Lecture Notes in Computer Science 4150

Commenced Publication in 1973
Founding and Former Series Editors:
Gerhard Goos, Juris Hartmanis, and Jan van Leeuwen

Marco Dorigo Luca Maria Gambardella
Mauro Birattari Alcherio Martinoli
Riccardo Poli Thomas Stützle (Eds.)

Ant Colony Optimization and Swarm Intelligence

5th International Workshop, ANTS 2006
Brussels, Belgium, September 4 – 7, 2006
Proceedings

Volume Editors

Marco Dorigo
Mauro Birattari
Thomas Stützle
IRIDIA, CoDE, Université Libre de Bruxelles
Av. F. Roosevelt 50, CP 194/6, 1050 Brussels, Belgium
E-mail: {mdorigo,mbiro,stuetzle}@ulb.ac.be

Luca Maria Gambardella
IDSIA, USI-SUPSI
Galleria 2, 6928 Manno-Lugano, Switzerland
E-mail: luca@idsia.ch

Alcherio Martinoli
SWIS, Ecole Polytechnique Fédérale de Lausanne
BC 210, Station 14, 1015 Lausanne, Switzerland
E-mail: alcherio.martinoli@epfl.ch

Riccardo Poli
University of Essex, Department of Computer Science
Wivenhoe Park, Colchester, CO4 3SQ, UK
E-mail: rpoli@essex.ac.uk

Library of Congress Control Number: 2006931489

CR Subject Classification (1998): F.2.2, F.1.1, G.1, G.2, I.2, C.2.4, J.1

LNCS Sublibrary: SL 1 – Theoretical Computer Science and General Issues

ISSN 0302-9743
ISBN-10 3-540-38482-0 Springer Berlin Heidelberg New York
ISBN-13 978-3-540-38482-3 Springer Berlin Heidelberg New York

Springer is a part of Springer Science+Business Media

springer.com

Typesetting: Camera-ready by author, data conversion by Scientific Publishing Services, Chennai, India
Printed on acid-free paper SPIN: 11839088 06/3142 5 4 3 2 1 0

Preface

ANTS – The International Workshop on Ant Colony Optimization and Swarm Intelligence is now at its fifth edition. The series started in 1998 with the organization of ANTS 1998. At that time the goal was to gather in a common meeting those researchers interested in ant colony optimization: more than 50 researchers from around the world joined for the first time in Brussels, Belgium, to discuss *ant colony optimization* and *swarm intelligence* related research. A selection of the best papers presented at the workshop was published as a special issue of the *Future Generation Computer Systems* journal (Vol. 16, No. 8, 2000). Two years later, ANTS 2000, organized again in Brussels, attracted more than 70 participants. The 41 extended abstracts presented as talks or posters at the workshop were collected in a booklet distributed to participants, and a selection of the best papers was published as a special section of the *IEEE Transactions on Evolutionary Computation* (Vol. 6, No. 4, 2002).

After these first two successful editions, it was decided to make of ANTS a series of biannual events with official workshop proceedings. The third and fourth editions were organized in September 2002 and September 2004, respectively. Proceedings were published by Springer within the *Lecture Notes in Computer Science* (LNCS) series.

The proceedings of ANTS 2002, LNCS Volume 2463, contained 36 contributions: 17 full papers, 11 short papers, and 8 extended abstracts, selected out of a total of 52 submissions. Those of ANTS 2004, LNCS Volume 3172, contained 50 contributions: 22 full papers, 19 short papers, and 9 extended abstracts, selected out of a total of 79 submissions.

Swarm intelligence is a rapidly growing field and the number of papers submitted to the 2006 edition reflects this growth: we received 115 submissions, a 45% increase with respect to the previous edition. Besides the higher number of researchers in the field, this increase can also be explained by the higher number of submitted papers that cover important subjects such as particle swarm optimization, swarm robotics, or ant-based clustering. ANTS is therefore slowly removing the initial bias towards ant colony optimization and becoming more and more the conference of the whole swarm intelligence community.

The higher number of submissions allowed us to increase the selective pressure: only 42% of the submitted papers was accepted for publication (i.e., 27 full papers and 23 short papers). This high selection threshold has made it possible to have a program of the highest standards. In addition to the accepted papers, a small number (12) of extended abstracts was selected for presentation: these are works that, although in a rather preliminary phase, show high potential and are therefore worth being discussed at the workshop.

To conclude this preface, we would like to thank all the people who helped in organizing ANTS 2006. We are very grateful to the authors who submitted their works; to the members of the International Program Committee and to the additional referees for their detailed reviews; to the IRIDIA people for their enthusiasm in helping with organizational matters; to the Université Libre de Bruxelles for providing rooms and logistic support; and, more generally, to all those contributing to the organization of the workshop. Finally, we would like to thank our sponsors: the IEEE Computational Intelligence Society, COMP2SYS,[1] *AntOptima*,[2] the Belgian National Funds for Scientific Research, and the French community of Belgium.

June 2006

Marco Dorigo
Luca M. Gambardella
Mauro Birattari
Alcherio Martinoli
Riccardo Poli
Thomas Stützle

[1] A Marie Curie Early Stage Training Site funded by the European Commission. More information is available at iridia.ulb.ac.be/comp2sys.

[2] More information is available at www.antoptima.com.

Organization

ANTS 2006 was organized by IRIDIA, Université Libre de Bruxelles, Belgium

Workshop Chairs

Marco Dorigo	IRIDIA, Université Libre de Bruxelles, Belgium
Luca M. Gambardella	IDSIA, USI-SUPSI, Manno, Switzerland

Technical Program Chairs

Alcherio Martinoli	EPFL, Lausanne, Switzerland
Riccardo Poli	University of Essex, UK
Thomas Stützle	IRIDIA, Université Libre de Bruxelles, Belgium

Publication Chair

Mauro Birattari	IRIDIA, Université Libre de Bruxelles, Belgium

Program Committee

Ashraf Abdelbar	American University in Cairo, Egypt
Carl Anderson	Qbit, LLC, Bethesda, MD, USA
Payman Arabshahi	University of Washington, Seattle, WA, USA
Tucker Balch	Georgia Institute of Technology, Atlanta, GA, USA
Tim Blackwell	Goldsmiths College, University of London, UK
Christian Blum	Universitat Politècnica de Catalunya, Spain
Eric Bonabeau	Icosystem Corporation, Cambridge, MA, USA
Jürgen Branke	University of Karlsruhe, Germany
Marco Chiarandini	University of Southern Denmark, Denmark
Maurice Clerc	Consultant, Groisy, France
Carlos Coello Coello	Instituto Politécnico Nacional, Mexico
Oscar Cordon	University of Granada, Spain
Jean-Louis Deneubourg	Université Libre de Bruxelles, Belgium
Gianni Di Caro	IDSIA, USI-SUPSI, Switzerland
Karl Doerner	Universität Wien, Austria
Kathryn Dowsland	University of Nottingham, UK

Hai-Bin Duan	Beihang University, P. R. China
Andries Engelbrecht	University of Pretoria, South Africa
Alex Freitas	University of Kent, UK
Caroline Gagné	Université du Québec à Chicoutimi, Canada
Deborah Gordon	Stanford University, Stanford, CA, USA
Walter Gutjahr	Universität Wien, Austria
Richard Hartl	Universität Wien, Austria
Marc Heissenbuettel	Swisscom Mobile Ltd, Switzerland
Beat Hirsbrunner	Universität Freiburg, Switzerland
Owen Holland	University of Essex, UK
Holger Hoos	University of British Columbia, Canada
Colin Johnson	University of Kent, UK
Jim Kennedy	Bureau of Labor Statistics, Washington D.C., USA
Franziska Klügl	Universität Würzburg, Germany
Joshua Knowles	University of Manchester, UK
William B. Langdon	University of Essex, UK
Vittorio Maniezzo	Università di Bologna, Italy
Monaldo Mastrolilli	IDSIA, USI-SUPSI, Manno, Switzerland
Ronaldo Menezes	Florida Tech, Melbourne, FL, USA
Daniel Merkle	Universität Leipzig, Germany
Peter Merz	Universität Kaiserslautern, Germany
Bernd Meyer	Monash University, Australia
Martin Middendorf	Universität Leipzig, Germany
Francesco Mondada	EPFL, Lausanne, Switzerland
Nicolas Monmarché	Université François Rabelais, Tours, France
Ann Nowé	Vrije Universiteit Brussel, Belgium
Luis Paquete	University of Algarve, Portugal
Rafael Stubs Parpinelli	Universidade do Estado de Santa Catarina, Brazil
Marc Reimann	Swiss Federal Institute of Technology, Switzerland
Andrea Roli	Università degli Studi "G. D'Annunzio", Italy
Martin Roth	Deutsche Telekom Laboratories, Germany
Erol Sahin	Middle East Technical University, Ankara, Turkey
Michael Sampels	IRIDIA, Université Libre de Bruxelles, Belgium
Giovanni Sebastiani	Ist. Applicazioni del Calcolo "Mauro Picone", Italy
Joerg Seyfried	Universität Karlsruhe, Germany
Mark Sinclair	Royal University of Phnom Penh, Cambodia
Christine Solnon	Université Claude Bernard, Lyon, France
William Spears	University of Wyoming, Laramie, WY, USA
Guy Theraulaz	Université Paul Sabatier, Toulouse, France
Alan Winfield	University of the West of England, UK
Jun Zhang	Sun Yat-Sen University, Guangzhou, P. R. China

Local Arrangements

Rehan O'Grady	IRIDIA, Université Libre de Bruxelles, Belgium
Carlotta Piscopo	IRIDIA, Université Libre de Bruxelles, Belgium

Additional Referees

Prasanna Balaprakash
Leonora Bianchi
Cecilia Di Chio
Frederick Ducatelle
Jens Gimmler
Oliver Korb
Manuel López-Ibáñez
Max Manfrin
David Martens
Chris Monson
Roberto Montemanni
Marco A. Montes de Oca
Paola Pellegrini
Andrea Rizzoli
Kevin Seppi

Sponsoring Institutions

AntOptima, Lugano, Switzerland
http://www.antoptima.com

COMP2SYS, Marie Curie Early Stage Training Site
http://iridia.ulb.ac.be/comp2sys

National Funds for Scientific Research, Belgium
http://www.fnrs.be

French Community of Belgium (through the research project ANTS)
http://www.cfwb.be

IEEE Computational Intelligence Society (as a technical co-sponsor)
http://www.ieee-cis.org

Table of Contents

Short Papers

Extended Abstracts

A Comparison of Particle Swarm Optimization Algorithms Based on Run-Length Distributions

Marco A. Montes de Oca, Thomas Stützle, Mauro Birattari, and Marco Dorigo

IRIDIA, CoDE, Université Libre de Bruxelles, Brussels, Belgium
{mmontes, stuetzle, mbiro, mdorigo}@ulb.ac.be

Abstract. In this paper we report an empirical comparison of some of the most influential Particle Swarm Optimization (PSO) algorithms based on run-length distributions (RLDs). The advantage of our approach over the usual report pattern (average iterations to reach a predefined goal, success rates, and standard deviations) found in the current PSO literature is that it is possible to evaluate the performance of an algorithm on different application scenarios at the same time. The RLDs reported in this paper show some of the strengths and weaknesses of the studied algorithms and suggest ways of improving their performance.

1 Introduction

Since the introduction of the first Particle Swarm Optimization (PSO) algorithm by Kennedy and Eberhart [1,2], many variants of the original algorithm have been proposed. The approach followed by many researchers to evaluate the performance of their variants has been to compare the proposed variant with the original version or, more recently, with the so-called canonical version [3]. In many cases, these new variants are reported to perform better, see for instance [4,5,6,7].

Unfortunately, since there are no cross-comparisons among variants, there is no general agreement on which PSO variant(s) could be considered the state-of-the-art in the field. The motivation for conducting the comparison reported in this paper is the identification of these variant(s). However, determining the state-of the-art algorithm is not a trivial task. In particular, one must be aware of the possible application scenarios in which a stochastic optimization algorithm may be used. Our main concern can be solution quality, time or both. Of course, in any case, the sooner we get a solution, the better. However, because of the stochastic nature of these algorithms, finding a high quality solution in a timely fashion only happens with a certain probability. Characterizing the distribution of this probability is the purpose of the run-time distribution. Formally, a stochastic optimization algorithm A applied to a problem Π will find a solution of quality q in time t with probability $P_{A,\Pi}(q,t) = P(RT_{A,\Pi} \leq t, SQ_{A,\Pi} \leq q)$, and the bivariate random variable $(RT_{A,\Pi}, SQ_{A,\Pi})$ describes the run-time and solution quality behavior of an algorithm A when applied to problem Π; the probability distribution of this random variable is also known as the run-time distribution of A on Π [8]. Since in continuous optimization we measure run-time in terms of

M. Dorigo et al. (Eds.): ANTS 2006, LNCS 4150, pp. 1–12, 2006.

the number of function evaluations, we talk of *run-length* distributions (RLDs) rather than *run-time* distributions. This is the approach followed in this paper.

An RLD completely characterizes the performance of a stochastic optimization algorithm on a particular problem, regardless of the actual application scenario in which we may be interested in. We say this because with an RLD we can estimate the probability of finding a solution of a certain quality given some time limit. This is the main reason why we chose to evaluate some of the most influential PSO algorithms using RLDs. As a bonus, an analysis based on RLDs allows the identification of some strengths and weaknesses of the studied algorithms and may also be used to design improved versions.

The rest of the paper is organized as follows. Section 2 briefly describes the PSO technique and the variants compared in this paper. Section 3 describes the experimental setup adopted for our comparison. Section 4 presents the development of the solution quality over time and the RLDs of the studied algorithms. Section 5 summarizes the main contributions and results presented in the paper.

2 Particle Swarm Optimization Algorithms

In the original PSO algorithm [1,2], a fixed number of solutions (called *particles* in a PSO context) are randomly initialized in a d-dimensional solution space. A particle i at time step t has a position vector $\boldsymbol{x}_i^t$ and a velocity vector $\boldsymbol{v}_i^t$. An objective function $f : S \rightarrow \Re$, with $S \subset \Re^d$, determines the quality of a particle's position, i.e., a particle's position represents a solution to the problem being solved. Each particle i has a vector $\boldsymbol{p}_i$ that represents its own best previous position that has an associated objective function value $pbest_i = f(\boldsymbol{p}_i)$. Finally, the best position the swarm has ever visited is stored in a vector $\boldsymbol{s}$ whose objective function value is $gbest = f(\boldsymbol{s})$.

The algorithm iterates updating the velocities and positions of the particles until a stopping criterion is met. The update rules are:

$$\boldsymbol{v}_i^{t+1} = \boldsymbol{v}_i^t + \varphi_1 \boldsymbol{U}_1(0,1) * (\boldsymbol{p}_i - \boldsymbol{x}_i^t) + \varphi_2 \boldsymbol{U}_2(0,1) * (\boldsymbol{s} - \boldsymbol{x}_i^t) , \tag{1}$$

$$\boldsymbol{x}_i^{t+1} = \boldsymbol{x}_i^t + \boldsymbol{v}_i^{t+1} , \tag{2}$$

where φ_1 and φ_2 are two constants called the *cognitive* and *social* acceleration coefficients respectively, $\boldsymbol{U}_1(0,1)$ and $\boldsymbol{U}_2(0,1)$ are two d-dimensional uniformly distributed random vectors in which each component goes from zero to one, and $*$ is an element-by-element vector multiplication operator.

The variants we compare in this study were selected either because they are among the most commonly used in the field or because they look very promising. In the following subsections, we describe them in more detail.

2.1 Canonical Particle Swarm Optimizer

Clerc and Kennedy [3] introduced a constriction factor into PSO to control the convergence properties of the particles. This constriction factor is added in Equation 1 giving

$$\boldsymbol{v}_i^{t+1} = \chi\left(\boldsymbol{v}_i^t + \varphi_1 \boldsymbol{U}_1(0,1) * (\boldsymbol{p}_i - \boldsymbol{x}_i^t) + \varphi_2 \boldsymbol{U}_2(0,1) * (\boldsymbol{s} - \boldsymbol{x}_i^t)\right) , \tag{3}$$

with

$$\chi = 2k / \left(\left|2 - \varphi - \sqrt{\varphi^2 - 4\varphi}\right|\right) , \tag{4}$$

where $k \in [0,1]$, $\varphi = \varphi_1 + \varphi_2$ and $\varphi > 4$. Usually, k is set to 1 and both φ_1 and φ_2 are set to 2.05, giving as a result χ equal to 0.729 [9,10]. This variant has been so widely used that it is known as the *canonical* PSO.

2.2 Time-Varying Inertia Weight Particle Swarm Optimizer

Shi and Eberhart [4,11] introduced the idea of a time-varying inertia weight. The idea was to control the diversification–intensification behavior of the original PSO. The velocity update rule is

$$\boldsymbol{v}_i^{t+1} = w(t)\boldsymbol{v}_i^t + \varphi_1 \boldsymbol{U}_1(0,1) * (\boldsymbol{p}_i - \boldsymbol{x}_i^t) + \varphi_2 \boldsymbol{U}_2(0,1) * (\boldsymbol{s} - \boldsymbol{x}_i^t) , \tag{5}$$

where $w(t)$ is the time-varying inertia weight which usually is linearly adapted from an initial value to a final one. In most cases, φ_1 and φ_2 are both set to 2.

There are two ways of varying the inertia weight in time: decreasingly (e.g., as in [4,12,11]) and increasingly (e.g., as in [13,14]). In this paper, we included both variants for the sake of completeness. Normally, the starting value of the inertia weight is set to 0.9 and the final to 0.4. Zheng et al. [13,14], use the opposite settings. In the results section, these variants are identified by Dec-IW and Inc-IW, respectively.

2.3 Stochastic Inertia Weight Particle Swarm Optimizer

Eberhart and Shi [15] proposed another variant in which the inertia weight is randomly selected according to a uniform distribution in the range [0.5,1.0]. This range was inspired by Clerc and Kennedy's constriction factor. In this version, the acceleration coefficients are set to 1.494 as a result of the multiplication $\chi \cdot \varphi_{1,2}$. Although this variant was originally proposed for dynamic environments, it has also been shown to be a competitive optimizer for static ones [16]. In the results section this variant is identified by Sto-IW.

2.4 Fully Informed Particle Swarm Optimizer

In the fully informed particle swarm (FIPS) proposed by Mendes et al. [7], a particle uses information from all its topological neighbors. This variant is based on the fact that Clerc and Kennedy's constriction factor does not enforce that the value φ should be split only between two attractors.

For a given particle, the way φ (i.e., the sum of the acceleration coefficients) is decomposed is $\varphi_k = \varphi/|\mathcal{N}|\ \forall k \in \mathcal{N}$ where $\mathcal{N}$ is the neighborhood of the particle. As a result, the new velocity update equation becomes

$$\boldsymbol{v}_i^{t+1} = \chi\left[\boldsymbol{v}_i^t + \sum_{k \in \mathcal{N}} \varphi_k \mathcal{W}(k) \boldsymbol{U}_k(0,1) * (\boldsymbol{p}_k - \boldsymbol{x}_i^t)\right] , \tag{6}$$

where $\mathcal{W}(k)$ is a weighting function.

2.5 Self-organizing Hierarchical Particle Swarm Optimizer with Time-Varying Acceleration Coefficients

The self-organizing hierarchical particle swarm optimizer with time-varying acceleration coefficients (HPSOTVAC) proposed by Ratnaweera et al. [16] drops the velocity term from the right side of Equation 5. If a particle's new velocity becomes zero (in any dimension), it is reinitialized to some value proportional to the maximum allowable velocity V_{max}. HPSOTVAC linearly adapts the value of the acceleration coefficients φ_1 and φ_2 to enforce the diversification behavior at the beginning of the run and the intensification behavior at the end. φ_1 is decreased from 2.5 to 0.5 and φ_2 increases from 0.5 to 2.5. Finally, the reinitialization velocity is also linearly decreased from V_{max} at the beginning of the run to $0.1 \cdot V_{max}$ at the end.

2.6 Adaptive Hierarchical Particle Swarm Optimizer

Proposed by Janson and Middendorf [17], the adaptive hierarchical PSO (AHPSO) is an example of a PSO with dynamic adaptation of the population topology. In AHPSO, the topology is a tree-like structure in which particles with a higher fitness evaluation are located in the upper nodes of the tree. At each iteration, a child particle updates its velocity considering its own previous best performance and the previous best performance of its parent. Additionally, before the velocity updating process takes place, the previous best fitness value of any particle is compared with that of its parent. If it is better, child and parent swap their positions in the hierarchy.

The branching degree of the tree is a factor that can balance the diversification-intensification behavior of the algorithm. To dynamically adapt the algorithm to the stage of the optimization process, the branching degree is decreased by k_{adapt} degrees until a certain minimum degree d_{min} is reached. This process takes place every f_{adapt} number of iterations. The parameters that control this process need to be tuned for each problem [17]. In our experiments, for the reasons explained in the next section, we set the initial branching factor to 20, parameters d_{min}, f_{adapt}, and k_{adapt} were set to 2, $1000 * m$, and 3 respectively, where m is the number of particles.

3 Experimental Setup

All the PSO variants described in the previous section were implemented for this comparison. To ensure the correctness of our implementations, we tested them on the same problems with the same parameters as reported in the literature[1]. To allow the comparison of the results with previous works, we used some of the most common benchmark functions in the PSO literature: Sphere, Rosenbrock, Rastrigin, Griewank, and Schaffer's F6 functions in 30 dimensions. The mathematical definition of these functions is readily available in the literature (cf. [10]).

[1] For space restrictions, we refer the interested reader to the following address: http://iridia.ulb.ac.be/supp/IridiaSupp2006-003/index.html

In our runs, these functions were shifted and biased exactly as specified in [18][2]. Because of this, our initializations are, in all cases, asymmetric with respect to the global optimum.

The reported results are based on 100 independent trials of 1 000 000 function evaluations. In our experiments, we used swarms of 20 particles using two different topologies: fully connected and ring with unitary neighborhood size. The results are organized by population topology. Both topologies included self-references (i.e., every particle is a neighbor to itself). This separation was needed to highlight the influence of the used topology in the behavior of the algorithms. Note that the AHPSO algorithm uses neither a fully connected topology nor a ring topology and therefore appears in both sets of results.

Before proceeding to the presentation of our results, it is worth noting that most PSO algorithms are not robust in their parameterization. For example, in the PSO variants based on a time-varying inertia weight, the slope of the increasing or decreasing inertia weight function is determined by the maximum number of function evaluations. Another problem (for comparison purposes) is that it is also possible to fine-tune the parameters of a variant to solve a particular problem. A possible solution to this problem is to fine-tune all variants for the problem at hand and proceed with the comparison; however, if our aim is to solve real-world problems which generally have a structure we do not know in advance, we need algorithms with a set of "normally good" parameters. For this reason, in this study each algorithm used the same parameterization across the benchmark problems. The actual values chosen for the parameters have already been mentioned in the preceding sections.

4 Results

Tables 1 and 2 show the average value and standard deviation of the number of function evaluations needed to achieve a certain solution quality with fully connected and ring topologies, respectively. For each function, there are three different solution qualities. The first one corresponds to the usual goal for that function (cf. [10]). The second and third can be considered medium and high solution qualities, respectively. The absolute values can be computed as follows: if, for example, the desired solution quality is 0.01% and the optimum is at -130.0, it means that the goal to reach is $-130 - (0.0001 \times -130.0) = -129.987$.

Most variants, most notably FIPS, are greatly affected in their performance by the used topology. With a fully connected topology, most of the tested variants reach the specified solution quality faster than with the ring topology. FIPS performs poorly with this topology: only in 4 out of 15 cases it reaches the specified solution quality. However, whenever it does, it is the fastest algorithm.

The data shown in Tables 1 and 2 should be taken *cum grano salis*. The averages and standard deviations reported there are computed over successful runs only. Since these data alone can be misleading, the median solution quality over time (not included in the paper due to space restrictions) is reported

[2] The values of the optima are specified in Tables 1 and 2.

in the already mentioned URL. However, the good performance of FIPS using the ring topology is confirmed by the median. FIPS is among the fastest variants.

HPSOTVAC is the only variant that is able to find the highest solution quality target in the Rastrigin function. HPSOTVAC succeeds at reaching the goal but spends many function evaluations to do that.

AHPSO performs relatively better than the other variants when they use the ring topology. This is expected since AHPSO adapts the population hierarchy from a highly connected one to a loosely connected one, so it exploits the benefits of converging faster at the beginning of the run. As seen from the results, fast

Table 1. Average value and standard deviation of the number of function evaluations needed to achieve a certain solution quality (S.Q.) using the fully connected topology. Only successful runs are considered. {f1=Griewank (optimum at -130.0), f2=Rastrigin (optimum at -330), f3=Rosenbrock (optimum at 390), f4=Schaffer's F6 (optimum at -300), f5=Sphere (optimum at -450)}.

Function	S.Q. (%)	AHPSO	Canonical	FIPS	Dec-IW	HPSOTVAC	Inc-IW	Sto-IW
f1	0.077	9641.3	**8345.8**	–	433036	25741.8	8365.5	9713.8
		1413	1271.4	–	24012.3	1647.1	852.8	1483.5
	0.01	11613.8	**9784.1**	–	442995	34474.3	10115.8	11806.7
		1508.1	1153	–	15996.2	7430.5	1274.6	1789
	0.001	12994.3	**11322.7**	–	453478	57830	41082.6	13306.2
		1481	2120.2	–	21515.5	61113.8	166025	2335.8
f2	30.30	4011	3836.6	**960**	364350	29852.8	53104.4	4820.7
		1478.6	1065.2	56.5	42590.1	21636.7	212169	1534.7
	15	6295	**5550**	–	427895	101322	516798	7758
		1123	1400.1	–	30835.2	44487.3	490241	2554.6
	1	–	–	–	–	**635060**	–	–
		–	–	–	–	133964	–	–
f3	25.64	108546	50148.6	–	492989	489920	**22381.2**	56747.8
		150900	58874	–	70214.1	258767	19505.5	105101
	10	153596	87750.5	–	548089	665170	**67398.8**	95153.8
		204835	90431.8	–	85071.9	205934	104077	148441
	1	353182	**168799**	–	653042	762488	288477	232858
		201800	97989.6	–	116809	218223	208034	152029
f4	3.3×10^{-6}	33636.3	21044.6	**5230**	115876	84893.3	72287.1	56837.6
		93014.2	44560.4	3921.3	24171.4	184640	196370	172404
	1×10^{-7}	34306.1	21478.2	**5540**	131253	86826	58508.4	57350.8
		93024.3	44536.7	3964.9	22963.3	184097	161070	172359
	1×10^{-8}	34540	21789.6	**5724**	140434	88095.1	58760.6	57698.4
		92995.1	44506.6	3932.1	19112.8	183701	161142	172325
f5	0.0022	11342.8	10913	–	433345	31210.4	**8135.1**	11127.8
		1386.8	2255.7	–	6885.3	1436.4	944.4	1317
	0.0002	14000.6	13122.4	–	446824	40913	**9772**	13680.4
		1468.2	2648	–	6707.7	1616.7	946	1919
	0.00001	15862.6	14955.8	–	456062	48546.6	**11147.4**	15469.6
		1506.1	2795.7	–	6346.5	1723	1152	1890.7

Table 2. Average value and standard deviation of the number of function evaluations needed to achieve a certain solution quality (S.Q.) using the ring topology. Only successful runs are considered. {f1=Griewank (optimum at -130.0), f2=Rastrigin (optimum at -330), f3=Rosenbrock (optimum at 390), f4=Schaffer's F6 (optimum at -300), f5=Sphere (optimum at -450)}.

Function	S.Q. (%)	AHPSO	Canonical	FIPS	Dec-IW	HPSOTVAC	Inc-IW	Sto-IW
f1	0.077	9641.3	12377.8	**8030.2**	456568	29862	17268.8	15530.2
		1413	849	957.8	11431.9	1538.6	1192.3	1538.2
	0.01	**11613.8**	15865.6	12454.5	499384	40342.2	29852.1	23091.9
		1508.1	4322.2	7074.7	68132.2	3660.7	50030.8	37164.4
	0.001	**12994.3**	32157.9	21422.5	536619	60952.1	61165.4	35051.8
		1481	62991	26631.3	90978.1	36481	115035	71418.5
f2	30.30	**4011**	30091.3	22599.6	360126	30052.8	34257.8	19767.8
		1478.6	89696.1	9998.2	63803.9	12946.6	142877	84546.4
	15	**6295**	–	136648	460690	117057	685132	206669
		1123	–	100821	59069.9	37364.8	429521	275732
	1	–	–	–	–	**811247**	–	–
		–	–	–	–	107569	–	–
f3	25.64	108546	**104684**	226828	518732	361997	127764	151060
		150900	130114	252722	70783.1	325196	154960	188680
	10	**153596**	189872	320153	605140	291822	173475	215458
		204835	216622	270742	124911	263481	166956	233277
	1	**353182**	426285	443895	734466	–	531935	458950
		201800	221124	242211	146611	–	270172	262995
f4	3.3×10^{-6}	33636.3	43691.4	40416.9	123649	126988	**28014**	40241.2
		93014.2	89702.5	90967.5	28220.1	204489	32195.1	79386.6
	1×10^{-7}	34306.1	44897.6	49353.5	139871	129022	**29321.8**	42340.2
		93024.3	89965.7	92344.4	27231.6	204003	32216.9	82079
	1×10^{-8}	34540	45482.8	54939.6	150797	130576	**29771.6**	43142.4
		92995.1	89856.7	92882.9	27877.7	203686	32327.5	82122.1
f5	0.0022	11342.8	13693.6	**8266**	459971	35518.2	14923	17148.6
		1386.8	572.3	480.3	9304.4	1382.3	894.8	1129.21
	0.0001	14000.6	16421.4	**9920.8**	477006	47480	18077.4	20660.8
		1468.2	663.3	491.7	8254.2	1469.7	956.3	1265.5
	0.00001	15862.6	18484.2	**11169.6**	488408	56889	20430.6	23289.4
		1506.1	684.3	523.2	7977.2	1923.1	1055	1344.9

convergence is somehow associated to the fully connected topology, or at least with a highly connected one.

The Canonical PSO and the Increasing Inertia Weight variants perform pretty well. With the fully connected topology, they are the best performers in 10 out of 15 cases. With the ring topology, this number drops to only 4. These variants clearly exploit the convergence properties of the fully connected topology.

In this paper we report *qualified* RLDs which are cross-sections along the computing time axis of the full joint distribution of the bivariate random variable $(RT_{A,\Pi}, SQ_{A,\Pi})$ described in Section 1. The interested reader is referred to [8]

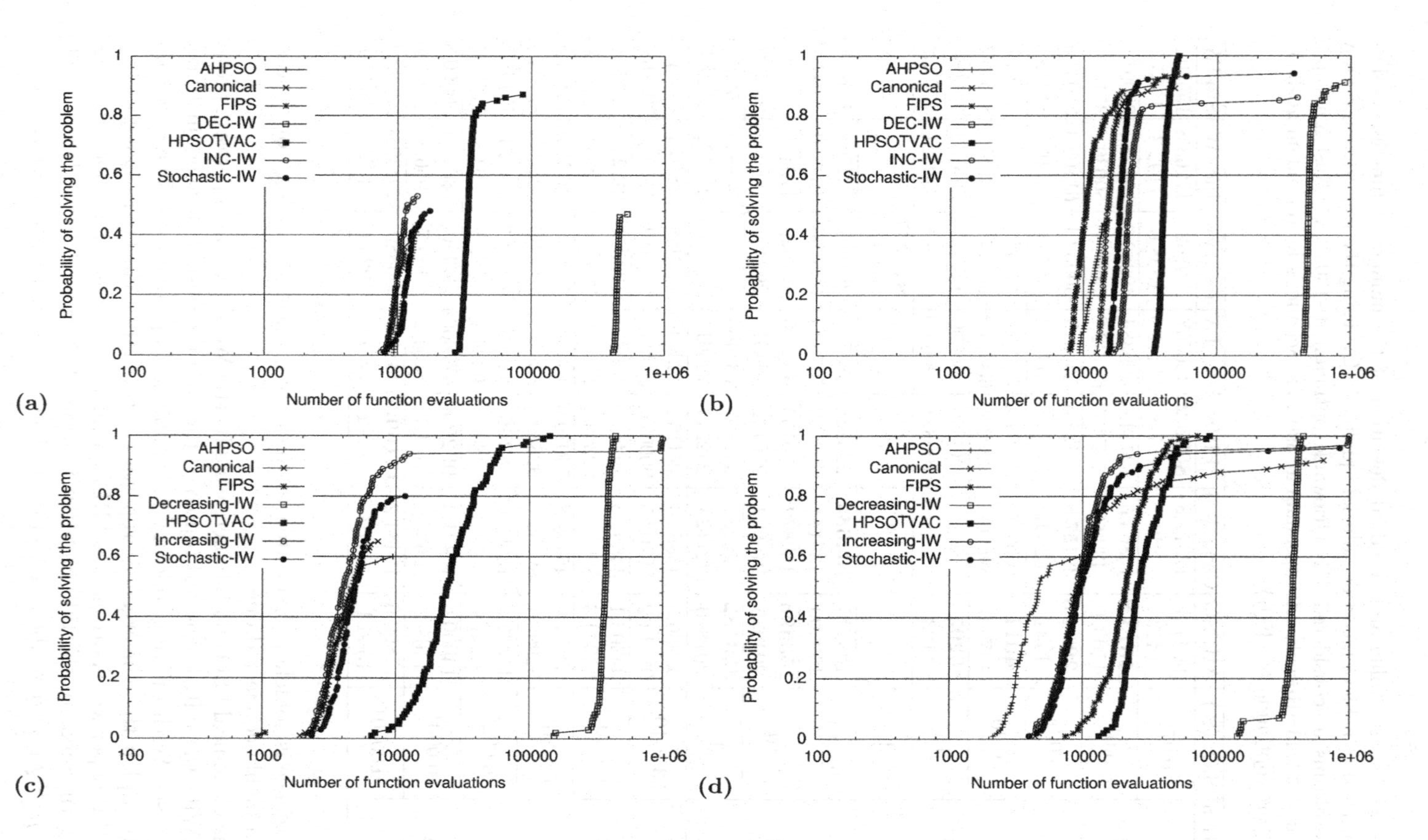

Fig. 1. Run-length distributions. In (a) and (b), the results obtained in Griewank function. In (c) and (d), the ones in Rastrigin.

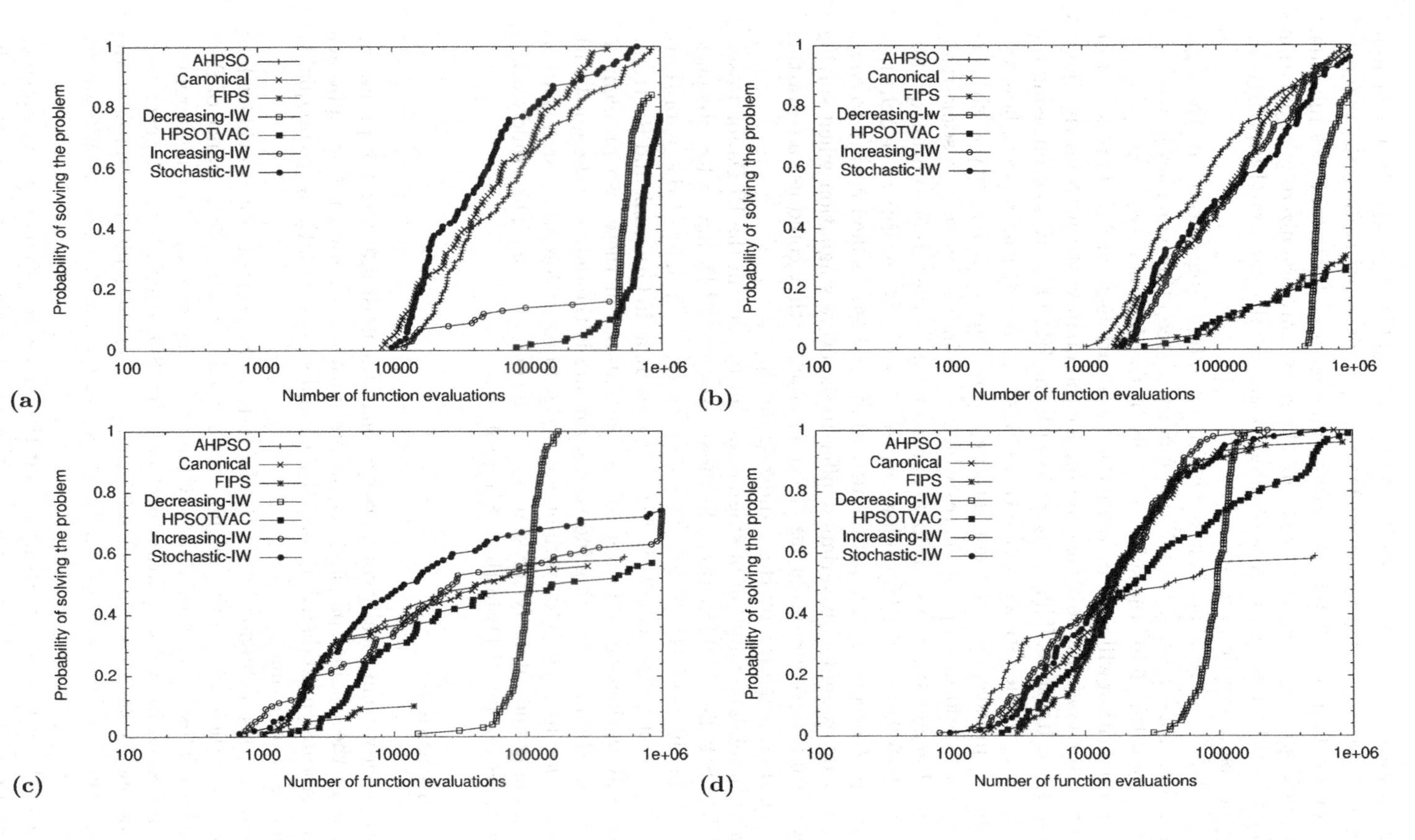

Fig. 2. Run-length distributions. In (a) and (b), the results obtained in Rosenbrock function. In (c) and (d), the ones in Schaffer's F6.

for more information about RLDs. Figures 1 and 2 show the RLDs in four benchmark functions. The shown RLDs correspond to solution qualities of 0.01% for Griewank function, 30% for Rastrigin function, 10% for Rosenbrock function, and 0.0001% for Schaffer's F6 function. The results are organized by population topology: on the left, the results obtained using a fully connected topology; on the right, using the ring topology.

The "slope" of the shown curves point out interesting features of the algorithms. If an RLD for a given solution quality is steep (but complete), it means that the algorithm finds the solution easily. If the demanded solution quality is high, the algorithm will need more function evaluations to find it. This will cause the curve to change its position (the higher the quality, the more to the right) and, possibly, its slope. This is the case with the HPSOTVAC variant using the ring topology in the Griewank function (Figure 1,(b)). It can be seen, however, that this is an exception and not the rule. Most variants have curves with low steepness or steep incomplete curves which is an indication that in some trials the algorithm gets stuck in some local optima far from the target solution quality.

An analysis based on RLDs allows us to measure the severity of search stagnation experienced by optimization algorithms and lets us devise ways to counteract it. For example, all variants suffer from severe stagnation when solving Griewank and Rastrigin problems. To counteract it, they could use a restarting mechanism as suggested by Hoos and Stützle [8].

Another related symptom of stagnation can be seen in the RLDs for Rosenbrock and Schaffer's F6 functions. In these cases, the RLDs have a low steepness which highlights the lack of diversification strategies in most of the algorithms. In these cases stagnation exists but is not as severe as in Griewank and Rastrigin.

The only variant that do not follow the pattern in these two problems is the one based on a decreasing inertia weight and is the only one designed with diversification in mind. This variant was designed to explore the search space at the beginning and intensify the search near the end of a run. This could explain the steepness of its RLDs in these two problems.

5 Conclusions

In this paper we empirically compared seven of the most influential or promising variants of the original particle swarm optimization algorithm. Our approach was to use run-length distributions (RLDs) and statistics of the solution quality development over time.

Regarding the behavior shown by the tested PSO variants, it is evident how important is the choice of the neighborhood topology in the performance of PSO algorithms. This is something already known in the field, but the measurement of its influence in the stagnation behavior of PSO algorithms had never been done before. With respect to our initial motivation, we limited ourselves to the comparison of some of the most influential variants, and from our results we did not find any dominant variant.

One of the advantages of RLDs is that they allow the evaluation of a stochastic optimization algorithm regardless of the actual application scenario it may be used in. Another advantage is that they allow the identification of some strengths and weaknesses of the studied algorithms that can be used to improve their performance. Future research will focus on exploiting the information provided by RLDs to the *engineering* of PSO variants. We sketched how this could be done.

Acknowledgments. This work was supported by the ANTS project, an *Action de Recherche Concertée* funded by the Scientific Research Directorate of the French Community of Belgium. Marco Montes de Oca acknowledges support from the Programme Alβan, the European Union Programme of High Level Scholarships for Latin America, scholarship No. E05D054889MX. Thomas Stützle and Marco Dorigo acknowledge suport from the Belgian National Fund for Scientific Research (FNRS), of which they are a Research Associate and a Research Director, respectively.

References

1. Kennedy, J., Eberhart, R.: Particle swarm optimization. In: Proceedings of IEEE International Conference on Neural Networks, Piscataway, NJ, IEEE Press (1995) 1942–1948
2. Eberhart, R., Kennedy, J.: A new optimizer using particle swarm theory. In: Proceedings of the 6th International Symposium on Micro Machine and Human Science, Piscataway, NJ, IEEE Press (1995) 39–43
3. Clerc, M., Kennedy, J.: The particle swarm–explosion, stability, and convergence in a multidimensional complex space. IEEE Transactions on Evolutionary Computation **6**(1) (2002) 58–73
4. Shi, Y., Eberhart, R.: A modified particle swarm optimizer. In: Proceedings of the 1998 IEEE World Congress on Computational Intelligence, Piscataway, NJ, IEEE Press (1998) 69–73
5. Kennedy, J.: Stereotyping: Improving particle swarm performance with cluster analysis. In: Proceedings of the 2000 IEEE Congress on Evolutionary Computation, Piscataway, NJ, IEEE Press (2000) 1507–1512
6. Fan, H.: A modification to particle swarm optimization algorithm. Engineering Computations **19**(8) (2002) 970–989
7. Mendes, R., Kennedy, J., Neves, J.: The fully informed particle swarm: Simpler, maybe better. IEEE Transactions on Evolutionary Computation **8**(3) (2004) 204–210
8. Hoos, H.H., Stützle, T.: Stochastic Local Search: Foundations and Applications. Morgan Kaufmann Publishers, San Francisco, CA, USA (2004)
9. Eberhart, R., Shi, Y.: Comparing inertia weights and constriction factors in particle swarm optimization. In: Proceedings of the 2000 IEEE Congress on Evolutionary Computation, Piscataway, NJ, IEEE Press (2000) 84–88
10. Trelea, I.C.: The particle swarm optimization algorithm: Convergence analysis and parameter selection. Information Processing Letters **85**(6) (2003) 317–325
11. Shi, Y., Eberhart, R.: Empirical study of particle swarm optimization. In: Proceedings of the 1999 IEEE Congress on Evolutionary Computation, Piscataway, NJ, IEEE Press (1999) 1945–1950
12. Shi, Y., Eberhart, R.: Parameter selection in particle swarm optimization. In: Proceedings of the 7th International Conference on Evolutionary Programming VII, LNCS Vol. 1447. Springer-Verlag, New York (1998) 591–600

13. Zheng, Y.L., Ma, L.H., Zhang, L.Y., Qian, J.X.: On the convergence analysis and parameter selection in particle swarm optimization. In: Proceedings of the 2003 IEEE International Conference on Machine Learning and Cybernetics, Piscataway, NJ, IEEE Press (2003) 1802–1807
14. Zheng, Y.L., Ma, L.H., Zhang, L.Y., Qian, J.X.: Empirical study of particle swarm optimizer with an increasing inertia weight. In: Proceedings of the 2003 IEEE Congress on Evolutionary Computation, Piscataway, NJ, IEEE Press (2003) 221–226
15. Eberhart, R., Shi, Y.: Tracking and optimizing dynamic systems with particle swarms. In: Proceedings of the 2001 IEEE Congress on Evolutionary Computation, Piscataway, NJ, IEEE Press (2001) 94–100
16. Ratnaweera, A., Halgamuge, S.K., Watson, H.C.: Self-organizing hierarchical particle swarm optimizer with time-varying acceleration coefficients. IEEE Transactions on Evolutionary Computation **8**(3) (2004) 240–255
17. Janson, S., Middendorf, M.: A hierarchical particle swarm optimizer and its adaptive variant. IEEE Transactions on Systems, Man and Cybernetics–Part B **35**(6) (2005) 1272–1282
18. Suganthan, P.N., Hansen, N., Liang, J.J., Deb, K., Chen, Y.P., Auger, A., Tiwari, S.: Problem definitions and evaluation criteria for the CEC 2005 special session on real-parameter optimization. Technical Report 2005005, Nanyang Technological University, Singapore and IIT Kanpur, India (2005)

A Framework and Model for Soft Routing: The Markovian Termite and Other Curious Creatures

Martin Roth

Deutsche Telekom Laboratories, Berlin, Germany
Martin.Roth02@telekom.de

Abstract. A theoretical framework and model is presented to study the self-organized behavior of probabilistic routing protocols for computer networks. Such soft routing protocols have attracted attention for delivering packets reliably, robustly, and efficiently. The framework supports several features necessary for emergent routing behavior, including feedback loops and indirect communication between peers. Efficient global operating parameters can be estimated without resorting to expensive monte-carlo simulation of the whole system. Key model parameters are routing sensitivity and routing threshold, or noise, which control the "randomness" of packet routes between source and destination, and a metric estimator. Global network characteristics are estimated, including steady state routing probabilities, average path length, and path robustness.

The framework is based on a markov chain analysis. Individual network nodes are represented as states. Standard techniques are used to find primary statistics of the steady state global routing pattern, given a set of link costs. The use of packets to collect information about, or "sample," the network for new path information is also reviewed. How the network sample rate influences performance is investigated.

1 Introduction

1.1 Overview

Adaptive behavior is one of the fundamental requirements of modern network routing protocols [1]. Deterministic approaches are able to perform well in relatively static networks by relying on traditional link state or distance vector shortest path algorithms. Difficulty arises under dynamic conditions when route costs are highly variable or the topology itself is unstable. Multipath routing is often cited as a solution, however its implementation is complicated by the need to manage additional routes; multipath routing does not fit elegantly into a traditional routing framework. Such a scenario is especially relevant in mesh and ad-hoc networks, where the communications characteristics of wireless links are subject to large and frequent variations. Probabilistic routing (here also known as soft routing or p.routing) is able to maintain route utility estimates on all routes simultaneously in one unified framework. Good paths are used proportionally

M. Dorigo et al. (Eds.): ANTS 2006, LNCS 4150, pp. 13–24, 2006.

more than bad ones, and all paths are successively refined based on simple local interactions between hosts. Adaptive end-to-end routing is an emergent property of the group. Soft routing is ultimately able to outperform deterministic routing (see Section 1.2 for details).

Probabilistic routing algorithms contain many parameters that influence their performance. Though there are several possible formulations of p.routing, each approach generally has three primary parameters. These are a sensitivity parameter, which controls the bias towards well performing links and paths, a noise parameter, which sets a lower bound on the entropy of the routing distribution, and an estimation parameter, which determines how different measurements of the network are combined to determine a single estimate of network state. Related issues include how often metric estimates are updated and how fast the network itself changes.

These parameters define interactions at a local level, such as at each network node, but it is unclear how they affect global behaviors, such as routing patterns. System properties of interest might include the expected per-packet path cost between a source and destination, or a robustness measure of the equilibrium routing solution. Extensive simulation can be used to calculate these characteristics, but an analytical model which can determine (possibly optimal) local parameter values based on required global behavior is missing. The framework presented here is the first step towards such a model.

1.2 Previous Work

Little work has been done to analytically model the performance of soft routing algorithms. The vast majority of the effort in the area has been simulation based. A variety of protocols have been proposed, primarily biologically inspired, including Q-Routing [2] (a reinforcement learning approach), Ant-Based Control (ABC) [3], AntNet [4], Cooperative Asymmetric Forwarding (CAF) [5], Probabilistic Emergent Routing Algorithm (PERA) [6], Ant-based Routing Algorithm (ARA) [7], Mobile Ant-Based Routing (MABR) [8], Multiple Ant Colony Optimization (MACO) [9], Termite [10], Ad-hoc Networking with Swarm Intelligence (ANSI) [11], AntHocNet [12], BeeAdHoc [13], and SAMPLE [14] (also a reinforcement learning approach). These protocols span the application space between wired and wireless ad-hoc or mesh networks, usually being compared to a well known deterministic protocol such as Open Shortest Path First routing (OSPF) [15] for wired networks or Ad-hoc On-demand Distance Vector routing (AODV) [16] for ad-hoc networks.

Strictly analytical work is scattered, and focuses on characterizing global behavior. [17] considers two different packet forwarding equations and shows that one type will end up using only the better of two links, regardless of their relative difference. The other method will split traffic across the two links proportionally to their utility. [18] develops a model for a node's estimate of the path utility to a destination given a metric update equation and its parameters. The model is used to explain differences in global routing performance when different metric updates are used, including a normalized exponential filter (also known as the pheromone update rule), an exponential filter, and a Dijkstra-inspired update. A scale free parameter is discovered which determines the maximum link estimate. [10] develops a heuristic for a good selection of the estimation parameter in the Termite p.routing algorithm based on the network correlation time. The results in this paper continue the analytical characterization effort by accounting for all fundamental aspects of soft routing algorithms.

Some effort towards parameter optimization has also been done in the related area of biologically inspired optimization (a form of stochastic gradient descent [19]), such as with Ant Colony Optimization (ACO) [20] and Particle Swarm Optimization (PSO) [21]. Approaches in this field center on classifying subproblems, and then running tests in order to find an optimal parameter space for each type of subproblem [22].

1.3 Structure of Paper

Section 2 gives an overview of how a generic probabilistic routing protocol works and introduces the p.routing framework in this light. The Termite protocol and a generic ACO based approach are used as examples. Section 3 develops a markov chain analysis of the framework, which is used to reveal various aspects of the performance of soft routing algorithms. This includes the equilibrium routing probabilities, expected path cost, and the frequency and fraction of links being tested. Section 4 gives simulation results and discussion based on the analysis methods presented in Section 3. Section 5 concludes the paper, in addition to providing avenues for future work.

2 Probabilistic Routing Framework

2.1 Framework Overview

A framework for probabilistic routing is presented. The Termite p.routing algorithm and a generic ACO based approach (hereafter referred to simply as ACO) are used as illustrative examples. Several proposed soft routing algorithms are based on ACO, including AntHocNet, ANSI, and ARA. The soft routing protocols proposed to date are essentially probabilistic distance vector protocols. There are two key components to any p.routing algorithm, the packet *forwarding equation* and the *metric estimator*. Each node estimates the cost to each destination through each neighbor based on routing data collected from received or overheard traffic. This information can be piggybacked on data packets, or found in control packets such as when using the forward/backward ant mechanism described in [4]. The network is continuously sampled for changes by each node in order to maintain up-to-date information. The estimated route utilities are used to generate a probability distribution from which a next hop for a specific neighbor can be randomly selected. Multipath routing is easily implemented since the utility of each neighbor to arrive at each destination is maintained by the metric estimator, and each neighbor is considered as a viable next hop by the forwarding equation. Soft routing is an application of dynamic optimization.

2.2 Packet Forwarding

The forwarding equation determines the next hop probability distribution for each packet, given its destination. The next hop is selected according to this distribution; this is per-packet probabilistic routing. The pmf, p, is a normalization of the current utility estimate, P, for each link to deliver a message to the destination. Normalization is an intuitive mechanism used to send more packets over good links, and fewer over lesser links. Two parameters influence the forwarding equation, including the *sensitivity*, $F \geq 0$, and *threshold*, $K \geq 0$. The sensitivity modulates the differences between link utilities, making the resulting probabilities more or less dependant on them. It controls how much better paths are used more than worse ones. The threshold parameter determines how good a link must be before it has a substantial impact on the routing distribution. This parameter balances the sensitivity by pushing the routing distribution closer towards uniform for large K. The threshold is alternatively known as *noise*, $0 \leq q \leq 1$, which takes a different form in the forwarding equation, but serves the same purpose. The balance between sensitivity and threshold (or noise) determines the tradeoff between network exploitation and exploration. The forwarding equation is denoted as the function $W(\mathsf{P}, K, F)$ or $W(\mathsf{P}, q, F)$, whichever is appropriate.

Routing metric estimates become stale as the network changes, requiring them to be constantly updated. The next hop distribution should reflect the uncertainty of the underlying route metric estimates at the current time. If an estimate is stale, then the corresponding link should be less probable. This intuition is included in the forwarding equation by multiplying the correlation between successive metric estimates, R, against the current estimate in the forwarding equation. The more time between estimate updates yields a lower correlation and ultimately a smaller routing probability for that link. The correlation function depends on the type of metric estimator used. Section 2.4 reviews estimate correlation in more detail.

Termite. Equation 1 shows the Termite forwarding equation. $P^n_{i,d}$ is the path utility estimate of node n using neighbor i to arrive at destination d. The normalization is with respect to all neighbors of n, $\mathcal{N}^n \subseteq \mathcal{V}$, where $\mathcal{V}$ is the set of all nodes in the network. The threshold must be set according to the expected range of the metric. $R(t - t^n_{i,d})$ is the correlation between the current estimate (the last sample of which generated at $t^n_{i,d}$) and the current time, t.

$$p^n_{i,d} = \frac{\left[P^n_{i,d} R\left(t - t^n_{i,d}\right) + K\right]^F}{\sum_{j \in \mathcal{N}^n} \left[P^n_{j,d} R\left(t - t^n_{j,d}\right) + K\right]^F} \tag{1}$$

ACO. A typical ACO type forwarding equation is shown in Equation 2. The noise parameter balances the routing distribution between a normalization of the link utilities and a uniform distribution across all outgoing links.

$$p_{i,d}^n = (1-q) \cdot \frac{\left[P_{i,d}^n R\left(t - t_{i,d}^n\right)\right]^F}{\sum_{j \in \mathcal{N}^n} \left[P_{j,d}^n R\left(t - t_{j,d}^n\right)\right]^F} + q \cdot \frac{1}{|\mathcal{N}^n|} \tag{2}$$

2.3 Metric Update

The metric update equation generates an estimate of the routing metric from each node to each destination through each neighbor. Current routing information is gathered either by proactively probing the network with control packets, or by passively collecting data from received and overheard packets. The returning information is treated as samples of a non-stationary stochastic process describing the change in link utilities. The samples are filtered to track the mean of the process, which is the estimate. Relevant parameters include the rate at which packets (those carrying routing information) are originated, λ, and the network correlation time, T, which is the period of time over which the network statistics are assumed to remain constant.

Of course different approaches must be taken to collect information about asymmetric or symmetric link metrics (assuming that all links are bidirectional). For this reason there exist mechanisms such as forward/backward ants (ACO) and data piggybacking (Termite), respectively. For the purposes of this analysis of the presented framework, how the information is collected is irrelevant, only that it is, and that the probe packets follow a routing rule defined by the forwarding equation.

There are two commonly used methods for estimating path utility based on samples of the network, both of which are basically low pass filters. Equation 3 shows the traditional exponential filter (also known as a pheromone filter). It requires little state but does not make an optimal estimate of the link utility. It includes information from all received samples in the current estimate (ie., it has an infinite impulse response). The time constant of the filter, τ, is characterized by the network correlation time, T. Here, $\gamma_{r,s}^n$ is the arrived utility update at node n from source node s over previous hop r. $P_{r,s}^n$ is the estimate at node n to get to the destination s, which is the source of the arriving packet, through the previous hop, r.

$$P_{r,s}^n \leftarrow P_{r,s}^n e^{-\left(t - t_{r,s}^n\right)\tau} + \gamma_{r,s}^n \left[1 - e^{-\left(t - t_{r,s}^n\right)\tau}\right] \tag{3}$$

Equation 4 shows the optimal path utility estimator in the form of a sliding window, or box, filter, with length equivalent to the network correlation time, T. Received utility updates are indexed as $\gamma_{r,s}^n[m]$, and corresponding arrival times as $t_{r,s}^n[m]$. The optimality of this filter is discussed in more detail in [10], however it is so because incoming network samples are assumed to be iid over a time period T, in which case an average is the best estimate of the process mean.

$$P_{r,s}^n \leftarrow \left|\{m : t - T \le t_{r,s}^n[m] \le t\}\right|^{-1} \sum_{m : t - T \le t_{r,s}^n[m] \le t} \gamma_{r,s}^n[m] \tag{4}$$

2.4 Sample Correlation

The value of the correlation function depends on the sort of filter that is used to estimate the path utility. Because different filters weight received packet values differently in order to generate their estimates, the correlations between successive estimates also differ. Equation 5 shows the output correlations for the exponential filter.

$$R_{exp}(\Delta t) = e^{-|\Delta t|\tau} \tag{5}$$

Equation 6 shows the output correlations of the box filter.

$$R_{\square}(\Delta t) = \begin{cases} 1 - \frac{|\Delta t|}{T} & , \quad |\Delta t| \leq T \\ 0 & , \quad |\Delta t| > T \end{cases} \tag{6}$$

Here, Δt is the elapsed time between estimates (packet arrivals). The functions are calculated according to standard methods by finding the inverse fourier transform of the power spectral density of a filtered uncorrelated random process [23].

3 Markovian Analysis

3.1 Steady State Routing Probabilities

With the general framework defined, it is now necessary to describe how to find the steady state routing solution based on a given forwarding equation and metric update scheme. The equilibrium solution can then be used to show how the local parameters affect the global routing pattern and how they may be adjusted in order to achieve a given performance level. Each node is represented as a state in a markov chain, and standard methods are used to find the statistics of paths from source to destination (represented as an absorbing state in the markov chain) [24].

First the average cost to a specific destination is calculated from any node. This determines the link utility estimates based on the routing probabilities. The per link packet arrival rate is then determined, which allows the average sample correlation to be set. An iterative algorithm is then shown which arrives at the equilibrium routing solution. First the average cost to a specific destination is calculated from any node. This determines the link utility estimates based on the routing probabilities. The per link packet arrival rate is then determined, which allows the average sample correlation to be set. An iterative algorithm is then shown which arrives at the equilibrium routing solution.

Average Cost to Destination. The average number of transitions from a source to the destination state is calculated. This models the average end-to-end path cost, and determines the link utility estimate for given routing probabilities. Suppose that the transition (or routing) probabilities for a network are given in a matrix p, shown in Equation 7.

$$\mathsf{p} = \begin{bmatrix} \mathsf{S} & \mathsf{T} \\ \mathsf{0} & \mathsf{I} \end{bmatrix} \tag{7}$$

In this matrix, S is a square $(\alpha - \beta) \times (\alpha - \beta)$ matrix representing the transition probabilities between the nonabsorbing states, T is a $(\alpha - \beta) \times \beta$ matrix representing the transition probabilities between the nonabsorbing states and the absorbing states, $\mathsf{0}$ is a $\beta \times (\alpha - \beta)$ matrix of zeros, and I is a $\beta \times \beta$ identity matrix representing the self-transition probabilities of the absorbing states. In this analysis, $\alpha = |\mathcal{V}|$ and $\beta = 1$, since there are $|\mathcal{V}|$ total nodes and only one destination node. The fundamental matrix of the system is Q, calculated in Equation 8. Here, $\mathsf{Q}_{i,j}$ describes the expected number of visits to state j, starting in state i, before arriving at an absorbing state. The exponent is the matrix inverse operator.

$$\mathsf{Q} = (\mathsf{I}_{(\alpha-\beta)\times(\alpha-\beta)} - \mathsf{S})^{-1} \tag{8}$$

The expected number of states visited when starting at state i before being absorbed (at the destination state) is the sum of each row of Q, and is calculated in Equation 9.

$$\mathsf{c}_i = \sum_{j \in \mathcal{V}-d} \mathsf{Q}_{i,j} \tag{9}$$

If the cost of each link is unity, such as with hop count, then c_i is also the expected cost from node i to the destination. In general, if the link costs are given in the matrix C such that $\mathsf{C}_{i,j}$ is the cost of the link from node i to j, then the cost from source to destination is calculated as in Equation 10.

$$\mathsf{c}_i = \sum_{j \in \mathcal{V}-d} \left[\mathsf{Q}_{i,j} \left(\sum_{k \in \mathcal{N}^j} \mathsf{C}_{j,k} \cdot \mathsf{P}_{j,k} \right) \right] \tag{10}$$

Equations 8 and 10 will be referred to as the function
$\mathsf{c} = avgCostToDestination(\mathsf{C}, \mathsf{p})$.

Per Link Packet Arrival Rates. The next step in calculating the equilibrium routing probabilities is to include the decay factor, R, since there is a finite packet arrival rate on each link. It is assumed that packets with routing information are sent between a single source and destination pair with independently and identically exponentially distributed interarrival times, with mean λ^{-1} seconds between packets. It is further assumed that as these packets make their way though the network towards the destination, that their arrival rate at any given node over any link is also independently distributed exponential with a link dependant mean sample rate, $\lambda_{i,j}$ packets per second. The packet arrival rate at node j from neighbor i when source s is sending a rate of λT packets per network correlation time unit to the destination, $(\lambda T)^s_{i,j}$, is shown in Equation 11.

$$(\lambda T)^s_{i,j} = \mathsf{Q}_{s,i} \cdot \mathsf{p}_{i,j} \cdot \lambda T \tag{11}$$

Expected Per Link Sample Correlation. The expected correlation between the current time and the previous estimate can then be calculated according to the filter shape used. Equation 12 shows the expected correlation for the exponential filter. The decay rate heuristic of $\tau = T^{-1}\ln(z)$ is used [10]. The constant z is small (ie., $z \approx 0.1$). $\Delta t \sim exp\left(\frac{1}{\lambda T}\right)$.

$$E\left[R_{exp}\left(\Delta t\right)\right] = \frac{\lambda T}{\lambda T - \ln(z)} \tag{12}$$

A similar analysis is shown in Equation 13 for the correlation of box filter.

$$E\left[R_{\square}\left(\Delta t\right)\right] = \begin{cases} \frac{\lambda T - 1}{\lambda T} & , \quad \lambda T \geq 1 \\ 0 & , \quad \lambda T < 1 \end{cases} \tag{13}$$

The critical parameter influencing the expected drop in correlation from one sample of the network to the next is the product λT. The units of this term my be thought of as *packets per network correlation time unit*, or *network samples per network correlation time unit*. Such language makes it clear that the probabilistic routing framework presented in this paper relies on samples of the network in order to make decisions, and that performance is improved (this term is used qualitatively at the moment) with higher sampling rates. This is especially so in the case of the box filter, where the correlation goes to zero when $|\Delta t| \geq T$, therefore a packet interarrival time of less than the network correlation time ($\lambda T \leq 1$) will leave (an average of) no information about the network at any node. This result is intuitive in the sense that if the network is sampled less often than the network itself changes, then the network will have never have current information about itself.

Steady State Algorithm. Algorithm 1 shows the how the steady state routing probabilities can be computed, based on a given source s and link costs. As long as certain parameter settings are avoided, such as $K = 0$, $q = 0$, or $F = \infty$, trivial routing solutions (ie., using only one path) are not generated.

Algorithm 1. Steady State Routing Probabilities with R

intialize p {random, nonzero, unity row sum}
while p not converged **do**
  $\mathsf{c} \leftarrow avgCostToDestination(\mathsf{C}, \mathsf{p})$ {Equations 8 and 10}
  for all $i, j \in (\mathcal{V} - d)$ **do**
    $\mathsf{P}_{i,j} \leftarrow (\mathsf{C}_{i,j} + \mathsf{c}_j)^{-1}$ {link utilities}
    $\mathsf{P}_{i,j} \leftarrow \mathsf{P}_{i,j} \cdot ER\left((\lambda T)^{s}_{i,j}\right)$
    {ER from Equation 12 or 13 depending on estimator}
    {$(\lambda T)^{s}_{i,j}$ from Equation 11}
  end for
  $\mathsf{p} = W(\mathsf{P}, K, F)$ {forwarding equation}
end while

4 Simulation Analysis

4.1 Analysis Overview

Using the previously described framework, this section examines the relationships between the sensitivity, threshold/noise, and sampling rate parameters on the equilibrium expected path cost between a source and destination. Their influence on each other and on the global routing performance is explained.

4.2 On Sensitivity and Noise

Figure 1 shows how the expected path cost between a source and destination vary with sensitivity and noise, using the Termite and ACO style forwarding equations. The network is randomly generated with 50 nodes, where node connectivity is determined by a uniform circular transceiver distance and link cost is inter-node distance squared, which is similar to an energy conservation metric. Transceiver range is set such that the average number of neighbors per node is eight. An infinite sample rate is assumed. The expected path cost approaches the minimum path cost as the routing sensitivity is increased. Because the threshold is under the influence of the sensitivity, the Termite forwarding equation will send all packets on the shortest path if the sensitivity is large enough. This is not the case for the ACO equation, where the noise and sensitivity are independent. For nonzero noise, there is a lower bound on the expected path cost which is strictly greater than the minimum path cost. The routing solution is tied to the network topology and associated link costs and source-destination distance.

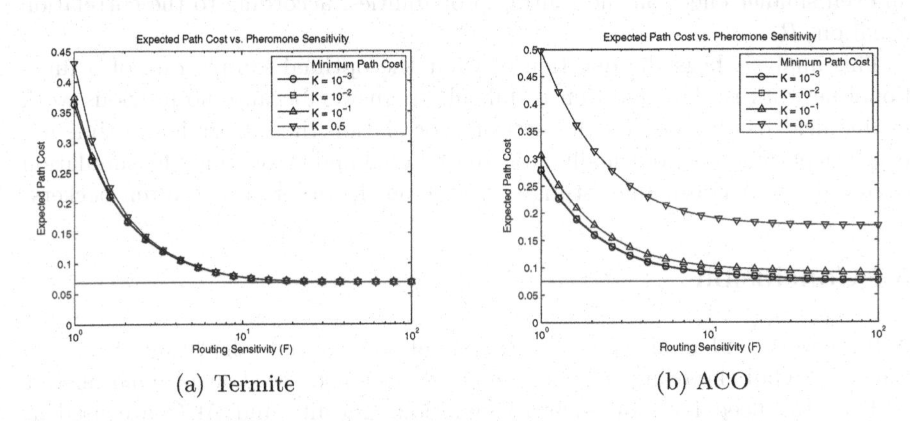

(a) Termite (b) ACO

Fig. 1. Path Cost Distribution vs. Routing Sensitivity (50 nodes)

4.3 Effect of Network Sample Rate

The network sample rate has a substantial effect on the performance of the routing algorithm. The samples are responsible for updating each node's estimate of how good each link is to arrive at each destination. Figure 2 shows the effect of the network sample rate on the expected path cost between source and destination. For clarity, the results are shown varied with F and λT, where $K = .1$ and $q = .05$ are held constant, using the box filter. "no λT" is equivalent to an infinite λT, as it removes the effect of the sample correlation function. A lower sample rate causes the algorithm to converge faster towards the shortest path than with a higher sample rate (assuming constant F). As good paths

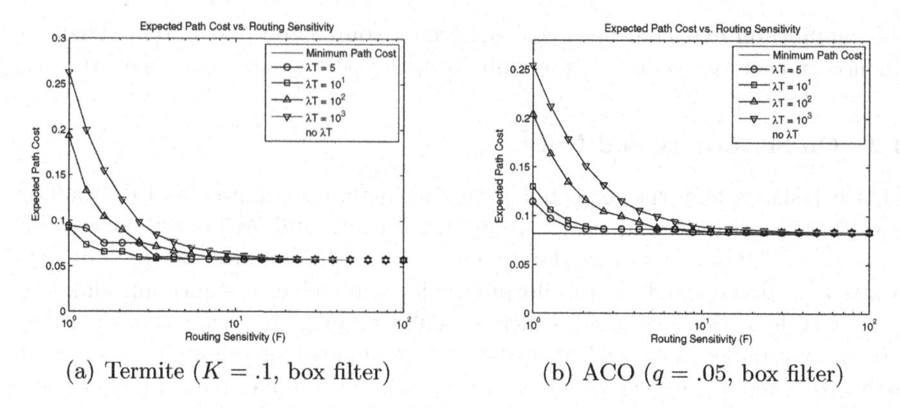

(a) Termite ($K = .1$, box filter)

(b) ACO ($q = .05$, box filter)

Fig. 2. Path Cost Distribution vs. Routing Sensitivity (50 nodes)

are found and their associated probabilities increase, the number of times that less attractive paths are sampled quickly drops (since only a limited number of samples are available). The infrequent arrival of new metric information on poor paths makes the currently stored information unreliable, the estimates then have an even smaller effect on the routing probabilities, according to the correlation equation, R.

This analysis begs the question of what the optimal sample rate of a functional network is. The question is difficult to answer because no previous work in this field has focused on the network correlation time. Correlation time dependant parameters are usually adjusted by trial and error to fit the simulation environment. A reasonable estimate *may be* on the order of one to ten packets.

5 Conclusion

A framework and model for the analysis of soft routing algorithms has been presented. Global routing behaviors can be determined based on local parameters without the need for monte-carlo simulation. Termite and ACO are used as examples to illustrate a two part framework including a forwarding equation and a metric updating equation. The continuous time metric update is reviewed, including the use of the filter correlation function to properly model the loss of metric information over time. An analysis of the filter correlation shows that the critical parameter for updating the network with new information is the network sampling rate, or the number of route metric updates sent per network correlation time.

A markov chain approach is used to find the steady state routing probabilities given the routing parameters and network costs. The expected path cost from source to destination can then be calculated. The cost is examined to reveal the tradeoffs between sensitivity, threshold, noise, and network sample rate in the context of average behavior and robustness against cost dynamics. A good parameter choice is a small threshold, a sensitivity proportional to the minimum path utility, and a sample rate proportional to the number of good paths between source and destination.

Future work should compare the results of the model analysis to monte-carlo simulation experiements. Steady state convergence time should also be considered, since dynamic networks may not offer enough time for the soft routing protocol to converge. The existence of an optimal sampling rate will also be explored, as the physical effects of a large sampling rate are detrimental to the network at large.

References

1. Tannenbaum A.: Computer Networks. Prentice Hall PTR (2002)
2. Boyan J., Littman M.: Packet Routing in Dynamically Changing Networks: A Reinforcement Learning Approach. In: Advances in Neural Information Processing Systems. Morgan Kaufmann, (1993)
3. Schoonderwoerd R., Holland O., Bruten J., Rothkrantz L.: Ant-Based Load Balancing In Telecommunications Networks. Adaptive Behavior (1996)
4. Di Caro G., Dorigo M.: Mobile Agents for Adaptive Routing. Technical Report, IRIDIA/97-12, Université Libre de Bruxelles, Belgium (1997)
5. Heusse M., Snyers D., Guérin S., Kuntz P.: Adaptive Agent-Driven Routing and Local Balancing in Communication Networks. ENST de Bretagne Technical Report RR-98001-IASC, (1997)
6. Baras J., Mehta H.: A Probabilistic Emergent Routing Algorithm for Mobile Adhoc Networks. In: WiOpt '03: Modeling and Optimization in Mobile, Ad-Hoc, and Wireless Networks (2003)
7. M. Günes, M. Kähmer, Bouazizi I.: Ant Routing Algorithm (ARA) for Mobile Multi-Hop Ad-Hoc Networks - New Features and Results. In: The Second Mediterranean Workshop on Ad-Hoc Networks (2003)
8. Heissenbüttel M., Braun T.: Ants-Based Routing in Large Scale Mobile Ad-Hoc Networks. Kommunikation in Verteilten Systemen (KiVS) (2003)
9. Sim K. M., Sun W. H.: Ant Colony Optimization for Routing and Load-Balancing: Survey and New Directions. IEEE Transactions on Systems, Man, and Cybernetics - Part A: Systems and Humans, **33**(5) (2003)
10. Roth M., Wicker S.: Termite: A Swarm Intelligent Routing Algorithm for Mobile Wireless Ad-Hoc Networks. Springer SCI Series: Swarm Intelligence and Data Mining, Springer (2005)
11. Rajagopalan S., Shen C.: A Routing Suite for Mobile Ad-hoc Networks using Swarm Intelligence. Unpublished (2004)
12. Ducatelle F., Di Caro G., Gambardella L. M.: Using Ant Agents to Combine Reactive and Proactive Strategies for Routing in Mobile Ad-Hoc Networks. Technical Report No. IDSIA-28-04-2004 (2004)
13. Wedde H., Farooq M.: The Wisdom of the Hive Applied to Mobile Ad-hoc Networks. IEEE Swarm Intelligence Symposium 2005 (SIS 2005) (2005)
14. Dowling J., Curran E., Cunningham R., Cahill V.: Using Feedback in Collaborative Reinforcement Learning to Adaptively Optimise MANET Routing. IEEE Transactions on Systems, Man and Cybernetics (Part A), Special Issue on Engineering Self-Orangized Distributed Systems, **35**(3) (2005) 360–372
15. http://www.faqs.org/rfcs/rfc2328.html

16. Perkins P., Royer E.: Ad-hoc On-demand Distance Vector. In: Proceedings of the 2nd IEEE Workshop on Mobile Computing Systems and Applications (1999)
17. Subramanian D., Druschel P., Chen J.: Ants and Reinforcement Learning: A Case Study in Routing in Dynamic Networks. In: Proceedings of the International Joint Conference on Artificial Intelligence (1997)
18. Roth M., Wicker S.: Asymptotic Pheromone Behavior in Swarm Intelligent MANETs: An Analytical Analysis of Routing Behavior. In: Sixth IFIP IEEE International Conference on Mobile and Wireless Communications Networks (MWCN) (2004)
19. Meuleau N., Dorigo M.: Ant Colony Optimization and Stochastic Gradient Descent. In: Artificial Life 8 (2002)
20. Dorigo M., Di Caro G.: Gambardella L. M., Ant Algorithms for Discrete Optimization. Artificial Life, **5**(2) (1999)
21. Eberhart R., Kennedy J.: Particle Swarm Optimization. In: Proceedings of the IEEE International Conference on Neural Networks (1995)
22. Gaertner D., Clark K.: On Optimal Parameters for Ant Colony Optimization algorithms. In: The International Conference on Artificial Intelligence (ICAI) (2004)
23. Brockwell P. J., Davis R. A.: Introduction to Time Series and Forecasting. Spinger-Verlag (2002)
24. Norris J.: Markov Chains. Cambridge University Press (1998)

A Stochastic Traffic Assignment Algorithm Based on Ant Colony Optimisation

Luca D'Acierno, Bruno Montella, and Fortuna De Lucia

Dept. of Transportation Engineering, 'Federico II' University, Naples, Italy
{dacierno, montella}@unina.it

Abstract. In this paper we propose a Stochastic User Equilibrium (SUE) algorithm that can be adopted as a model, known as a simulation model, that imitates the behaviour of transportation systems. Indeed, analyses of real dimension networks need simulation algorithms that allow network conditions and performances to be rapidly determined. Hence, we developed an MSA (*Method of Successive Averages*) algorithm based on the Ant Colony Optimisation paradigm that allows transportation systems to be simulated in less time but with the same accuracy as traditional MSA algorithms. Finally, by means of Blum's theorem, we stated theoretically the convergence of the proposed ACO-based algorithm.

1 Introduction

In design problems or in real-time management of transportation systems, it is necessary to have a simulation model that allows network performances and features to be defined for each alternative project or each management strategy. In analyses of real dimension networks, it is important that simulation models allow solutions to be obtained swiftly such that it is possible to explore a large number of alternative projects or simulate beforehand consequences of a strategy in terms of future (minutes or hours) network conditions. Most simulation algorithms used in the case of steady-state conditions (assumption of inter-period and intra-period stationarity) are based on the calculation of a sequence of network loading (assignment with a fixed-point approach, as proposed by [1]).

In this paper, we verify the possibility of developing a meta-heuristic algorithm that allows network flows to be calculated more quickly than by using traditional algorithms proposed in the literature. In particular, we steered our research into ant-based algorithms. These algorithms, developed about a decade ago (first papers were [2], [3], [4]) and based on the food source search of ant colonies, have shown their efficiency in terms of calculation time in many cases, such as in travelling salesman problems, quadratic assignment problems, job/shop scheduling, vehicle routing, and dynamic problems. An extended overview of ant-based algorithms can be found in [5].

In the literature, ACO algorithms have been developed for solving problems with many local optima where the initial value of variables could influence final

M. Dorigo et al. (Eds.): ANTS 2006, LNCS 4150, pp. 25–36, 2006.

solutions. Instead, in the case of the road traffic assignment problem the solution is unique (as shown in Sect. 3) and traditional algorithms yield solutions regardless of the initial values. Therefore, the aim of developing an ACO algorithm for solving the traffic assignment problem is related to the need to provide more efficient algorithms in terms of calculation times.

In the case of transportation systems, an algorithm based on the Ant Colony approach for solving a network design problem, consisting of choosing among a set of alternative projects, was proposed by [6]. In this case the problem was formulated as a bi-level programming problem where the upper level (the network design problem) was organised as an ACO-based algorithm and the lower level (the traffic assignment problem) as a traditional (non-ACO-based) algorithm.

An assignment equilibrium algorithm based on ACO that tends to load mainly minimum cost paths (deterministic approach) was proposed by [7]; the same authors compared its performance with that of Frank and Wolfe's algorithm ([8]), showing the efficiency of the ACO-based approach in simulation problems. Therefore, this paper can be considered an extension of ACO-based traffic assignment algorithms in the case of a stochastic approach.

An important aspect concerning the development of solution algorithms is the theoretical proof of convergence. Nevertheless, initially most of the literature analysed convergence properties only from a numerical point of view. However, in recent papers (such as [9], [10], [11],[12]) convergence is stated for some classes of ACO-based algorithms. In this context, the aim of the paper is to develop an ACO-based algorithm to solve the Stochastic User Equilibrium (SUE) problem and prove its convergence and efficiency. The paper is organised as follows: Sect. 2 describes the problem of traffic assignment and its analytical formulation; Sect. 3 proposes an overview of theoretical properties and solution algorithms of the traffic assignment problem; Sect. 4 shows the proposed model and the first results are summarised in Sect. 5; finally, Sect. 6 concludes and comments on prospects for future research.

2 The Traffic Assignment Problem

A transportation network can be defined as a *graph* whose links are associated functions called *link cost functions*. A graph, indicated as $G = (N, L)$, can be defined as an ordered pair of sets: the set of *nodes*, indicated as N, and the set of *links*, indicated as L, which is a set of pairs of nodes belonging to N ($L \subseteq N \times N$). It is possible to define a set N' of centroid nodes that are nodes belonging to N ($N' \subseteq N$) representing the beginning and/or the end of all trips simulated on the network. Under the assumption of intra-period stationarity, it is possible to associate to each link $l \in L$ a quantity called *link flow*, f_l, that expresses the average number of homogeneous units using link l in a time unit. In general, link flows can be user flows (number of travellers utilising the links considered) as well as vehicle flows (number of cars or buses moving on the links).

In a graph, a *path* may be defined, indicated with k, as an ordered sequence of consecutive links connecting two centroid nodes such that the final node of a link coincides with the initial node of the following link. In particular, in traffic assignment problems only acyclic paths are considered.

In general, the cost function of generic link $l \in L$, indicated as c_l, can be expressed as a non-negative function of link flows of the network, that is:

$$c_l = c_l(\boldsymbol{f}) \tag{1}$$

where $\boldsymbol{f}$ is the vector of link flows f_l, of dimensions $(n_{Links} \times 1)$, and $|L| = n_{Links}$. Moreover, it is possible to define a *path cost function*, C_k, that is equal to the sum of link costs of all links belonging to path k plus a term of *non-additive costs*, C_k^{NA} , i.e. costs that depend only on the path (such as road tolls at motorway entrance/exit points), that is:

$$C_k = C_k(\boldsymbol{f}) = \sum\nolimits_{l \in k} c_l(\boldsymbol{f}) + C_k^{NA} \tag{2}$$

It is worth noting that C_k, as defined above, is a non-negative function that refers only to acyclic paths. Let $\boldsymbol{A}$ be a binary matrix, called a *link-path incidence matrix*, of dimensions $(n_{Links} \times n_{Paths})$ where n_{Paths} is the number of paths of the considered graph, whose generic element $a_{l,k}$ is equal to 1 if link l belongs to path k and 0 otherwise. Equation (2) can be expressed as:

$$\boldsymbol{C} = \boldsymbol{A^T c}(\boldsymbol{f}) + \boldsymbol{C^{NA}} \tag{3}$$

where $\boldsymbol{C}$ is the vector of path costs and $\boldsymbol{C^{NA}}$ is the vector of non-additive path costs, both of dimensions $(n_{Paths} \times 1)$. The above equation is also known as the *supply model* of the considered network.

Most mathematical models simulating user choices are based on random utility theory and hypothesise that users are rational decision-makers maximising utility relative to their choices. In particular, the *perceived utility* (that is the utility perceived by each single user moving along path k), U_k, can be expressed as the sum of a *systematic utility*, V_k, which represents the mean value of the utility perceived by all decision-makers having the same choice context, and a *random residual*, ε_k, which represents the difference between the perceived and systematic utility, that is:

$$U_k = V_k + \varepsilon_k \tag{4}$$

With the above assumption, the probability of choosing path k among all feasible paths, all joining centroid nodes o and d, can be expressed as the probability that the perceived utility of path k is greater than that of all other available alternatives, that by means of (4), can be expressed as:

$$p[k \mid od] = Prob[U_k > U_h \qquad \forall h \neq k \in I_{od}] \tag{5}$$

where I_{od} is the set of all available paths that join origin node o with destination node d, and obviously $k \in I_{od}$. Moreover, it may be stated that, if the random residuals are independently and identically distributed according to a Gumbel random variable of zero mean and parameter θ (where the variance of the Gumbel variable is equal to $\pi^2\theta^2/6$), (5) can be expressed as:

$$p[k \mid od] = \exp(V_k/\theta) / \sum\nolimits_{h \in I_{od}} \exp(V_h/\theta) \tag{6}$$

The flow on path k, F_k, can be expressed as the travel demand flow between centroid node o and d, d_{od}, multiplied by the probability of choosing path k, that is:

$$F_k = p\,[k \mid od]\, d_{od} \tag{7}$$

Moreover, link flow of generic link l is equal to the sum of all path flows that utilise link l, that is:

$$f_l = \sum\nolimits_{k:l\in k} F_k \tag{8}$$

With the assumption that the systematic utility is equal to the opposite of the path cost ($V_k = -C_k$), it is possible to define a link flow function, called *network loading function*, that expresses link flows as a function of path costs, that is:

$$\boldsymbol{f} = \boldsymbol{AP}\,(-\boldsymbol{C})\,\boldsymbol{d} \tag{9}$$

where $\boldsymbol{d}$ is the vector of demand flows, of dimensions $(n_{Pairs} \times 1)$, n_{Pairs} is the number of available origin-destination pairs, and $\boldsymbol{P}$ is the path choice probability matrix, known as the *path choice map*, of dimensions $(n_{Paths} \times n_{Pairs})$, whose generic element $p_{k,od}$ expresses the probability that users travelling between origin-destination pair od choose path k (this term has been expressed in 5 as $p\,[k \mid od]$), while $p_{k,od} = 0$ if path k is not starting from o and/or ending in d. The above equation is also known as the *demand model* of the considered network.

Combining demand model (9) with supply model (3), the demand-supply interaction model, known as the *assignment model*, is obtained as a fixed-point model:

$$\boldsymbol{f}^* = \boldsymbol{AP}\left(-\boldsymbol{A}^T\boldsymbol{c}\,(\boldsymbol{f}^*) - \boldsymbol{C}^{NA}\right)\boldsymbol{d} \tag{10}$$

where the equilibrium solution $\boldsymbol{f}^*$ is a link flow vector that yields costs that generate a network loading vector that is equal to term $\boldsymbol{f}^*$.

The traffic assignment problem with the assumption of the rational decision-maker, expressed by (5), is known also as the *Stochastic User Equilibrium* (SUE) problem.

3 Theoretical Properties and Classical Solution Algorithms of the Traffic Assignment Problem

In the literature, two papers ([1] and [13]) state that the fixed-point problem, expressed by (10), has at least one solution if:

1. choice probability functions, $\boldsymbol{P}\,(-\boldsymbol{C})$, are continuous;
2. link cost functions, $\boldsymbol{c}\,(\boldsymbol{f})$, are continuous;
3. each OD pair is connected (i.e. $I_{od} \neq \Phi \quad \forall od$).

The first two conditions are verified by almost all functions proposed in the literature, and the third is related to the network framework that generally satisfies that condition. Moreover [1] and [13] state that the above fixed-point problem has at most one solution if:

1. route choice models are expressed by strictly increasing functions with respect to systematic utilities, that is:

$$\left[\boldsymbol{P}(\boldsymbol{V}') - \boldsymbol{P}(\boldsymbol{V}'')\right]^{\boldsymbol{T}} (\boldsymbol{V}' - \boldsymbol{V}'') > 0 \qquad \forall \boldsymbol{V}' \neq \boldsymbol{V}'' \tag{11}$$

   where $\boldsymbol{V}$ is the systematic utility vector, of dimensions $(n_{Paths} \times 1)$, whose k-th element is term V_k;
2. cost functions are expressed by monotone non-decreasing functions with respect to link flows, that is:

$$\left[\boldsymbol{c}(\boldsymbol{f}') - \boldsymbol{c}(\boldsymbol{f}'')\right]^{\boldsymbol{T}} (\boldsymbol{f}' - \boldsymbol{f}'') \geq 0 \qquad \forall \boldsymbol{f}', \boldsymbol{f}'' \in \boldsymbol{S_f} \tag{12}$$

   where $\boldsymbol{S_f}$ is the feasibility set of vectors $\boldsymbol{f}$.

It may be stated that the first condition is always satisfied if path choice models belong to Logit or Probit families; therefore, (6), known as Multinomial Logit, allows (11) to be verified. Besides, (12) is always satisfied by almost all functions proposed in the literature. Therefore with the above assumptions, we may state that the fixed-point solution exists and is unique.

An extension of Blum's theorem ([14]) was proposed by [1], where it was shown that a convergent solution algorithm for solving the fixed-point problem of a function $\boldsymbol{y} = \boldsymbol{\lambda}(\boldsymbol{x})$ with:

- $\boldsymbol{x} \in \boldsymbol{S_x}$ and $\boldsymbol{y} \in \boldsymbol{S_x}$ where $\boldsymbol{S_x}$ is non-empty, compact and convex set;
- a unique fixed-point $\boldsymbol{x}^* = \boldsymbol{\lambda}(\boldsymbol{x}^*)$;
- a function $\varphi(\boldsymbol{x}) \geq 0 \quad \forall \boldsymbol{x} \in \boldsymbol{S_x}$ where $\varphi(\boldsymbol{x})$ is continuous with first $\nabla\varphi(\boldsymbol{x})$ and second $\nabla^2\varphi(\boldsymbol{x})$ derivative continuous;
- $\nabla^2\varphi(\boldsymbol{x})^{\boldsymbol{T}} [\boldsymbol{\lambda}(\boldsymbol{x}) - \boldsymbol{x}] < 0 \quad \forall \boldsymbol{x} \in \boldsymbol{S_x}, \boldsymbol{x} \notin \boldsymbol{S_{\tilde{x}}}$ and $\nabla^2\varphi(\tilde{\boldsymbol{x}})^{\boldsymbol{T}} [\boldsymbol{\lambda}(\tilde{\boldsymbol{x}}) - \tilde{\boldsymbol{x}}] = 0$ $\forall \tilde{\boldsymbol{x}} \in \boldsymbol{S_{\tilde{x}}}$ where $\boldsymbol{S_{\tilde{x}}} \subseteq \boldsymbol{S_x}$ and $\boldsymbol{x}^* \in \boldsymbol{S_{\tilde{x}}}$;
- $|\varphi(\boldsymbol{x}) - \varphi(\boldsymbol{x}^*)| > 0 \quad \forall \boldsymbol{x} \in \boldsymbol{S_x}, \boldsymbol{x} \notin \boldsymbol{S_{\tilde{x}}}$ and $|\varphi(\tilde{\boldsymbol{x}}) - \varphi(\boldsymbol{x}^*)| = 0 \quad \forall \tilde{\boldsymbol{x}} \in \boldsymbol{S_{\tilde{x}}}$;
- $\boldsymbol{x}^{\boldsymbol{T}} \nabla^2\varphi(\boldsymbol{x}')\, \boldsymbol{x} = M < +\infty \quad \forall \boldsymbol{x}, \boldsymbol{x}' \in \boldsymbol{S_x}$;

can be formulated by the following recursive equation:

$$\boldsymbol{x}^{t+1} = \boldsymbol{x}^t + \mu_t \left(\boldsymbol{\lambda}\left(\boldsymbol{x}^t\right) - \boldsymbol{x}^t\right) \qquad \text{with } \boldsymbol{x}^t \in \boldsymbol{S}_x \tag{13}$$

if the sequence $\{\mu_t\}_{t>0}$ satisfies the following conditions:

$$\sum\nolimits_{t>0} \mu_t = +\infty \qquad \sum\nolimits_{t>0} (\mu_t)^2 = M < +\infty \tag{14}$$

Moreover, if the sequence $\{\mu_t\}_{t>0}$ satisfies the condition:

$$\mu_t \in \left]0, 1\right] \tag{15}$$

then the elements of the sequence described by (13) belong to set $\boldsymbol{S_x}$, which is convex. A sequence $\{\mu_t\}_{t>0}$ which satisfies both (14) and (15) is given by $\{\mu_t = 1/t\}_{t>0}$ such that (13) becomes:

$$\boldsymbol{x}^{t+1} = \boldsymbol{x}^t + (1/t)\left(\boldsymbol{\lambda}\left(\boldsymbol{x}^t\right) - \boldsymbol{x}^t\right) = (1 - 1/t)\,\boldsymbol{x}^t + \boldsymbol{\lambda}\left(\boldsymbol{x}^t\right)/t \tag{16}$$

With the above assumption we may develop a solution algorithm known as the *Method of Successive Averages* (MSA). In particular, [1] proposed a *Flow Averaging* algorithm (MSA-FA) and a *Cost Averaging* algorithm (MSA-CA) based respectively on the following sequences:

$$\boldsymbol{f}^{t+1} = \boldsymbol{f}^t + (1/t)\left(\boldsymbol{f}\left(\boldsymbol{c}\left(\boldsymbol{f}^t\right)\right) - \boldsymbol{f}^t\right) \in \boldsymbol{S_f} \qquad \text{with } \boldsymbol{f}^1 \in \boldsymbol{S_f} \tag{17}$$

$$\boldsymbol{c}^{t+1} = \boldsymbol{c}^t + (1/t)\left(\boldsymbol{c}\left(\boldsymbol{f}\left(\boldsymbol{c}^t\right)\right) - \boldsymbol{c}^t\right) \in \boldsymbol{S_c} \text{ with } \boldsymbol{c}^1 = \boldsymbol{c}\left(\boldsymbol{f}^1\right) \in \boldsymbol{S_c};\ \boldsymbol{f}^1 \in \boldsymbol{S_f} \tag{18}$$

where $\boldsymbol{f}(\cdot)$ is the network loading function described by (9). Moreover, in order to prove the convergence of algorithms, assuming that solution existence and uniqueness conditions hold, it is necessary to verify that link cost functions have a symmetric continuous Jacobian $Jac\left[\boldsymbol{c}(\boldsymbol{f})\right]$ over set $\boldsymbol{S_f}$, for MSA-FA and choice map functions, which are expressed by (9), are additive and continuous with continuous first derivative for algorithm MSA-CA. Indeed, [1] shows that in the case of algorithm MSA-FA with the assumption that $\boldsymbol{\lambda}(\cdot) = \boldsymbol{f}(\boldsymbol{c}(\boldsymbol{f}))$ and $\varphi(\cdot) : \nabla\varphi(\cdot) = \boldsymbol{c}(\boldsymbol{f}) - \boldsymbol{c}^*$ where $\boldsymbol{c}^* = \boldsymbol{c}(\boldsymbol{f}^*)$, it is possible to state that the Jacobian condition allows all convergence requirements to be satisfied. Likewise, in the case of algorithm MSA-CA, [1] shows that the assumption should be $\boldsymbol{\lambda}(\cdot) = \boldsymbol{c}(\boldsymbol{f}(\boldsymbol{c}))$ and $\varphi(\cdot) : \nabla\varphi(\cdot) = \boldsymbol{f}^* - \boldsymbol{f}(\boldsymbol{c})$ where $\boldsymbol{f}^* = \boldsymbol{f}(\boldsymbol{c}^*)$. Importantly, these conditions are generally satisfied by almost all functions proposed in the literature. For both algorithms, the termination test is:

$$\left|\left(f_l\left(\boldsymbol{c}\left(\boldsymbol{f}^{t-1}\right)\right) - f_l^{t-1}\right)/f_l^{t-1}\right| < \varepsilon \qquad \forall l \in L \tag{19}$$

where f_l^{t-1} is the flow of link l at iteration $(t-1)$-th, $f_l\left(\boldsymbol{c}\left(\boldsymbol{f}^{t-1}\right)\right)$ is the network loading flows associated to link l at iteration t-th and ε is the algorithm threshold.

A network loading algorithm, known as *Dial's algorithm*, was proposed by [15]. This algorithm allows function $\boldsymbol{y} = \boldsymbol{f}(\boldsymbol{c})$ to be calculated with the use of the path choice model described by (6), for managing real size networks based on three phases: path generation, weight calculation and network loading.

In the first phase, a shortest tree algorithm from each destination node d is performed in order to associate to each node i a value $Z_{d,i}$ that expresses the cost of reaching the destination node d. A link $l = (i,j)$ can be considered for a feasible path only if $Z_{d,i} > Z_{d,j}$. Paths generated with this approach are called *Dial efficient paths*. In the second phase, for each destination d, a weight is associated to each link $l = (i,j)$, indicated as $w_{d,(i,j)}$, and to each node i, $W_{d,i}$, starting from the destination node d, with $W_{d,d} = 1$, and continuing with the other nodes by increasing $Z_{d,i}$ values, such that:

$$w_{d,(i,j)} = \begin{cases} \exp\left(\frac{-c_{(i,j)}}{\theta}\right) W_{d,j} & \text{if } Z_{d,i} > Z_{d,j} \\ 0 & \text{if } Z_{d,i} \le Z_{d,j} \end{cases} \qquad W_{d,i} = \sum_{(i,h) \in FS(i)} w_{d,(i,h)} \tag{20}$$

where $FS(i)$ is the set of links belonging to the forward star of node i. With the assumption that the probability of choosing link l, with $l = (i,j)$, at diversion node i is equal to the ratio of Dial weights, that is:

$$p\left[l \mid i\right] = w_{d,(i,j)}/W_{d,j} \tag{21}$$

we may develop the third phase that consists, for each destination d, in associating a term to each link $l = (i, j)$, indicated as $e_{d,(i,j)}$, and to each node i, $E_{d,i}$. These terms can be determined by setting for each origin o the value of $E_{d,o}$ equal to d_{od}. Then, starting from the centroid node with the maximum value of $Z_{d,i}$ and continuing with the other nodes by decreasing $Z_{d,i}$ values, we apply the following relations:

$$e_{d,(i,j)} = E_{d,i}\, p\left[l \mid i\right] = E_{d,i}\, w_{d,(i,j)} / W_{d,j} \qquad E_{d,j} = \sum\nolimits_{(h,j) \in BS(j)} e_{d,(h,j)} \tag{22}$$

where $BS\left(j\right)$ is the set of links belonging to the backward star of node j.

Finally, network loading flows can be calculated as:

$$f_{l=(i,j)} = \sum\nolimits_{d} e_{d,(i,j)} \tag{23}$$

It is worth noting that on applying Dial's algorithm for calculating network loading flows, it is necessary to implement the path generation phase (phase 1) only once to ensure the convergence of MSA algorithms. Indeed, as shown by [16], if this phase were performed at each iteration, the set of Dial efficient paths could change at each iteration and therefore (11), which is a necessary condition for verifying convergence of solution algorithms, could not be verified.

4 The Proposed Assignment Algorithm

In this paper, we propose an ACO solution algorithm for the fixed-point problem (10) based on the following assumptions:

- for each *od* pair there is an ant colony with its nest (centroid o) and its food source (centroid node d). Since every colony has a distinctive kind of pheromone, ants can recognise only paths utilised by the same colony. This hypothesis was formerly introduced by [7];
- the initial *intensity of the pheromone trail* on each link l, associated to ant colony *od*, indicated as $\tau^0_{od,l}$, is a function of path costs, that is:

$$\tau^0_{od,l} = \sum\nolimits_{k:l \in k} T^0_{od,k} \qquad \text{with } T^0_{od,k} = \begin{cases} \exp\left(-C^0_k/\theta\right) & \text{if } k \in I_{od} \\ 0 & \text{if } k \notin I_{od} \end{cases} \tag{24}$$

  where $C^0_k = C_k\left(\boldsymbol{f}^0\right)$ and $\boldsymbol{f}^0 \in \boldsymbol{S_f}$;
- the probability of choosing link l, with $l = (i, j)$, at diversion node i (known in the literature as *transition probability*), at iteration t, can be expressed as:

$$p^t\left[l \mid i\right] = \tau^t_{od,l} \Big/ \sum\nolimits_{l' \in FS(i)} \tau^t_{od,l'} \tag{25}$$

  where $FS\left(i\right)$ is the set of links belonging to the forward star of node i. In this case, the *visibility term* (generally indicated as $\eta_{od,l}$) is equal to 1;

– the 'ant' (or 'vehicle') flow on the path k is equal to:

$$F_{od,k}^{t} = d_{od} \prod\nolimits_{l \in k, i \in k} p^{t} [l \mid i] \tag{26}$$

It may be stated that (26) is equal to the flow expression proposed by [7];

– the *updating of the pheromone trail* can be expressed as:

$$\tau_{od,l}^{t+1} = (1 - \rho)\, \tau_{od,l}^{t} + \rho \Delta \tau_{od,l}^{t+1} \tag{27}$$

where *evaporation coefficient* ρ is variable and equal to $1/t$. The variability of this coefficient was introduced by [17], even though the expression was different;

– each ant is provided with a memory that stores the sequences of used links. This property allows ants to update the pheromone trail at each iteration only if they are the first to arrive, after reaching the food source, at their nest;

– the increase in the pheromone trail is based on a global approach, that is all links are simultaneously updated. This assumption, in combination with the ant memory property, can be expressed by a function of path costs, that is:

$$\Delta \tau_{od,l}^{t} = \sum\nolimits_{k: l \in k} \Delta T_{od,k}^{t} \quad \text{with } \Delta T_{od,k}^{t} = \begin{cases} \exp\left(-C_{k}^{t}/\theta\right) & \text{if } k \in I_{od} \\ 0 & \text{if } k \notin I_{od} \end{cases} \tag{28}$$

The usefulness of a global approach in transportation problems was highlighted by [6].

With the above hypotheses, it may be stated that the application of the proposed ACO algorithm in the case of a transportation network is equivalent to the application of an MSA algorithm where the successive averages are applied to weights of Dial's algorithm, that is:

$$w_{d,(i,j)}^{t+1} = w_{d,(i,j)}^{t} + \frac{\Delta w_{d,(i,j)}^{t+1} - w_{d,(i,j)}^{t}}{k} \qquad W_{d,i}^{t+1} = W_{d,i}^{t} + \frac{\Delta W_{d,i}^{t+1} - W_{d,i}^{t}}{k} \tag{29}$$

Indeed, in the simple case of a single *od* pair joined by only two paths, four cases may be found: the two paths do not have any link in common; the two paths share a part between a generic node and the destination node; the two paths share a part between the origin node and a generic node; the two paths share two parts: the first between the origin node and a generic node and the second between a generic node and the destination node.

In the first case, the two paths have only origin and destination nodes in common, hence the diversion node is the origin node and the diversion node probabilities are equal because $\tau_{od,l=(o,i)}^{t} = w_{od,l=(o,i)}^{t}$ for each link leaving origin node o. In the second case, the diversion node is again the origin one, hence this case is similar to the previous one. In the third case, the diversion node is the node at the end of the common part. Hence, if $C_{k_1}^{t}$ and $C_{k_1}^{t}$ are path costs respectively of the first and second path, and $C^t_{k^*}$ is the path cost of the common part, the Ant diversion probability is equal to:

$$\begin{aligned} p^t_{Ant}[k_1] &= 1/\left(1+\exp\left(-\left(C^t_{k_2}-C^t_{k_1}\right)/\theta\right)\right) = \\ &= 1/\left(1+\exp\left(-\left(\left(C^t_{k^*}+C^t_{k_2-k^*}\right)-\left(C^t_{k^*}+C^t_{k_1-k^*}\right)\right)/\theta\right)\right) = \\ &= 1/\left(1+\exp\left(-\left(C^t_{k_2-k^*}-C^t_{k_1-k^*}\right)/\theta\right)\right) \end{aligned} \quad (30)$$

whereas the Dial diversion probability is equal to:

$$p^t_{Dial}[k_1] = 1/\left(1+\exp\left(-\left(C^t_{k_2-k^*}-C^t_{k_1-k^*}\right)/\theta\right)\right) \quad (31)$$

which is equal to the Ant one. Hence, also in this case, the two approaches provide the same probabilities. Finally, the last case is similar to the third and therefore provides the same results.

Similarly, the two approaches may be stated as providing the same probability even when the number of paths, joining the same OD pair, is greater than two. Therefore it is possible to state the perfect equivalence of diversion node probabilities between the Ant approach (25) and the Dial approach (21).

With the use of the extension of Blum's theorem ([14]) proposed by [1], we may state the convergence of the proposed ACO-based MSA algorithm, assuming that existence and uniqueness conditions hold, if $Jac[\boldsymbol{c}(\boldsymbol{\tau})]$ is symmetric and continuous. Indeed, this condition satisfies all hypotheses of Blum's theorem in the case of the fixed-point problem $\boldsymbol{\tau} = \boldsymbol{\tau}(\boldsymbol{c}(\boldsymbol{f}(\boldsymbol{\tau})))$ where $\boldsymbol{\tau}$ is a vector whose generic element is the pheromone trail $\tau_{od,l}$ (or equivalently $w_{od,l}$), of dimensions $((n_{Pairs}\cdot n_{Links})\times 1)$, and with the assumption that $\boldsymbol{\lambda}(\cdot) = \boldsymbol{\tau}(\boldsymbol{c}(\boldsymbol{f}(\boldsymbol{\tau})))$ and $\varphi(\cdot):\nabla\varphi(\cdot) = \boldsymbol{c}(\boldsymbol{f}(\boldsymbol{\tau}))-\boldsymbol{c}^*$ where $\boldsymbol{c}^* = \boldsymbol{c}(\boldsymbol{f}(\boldsymbol{\tau}^*))$. The proof of convergence in this case is similar to that proposed by [1], hence for brevity it is not reported in this paper. However, sufficient conditions to verify the $Jac[\boldsymbol{c}(\boldsymbol{\tau})]$ hypothesis are that link cost functions are separable (i.e. $c_l = c(f_l)$) and have a symmetric and continuous Jacobian $Jac[\boldsymbol{c}(\boldsymbol{f})]$ over set $\boldsymbol{S_f}$, and choice map functions, which are expressed by (9), are additive and continuous with the continuous first derivative. Indeed, since $Jac[\boldsymbol{c}(\boldsymbol{\tau})] = Jac[\boldsymbol{c}(\boldsymbol{f})]\,Jac[\boldsymbol{f}(\boldsymbol{\tau})]$ the condition on $Jac[\boldsymbol{c}(\boldsymbol{\tau})]$ is verified. Moreover, the above conditions are generally satisfied by almost all functions proposed in the literature and therefore convergence of the proposed algorithm may be postulated.

5 First Results

In order to verify the efficiency of the proposed MSA algorithm based on Ant Colony Optimisation, it was applied to simulate traffic conditions in the case of two Italian real dimension networks: the network of Salerno (a city of about 140,000 inhabitants) and the network of Naples (a city of about 1,000,000 inhabitants). Table (1) shows features of the analysed networks, whereas Tables (2) and (3), as well as Figs. (1) and (2), indicate algorithm performances. In both networks, the proposed algorithm (indicated as MSA-ANT) is shown to provide

Table 1. Network features

City name	*Number of links*	*Number of nodes*	*Number of centroid nodes*	*Number of OD pairs*	*Number of peak-hour trips*
Salerno	1,133	529	62	3,844	21,176
Naples	5,750	3,075	167	26,916	118,764

Table 2. Algorithm performance on the Salerno network

Algorithm name	*Number of interations*	*Convergence* $> 90.0\%$	*Calculation time* [*s*]	*Solution error*
$MSA - FA$	54	12	29	Reference
$MSA - CA$	7	5	5	0.004 %
$MSA - ANT$	5	3	4	0.004 %

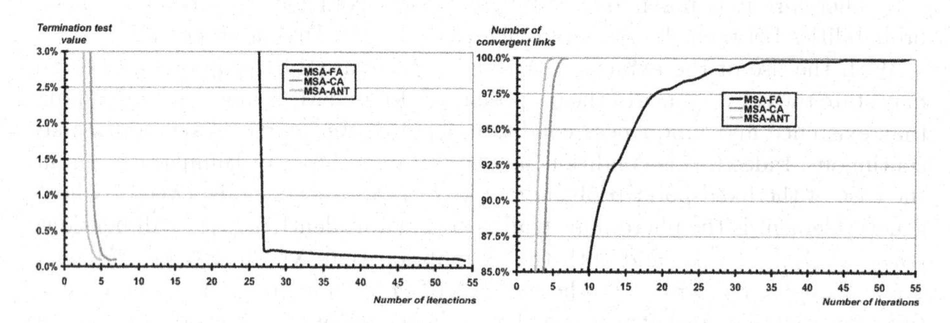

Fig. 1. Algorithm performance on the Salerno network

Table 3. Algorithm performance on the Naples network

Algorithm name	*Number of interations*	*Convergence* $> 90.0\%$	*Calculation time* [*min*]	*Solution error*
$MSA - FA$	59	11	42.15	Reference
$MSA - CA$	14	4	11.26	0.013 %
$MSA - ANT$	9	4	7.65	0.006 %

the same solution as the traditional algorithm (algorithm thresholds being equal to 0.1% and 1.0% respectively in the case of Salerno and Naples) in lower calculation times. In particular, calculation times of the proposed algorithm are lower than 86.2% in the first network and 81.9% in the second network with respect to the case of algorithm MSA-FA. Instead, these values become 20.0% and 32.1% with respect to the case of algorithm MSA-CA. In terms of number of iterations, the proposed algorithm requires 90.7% and 84.7% less with respect

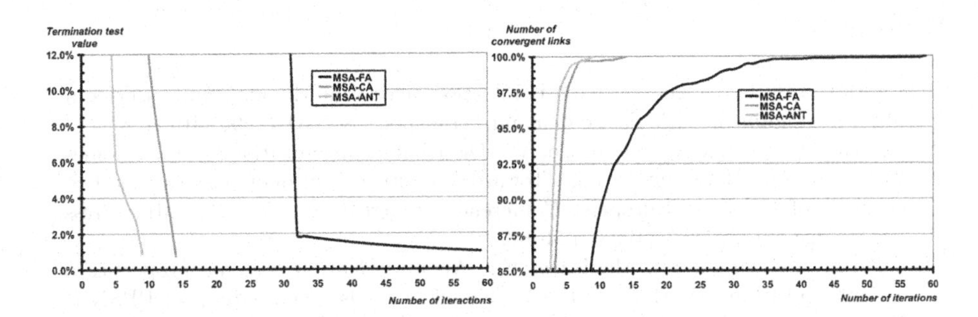

Fig. 2. Algorithm performance on the Naples network

the algorithm MSA-FA; these values become 28.6% and 35.7% in the case of algorithm MSA-CA. Finally, if we accept that the convergence of algorithm is achieved when only 90% of links satisfy termination test (19) then the reductions in number of iterations would be 75.0% and 63.6% in the case of algorithm MSA-FA, and 40.0% and 0.0% in the case of algorithm MSA-CA.

6 Conclusions and Research Prospects

In this paper we proposed an ACO algorithm that can be utilised for solving the road traffic assignment problem. In particular, we showed that the proposed algorithm has an MSA framework where averages are applied to Dial's weights (as shown in 29) and stated the perfect equivalence in terms of path choice behaviours between artificial ants (with the proposed approach) and road users (simulated with the traditional algorithms of traffic assignment). Moreover we stated theoretically the convergence of the proposed MSA-ANT algorithm by means of the extension of Blum's theorem ([14]) proposed by [1], and numerically its efficiency in terms of calculation time with respect to traditional MSA algorithms. Hence, the proposed algorithm can be considered an extension of ACO-based assignment algorithms in the case of a stochastic assumption on the path choice model (the previous paper [7] was based on a deterministic approach). Moreover, theoretical proof on convergence overcomes limitations resulting from the use of numerical proof in [7]. Finally, the new MSA-ANT could be used to develop an ACO-based algorithm also in the lower level of paper [6] in order to speed up solution search.

In terms of future research, we propose an extension of the proposed algorithm in the case of more complex path choice models (such as C-Logit and Probit) and in the case of preventive-adaptive choice behaviour that is typical of mass transit system users (i.e. the hyper-path choice approach proposed by [18]). Finally, we advocate using the proposed algorithm as a simulation model for imitating the behaviour of transportation systems in network design problems or in real-time management in order to highlight the advantages of the ACO approach.

References

1. Cantarella, G.E.: A general fixed-point approach to multimodal multi-user equilibrium assignment with elastic demand. Transport. Sci. **31** (1997) 107–128
2. Colorni, A., Dorigo, M., Maniezzo, V.: Distributed optimization by ant colonies. In Varela, F.J., Bourgine, P., eds.: Toward a practice of autonomous systems: Proceedings of the First European Conference on Artificial Life. The MIT Press, Cambridge (MA), USA (1992) 134–142
3. Colorni, A., Dorigo, M., Maniezzo, V.: An investigation of some properties of an ant algorithm. In Männer, R., Manderick, B., eds.: Proceedings of PPSN II. Elsevier, Amsterdam (1992) 509–520
4. Dorigo, M.: Optimization, learning and natural algorithms (in Italian). PhD thesis, Department of Electronics, Polytechnic of Milan, Italy (1992)
5. Dorigo, M., Stützle, T.: Ant colony optimization. The MIT Press, Cambridge (MA), USA (2004)
6. Poorzahedy, H., Abulghasemi, F.: Application of ant system to network design problem. Transportation **32** (2005) 251–273
7. Mussone, L., Ponzi, M., Matteucci, M.: Ant colony optimization for stochastic user equilibrium. In Bifulco, G.N., ed.: I sistemi stradali di trasporto nella societá dell'informazione. Aracne Editrice, Rome, Italy (2005) 175–199
8. Frank, M., Wolfe, P.: An algorithm for quadratic programming. Nav. Res. Log. Quart. **3** (1956) 95–110
9. Gutjahr, W.J.: A graph-based ant system and its convergence. Future Gener. Comp. Sy. **16** (2000) 873–888
10. Gutjahr, W.J.: A converging ACO algorithm for stochastic combinatorial optimization. In Albrecht, A., Steinhöfel, F., eds.: Stochastic algorithms: Foundations and applications. Volume 2827 of LNSC. Springer-Verlag, Berlin (2003) 10–25
11. Chang, H.S., Gutjahr, W.J., Yang, J., Park, S.: An ant system based approach to markov decision processes. In: Proceedings of the 2004 American Control Conference, Boston (MA), USA (2004) 3820–3825
12. Dorigo, M., Blum, C.: Ant colony optimization theory: a survey. Theor. Comput. Sci. **344** (2005) 243–278
13. Cascetta, E.: Transportation systems engineering: theory and methods. Kluwer Academic Publisher, Dordrecht, The Netherlands (2001)
14. Blum, J.R.: Multidimensional stochastic approximation methods. Ann. Math. Stat. **25** (1954) 737–744
15. Dial, R.B.: A probabilistic multipath traffic assignment model which obviates path enumeration. Transport. Res. **5** (1971) 83–111
16. D'Acierno, L., Montella, B., Gallo, M.: Multimodal assignment to congested networks: fixed-point models and algorithms. In: Proceedings of European Transport Conference 2002, Cambridge, United Kingdom (2002)
17. Li, Y., Gong, S.: Dynamic ant colony optimisation for TSP. Int. J. Adv. Manuf. Tech. **22** (2003) 528–533
18. Nguyen, S., Pallottino, S., Gendreau, M.: Implicit enumeration of hyperpaths in a logit model for transit networks. Transport. Sci. **32** (1998) 54–64

An Analysis of the Different Components of the AntHocNet Routing Algorithm

Frederick Ducatelle, Gianni A. Di Caro, and Luca Maria Gambardella*

Istituto "Dalle Molle" di Studi sull'Intelligenza Artificiale (IDSIA)
Manno-Lugano, Switzerland
{frederick, gianni, luca}@idsia.ch

Abstract. Mobile ad hoc networks are a class of highly dynamic networks. In previous work, we developed a new routing algorithm, called AntHocNet, for these challenging network environments. AntHocNet has been designed after the Ant Colony Optimization (ACO) framework, and its general architecture shares strong similarities with the architectures of typical ACO implementations for network routing. On the other hand, AntHocNet also contains several elements which are new to ACO routing implementations, such as the combination of ant-based path sampling with a lightweight information bootstrapping process, the use of both reactive and proactive components, and the use of composite pheromone metrics. In this paper we discuss all these elements, pointing out their general usefulness to face the multiple challenges of mobile ad hoc networks, and perform an evaluation of their working and effect on performance through extensive simulation studies.

1 Introduction

Mobile Ad Hoc Networks (MANETs) [1] are networks in which all nodes are mobile and communicate with each other via wireless connections. Nodes can join or leave at any time. There is no fixed infrastructure. All nodes are equal and there is no central control or overview. There are no designated routers: nodes serve as routers for each other, and data packets are forwarded from node to node in a multi-hop fashion. MANETs are useful to bring wireless connectivity in infrastructureless areas or to provide instantaneous connectivity free of charge inside specific user communities and/or geographic areas. However, the control of a MANET is very challenging. Its topology and traffic patterns are defined by the present users, their positions and radio ranges. The effectively available bandwidth is defined by the characteristics of the wireless signal between the nodes, and by the amount of simultaneous contention to access the shared wireless medium. Due to the mobility and the constant arrival/departure of users,

* This work was supported by the Future & Emerging Technologies unit of the European Commission through project "BISON: Biology-Inspired techniques for Self-Organization in dynamic Networks" (IST-2001-38923) and by the Swiss Hasler Foundation through grant DICS-1830.

M. Dorigo et al. (Eds.): ANTS 2006, LNCS 4150, pp. 37–48, 2006.

all these characteristics, which make up the mode of the network, change over time, and different modes can coexist in different parts of the network.

Routing is at the core of the functioning of a MANET, and the challenges mentioned above call for a fully adaptive, multi-modal routing controller. We believe that the multidimensional complexity of the task makes it necessary to include multiple learning, adaptive, and behavioral components in the design of the routing algorithm. This is the approach we followed in *AntHocNet* [2,3,4,5,6]. It combines *Monte Carlo path sampling and learning using ant agents*, which is characteristic of the ACO framework [7], with an *information bootstrapping process*, which is typical for dynamic programming and some reinforcement learning approaches [8]. Operating the two learning mechanisms at different speeds allows to obtain an adaptivity, robustness and efficiency which neither of the subcomponents could offer on its own. Moreover, the use of both proactive and reactive behaviors allows to both anticipate and respond in timely fashion to sudden disruptive events. AntHocNet's design also includes the use of multiple metrics (e.g., number of hops and signal quality) in the definition of the pheromone variables used to guide ant decisions.

AntHocNet was inspired by previous work on ACO routing for wired networks [9], but its composite design represents a departure from previous instances of ACO algorithms for routing. The effective integration of a bootstrapping-based mechanisms within a typical ACO architecture is in fact an approach which is innovative in general terms, and not only concerning the application of ACO to network problems. Furthermore, the way we integrated the two mechanisms in AntHocNet is rather general and could be applied with success also to different application domains. Moreover, while the combined use of reactive and proactive ant generation, as well as the use of a composite pheromone metric, are not totally a novelty in ACO routing algorithms, the way these schemes are used is original and general (e.g., see [10,11,12] for examples of other ACO algorithms for MANETs).

The purpose of this paper is to report an experimental analysis of the role and effect of these different design components of the algorithm. In particular, we show the effect on performance of using bootstrapping and proactive components, of adopting different choices for the composite pheromone metric used to guide ant activities, and of selecting different levels of exploration. Even if this *sensitivity analysis* is specific for AntHocNet, we believe that to a certain extent the reported results can also provide general insights about all the considered issues, and in particular about the integration of ant-based path sampling with pheromone bootstrapping mechanisms.

The general effectiveness of AntHocNet's integrated approach was assessed in a number of papers [2,3,4,5,6] and a technical report [13]. Over a wide range of scenarios with different characteristics in terms of mobility, data traffic load, modality, etc., AntHocNet always showed excellent performance compared to state-of-the-art MANET routing algorithms such as AODV [14] and OLSR [15]. In this paper we focus on the sensitivity analysis and we do not report any further results concerning AntHocNet's general performance.

The rest of the paper is organized as follows. In Section 2 we provide a concise description of AntHocNet (for more details the reader can consult the mentioned references). Section 3 describes the experimental methodology and the general characteristics of the simulation environment. In Section 4 and its subsections, we report the results of the experimental analysis and discuss them.

2 The AntHocNet Routing Algorithm

In MANET jargon AntHocNet is termed a *hybrid* algorithm since it makes use of both reactive and proactive strategies to establish routing paths. It is *reactive* in the sense that a node only starts gathering routing information for a specific destination when a local traffic session needs to communicate with the destination and no routing information is available. It is *proactive* because as long as the communication starts, and for the entire duration of the communication, the nodes proactively keep the routing information related to the ongoing flow up-to-date with network changes. In this way both the costs and the number of paths used by each running flow can reflect the actual status of the network, providing an optimized network response. The reactive component of the algorithm deals with the phase of *path setup* and is totally based on the use of *ACO ant agents* to find a good initial path. Routing information is encoded in node *pheromone tables*. The proactive component implements *path maintenance and improvement*, proactively adapting during the course of a session the paths the session is using to network changes. Path maintenance and improvement is realized by a combination of ant path sampling and slow-rate *pheromone diffusion*: the routing information obtained via ant path sampling is spread between the nodes of the MANET and used to update the routing tables according to a bootstrapping scheme that in turn provide main guidance for the ant path exploration. *Link failures* are dealt with using a local path repair process or via explicit notification messages. *Stochastic decisions* are used both for ant exploration and to distribute data packets over multiple paths.

In the following we provide a concise description of each of these components. The component dealing with link failures is not described since it is neither central for the algorithm nor relevant for the sensitivity analysis reported here.

2.1 Metrics for Path Quality and Pheromone Tables

Paths are implicitly defined by tables of pheromone variables playing the role of routing tables. An entry $\mathcal{T}^i_{nd} \in \mathbb{R}$ of the pheromone table $\mathcal{T}^i$ at node i contains a value indicating the estimated goodness of going from i over neighbor n to reach destination d. Since AntHocNet only maintains information about destinations which are active in a communication session, and the neighbors of a node change continually, the filling of the pheromone tables is sparse and dynamic. Several different metrics, such as *number of hops, end-to-end delay, signal quality, congestion, etc.*, can be used to define the goodness of a path. AntHocNet makes use of a combination of these metrics to define the pheromone variables.

The effect of different combinations of metrics is studied in Subsection 4.1.

2.2 Reactive Path Setup

When a source node s starts a communication session with a destination node d, and it does not have routing information for d, it broadcasts a *reactive forward ant.* At each node, the ant is either unicast or broadcast, according to whether or not the current node has pheromone information for d. If information is available, the ant chooses its next hop n with the probability P_{nd} which depends on the relative goodness of n as a next hop, expressed in the pheromone variable $\mathcal{T}^i_{nd}$:

$$P_{nd} = \frac{(\mathcal{T}^i_{nd})^\beta}{\sum_{j \in \mathcal{N}^i_d} (\mathcal{T}^i_{jd})^\beta}, \quad \beta \geq 1, \tag{1}$$

where $\mathcal{N}^i_d$ is the set of neighbors of i over which a path to d is known, and β is a parameter which controls the exploratory behavior of the ants. If no pheromone is available, the ant is broadcast. Since it is quite likely that somewhere along the path no pheromone is found, in the experiments we normally use a high value of β to avoid excessive ant proliferation. Due to subsequent broadcasts, many duplicate copies of the same ant travel to the destination. A node which receives multiple copies of the same ant only accepts the first and discards the other. This way, only one path is set up initially. During the course of the communication session, more paths are added via the proactive path maintenance and exploration mechanism discussed in the next subsection.

Each forward ant keeps a list of the nodes it has visited. Upon arrival at the destination d, it is converted into a *backward ant*, which travels back to the source retracing the path. At each intermediate node i, coming from neighbor n, the ant updates the entry $\mathcal{T}^i_{nd}$ in the i's pheromone table. The way the entry is updated depends on the path quality metrics used to define pheromone variables. For instance, if the pheromone is expressed using the number of hops as a measure of goodness, at each hop the backward ant increments an internal hop counter and uses the inverse of this value to locally assign the value τ^i_d which is used to update the pheromone variable $\mathcal{T}^i_{nd}$ as follows: $\mathcal{T}^i_{nd} = \gamma \mathcal{T}^i_{nd} + (1-\gamma)\tau^i_d$, $\gamma \in [0,1]$. For different metrics, the calculation of τ^i_d is more complex but follows the same logic. For instance, if delay is used, the ant needs to incrementally calculate at each node a robust estimate of the expected delay to reach the destination.

2.3 Proactive Path Maintenance and Exploration

During the course of a communication session, source nodes periodically send out *proactive forward ants* to update the information about currently used paths and try to find new and better paths. They follow pheromone and update pheromone tables in the same way as reactive forward ants. Such continuous proactive sampling of paths is the typical mode of operation in ACO routing algorithms. However, the ant sending frequency needed to faithfully keep track of the constant network changes is in general too high for the available bandwidth. Moreover, to find entirely new paths, excessive blind exploration through random walks or broadcasts would be needed. Therefore, we keep ant sending rate low but ant actions are integrated with a lightweight process combining *pheromone diffusion and bootstrapping.* This process provides a second way of updating pheromone on existing paths, and can give information to guide exploratory ant behavior.

Pheromone diffusion is implemented using *beacon messages* broadcast periodically and asynchronously by the nodes to all their neighbors. In these messages, the sending node n places a list of destinations it has information about, including for each destination d its best pheromone value $\mathcal{T}^n_{m^*d}$. A node i receiving the message from n first of all registers that n is its neighbor. Then, for each destination d listed in the message, it derives an estimate of the goodness of going from i to d over n, combining the cost of hopping from i to n with the reported pheromone value $\mathcal{T}^n_{m^*d}$. We call the obtained estimate $\mathcal{B}^i_{nd}$ *bootstrapped pheromone*, since it is built up bootstrapping on the value of the path goodness estimate received from an adjacent node. The bootstrapped pheromone can in turn be forwarded in the next message sent out by n, giving rise to a bootstrapped pheromone field over the MANET. This is the typical way of calculating estimates followed by all approaches based on *dynamic programming* such the class of distributed Bellman-Ford routing algorithms [16] and the reinforcement learning algorithms derived from Q-learning [8]. However, due to the slow multi-step forwarding, bootstrapped pheromone does not provide the most accurate view of the current situation and has difficulty to properly track the distributed changes in the network. Generally speaking, bootstrapping alone is not expected to work effectively in highly non-stationary environments. However, here the bootstrapped pheromone is obtained via a lightweight, efficient process, and is complemented by the explicit Monte Carlo path sampling and updating done by the ants. In this way we have two complementary updating frequencies in the path updating process. Bootstrapped pheromone is used directly for the *maintenance* of existing paths. That is, if an entry $\mathcal{T}^i_{nd}$ is present in the routing table, $\mathcal{B}^i_{nd}$ is used as a replacement of it. For path *exploration*, bootstrapped pheromone is used indirectly. If i does not yet have a value for $\mathcal{T}^i_{nd}$ in its routing table, $\mathcal{B}^i_{nd}$ could indicate a possible new path from i to d over n. However, this path has never been sampled explicitly by an ant, and due to the slow multi-step process it could contain undetected loops or dangling links. It is therefore not safe to use for data forwarding before being checked. This is the task of proactive forward ants, which use both the regular and the bootstrapped pheromone on their way to the destination. This way, promising pheromone is investigated, and can be turned into a regular path available for data. This increases the number of paths available for data routing, which grows to a full mesh, and allows the algorithm to exploit new opportunities in the ever changing topology.

The effect of varying the number of bootstrapping entries in the beacon messages, the sending rate of proactive ants, and their β exploration exponent is studied respectively in Subsections 4.3, 4.2, and 4.4.

2.4 Stochastic Data Routing

Nodes in AntHocNet *forward data stochastically* according to the pheromone values. When a node has multiple next hops for the destination d of the data, it randomly selects one of them, with probability P_{nd}. P_{nd} is calculated like for reactive forward ants, using Eq. 1. According to this strategy, a mesh of multiple paths is made available to route data. The number of paths to use is automatically and dynamically selected in function of their estimated quality. The effect of varying β in Eq. 1 for data forwarding is studied in Subsection 4.4.

3 Experimental Methodology and Characteristics of the Simulation Environment

For the study of the performance of AntHocNet, we use simulation experiments. This is the most common approach in MANETs, since the complexity of this kind of networks makes analytical evaluations difficult and limited in scope, while the high costs involved in purchasing and configuring hardware limit the use of real testbeds. As simulation software, we use QualNet [17], a commercial simulation package which comes with correct implementations of the most important protocols at all layers of the network protocol stack.

All the simulation tests reported in this paper last 900 seconds, and each data point represents the average taken over 10 runs using different random seeds. The tests are carried out in open space scenarios (see [13] for evaluations of AntHocNet in a structured urban scenario). 100 nodes move in an area of $2400 \times 800\ m^2$. The movements of the nodes are defined according to the random waypoint mobility model [18]. Nodes pick a random destination point in the area, and move to that point with a randomly chosen speed. Upon arrival, they stay put for a fixed pause time, after which a new random destination and speed are chosen. In our experiments, the node speed is chosen uniformly between 0 and 10 m/s, unless stated otherwise. The pause time is always 30 seconds.

Radio signal propagation is modeled with the two-ray ground reflection model, which considers both the direct and the ground reflection path [19]. The transmission range of each node is 250 meters. At the physical and medium access control layers of the network protocol stack, we use the IEEE 802.11b protocol in DCF function with 2 Mbits/s bandwidth. At the application layer, data traffic is generated by 20 constant bit rate (CBR) sources, sending packets of 64 bytes. The CBR sessions start at a random time between 0 and 180 seconds after the start of the simulation, and go on until the end of the simulation. The data rate is 4 packets per second, unless stated differently. CBR uses the UDP protocol at the transport layer. All these settings reflect choices widely adopted in MANET research. Concerning the AntHocNet parameters, if not stated differently, the value of β in Eq. 1 is set to 20, the maximum number of entries in the pheromone diffusion messages is set to 10, and the sending interval for the proactive ants is 2 seconds.

To evaluate the performance of the routing algorithms, we measure the *average end-to-end delay* for data packets and the *ratio of correctly delivered versus sent packets* (delivery ratio). These are standard measures of effectiveness in MANETs. Other metrics which we consider are *delay jitter* and *routing overhead.* Delay jitter is the average difference in inter-arrival time between packets. This is an important metric in quality-of-service networks and in MANETs provides also a measure of the stability of the algorithm's response to topological changes. Routing overhead is a measure of efficiency. We calculate it as the number of control packet transmissions (counting every hop) per data packet delivered. Due to space limitations we do not show the results for jitter and overhead. In any case, for all the tests of Section 4, the results for jitter and overhead always follow the same trends of those for delivery ratio and end-to-end delay.

4 Experimental Analysis of the Different Components of AntHocNet

In this section, we study how the different components influence the performance of the algorithm. In 4.1, we investigate the use of different path evaluation metrics for the definition of pheromone values. In 4.2 we study the effect of the frequency with which proactive ants are sent out. In 4.3, we study the importance of the amount of information transmitted in the pheromone diffusion messages and made available for bootstrapping. Finally, in 4.4, we investigate the effect of allowing different levels of exploration in proactive forward ants and for data.

4.1 Using Different Optimization Metrics for Pheromone Definition

A pheromone value encodes the adaptive estimated goodness of a routing decision. Different path evaluation metrics can be used to measure this goodness. We investigate the use of path length in number of hops (referred to as "hops" in Figure 1), path end-to-end delay ("delay"), number of hops combined with the quality of each hop in terms of signal-to-noise ratio ("snr"), and number of hops combined with end-to-end delay ("hops+delay"). We also plot results for a version of AntHocNet where proactive learning is switched off ("no proactivity") and the "snr" metric is used. Figures 1a and 1b show respectively the delivery ratio and the average end-to-end delay of the different versions of the algorithm.

A first clear result is that the metric combining hops and signal-to-noise ratio is the most effective one. On the other hand, the optimization with respect to the sole number of hops produces the worst results. Even worse than using no proactivity at all with the "snr" metric. Similar results have been reported in the MANET community [20]: paths with a low number of hops usually use long hops, which necessarily have lower signal strength, and can therefore easily loose connection. This diagnosis is confirmed by the fact that combining number of hops and signal-to-noise ratio leads to large improvement in performance. Finally, we point out that also the use of the end-to-end delay alone or together with hops does not give good results. This is partly due to the backoff mechanisms at the MAC layer, that make the single node experiencing fluctuating delays accessing the shared wireless channel. This generates large variations in terms of end-to-end delay. Under these conditions it becomes hard to learn estimates of path latencies which are robust enough to rank the quality of the different paths.

These results indicate that it is crucial to choose a good path evaluation metric to use in the pheromone model, based on knowledge of the specific network environment. A composite metric is the right way to go.

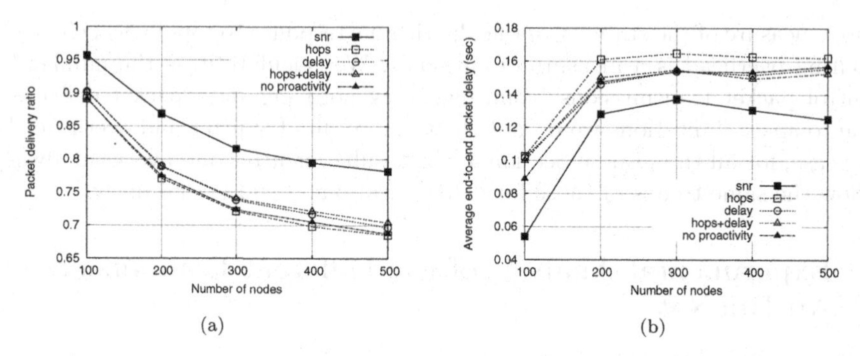

Fig. 1. (a) Delivery ratio and (b) average end-to-end delay for AntHocNet using different optimization metrics in scenarios with increasing nodes and constant density

4.2 Varying the Proactive Ant Send Interval

The proactive ant send interval is the time between successive proactive ants in the proactive path maintenance and improvement phase. It defines how often the algorithm looks for path improvements, and therefore how quickly it can adapt to changes. We made tests with send intervals of 0.5, 1, 2, 5, 10, 20, and 50 seconds. The test scenario is the basic scenario described in Section 3. The different curves in Figures 2a and 2b represent tests using different maximum node speeds (so changing the network mobility).

All tests show a similar pattern. A too low ant send interval leads to bad performance, because the network gets flooded by ants. At 2 seconds, there seems to be an optimal send interval. For frequencies lower than that, the performance decays because the algorithm is not sending enough ants to keep up with the changes in the network. For low speeds, this decay is slower since the network changes less fast. However, it is interesting to note that the best send interval value is independent of the node speed. We also did some tests keeping the speed constant on 10 m/s, and varying the data traffic load (not presented here due to space limitations). Although higher traffic load could be expected to leave less space for ants, also there the best ant send interval was always around 2 seconds. The general effect of the use of proactive actions was also shown in previous Figure 1 where it is evident the significant decrease in performance when the proactive mechanisms are switched off (using "snr" metric) with respect to the case of using the same metric but switching proactivity on.

4.3 Varying the Number of Entries in Pheromone Diffusion Messages

The number of maximum entries in the pheromone diffusion messages defines how much pheromone information is spread at each step of the information bootstrapping process. Concretely, a low number of entries spreads little information and determines also a slow running of the bootstrapping process. 0 entries is the extreme case where the supporting pheromone diffusion function is disabled and the proactive ants get no guidance. We made tests using 0, 2, 5, 10 and 20

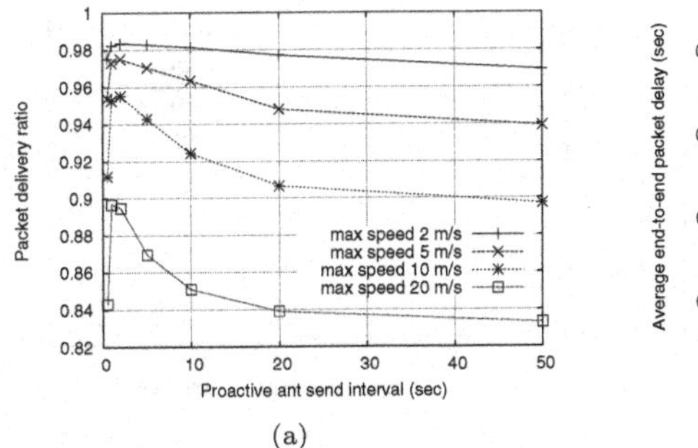

(a)

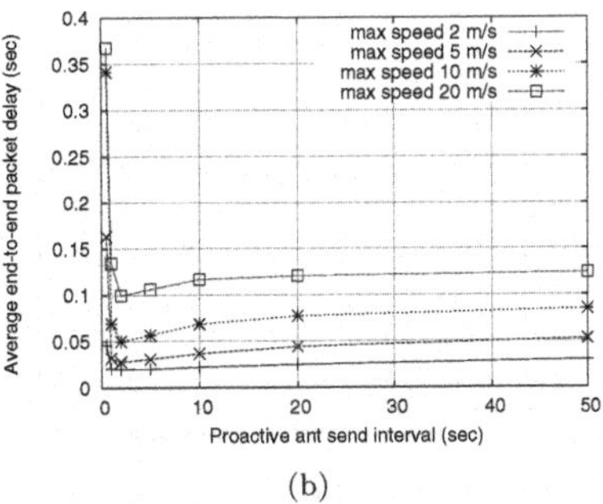

(b)

Fig. 2. (a) Delivery ratio and (b) average end-to-end delay for AntHocNet increasing the proactive ant send interval in scenarios with different node speeds

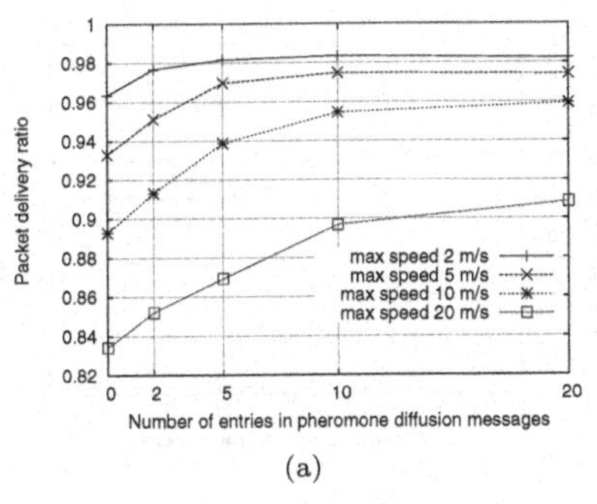

(a)

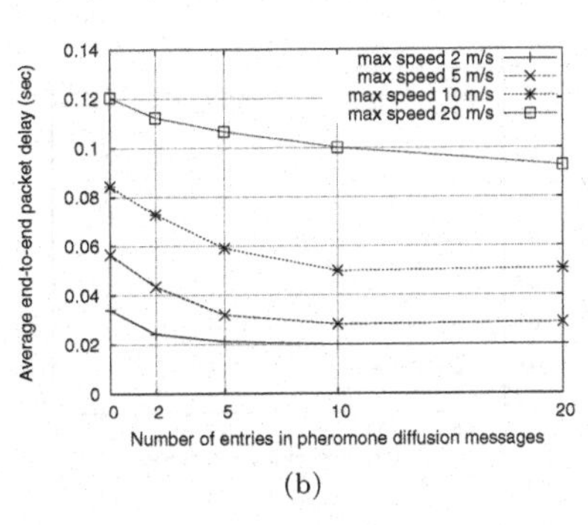

(b)

Fig. 3. (a) Delivery ratio and (b) average end-to-end delay for AntHocNet varying the number of entries in pheromone diffusion messages for scenarios with different speeds

entries. The destinations whose routing information is included in each message are selected randomly out of the known destinations stored in the pheromone table. The test scenario is the basic scenario described in Section 3, with 20 active CBR sessions. Like in Subsection 4.2, we report results for different maximum node speeds. Figure 3 reports delivery ratio and average delay.

The graphs show the importance of the supporting pheromone diffusion process: giving more efficiency to this process allows for better performance. Moreover, for the tested sizes the benefit of the increase in transmitted information is still greater than the negative impact due to the generation of larger messages. As was observed in 4.2, the importance of the efficiency of the learning process decreases for slower changing networks.

4.4 Varying the Routing Exponent for Ants and Data

The exponent β in Equation 1 defines the amount of exploration allowed to the ants during their path search phase. As previously mentioned, for the reactive forward ants β is set to a quite high value (≈ 20) in order to reduce counter-productive ant proliferation and the establishment of sub-optimal paths at the start of a new data session. On the other hand, proactive ants are meant to explore and check the path improvements suggested by bootstrapped pheromone.

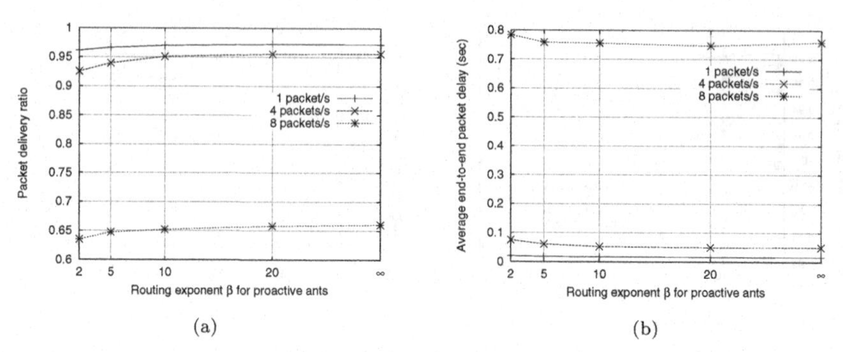

Fig. 4. (a) Delivery ratio and (b) average end-to-end delay for AntHocNet increasing the proactive ant routing exponent in scenarios with different data traffic send rates

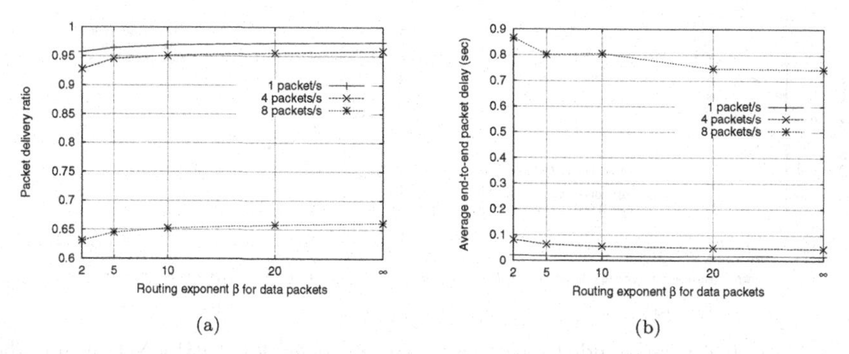

Fig. 5. (a) Delivery ratio and (b) average end-to-end delay for AntHocNet increasing the data packet routing exponent in scenarios with different data traffic send rates

Therefore, we studied the effect of varying the degree of exploration allowed to the proactive forward ants by considering β values of 2, 5, 10, 20 and also the case of deterministic choice of the best path. We considered scenarios with data rates of 1, 4 and 8 packets per second for the 20 CBR sessions.

The results reported in Figure 4 shows a large difference in performance for the cases of 1 and 4 packets and that of 8 packets per second. Nevertheless, an equivalent performance response with respect to β is evidenced in all the three different scenarios. Performance increases reducing exploration. The difference between the extreme cases of $\beta = 2$ and deterministic choices (indicated with ∞ in the plots), is small but clear. These results show that in MANETs, exploration at the level of the ants do not really pay back due to constant changes and strong bandwidth limitations that limit the frequency of ant generation and the number of different paths that can be effectively explored. We observed a similar behavior also increasing the degree of exploration in reactive forward ants. On the other hand, in AntHocNet path exploration is implicitly carried out with the low-overhead mechanisms of pheromone diffusion and bootstrapping. New paths are built in a multi-step fashion by bootstrapping on the pheromone information received from the neighbors. Results in Figure 4 says that the best performance is obtained when proactive ants are used to test only the best path indicated by the current combination of regular and bootstrapped pheromone.

Analogous results are reported in Figure 5, that shows the effect of varying the value of the exponent β for data packets. In this case, β controls the amount of multiple paths used to spread data. The results suggest that the best choice is to adopt a deterministic greedy policy for next hop selection. On the other hand, this does not imply that a single path is used to forward the data packets of a same traffic flow. We experimentally observed that multiple paths are actually used even when a deterministic policy is adopted, as the result of the continual proactive updating and addition/removal of paths made available to the running flows. However, the results indicate that an excessive use of multiple paths can easily bring performance down. This is due to the fact that if two paths simultaneously used for the same flow are not radio-disjoint they will interfere with a consequent degradation of performance.

5 Conclusions

MANETs are extremely dynamic network environments. Their multi-modality represents an important challenge for algorithms at all levels of the network protocol stack, and specifically for routing. We addressed these challenges with AntHocNet, a routing algorithm designed after ACO ideas. AntHocNet was introduced in previous work and showed superior performance compared to other state-of-the-art algorithms over a wide range of MANET simulation scenarios [2,3,4,5,6,13]. In this paper we discussed AntHocNet emphasizing its innovative design, especially with respect to previous ACO algorithms for routing. In particular, we pointed out the fact that AntHocNet is based on the integration of reactive and proactive components, and on the integration of the typical ACO path sampling mechanism with the learning of routing information using an information bootstrapping process. In a detailed experimental study we have investigated the role and the importance of these and other different components of the algorithm, studying their effect on the overall performance. The effectiveness of the use of a composite design to deal with the multiple challenges of MANETs, and in particular the effectiveness of combining ant-based Monte Carlo sampling with pheromone bootstrapping, has been confirmed by the experimental results. Moreover, the experimental analysis has pointed out the need for adopting low-overhead and low-interference strategies for exploration and data forwarding, as well as the importance of defining a composite pheromone metric taking into account different multiple aspects of the network environment.

References

1. Royer, E., Toh, C.K.: A review of current routing protocols for ad hoc mobile wireless networks. IEEE Personal Communications (1999)
2. Di Caro, G., Ducatelle, F., Gambardella, L.: AntHocNet: an adaptive nature-inspired algorithm for routing in mobile ad hoc networks. European Transactions on Telecommunications, *Special Issue on Self Organization in Mobile Networking* **16**(5) (2005) 443–455

3. Ducatelle, F., Di Caro, G., Gambardella, L.: Using ant agents to combine reactive and proactive strategies for routing in mobile ad hoc networks. Int. Journal of Computational Intelligence and Applications (IJCIA), *Special Issue on Nature-Inspired Approaches to Networks and Telecommunications* **5**(2) (2005) 169–184
4. Di Caro, G., Ducatelle, F., Gambardella, L.: Swarm intelligence for routing in mobile ad hoc networks. In: Proceedings of the 2005 IEEE Swarm Intelligence Symposium (SIS). (2005)
5. Ducatelle, F., Di Caro, G., Gambardella, L.: Ant agents for hybrid multipath routing in mobile ad hoc networks. In: Proceedings of the Second Annual Conference on Wireless On demand Network Systems and Services (WONS), St. Moritz, Switzerland (2005)
6. Di Caro, G., Ducatelle, F., Gambardella, L.: AntHocNet: an ant-based hybrid routing algorithm for mobile ad hoc networks. In: Proceedings of Parallel Problem Solving from Nature (PPSN) VIII. Volume 3242 of Lecture Notes in Computer Science., Springer-Verlag (2004) 461–470 (Conference best paper award).
7. Di Caro, G.: Ant Colony Optimization and its application to adaptive routing in telecommunication networks. PhD thesis, Faculté des Sciences Appliquées, Université Libre de Bruxelles, Brussels, Belgium (2004)
8. Sutton, R., Barto, A.: Reinforcement Learning: An Introduction.MITPress (1998)
9. Di Caro, G., Dorigo, M.: AntNet: Distributed stigmergetic control for communications networks. J. of Artificial Intelligence Research (JAIR) **9** (1998) 317–365
10. Shen, C.C., Jaikaeo, C., Srisathapornphat, C., Huang, Z., Rajagopalan, S.: Ad hoc networking with swarm intelligence. In: Proceedings of ANTS 2004, Fourth International Workshop on Ant Algorithms. LNCS, Springer-Verlag (2004)
11. Baras, J.S., Mehta, H.: A probabilistic emergent routing algorithm for mobile ad hoc networks. In: WiOpt03: Modeling and Optimization in Mobile, Ad Hoc and Wireless Networks. (2003)
12. Günes, M., Kähmer, M., Bouazizi, I.: Ant-routing-algorithm (ARA) for mobile multi-hop ad-hoc networks - new features and results. In: Proceedings of the 2nd Mediterranean Workshop on Ad-Hoc Networks (Med-Hoc-Net'03), Mahdia, Tunisia (2003)
13. Di Caro, G., Ducatelle, F., Gambardella, L.: Studies of routing performance in a city-like testbed for mobile ad hoc networks. Technical Report 07-06, IDSIA, Lugano (Switzerland) (2006)
14. Perkins, C., Royer, E.: Ad-hoc on-demand distance vector routing. In: Proc. of the 2nd IEEE Workshop on Mobile Computing Systems and Applications. (1999)
15. Clausen, T., Jacquet, P., Laouiti, A., Muhlethaler, P., Qayyum, A., Viennot, L.: Optimized link state routing protocol. In: Proceedings of IEEE INMIC. (2001)
16. Bertsekas, D., Gallager, R.: Data Networks. Prentice–Hall (1992)
17. Scalable Network Technologies, Inc. Culver City, CA, USA: QualNet Simulator, Version 3.8. (2005) `http://www.scalable-networks.com`.
18. Johnson, D., Maltz, D.: Dynamic Source Routing in Ad Hoc Wireless Networks. In: Mobile Computing. Kluwer (1996) 153–181
19. Rappaport, T.Wireless communications, principles and practicePrenticeHall (1996)
20. De Couto, D., Aguayo, D., Chambers, B., Morris, R.: Performance of multihop wireless networks: Shortest path is not enough. In: Proceedings of the First Workshop on Hot Topics in Networks (HotNets-I), ACM SIGCOMM (2002)

An Energy-Efficient Ant-Based Routing Algorithm for Wireless Sensor Networks

Tiago Camilo[1], Carlos Carreto[2], Jorge Sá Silva[1], and Fernando Boavida[1]

[1] Laboratory of Communications and Telematics
University of Coimbra, Coimbra, Portugal
{tandre, sasilva, boavida}@dei.uc.pt
[2] Escola Superior de Tecnologia e Gestão
Instituto Politécnico da Guarda, Guarda, Portugal
ccarreto@ipg.pt

Abstract. Wireless Sensor Networks are characterized by having specific requirements such as limited energy availability, low memory and reduced processing power. On the other hand, these networks have enormous potential applicability, e.g., habitat monitoring, medical care, military surveillance or traffic control. Many protocols have been developed for Wireless Sensor Networks that try to overcome the constraints that characterize this type of networks. Ant-based routing protocols can add a significant contribution to assist in the maximisation of the network lifetime, but this is only possible by means of an adaptable and balanced algorithm that takes into account the Wireless Sensor Networks main restrictions. This paper presents a new Wireless Sensor Network routing protocol, which is based on the Ant Colony Optimization metaheuristic. The protocol was studied by simulation for several Wireless Sensor Network scenarios and the results clearly show that it minimises communication load and maximises energy savings.

1 Introduction

Identified as one of the most important technologies of the XXI century, Wireless Sensor Networks (WSNs), are becoming the next step in information revolution [1]. This enhancement was only possible due to the recent advances in electronic sensors, communication technologies and computation algorithms; however, because of their novelty, WSNs present new challenges compared to custom wireless networks. Although they can be considered ad hoc networks, WSNs present unique characteristics mainly due to their component devices, the sensor nodes.

A sensor node, typically, contains signal-processing circuits, micro-controllers and a wireless transmitter/receiver antenna, and is characterized by limited resources: low memory, reduced power battery and limited processing capabilities. Sink-nodes are the devices responsible for managing the communication from the sensor network to the base station, normally located in the wired network where the observer keeps record of the sensor data. After receiving packets, sink-nodes can send them to the base station if it is located inside the communication range, or send them to other sink-nodes, through known ad hoc techniques. Furthermore,

M. Dorigo et al. (Eds.): ANTS 2006, LNCS 4150, pp. 49–59, 2006.

sink-nodes have distinctive characteristics when compared to typical sensor-nodes, such as more energy capacity, more processing power and more memory, which makes them perfect to perform high demand processing and storing tasks.

Potential WSNs applications include security, traffic control, industrial and manufacturing automation, medical or animal monitoring, and many more. This wide applicability range forces WSN protocols to become application-based, meaning that it is not feasible to build a WSN algorithm that fulfils all application requirements. Instead it is important to build generic algorithms that somehow can be adapted to some application requirements and at the same time prolong the network lifetime as long as possible. The lifetime of a sensor network can be measured based on generic parameters, such as the time when half of the sensor nodes lose their transmitting capability, or through specific metrics of each application, e.g. minimum delay.

This paper presents a new communication protocol for WSNs called energy-efficient ant-based routing algorithm (EEABR), which is based on the Ant Colony Optimization (ACO) metaheuristic [13]. EEABR uses a colony of artificial ants that travel through the WSN looking for paths between the sensor nodes and a destination node, that are at the same time short in length and energy-efficient, contributing in that way to maximise the lifetime of the WSN. Each ant chooses the next network node to go to with a probability that is a function of the node energy and of the amount of pheromone trail present on the connections between the nodes. When an ant reaches the destination node, it travels backwards trough the path constructed and updates the pheromone trail by an amount that is based on the energy quality and the number of nodes of the path. After some iterations the EEABR protocol is able to build a routing tree with optimized energy branches.

In this paper we do not consider energy saving techniques based on the management of the node status [12]. These techniques are normally implemented in physical and access layers, and allow turning nodes from sleep mode to transmitting/receiving mode.

The remainder of this paper is organized as follows. Section 2 describes the state-of-the-art of WSN protocols; wellknow algorithms are described as well as some approaches that combine ant-based algorithm with such networks. In Section 3 the EEABR protocol is described, in conjunction with two other approaches. Section 4 presents the studies performed to evaluate the proposed protocol; these simulation environments try to emulate real WSN deployment, so that real sensor characteristics can be studied. Conclusions and topics for further work are presented in Section 5.

2 Related Work

Wireless sensor networks can be considered, as mentioned before, ad-hoc networks. However, protocols for mobile ad hoc networks (MANETs) do not offer some of the sensor networks requirements: sensors typically have low power battery, low memory, and the routing tables grow up with the network length and do not support diffusion communication. These are the main reasons why it is necessary to design new protocols, built on the most important criterion of energy-efficiency.

Low Energy Adaptive Clustering Hierarchy (LEACH), described in [2], is probably one of the more referenced protocols in the sensor networks area. It is a powerful, efficient protocol created to be used in sensor networks with continued data flowing (unstopped sensor activity). This is a protocol that uses a hierarchical topology, randomly creates cluesterheads, and presents data aggregation mechanisms.

Power-Efficient GAthering inSensor Information Systems (PEGASIS), is a recently developed protocol, which is similar to LEACH but that requires less energy per round [3]. In PEGASIS, a chain is created so that each node receives aggregate information and forwards it to a nearby neighbour. It presents mechanisms that allow the variation of radio communications energy parameters. Compared to LEACH, the PEGASIS protocol obtains up to 100% of energy cost improvement per round [4]. However these two protocols are not suitable for mobility, and both assume that data packets can be aggregated at clusterheads.

Direct Diffusion (DD) [5] is a data-centric protocol, which addresses nodes by the monitored data instead of their network addresses. In this protocol the application is responsible to query the network for a specific phenomenon value. Sensor nodes that satisfy the specific query start transmitting their data. Based on sink-nodes requests this protocol does not consider the node's available energy when building their flood-based routing scheme.

Jeon et al. [6] proposed an energy-efficient routing protocol that tries to manage both delay and energy concerns. Based on AntNet protocol [7], this algorithm uses the concept of ant pheromone to produce two prioritized queues, which are used to send differentiated traffic. However, such approach can be infeasible in current sensor nodes due to the memory required to save both queues. This can be even more problematic if the sensor network is very populated, since the routing table on each device depends on the number of neighbours.

Zhang et al. [8], study three distinct Ant-based algorithms for WSNs. However, the authors only focus on the building of an initial pheromone distribution, good at system start-up.

Finally in [9], the authors present an ant colony algorithm for Steiner Trees which can be ported to WSNs routing. However, no changes are considered regarding the specific WSNs requirements and also no considerations are made regarding the energy management essential to the WSNs performance.

The ant-based algorithms presented above assume that communication between sensor nodes (end-to-end) is required by the WSN application, and build their algorithms based on such assumption. However this is not the case in most WSNs scenarios, where the hop-by-hop or single hop communication is performed from source node (sensor node) to sink node, which is responsible to collect sensor data from the network. This node presents different characteristics compared with normal sensor nodes (more energy, more memory and more processing power), and such differences are not considered in the referred algorithms.

3 Energy-Efficient Ant-Based Routing Algorithm

Whenever a WSN protocol is designed, it is important to consider the energy efficiency of the underlying algorithm, since this type of networks have strict power requirements. In this section we describe a new energy-constrained protocol, the EEABR protocol, which is based on the Ant Colony Optimization heuristic and is focused on the main WSNs constraints.

On such networks deployed in real environment it is important to point out that sensor nodes may not have energy replenishment capabilities. This assumption forces the use of energy-efficient algorithms in order to maximize the network's life time. In contrast, in timely delivery packet networks, a routing algorithm attempts to find the shortest path between two distinct devices (source and receiver), which can be easily done by choosing the path with less communication hops. In WSNs such requirements are relegated to second plane, since quality of service and service awareness are not as important as in normal MANETs, where running protocols required low communication delays.

The remainder of this section summarizes the idea behind EEABR. First, a basic ant-based routing algorithm for WSNs is presented that describes the adaptation of the ACO metaheuristic to solve the routing problem in WSNs. Next, an improved algorithm is presented that reduces the memory used in the sensor nodes and also considers the energy quality of the paths found by the ants. Finally the EEABR protocol is presented and further improvements are described to reduce the communication load and the energy spent with communications.

3.1 Basic Ant Based Routing for WSNs

The ACO metaheuristic has been applied with success to many combinatorial optimisation problems [13]. Its optimization procedure can be easily adapted to implement an ant based routing algorithm for WSNs. A basic implementation of such algorithm can be informally described as follows.

1. At regular intervals, from every network node, a forward ant k is launched with the mission to find a path until the destination. The identifier of every visited node is saved onto a memory M_k and carried by the ant.
2. At each node r, a forward ant selects the next hop node using the same probabilistic rule proposed in the ACO metaheuristic:

$$p_k(r,s) = \begin{cases} \dfrac{[T(r,s)]^\alpha [E(s)]^\beta}{\sum\limits_{u \notin M_k} [T(r,s)]^\alpha [E(s)]^\beta} & \text{if } s \notin M_k \\ 0 & \text{otherwise} \end{cases} \quad (1)$$

where $p_k(r,s)$ is the probability with which ant k chooses to move from node r to node s, T is the routing table at each node that stores the amount of pheromone trail on connection (r,s), E is the visibility function given by $\frac{1}{(C-e_s)}$ (C is the initial energy level of the nodes and e_s is the actual energy level of node s), and α and β are parameters that control the relative importance of trail versus visibility. The selection probability is a trade-off between visibility (which says that nodes with more energy should be chosen with high probability) and actual trail intensity (that says that if on connection (r, s) there has been a lot of traffic then it is highly desirable to use that connection.

3. When a forward ant reaches the destination node, it is transformed in a backward ant which mission is now to update the pheromone trail of the path it used to reach the destination and that is stored in its memory.
4. Before backward ant k starts its return journey, the destination node computes the amount of pheromone trail that the ant will drop during its journey:

$$\Delta T_k = \frac{1}{N - Fd_k} \tag{2}$$

where N is the total number of nodes and Fd_k is the distance travelled by the forward ant k (the number of nodes stored in its memory).

5. Whenever a node r receives a backward ant coming from a neighbouring node s, it updates its routing table in the following manner:

$$T_k\,(r, s) = (1 - \rho)\, T_k\,(r, s) + \Delta T_k \tag{3}$$

where ρ is a coefficient such that $(1 - \rho)$ represents the evaporation of trail since the last time $T_k\,(r, s)$ was updated.

6. When the backward ant reaches the node where it was created, its mission is finished and the ant is eliminated.

By performing this algorithm several iterations, each node will be able to know which are the best neighbours (in terms of the optimal function represented by (2)) to send a packet, towards a specific destination.

3.2 Improved Ant Based Routing for WSNs

In this section we propose two improvements in the basic ant-based routing algorithm described in the previous section in order to reduce the memory used in the sensor nodes and also to consider the energy quality of the paths found by the ants.

In the basic algorithm the forward ants are sent to no specific destination node, which means that sensor nodes must communicate with each other and the routing tables of each node must contain the identification of all the sensor nodes in the neighbourhood and the correspondent levels of pheromone trail. For large networks, this can be a problem since nodes would need to have big amounts of memory to save all the information about the neighbourhood. Nevertheless, the algorithm can be easily changed to save memory. If the forward ants are sent directly to the sink-node, the routing tables only need to save the neighbour nodes that are in the direction of the sink-node. This considerably reduces the size of the routing tables and, in consequence, the memory needed by the nodes.

As described in the Introduction, sensor nodes are devices with a very limited energy capacity. This means that the quality of a given path between a sensor node and the sink-node, should be determined not only in terms of the distance (number of nodes of the path), but also in terms of the energy level of that path. For example, it would be preferable to choose a longer path with high energy level than a shorter path with very low energy levels.

To consider the energy quality of the paths on the basic algorithm a new function is proposed to determine the amount of pheromone trail that the backward ant will drop during its returning journey:

$$\Delta T_k = \frac{1}{C - \left(\operatorname{avg}(E_k) - \frac{1}{\min(E_k)}\right)} \tag{4}$$

where E_k is a new vector carried by forward ant k with the energy levels of the nodes of its path, C is the initial energy level of the nodes, $\operatorname{avg}(E_k)$ is the average of the vector values and $\min(E_k)$ is the minimum value of the vector.

3.3 Energy-efficient Ant Based Routing for WSNs

In this section we propose further improvements in the routing algorithm described in the previous section in order to reduce the communication load related to the ants and the energy spent with communications. We also propose new functions to update the pheromone trail.

It has been proved that the tasks performed by the sensor nodes that are related with communications (transmitting and receiving data), spend much more energy than those related with data processing and memory management [10,11]. Since one of the main concerns in WSNs is to maximise the lifetime of the network, which means saving as much energy as possible, it would be preferable that the routing algorithm could perform as much processing as possible in the network nodes, than transmitting all data through the ants to the sink-node to be processed there. In fact, in huge sensor networks where the number of nodes can easily reach more than 1000 units, the memory of the ants would be so big that it would be unfeasible to send the ants through the network.

To implement these ideas, the memory M_k of each ant is reduced to just two records, the last two visited nodes. Since the path followed by the ants is no more in their memories, a memory must be created at each node that keeps record of each ant that was received and sent. Each memory record saves the previous node, the forward node, the ant identification and a timeout value. Whenever a forward ant is received, the node looks into its memory and searches the ant identification for a possible loop. If no record is found, the node saves the required information, restarts a timer, and forwards the ant to the next node. If a record containing the ant identification is found, the ant is eliminated. When a node receives a backward ant, it searches its memory to find the next node to where the ant must be sent. The timer is used to delete the record that identifies the backward ant, if for any reason the ant does not reach that node within the time defined by the timer.

The vector E_k was erased from the forward ants k, that now only carry the average energy till the current node ($E\text{avg}_k$), and the minimum energy level registered ($E\min_k$). These values are updated by each node that receives the forward ants.

When the forward ant reaches the sink-node these values are used to calculate the amount of pheromone trail used by the corresponding backward ant:

$$\Delta T_k = \frac{1}{C - \left[\frac{E\min_k - Fd_k}{E\text{avg}_k - Fd_k}\right]} \tag{5}$$

With these changes it is possible to reduce the ant's length by ≅700%, and save on each ant hop the transmission of ≅250 bytes. This is a significant achievement, since it allows the saving of precious energy levels on sensor nodes.

Calculating ΔT_konly as a function of the energy levels of the path, as it is done in (4), can bring no optimized routes, since a path with 15 nodes can have the same energy average as a path with only 5 nodes. Therefore ΔT_kmust be calculated as a function of both parameters: the energy levels and the length of the path. This can be achieved by introducing the parameter Fd_k in the (5), which represents the number of nodes that the forward ant k has visited.

The equation used to update the routing tables at each node is now changed to:

$$T_k(r,s) = (1-\rho)\,T_k(r,s) + \left[\frac{\Delta T_k}{\varphi B d_k}\right] \tag{6}$$

where φ is a coefficient and Bd_k is the travelled distance (the number of visited nodes), by backward ant k until node r. These two parameters will force the ant to loose part of the pheromone strength during its way to the source node. The idea behind this behaviour is to build a better pheromone distribution (nodes near the sink-node will have more pheromone levels) and will force remote nodes to find better paths. Such behaviour is extremely important when the sink-node is able to move, since the pheromone adaptation will be much quicker.

4 Experimental Results

In this section we present the experimental results obtained for the three algorithms described in section 3: the basic ant-based routing algorithm (BABR), described in section 3.1, the improved ant-based routing algorithm (IABR), presented in section 3.2, and the energy-efficient ant-based routing algorithm (EEABR), presented in section 3.3. The algorithms were tested using the well known ns-2 simulator [14], with the two-ray ground reflection model.

To better understand the differences between the three algorithms, three distinct scenarios were used, each one trying to represent real WSN deployment environments, as well as possible. On all scenarios the nodes were deployed in random fashion, since in real sensor networks the device deployment, in general, cannot be controlled by an operator due to the environment characteristics. The number of deployed sensor nodes varied between 10 and 100 nodes. In terms of

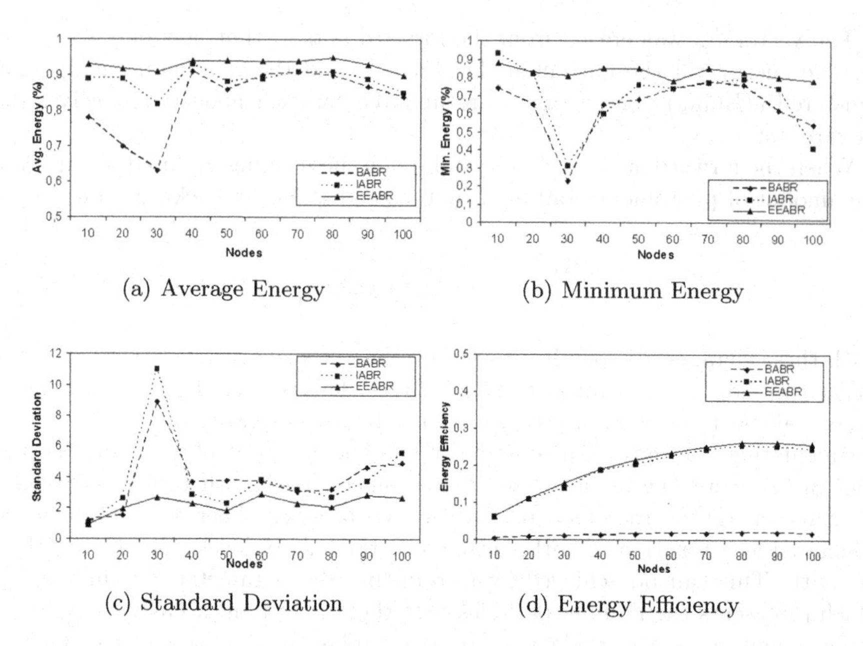

(a) Average Energy (b) Minimum Energy

(c) Standard Deviation (d) Energy Efficiency

Fig. 1. Performance in sensor network with static phenomenon

simulated area it also varied, forcing the connectivity between all nodes, from 200x200 m^2 (10 nodes), 300x300 m^2 (20 nodes), 400x400 m^2 (30 nodes), 500x500 m^2 (40 nodes) and 600x600 m^2 when 50, 60, 70, 80, 90 and 100 nodes were used. For each environment four metrics were used to compare the performance of the algorithms: the Average Energy, which gives the average of energy of all nodes at the end of simulation; the Minimum Energy, which gives the lowest energy amount of all nodes; the Standard Deviation, which gives the average variance between energy levels on all nodes; and finally the Energy Efficiency, which gives the ratio between total consumed energy and the number of packets received by the sink-node.

The first scenario simulates a static WSN where the sensor nodes were randomly deployed with the objective to monitor a static phenomenon. The location of the phenomenon and the sink-node are not known. Nodes are responsible to monitor the phenomenon and send the relevant sensor data to the sink-node. In this peculiar scenario the nodes near the phenomenon will be affected in terms of energy consumption, since they will be forced to periodically transmit data. Figure 1 presents the results of the simulation for the studied parameters. In the majority of the scenarios (from 10 till 100 nodes) the EEABR protocol gives the best results. In Figure 1b) the minimum energy in both protocols, BABR and IABR, present a very low value when the network has 30 nodes, however in the EEABR protocol this does not happen. This behaviour is also visible in Figure 1c) where the standard deviation shows us the same distinctive values. This behaviour can be explained considering the used network topology, where there exist few communication paths from source to the sink-node.

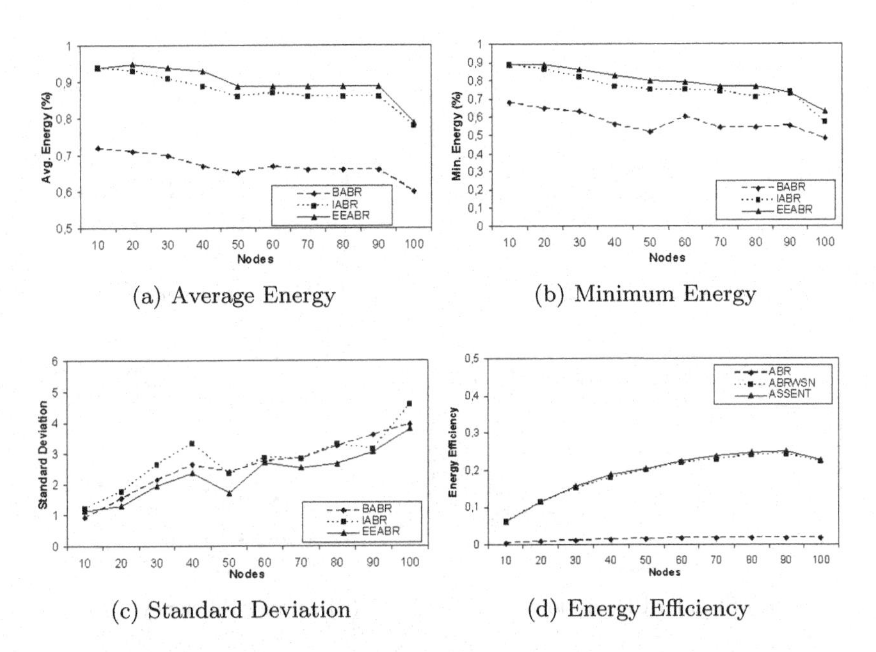

(a) Average Energy

(b) Minimum Energy

(c) Standard Deviation

(d) Energy Efficiency

Fig. 2. Performance in sensor network with mobile phenomenon

The results illustrated in Figure 2 correspond to the second scenario, where the phenomenon is mobile. Comparing with results from previous scenarios, the phenomenon mobility decreases the performance of the algorithm, which is understandable and expected since more nodes become sources of data packets, increasing the number of packets in the network. Once again the EEABR protocol presents the best results when compared to the others protocols, but results can easily be compared to scenarios where all environment variables are static.

The final study simulates a mesh sensor network. These networks are composed of several nodes with different capabilities. On each network three energy levels were used: 50, 30 and 20 joules. These levels were uniformly distributed over the nodes. Figure 3 shows the simulation results. The EEABR protocol had better final results compared to the previous studies. This can be explained by the adaptability of the protocol, which efficiently tries to balance the energy levels on all nodes. This conclusion is more evident in Figure 3d). When compared with the other algorithms the EEABR presents a significant reduction in relation to the standard deviation. In terms of average energy levels the EEABR always presents the best results. When compared to the IABR the difference between the average values varied between 3% and 10%, and when compared with BABR varied between 17% and 25%. In terms of the minimum energy of the nodes at the end of the simulation, no algorithm could avoid the existence of "dead" nodes, however BABR and IABR presented two "dead" nodes contrasting to only one presented by the EEABR protocol. This is due to the random node distribution, where only two nodes were responsible to provide connectivity between the source and the sink-node, since the phenomenon was static.

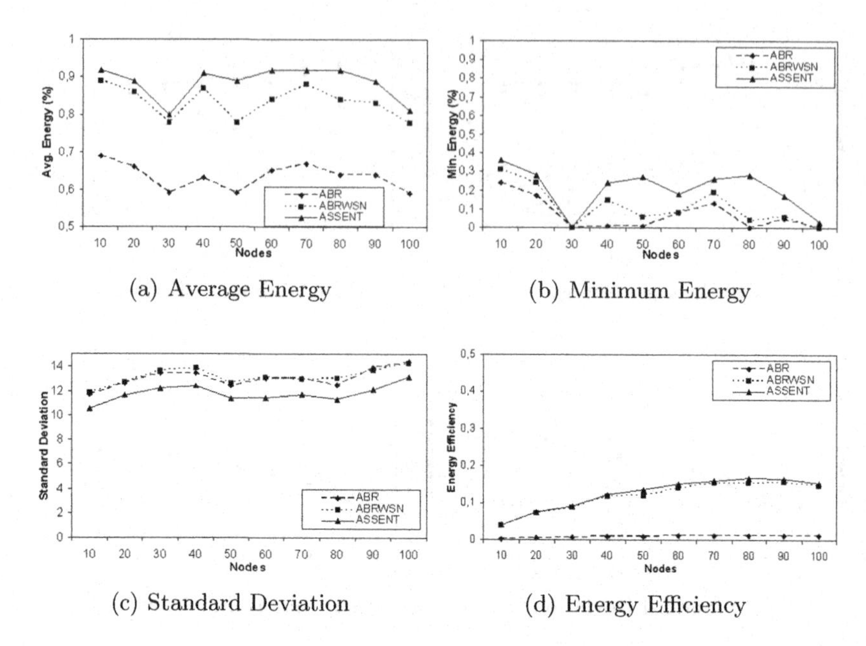

(a) Average Energy (b) Minimum Energy

(c) Standard Deviation (d) Energy Efficiency

Fig. 3. Performance in sensor network with different initial energy levels

In relation to energy efficiency, the results were very similar in all scenarios. EEABR and IABR present the best results, which are also very similar because both algorithms are energy-aware. However, in terms of the other parameters, the difference between both protocols became higher, meaning EEABR performance is better since it significantly reduces the energy consumed in communications. On the other hand, the BABR algorithm presents the worst results for all the studied parameters, although in some cases it reaches the same values as the IABR protocol, due to the inefficiency of the IABR in reducing the overhead in exchange messages.

5 Conclusions

In this paper we studied the application of the Ant Colony Optimization metaheuristic to solve the routing problem in wireless sensor networks. A basic ant-based routing algorithm was proposed, and several improvements, inspired by the features of wireless sensor networks (low energy levels, low processing and memory capabilities), were considered and implemented. The resulting routing protocol, called Energy-Efficient Ant Based Routing (EEABR), uses "lightweight" ants to find routing paths between the sensor nodes and the sink nodes, which are optimised in terms of distance and energy levels. These special ants minimise communication loads and maximise energy savings, contributing to expand the lifetime of the wireless network. The experimental results showed that the algorithm leads to very good results in different WSN scenarios.

As future work we intend to study the initialization method to populate the routing tables with initial pheromone levels. As shown in the literature [8], such mechanisms can increase even more the efficiency of the networks. Another approach to be studied is the integration of multiple sink-nodes.

Acknowledgments. The work presented in this paper is partially financed by the Portuguese Foundation for Science and Technology, FCT through the 6Mnet POSI/REDES/44089/2002 project.

This work has been partly supported by the European Union under the E-Next FP6-506869 NoE.

References

1. Estrin, D., et al.: Embedded, Everywhere: A research Agenda for Network Systems of Embedded Computers, National Research Council Report, 2001
2. Handy, M., Haase, M., Timmermann, D.: Low Energy Adaptive Clustering Hierarchy with Deterministic Cluster-Head Selection, 4th IEEE International Conference on Mobile and Wireless Communications Networks, Stockholm, 2002
3. Lindsey, S., Raghavendra, C.: PEGASIS: Power Efficient GAthering in Sensor Information Systems, ICC, 2001
4. Lindsey, S., Raghavendra, C., Sivalingam, K.: Data Gathering in Sensor Networks using the EnergyDelay Metric, 2000
5. Intanagonwiwat, C., Govindan, R., Estrin, D.: Directed Diffusion: a scalable and robust communication paradigm for sensor networks, ACM Press, 2000
6. Jeon, P., Rao, R., Kesidis, G.: Two-Priority Routing in Sensor MANETs Using Both Energy and Delay Metrics, in preparation, 2004
7. Di Caro G., Dorigo M.: AntNet: Distributed Stigmergetic Control for Communications Networks, Journal of Artificial Intelligence Research (JAIR), Vol. 9, Pag. 317-365, 1998
8. Zhang, Y., Kuhn, L., Fromherz, M.: Improvements on Ant Routing for Sensor Networks, In: Ants 2004, Int. Workshop on Ant Colony Optimization and Swarm Intelligence, Sept. 2004
9. Singh, G., Das, S. Gosavi, S., Pujar, S.: Ant Colony Algorithms for Steiner Trees: An Application to Routing in Sensor Networks, Recent Developments in Biologically Inspired Computing, Eds. L. N. de Castro, F. J. von Zuben, Idea Group Publishing, pp. 181-206, 2004
10. Zuniga, M. Z.; Krishnamachari, B.: Integrating Future Large-Scale Wireless Sensor Networks with the Internet, Department of Electrical Engineering, UNiversity of Southern California, 2002
11. Alonso, J., Dunkels, A., Voigt. T,: Bounds on the energy consumption of routings in wireless sensor nodes, WiOpt'04: Modeling and Optimization in Mobile, Ad Hoc and Wireless Networks, Cambridge, UK, March 2004
12. Ye, W., Heidemann, J.: Medium Access Control in Wireless Sensor Networks, in Wireless Sensor Networks, Kluwer Academic Publishers, 2004
13. Dorigo, M., Stützle, T.: Ant Colony Optimization, MIT Press, 2004.
14. Network Simulator-2: http://www.isi.edu/nsnam/ns/

An Enhanced Aggregation Pheromone System for Real-Parameter Optimization in the ACO Metaphor

Shigeyoshi Tsutsui

Hannan University, Matsubara, Osaka, Japan
tsutsui@hannan-u.ac.jp

Abstract. In previous papers we proposed an algorithm for real parameter optimization called the Aggregation Pheromone System (APS). The APS replaces pheromone trails in traditional ACO with aggregation pheromones. The pheromone update rule is applied in a way similar to that of ACO. In this paper, we proposed an enhanced APS (eAPS), which uses a colony model with *units*. It allows a stronger exploitation of better solutions found and at the same time it can prevent premature stagnation of the search. Experimental results showed eAPS has higher performance than APS. It has also shown that the parameter settings for eAPS are more robust than for APS.

1 Introduction

Ant colony optimization (ACO) has been applied with great success to a large number of discrete optimization problems [1,2,3,4,5,6,7]. However, up to now, few ant-colony based approaches for continuous optimization have been proposed in the literature. The first such method, called Continuous ACO (CACO), was proposed in [8]. CACO combines ACO with a real-coded GA, with the ACO approach being largely devoted to local search. CACO was extended in [9,10]. In [11], the behavior of an ant called *Pachycondyla Apicalis* was introduced as the base metaphor of the algorithm (API) to solve continuous problems. The Continuous Interacting Ant Colony (CIAC) in [12] also introduces an additional direct communication scheme among ants.

In contrast to the above studies, studies that have a purely pheromone based method were proposed independently in [13,14]. In [13], the pheromone intensity is represented by a single normal probability distribution function. The center of the normal distribution is updated at each iteration to the position which has the best functional value. The variance value is updated according to the current ant distribution. In [14], a mixture of normal distributions was introduced to solve multimodal functions. However, both of the above two approaches use marginal distribution models and the pheromone update rules used are quite different than those of the original ACO methods.

In [15,16], we proposed a model called the Aggregation Pheromone System (APS). The APS replaces pheromone trails in traditional ACO with aggregation pheromones and uses them as the base metaphor of the algorithm. The

M. Dorigo et al. (Eds.): ANTS 2006, LNCS 4150, pp. 60–71, 2006.

pheromone update rule is applied in a way similar to that of AS [1], and as a result, the aggregation pheromone density eventually becomes a mixture of multivariate normal distributions. APS could solve even hard problems which have a tight linkage among parameters. However, this model was sensitive to control parameters.

In this paper, we propose an enhanced APS (eAPS) which is more robust to the control parameters and has higher performance than APS. This is achieved by applying stronger exploitation of better solutions found during the search and by using a colony model with *units*. The remainder of this paper is organized as follows. Section 2 gives a brief review of APS. Section 3 describes how the eAPS is structured. Section 4 presents experimental results. Finally, Section 5 concludes the paper.

2 A Review of Aggregation Pheromone System (APS)

In the real world, *aggregation pheromones* are used by some insect species to communicate with members of their community to share information about the location of food, safe shelter, potential mates, or enemies. Many types of function of aggregation behavior have been observed. These include foraging-site marking, mating, finding shelter, and defense. When they have found safe shelter, cockroaches produce a specific pheromone in their excrement which attracts other members of their species [17].

The difference between ACO and APS is in how the pheromones function in the search space [15,16]. In ACO, pheromone intensity is defined as a trail on a node or an edge between nodes of a given sequencing problem. On the other hand, in APS, the aggregation pheromone density is defined by a continuous density function in the search space X in R^n.

Fig.1 shows the colony model of the previously proposed APS. Here, the colony consists of a population of m individuals. The aggregation pheromone density $\tau(t,x)$ is defined in the search space X in R^n. Initially (t=0), the aggregation pheromone is distributed uniformly. The initial pheromone density is used to generate the first m individuals; therefore, the initial individuals are generated randomly with a uniform distribution over all possible individuals. Next, the individuals are evaluated and aggregation pheromone is emitted by m individuals depending on their functional values. Then the pheromone density for the next iteration $\tau(t+1,x)$ is obtained and m individuals for next iteration are generated according to $\tau(t+1,x)$.

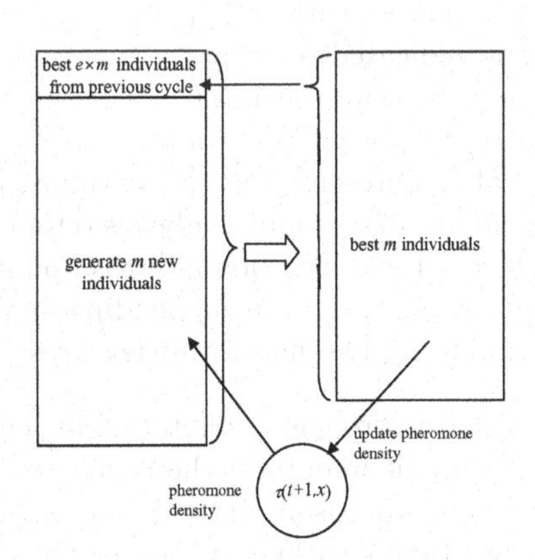

Fig. 1. Colony model of the APS

Here the best individuals of $e \times m$ from the previous population are added to the newly generated solutions (the value of e should be rather small, e.g. $e = 0.1$). The best m individuals are selected from the combined population of $(e+1) \times m$ to create the next population of m individuals. Setting $t \leftarrow t+1$, this iteration continues until the predetermined termination criteria are satisfied. Note here that in APS, the best $e \times m$ individuals are retained as candidate individuals for the next iteration so that APS can perform a stronger exploitation of best solutions found in the search.

3 Enhanced Aggregation Pheromone System (eAPS)

3.1 The Model of the eAPS

In eAPS, we use a colony model as shown in Fig.2. The colony model consists of m *units*, and each unit consists of only one individual. At iteration $t+1$, a new individual $x^{new}_{i,t+1}$ ($i = 1, 2, \ldots, m$) is generated according to the pheromone density $\tau(t+1, x)$. It is then compared with the existing individual $x^*_{i,t}$ in the unit. If $x^{new}_{i,t+1}$ is better than $x^*_{i,t}$, it is replaced with $x^{new}_{i,t+1}$ and is indicated as $x^*_{i,t+1}$ (here note that the notation of individual$x^{new}_{i,t}$ (or $x^*_{i,t}$) also represents the vector value of the individual). Thus in eAPS, the best individual of each unit is always retained. After the best individual for each unit is obtained and iteration index is updated, aggregation pheromone is emitted by all $x^*_{i,t}$ (i =1, 2, ..., m) depending on their functional values. This colony model has following two important features:

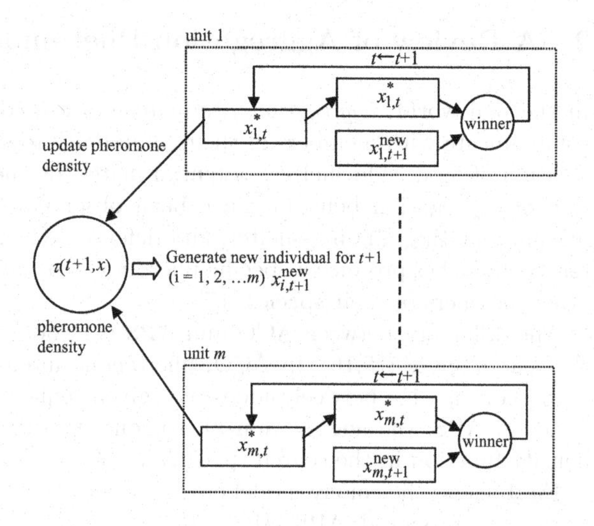

Fig. 2. Colony model of the eAPS

1. It has a stronger exploitation of the best individuals than APS in Fig. 1 since each unit maintains the best individual found so far in the unit.
2. The comparison method in each unit is similar to tournament selection in genetic algorithms (GAs), but it differs from traditional tournament selection because the comparison is performed only between existing and newly generated individuals. However, just as tournament selection can maintain diversity of a population in GAs, this colony model can also maintain the diversity of $x^*_{i,t}$.

3.2 Updating the Pheromone Intensity and Sampling New Individuals

Although the methods of updating the pheromone intensity and sampling new individuals in eAPS are basically the same as with APS, here we describe them in a more schematic manner.

3.2.1 Probability Density Function of Pheromone

In each iteration, $x_{i,t}^{new}$ (i=1,2, ..., m) individuals are generated based on the density of the aggregation pheromone. Individuals are attracted more strongly to positions where the pheromone density is higher, whereas they are attracted less strongly to positions where the density is lower. More specifically, eAPS generates new individuals using probabilities proportional to the values of the aggregation density function. Let $\tau(t, x)$ be the density function of the aggregation pheromone in iteration t in search space X. Initially (t=0), the aggregation pheromone is distributed uniformly, that is, $\tau(0, x) = c$ where c is an arbitrary positive constant. The probability density function used to generate new candidate solutions at iteration t, denoted by $p_\tau(t,x)$, is defined as

$$p_\tau(t, x) = \frac{\tau(t, x)}{\int_X \tau(t, x)dx}. \tag{1}$$

Note that each individual $x_{i,t}^*$ emits aggregation pheromone in its neighborhood. The intensity of aggregation pheromone emitted by individual $x_{i,t}^*$ is based on the following properties:

1. The intensity of the pheromone emitted by $x_{i,t}^*$ depend on its functional value $f(x_{i,t}^*)$ – the higher the functional value of $x_{i,t}^*$, the higher the pheromone intensity emitted by $x_{i,t}^*$. We use ranking based on functional value to control pheromone intensity. The best individual has a rank m, whereas the worst individual has a rank 1.
2. The intensity emitted by $x_{i,t}^*$ should decrease with distance so that the pheromone intensity of the points that are closer to $x_{i,t}^*$ is increased more than the pheromone intensity of the points that are further from $x_{i,t}^*$.
3. In order to alleviate the effects of linear transformations of the search space, the intensity should consider the overall distribution of individuals in each unit. To achieve this, the pheromone intensity emitted by $x_{i,t}^*$ was chosen to be a Gaussian distribution with covariance equal to the covariance of $x_{i,t}^*$ (i=1, 2, ..., m).

Let us represent the rank value of individual $x_{i,t}^*$ at iteration t by $r(t, x_{i,t}^*)$. Then, the density of the pheromone intensity emitted by $x_{i,t}^*$ at point x is given by a scaled multivariate normal distribution as

$$\Delta\tau'(t, x_{i,t}^*, x) = C\frac{(r(t, x_{i,t}^*))^\alpha}{\sum_{j=1}^{m} j^\alpha} N(x;\ x_{i,t}^*, \beta^2\Sigma_t), \tag{2}$$

where $N(x; x_{i,t}^*, \beta^2\Sigma_t)$ is a multivariate normal distribution function (its center is located at $x_{i,t}^*$), Σ_t is the covariance matrix of all individuals of units at iteration

t, α (α >0) is a parameter for adjusting the relative importance of rank, β (β>0) is a scaling factor, and C is the total pheromone intensity emitted from all units at iteration t. The total aggregation pheromone density emitted by the m individuals $x^*_{i,t}(i = 1, 2, ..., m)$ at iteration t is then given by

$$\Delta\tau(t,x) = \sum_{i=1}^{m} \Delta\tau'(t, x^*_{i,t}, x). \quad (3)$$

Note that C is assumed to be constant for all t, that is,

$$\int_X \Delta\tau(t,x)dx = C \text{ for } t \geq 0. \quad (4)$$

The sampling method in 3.2.2 is based on the assumptions of Eq. 4. The scaling factor β controls exploration by reducing the amount of variation as the colony converges to the optimum. The total aggregation pheromone density is updated according to the following formula as in ACO:

$$\tau(t+1,x) = \rho \cdot \tau(t,x) + \Delta\tau(t,x), \quad (5)$$

where ρ ($0 \leq \rho < 1$) controls the evaporation rate.

After the pheromone updating is performed, a new individual $x^{new}_{i,t+1}$ is generated based on the new pheromone density $\tau(t+1,x)$, $x^*_{i,t+1}$ is obtained by comparison between $x^*_{i,t}$ and $x^{new}_{i,t+1}$ at each unit i, and the next iteration of eAPS is performed. Since the pheromone density updates increase pheromone intensity near individuals with better functional value, the pheromone density in promising regions of the search space is expected to increase over successive iterations, eventually converging to global optima.

3.2.2 Sampling New Individuals

As described in 3.2.1, new individuals are sampled with probabilities proportional to the aggregation density $\tau(t+1,x)$. To perform the sampling, we need to obtain the probability density function $p_\tau(t+1,x)$ from $\tau(t+1,x)$. Since $\tau(t+1,x)$ in Eq. 5 can be rewritten as

$$\tau(t+1,x) = \rho^{t+1}\tau(0,x) + \sum_{h=0}^{t} \rho^h \Delta\tau(t-h,x). \quad (6)$$

$p_\tau(t+1,x)$ is obtained from Eqs. 1, 4, and 6 as

$$p_\tau(t+1,x) = \frac{\rho^{t+1}}{\sum_{k=0}^{t+1}\rho^k} \cdot \frac{\tau(0,x)}{C} + \sum_{h=0}^{t} \frac{\rho^h}{\sum_{k=0}^{t+1}\rho^k} \cdot \frac{\Delta\tau(t-h,x)}{C}. \quad (7)$$

Here note that we assume $\rho^0 = 1$ for $0 \leq \rho < 1$ for the sake of convenience.

In general, if a probability density function $f(x)$ can be written as a mixture of other probability density functions

$$f(x) = p_1 f_1(x) + p_2 f_2(x) + \cdots + p_S f_S(x) \quad (8)$$

with $p_1 + p_2 + ... + p_S = 1$, then the sampling of each point according to the distribution defined by $f(x)$ proceeds as follows:

1. Choose which mixture component $f_s(x)$ should be used (a mixture component $f_s(x)$ is chosen with probability p_s).
2. Generate a random point according to the density function of the chosen component.

Since $p_\tau(t+1,x)$ is indeed a mixture distribution of $t+1$ multivariate Gaussian distributions and one uniform distribution, we can sample it using the standard sampling procedure for mixture distributions where the probability of the s-th component of the mixture is

$$p_s = \rho^s \Big/ \sum_{k=0}^{t+1} \rho^k \tag{9}$$

and the s-th mixture component is given by

$$f_s(x) = \begin{cases} \Delta\tau(t-s,x)/C \text{ if } s \leq t \\ \tau(0,x)/C \text{ otherwise.} \end{cases} \tag{10}$$

This sampling is called *iteration sampling.* Sampling according to the last mixture component $f_{t+1}(x)$ is simple because $f_{t+1}(x)$ is a constant and thus represents a uniform distribution over the entire search space. The remaining components, $f_s(x)$, where $s \leq t$, are mixture distributions of m components each:

$$\frac{\Delta\tau(t-s,x)}{C} = \sum_{i=1}^{m} \frac{(r(t-s, x^*_{i,t-s}))^\alpha}{\sum_{j=1}^{m} j^\alpha} N(x;\ x^*_{i,t-s}, \beta^2 \Sigma_{t-s}). \tag{11}$$

Sampling according to $f_s(x)$, where $s \leq t$, can be done in a similar way, i.e., by using the sampling procedure for mixture distributions where the probability of the i-th component $N(x;\ x^*_{i,t-s}, \beta^2 \Sigma_{t-s})$ of Eq. 11 is

$$p_i = (r(t-s, x^*_{i,t-s}))^\alpha \Big/ \sum_{j=1}^{m} j^\alpha. \tag{12}$$

We call this sampling *rank sampling.* Since each component is a normal distribution, it can be sampled using Cholesky decomposition [18].

To perform the sampling based on Eq. 7, using the above sampling method, we need a large amount of memory to store $x^*_{r,t-h}$ vector values and the covariance matrix $\beta^2 \Sigma_{t-h}$ when the iteration index t becomes large. In this case $\rho^h \rightarrow 0$ for large h since $\rho < 1$. Thus, we can limit the maximum number of iterations to keep data up to a constant H. For $t \geq H$ we need to use slightly modified probability function $p_\tau(t+1,x)$ from Eq. 7.

Although at each iteration sampling is performed probabilistically according to Eq. 7 (or its modification for $t \geq H$), we also introduce a perturbation, resulting from conflict among individuals, or environmental disturbances. Perturbation works to perform the same function as mutation in evolutionary algorithms. The perturbation rate is represented by P_{rate}. The pseudo-code of eAPS is shown in Fig. 3.

1. $t \leftarrow 0$
2. Set the initial pheromone density $\tau(0, x)$ uniformly
3. Sample two individuals randomly for each unit i
4. Obtain the rank number $r(t, x^*_{i,t})$ for each individual $x^*_{i,t}$ (m for the best, 1 for the worst 1) in the colony
5. Compute the covariance matrix Σ_t of m individuals of $x^*_{i,t}$ of i=1,2,..., m
6. Update the pheromone density $\tau(t+1, x)$ according to Eq. 5
7. Sample new individual $x^{new}_{i,t+1}$ according to Eq. 7 for i=1,2,..., m
8. Apply perturbation to $x^{new}_{i,t+1}$ with rate of P_{rate} for i=1,2,..., m
9. Compare $x^{new}_{i,t+1}$ and $x^*_{i,t}$ and set the best one as $x^*_{i,t+1}$ for i=1,2,..., m
10. $t \leftarrow t+1$
11. If the termination criteria are met, terminate the algorithm. Otherwise, go to 4

Fig. 3. The pseudo-code of eAPS

3.2.3 The Computational Complexity

Here, we consider the computational complexity of the eAPS. Let n be the problem size. First, let us consider the computational complexity of updating the pheromone intensity described in 3.2.1. Comparing $x^{new}_{i,t+1}$ and $x^*_{i,t}$ for m units is simply performed with O(m). To give the rank number for each individual $x^*_{i,t}$, we use a quick sort with a complexity of O($m\times\log(m)$). Computing the covariance matrix Σ_t of $m$individuals is performed with O($m \times n^2$). Next, the computational complexity of the three samplings to generate m individual described in 3.2.2 is as follows: The iteration sampling is simply performed with O($t \times m$) for $t < H$ or O($H \times m$) for $t \geq H$. The rank sampling is also simply performed with O(m^2). Cholesky decomposition is performed with O(n^3) and the sampling after the Cholesky decomposition is O($m \times n^2$).

Thus the computation time of the algorithm is mainly occupied by computing the covariance matrix (O($m \times n^2$)), the Cholesky decomposition (O(n^3)) and the sampling after the Cholesky decomposition (O($m \times n^2$)). However, since in real-world parameter optimization a function evaluation usually needs a much larger computational time compared to the evolutionary algorithm run time, the number of function evaluations is the more critical issue. In Section 4, we mainly evaluate eAPS using the number of function evaluations.

4 Experimental Study

In this section, the search capability and the characteristics of eAPS are studied in comparison with APS using test functions which are commonly used in the evolutionary computation community.

4.1 Experimental Methodology

We used the following four test functions: the Ellipsoidal function ($F_{Ellipsoidal}$) [19], the Ridge function (F_{Ridge}), the Rosenbrock function ($F_{Rosenbrock}$), and the Rastrigin function ($F_{Rastrigin}$).

$$F_{Ellipsoidal} = \sum_{i=1}^{n} ix_i^2, \; (-3.12 \le x_i < 7.12) \tag{13}$$

$$F_{Ridge} = \sum_{i=1}^{n} \left(\sum_{j=1}^{i} x_j \right)^2, \; (-44 \le x_i < 84) \tag{14}$$

$$F_{Rosenbrock} = \sum_{i=2}^{n} (100(x_1 - x_i^2)^2 + (x_i - 1)^2), \; (-2.048 \le x_i < 2.048) \tag{15}$$

$$F_{Rastrigin} = 10n + \sum_{i=1}^{n} (x_i^2 - 10\cos(2\pi x_i)) \;, \; (-3.12 \le x_i < 7.12) \tag{16}$$

F_{Ridge} has a linkage among variables. $F_{Rosenbrock}$ has a strong linkage among variables. $F_{Ellipsoidal}$ and $F_{Rastrigin}$ have no linkage among variables. $F_{Rosenbrock}$, $F_{Ellipsoidal}$, and F_{Ridge} are unimodal, and $F_{Rastrigin}$ is multimodal. $F_{Ellipsoidal}$, F_{Ridge}, and $F_{Rastrigin}$ have their global optima at (0, 0, ..., 0). To avoid sampling bias as discussed in [19], search space definitions are altered from the original definitions.

We evaluated the algorithms by measuring *#OPT* (the number of runs in which algorithms succeeded in finding the global optimum) and *MNE* (the mean number of function evaluations to find the global optimum in those runs where it did find the optimum). Problem size $n = 20$ was used for all test functions. We assumed the solution to be successfully detected if the functional value was within $n \times 10^{-6}$ of the actual optimum value. 20 runs were performed in each setting. Each run continued until the global optimum was found or a maximum of 500,000 evaluations was reached. There are four common parameters for both eAPS and APS: m, α, β, and ρ. APS contains an extra parameter e which determines the number of best individuals passed to the next iteration by $e \times m$ (see Fig. 1). In all experiments, the same unit size of $m= 120$ was used. Perturbation was applied to each solution with $P_{rate} = 0.0005$ by adding a randomly generated number from a zero-mean, unit-normal distribution. For the value of H, H=100 was used. Here the remaining parameters were set as follows. For the APS, the following parameters were used in the experiments: $\beta = 0.6$, $\alpha = 4$, $\rho = 0.9$, and $e = 0.1$. For eAPS, the following parameters were used in the experiments: $\beta = 1.0$, $\alpha = 32$, and $\rho = 0.2$, except for the value of β of $F_{Rastrigin}$. For $F_{Rastrigin}$, $\beta = 0.6$ was used. These values were obtained by tuning, using test function $F_{Rosenbrock}$. $F_{Rosenbrock}$ was chosen because it is one of the hardest problems to solve due to a strong linkage among variables. For a comparison with a real-coded GA (RGA), the results were also compared to the results obtained with SPX crossover [20], a typical crossover operator for RGAs, with the population size tuned to 300. Parameters for SPX were also tuned using $F_{Rosenbrock}$.

4.2 Results

Table 1 shows the results of eAPS, APS, and RGA using the parameter values described in Section 4.1. Among these three, eAPS showed the best results with the smallest values of *MNE*. For examples, values of MNE on $F_{Rosenbrock}$ are 61, 768.7,126,994.4, and 255, 483.5 for eAPS, APS, and RGA, respectively. Values of *#OPT* on $F_{rastrigin}$ are 20, 16, and 17 for eAPS, APS, and RGA, respectively. (*Time* indicates the average time for successful runs in seconds on a 2.8G Hz Pentium 4 with 512MB main memory. The code is written in Java). Fig. 4 shows the convergence processes of eAPS, APS, and RGA on $F_{Rosenbrock}$. Each dot sequence indicates a change of functional values over 20 runs with each algorithm. We can see that eAPS converged more rapidly than APS and RGA.

Table 1. Results with default values

Algorithm		$F_{Ellipsoidal}$	F_{Ridge}	$F_{Rosenbrock}$	$F_{Rastrigin}$
eAPS	*#OPT*	20/20	20/20	20/20	20/20
	MNE	**35927.7**	**43043.5**	**61768.7**	**239364.5**
	STD	1137.9	1510.7	4169.6	55065.9
	Time(sec)	1.8	1.4	2.7	9.7
APS	*#OPT*	20/20	20/20	20/20	16/20
	MNE	70001.4	83730.6	126994.4	412560.9
	STD	1521.5	1539.9	7325.7	45166.8
	Time(sec)	3.1	3.8	5.5	17.4
RGA	*#OPT*	20/20	20/20	20/20	17/20
	MNE	113076.9	137996.0	255483.5	406759.0
	STD	1716.2	137996.0	21392.3	42639.6
	Time(sec)	8.5	7.8	10.6	30.0

In addition to examining the performance results with fixed parameter values, we analyzed the sensitivity of eAPS and APS to changing parameters ρ and α. To make the analysis feasible, the sensitivity of eAPS and APS with respect to only one parameter at a time is analyzed, while all the remaining parameters being kept constant.

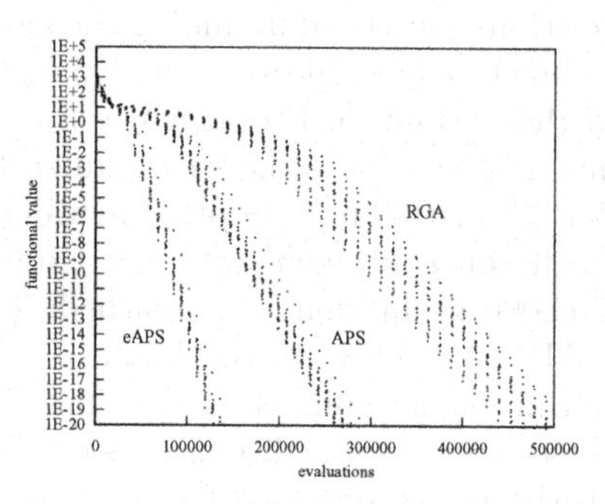

Fig. 4. Convergence processes on $F_{Rosenbrock}$

Fig. 5 shows the variation of *MNE* and *#OPT* with ρ for the function $F_{Ellipsoidal}$ (unimodal, no linkage among variables) and $F_{Ellipsoidal}$ (unimodal, strong linkage among variables). The first figure shows the variation with eAPS and the second one with APS. To explore the effect of the evaporation coefficient ρ, we varied the value of ρ over the ranges of [0, 0.9] for eAPS and [0.5, 0.95] for APS. We can see that eAPS is robust to the variation of ρ, but APS is very sensitive to this value. In eAPS, the best individuals are maintained in each unit. This works as a kind of memory of past searches. This is the reason why eAPS is robust to variation of ρ. However, on $F_{Rosenbrock}$ which has a strong linkage among variables, ρ still plays an important role. On the other hand, on $F_{Ellipsoidal}$ which has no linkage among variables, $\rho = 0$ showed the best performance. In APS, the appropriate choice of ρ value becomes important for APS to perform well.

To see the effect of the parameter α, which adjusts the relative importance of rank in determining the pheromone emitted by an individual, we tested α in the range of [4, 64] on the same functions (Fig. 6). With larger values of α,higher ranked individuals emit pheromone at increasing rates. With larger values of α, the performance of eAPS increased, as seen in the results with $F_{Ellipsoidal}$. However, as seen in the results with $F_{Rosenbrock}$, which has a strong linkage among variables, larger α values affect convergence. Again in APS, varying α has a strong effect on performance, especially on $F_{Rosenbrock}$.

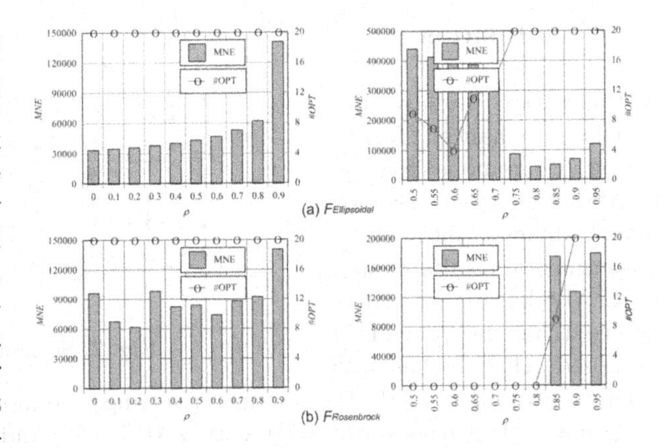

Fig. 5. Variation of MNE and $\#OPT$ with ρ

In the above experiment, we analyzed the algorithms with problem size (n) fixed at 20. Here we focus on the scale-up behavior of eAPS and APS using function $F_{Rosenbrock}$ with the number of variables increasing from 10 to 40 with a step size of 10. We ran the experiment for each number of variables (n) increasing the values of m starting at 100 with a step size of 20 until the optimal solution was obtained 5 times in 5 runs. The values of the other control parameters were the same as those described in Section 4.1. Fig. 7 shows changes of $MNE$and m. The MNE increased at rate of almost $O(n^{2.6})$ and m increased almost linearly. However, we need to undertake a more intensive study on the scaling behavior of the algorithms before drawing firm conclusions.

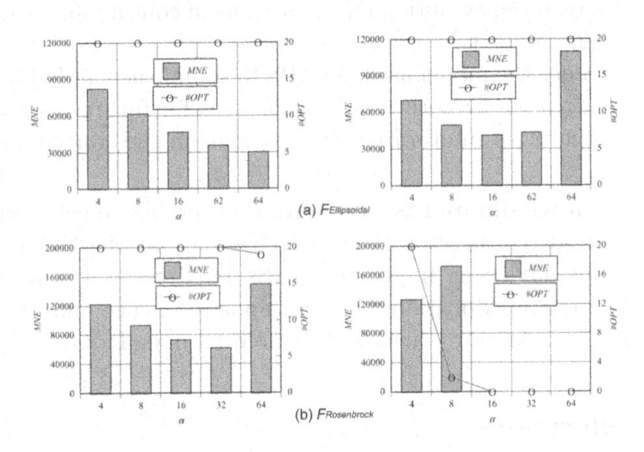

Fig. 6. Variation of MNE and $\#OPT$ with α

5 Conclusions

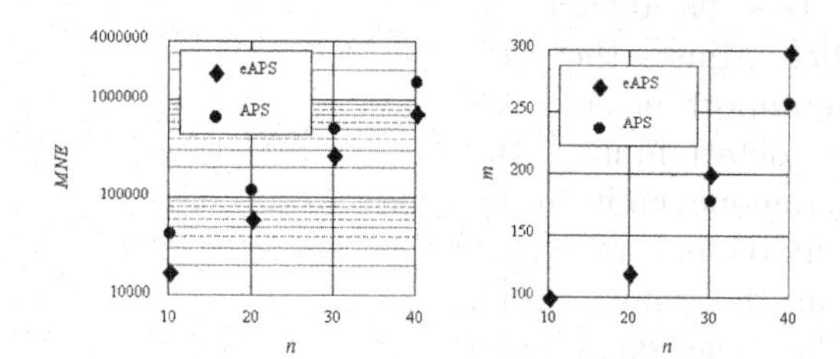

Fig. 7. Scale-up behavior

In this paper, we have proposed an enhanced Aggregation Pheromone System (eAPS), which uses a colony model divided into units. Experimental results showed that eAPS has higher performance than the previously proposed APS. It has also been shown that in eAPS the parameter settings are more robust.

Although the results obtained with eAPS on standard test problems are promising, we do not claim that eAPS is the most efficient algorithm for solving the test problems. However, it provides an alternative approach to the use of probabilistic model building and sampling in evolutionary algorithms [21] for solving real-valued problems. There are many important topics for future research:

1. It is important to identify classes of problems for which eAPS outperforms advanced evolutionary algorithms in continuous domains, such as, CMA-ES) [22] and PCX [19].
2. Since the model used in eAPS to sample new individuals can contain multiple attractors, another interesting area for future research is to extend eAPS to deal with multi-objective problems in the continuous domain.

Acknowledgements. The author would like to acknowledge Dr. Marco Dorigo, the research director of the IRIDIA Lab. at the Université Libre de Bruxelles, for his advice on improving APS. This research is partially supported by the Ministry of Education, Culture, Sports, Science and Technology of Japan under Grant-in-Aid for Scientific Research number 16500143.

References

1. Dorigo, M., Maniezzo, V., Colorni, A.: Ant System: Optimization by a colony of cooperating agents. IEEE Trans. on SMC-Part B **26**(1) (1996) 29–41
2. Dorigo, M., Gambardella, L.M.: Ant colony system: A cooperative learning approach to the traveling salesman problem. IEEE Trans. on EC **1**(1) (1997) 53–66
3. Dorigo, M., Stützle, T.: Ant Colony Optimization. MIT press, Massachusetts (2004)
4. Stützle, T., Hoos, H.: Max-min ant system. Future Generation Computer Systems **16**(9) (2000) 889–914
5. Maniezzo, V.: Exact and approximate nondeterministic tree-search procedures for the quadratic assignment problem. Research Report CSR 98-1 (1998)
6. Forsyth, P., Wren, A.: An ant system for bus driver scheduling. Proc. of the 7th Int. Workshop on Computer-Aided Scheduling of Public Transport (1997)

7. Bullnheimer, B., Hartl, R.F., Strauss, C.: Applying the Ant System to the Vehicle Routing Problem. Meta-heuristics: Advances and trends in local search paradigms for optimization, voss, s., et al (eds.) edn. Kluwer (1999)
8. Bilchev, G., Parmee, I.C.: The ant colony metaphor for searching continuous design spaces. Proc. of the AISB Workshop on Evo. Comp. (1995) 24–39
9. Mathur, M., Karle, S.B., Priye, S., Jyaraman, V.K., Kulkarni, B.D.: Ant colony approach to continuous function optimization. Ind. Eng. Chem. Res **39** (2000) 3814–3822
10. Wodrich, M., Bilchev, G.: Cooperative distribution search: the ant't way. Control Cybernetics **3** (1997) 413–446
11. Monmarch'e, N., Venturini, G., Slimane, M.: On how pachycondyla apicalis ants suggest a new search algorithm. Future Generation Computer Systems **16**(8) (2000) 937–946
12. Dreo, J., Siarry, P.: A new ant colony algorithm using the heterarchical concept aimed at optimization of multiminima continuous functions. Proc. of the Third Int. Workshop on Ant Algorithms (ANTS 2002) (2002) 216–221
13. Pourtakdoust, S.H., Nobahari, H.: An extension of ant colony system to continuous optimization problems. Proc. of Fourth Int. Workshop on Ant Colony Optimization and Swarm Intelligence (ANTS 2004) (2004) 294–301
14. Socha, K.: ACO for continuous and mixed-variable optimization. Proc. of Fourth Int. Workshop on Ant Colony Optimization and Swarm Intelligence (ANTS 2004) (2004) 25–36
15. Tsutsui, S.: Ant colony optimisation for continuous domains with aggregation pheromones metaphor. Proc. of the fifth Int. Conf. on Recent Advances in Soft Computing (2004) 207–212
16. Tsutsui, S., Pelikan, M., Ghosh, A.: Performance of aggregation pheromone system on unimodal and multimodal problems. Proc. of the 2005 IEEE CEC (2005) 880–887
17. Bell, W.J., Car, R.T.: Chemical ecology of insects. Journal of Applied Ecology **22**(1) (1985) 299–300
18. Schatzman, M., Taylor, J., Schatzman, M.: Numerical Analysis: A Mathematical Introduction. Oxford Univ Press (2002)
19. Deb, K., Anand, A., Joshi, D.: A computationally efficient evolutionary algorithm for real-parameter optimization. Evolutionary Computation **10**(4) (2002) 371–395
20. Higuchi, T., Tsutsui, S., Yamamura, M.: Theoretical analysis of simplex crossover for real-coded genetic algorithms. Proc. of the PPSN VI (2000) 365–374
21. Pelikan, M., Goldberg, D.E., Lobo, F.: A survey of optimization by building and using probabilistic models. Computational Optimization and Applications **21**(1) (2002) 5–20
22. Hansen, N., Ostermeier, A.: Adapting arbitrary normal mutation distributions in evolution strategies: The covariance matrix adaptation. Proc. of the 1996 CEC (1996) 312–317

An Estimation of Distribution Particle Swarm Optimization Algorithm

Mudassar Iqbal[1] and Marco A. Montes de Oca[2]

[1] Computing Laboratory, University of Kent, Canterbury, United Kingdom
mi26@kent.ac.uk
[2] IRIDIA, CoDE, Université Libre de Bruxelles, Brussels, Belgium
mmontes@ulb.ac.be

Abstract. In this paper we present an estimation of distribution particle swarm optimization algorithm that borrows ideas from recent developments in ant colony optimization which can be considered an estimation of distribution algorithm. In the classical particle swarm optimization algorithm, particles exploit their individual memory to explore the search space. However, the swarm as a whole has no means to exploit its collective memory (represented by the array of previous best positions or pbests) to guide its search. This causes a re-exploration of already known bad regions of the search space, wasting costly function evaluations. In our approach, we use the swarm's collective memory to probabilistically guide the particles' movement towards the estimated promising regions in the search space. Our experiments show that this approach is able to find similar or better solutions than the canonical particle swarm optimizer with fewer function evaluations.

1 Introduction

The first Particle Swarm Optimization (PSO) algorithm was introduced by Kennedy and Eberhart [1,2]. It is a population-based optimization algorithm inspired by the social behavior of birds and, like other algorithms of its kind, it is initialized with a population of complete solutions (called particles) randomly located in a d-dimensional solution space. An objective function $f : S \rightarrow \Re$ where $S \subset \Re^d$, determines the quality of a particle's position, that is, a particle's position represents a solution to the problem being solved. A particle i at time step t has a position vector $\boldsymbol{x}_i^t$ and a velocity vector $\boldsymbol{v}_i^t$. Another vector $\boldsymbol{pbest}_i$ stores the position in which it has received the best evaluation of the objective function. This vector is updated every time the particle finds a better position. Finally, the best vector $\boldsymbol{pbest}$ (i.e., the one with the best objective function value) of any particle belonging to the "neighborhood" of particle i is stored in vector $\boldsymbol{s}_i$. If the neighborhood of particle i is the whole swarm, then $\boldsymbol{s}_i$ is the best solution found so far.

The algorithm iterates updating particles' velocity and position until a stopping criterion is met, usually a sufficiently good solution value or a maximum number of iterations or function evaluations. The update rules are:

M. Dorigo et al. (Eds.): ANTS 2006, LNCS 4150, pp. 72–83, 2006.

$$\boldsymbol{v}_i^{t+1} = \boldsymbol{v}_i^t + \varphi_1 \boldsymbol{U}_1(0,1) * (\boldsymbol{pbest}_i - \boldsymbol{x}_i^t) + \varphi_2 \boldsymbol{U}_2(0,1) * (\boldsymbol{s}_i - \boldsymbol{x}_i^t), \quad (1)$$

$$\boldsymbol{x}_i^{t+1} = \boldsymbol{x}_i^t + \boldsymbol{v}_i^{t+1}, \quad (2)$$

where φ_1 and φ_2 are two constants called the *cognitive* and *social* coefficients respectively, $\boldsymbol{U}_1(0,1)$ and $\boldsymbol{U}_2(0,1)$ are two d-dimensional uniformly distributed random vectors (generated every iteration) in which each component goes from zero to one, and $*$ is an element-by-element vector multiplication operator.

Clerc and Kennedy [3] introduced the concept of *constriction* in PSO. Since it is based on a rigorous analysis of the dynamics of a simplified model of the original PSO, it became highly influential in the field to the point that it is now referred to as the *canonical* PSO. The difference with respect to the original PSO is the addition of a constriction factor in Equation 1. The modified velocity update rule becomes

$$\boldsymbol{v}_i^{t+1} = \chi(\boldsymbol{v}_i^t + \varphi_1 \boldsymbol{U}_1(0,1) * (\boldsymbol{pbest}_i - \boldsymbol{x}_i^t) + \varphi_2 \boldsymbol{U}_2(0,1) * (\boldsymbol{s}_i - \boldsymbol{x}_i^t)), \quad (3)$$

with

$$\chi = \frac{2k}{\left|2 - \varphi - \sqrt{\varphi^2 - 4\varphi}\right|}, \quad (4)$$

where $k \in [0,1]$, $\varphi = \varphi_1 + \varphi_2$ and $\varphi > 4$. Usually, k is set to 1 and both φ_1 and φ_2 are set to 2.05, giving as a result χ equal to 0.729 [4,5]. This is the PSO version we use in our comparisons[1].

From Equations 1 and 3, it is clear that the behavior of every particle is partially determined by its previous experience (through vector $\boldsymbol{pbest}_i$). This memory allows a particle to search somewhere around its own previous best position and the best position ever found by a particle in its neighborhood. However, during a search different particles move and test (i.e., evaluate the objective function) over and over again the same, or approximately the same, region in the search space without any individual improvement. While this is part of the search process and allows the swarm to explore the search space, it is also a waste of computing power when the explored regions have been visited before by the swarm without success. This happens because the swarm as a single entity does not learn.

In this paper, we present a generic extension to the PSO paradigm that allows a particle swarm to estimate the distribution of promising regions—and thus "learn" from previous experience—of the fitness landscape by exploiting the information it gains during the optimization process. This distribution is in turn used to try to keep the particles within the promising regions. It is a modular extension that can be used in any PSO variant that uses a position update rule based on previously found solutions. The estimation of the distribution is done by means of a mixture of normal distributions taking into account the set of *pbest* vectors. It borrows some ideas from recent developments in Ant

[1] Since there is no agreement about which algorithmic variant can be considered the state-of-the-art in the PSO field, we decided to use this version as our reference.

Colony Optimization (ACO) [6] in which an archive of solutions is used to select the next point to explore in the search space. The underlying assumption of independence between variables common to many Estimation of Distribution Algorithms (EDAs) for continuous optimization problems (see [7]) is also present in this work.

The rest of the paper is organized as follows. Section 2 presents some background information on the class of estimation of distribution optimization algorithms to which our proposed algorithm belongs. Section 3 presents in detail the estimation of distribution particle swarm optimizer proposed in this paper. In section 4 we describe the experimental setup we used to assess the performance of our proposed algorithm. Section 5 presents our empirical results along with some discussion and finally, in section 6, we conclude.

2 Estimation of Distribution Optimization Algorithms

Evolutionary Algorithms that use information obtained during the optimization process to build probabilistic models of the distribution of good regions in the search space and that use these models to generate new solutions are called Estimation of Distribution Algorithms (EDAs) [7]. The fully joint probability distribution characterizes the problem being solved. Depending on whether there is *a priori* knowledge about the underlying distribution or not, one can either use a suitable parameterization to get fast convergence rates or use machine learning methods to approximate this unknown distribution, respectively. The latter case is the most commonly found in practice.

EDAs differ in three aspects: (i) in the way they gather information during the optimization process, (ii) in the way they use the gathered information to build probabilistic models, and (iii) in the way they use these models to generate new solutions. An experimental comparison of some of the best known EDAs has been done by Kern et al. [8].

A pseudo-code view of the algorithmic structure behind most EDAs can be seen in Algorithm 1. An EDA starts with a solution population $\mathbf{X}^0$ and a solution distribution model $\mathcal{P}^0$. The main loop consists of four principal stages. The first stage is to select the best individuals (according to some fitness criteria f) from the population. These individuals are used in a second stage in which the solution distribution model $\mathcal{P}^t$ is updated or recreated. The third stage consists of sampling the updated solution distribution model $\mathcal{P}^{t+1}$ to generate new solutions $\mathbf{X}^{t+1}_{offspring}$. The last stage involves the base population $\mathbf{X}^t_{base}$, the new solutions and the fitness criteria. The end result is a new base population and the process starts over again until the stopping criteria are satisfied.

There has been a growing interest for EDAs in the last years and there are now some hybrid approaches. One of them is our proposed algorithm. It works as a canonical PSO but uses information gathered during the optimization process to keep the particles within the promising regions so that it does not waste function evaluations. We detail our algorithm in the next section. For a comprehensive presentation of the EDA field see the work of Larrañaga and Lozano [9].

Algorithm 1. Algorithmic structure of EDAs.

/* Initialization */
Initialize population of solutions $\mathbf{X}^0_{base}$ and solution distribution model $\mathcal{P}^0$

/* Main Loop */
while Stopping criteria are not satisfied **do**
  $\mathbf{X}^t_{parent} = select(\mathbf{X}^t_{base}, f)$ /* Selection */

  $\mathcal{P}^{t+1} = estimate(\mathbf{X}^t_{parent}, \mathcal{P}^t)$ /* Estimation */

  $\mathbf{X}^{t+1}_{offspring} = sample(\mathcal{P}^{t+1})$ /* Sampling */

  $\mathbf{X}^{t+1}_{base} = replacement(\mathbf{X}^{t+1}_{offspring}, \mathbf{X}^t_{base}, f)$ /* Replacement */

  $t = t + 1$
end while

3 Estimation of Distribution Particle Swarm Optimization Algorithm

PSO algorithms are considered to be part of the emerging field of *Swarm Intelligence* [10,11]. Swarm Intelligence is the discipline that studies natural and artificial systems comprised of multiple simple entities that collectively exhibit adaptive behaviors. Some examples of natural swarm intelligent systems are ant colonies, slime molds, bee and wasp swarms.

Besides PSO, the other prominent representative of artificial swarm intelligent systems is Ant Colony Optimization (ACO) [6]. ACO is usually used for solving combinatorial optimization problems. In ACO, artificial ants build solutions incrementally selecting one solution component at a time. The probabilistic selection is biased by a trail of "pheromone" deposited by other ants in previous iterations of the algorithm. The amount of pheromone is proportional to the quality of the complete solutions, so that ants will prefer to choose solution components that are known to yield good solutions. In fact, the role of the so-called pheromone matrix is to approximate the distribution of good solutions in the search space. Seen from this point of view, ACO is an EDA.

A recent development of ACO that extends it to continuous optimization is called ACO_R [12,13]. ACO_R approximates the joint probability distribution, one dimension at a time, by using mixtures of weighted Gaussian functions. The weights try to represent the quality of different search regions. This allows the algorithm to deal with multimodal functions. Figure 1 illustrates the idea of approximating the distribution of good regions in a single dimension using a mixture of weighted Gaussian functions.

The source of information to parameterize these univariate distributions is an archive of solutions of size k. The i-th component of the l-th solution is denoted by s^i_l. For an n-dimensional problem, $1 \leq i \leq n$ and $1 \leq l \leq k$. For each

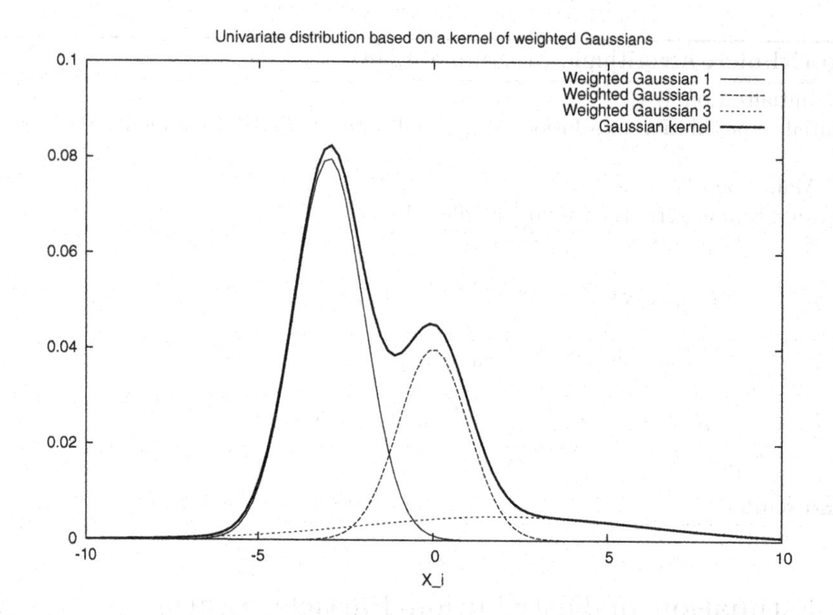

Fig. 1. Mixture of weighted one-dimensional Gaussian functions used to approximate two promising (but in a different degree) search regions

dimension i, the vector $\boldsymbol{\mu}_i = < s_1^i, \ldots, s_k^i >$ is the vector of means that is used to model the univariate probability distribution of the i-th dimension. The vector of weights $\boldsymbol{w} = < w_1, \ldots, w_k >$ is the same across all dimensions because it is based on the relative quality of the complete solutions. Every iteration, after the solutions are ranked, the weights are determined by

$$w_l = \frac{1}{qk\sqrt{2\pi}} e^{-\frac{(l-1)^2}{2(qk)^2}}, \tag{5}$$

where q is a parameter that determines the degree of preferability of good solutions. With a small q the best solutions are strongly preferred to guide the search.

Since ACO$_R$ samples the mixture of Gaussians, it has to first select one of the Gaussian functions from the kernel. The selection is done probabilistically. The probability of choosing the l-th Gaussian function is proportional to its weight and it is computed using

$$p_l = \frac{w_l}{\sum_{j=1}^{k} w_j}. \tag{6}$$

Then, ACO$_R$ computes the standard deviation of the chosen Gaussian function as

$$\sigma_l^i = \xi \sum_{j=1}^{k} \frac{|s_j^i - s_l^i|}{k-1}, \tag{7}$$

where ξ is a parameter that allows the algorithm to balance its exploration–exploitation behaviors. ξ has the same value for all the dimensions. Having computed all the needed parameters, ACO_R samples the Gaussian function to generate a new solution component. The process is repeated for every dimension, for every ant until a stopping criterion is met.

This fast presentation of ACO_R was needed to introduce our Estimation of Distribution Particle Swarm Optimization (EDPSO) algorithm. The reason is that EDPSO borrows some ideas from ACO_R. First, the set of *pbest* vectors plays the role of the solution archive in ACO_R. In EDPSO, k (i.e., the size of the solution archive) is equal to the number of particles. The dynamics of the algorithm, however, is somewhat different. EDPSO works as a canonical PSO as described in section 1 but with some modifications: after the execution of the velocity update rule shown in Equation 3, EDPSO selects one Gaussian function just as ACO_R does. Then, the selected Gaussian function is evaluated (not sampled) to probabilistically move the particle to its new position. If the movement is successful, the algorithm continues as usual, but if the movement is unsuccessful, then the selected Gaussian function is sampled in the same way as in ACO_R. The result is a "hybrid" algorithm that explores the search space using the PSO dynamics but when this approach fails (i.e., when a particle's tendency is to move far away from good solutions) a direct sampling of the probability distribution is used instead. It is important to mention that when the selected Gaussian function is evaluated, we use an *unscaled* version of it, so that its range is [0,1] (i.e., a true probability). A pseudo-code version of EDPSO can be seen in Algorithm 2.

4 Experimental Setup

To evaluate the performance of EDPSO we used the most commonly used benchmark functions in the PSO literature (see [14] for details). We have compared our algorithm with the canonical PSO as described in section 1. Table 1 shows the initialization ranges and the goals that had to be achieved by the algorithms in terms of solution quality, although this goal was not used as a stopping criterion. We ran 30 independent runs for each function in 30,40 and 50 dimensions for a maximum of 120 000, 160 000, and 200 000 function evaluations respectively. The number of particles was equal to 40.

Table 1. Parameters for the test functions

Function	Initialization range	Goal
Sphere	$[-100, 100]^D$	0.01
Rosenbrock	$[-30, 30]^D$	100
Rastrigin	$[-5.12, 5.12]^D$	100
Griewank	$[-600, 600]^D$	0.1
Ackley	$[-32, 32]^D$	0.1

Algorithm 2. Pseudocode version of the EDPSO algorithm

```
/* Initialization. k is the number of particles,
and n is the dimension of the problem */
for i = 1 to k do
  Create particle i with random position and velocity
end for
Initialize gbest and all pbest_i to some sensible values /* To a sufficiently large number,
for example, if we want to minimize a function */

/* Main Loop */
t = 0
while gbest is not good enough or t < t_max do
  /* Evaluation Loop */
  for i = 1 to k do
    if f(x_i) is better than pbest_i then
      p_i = x_i
      pbest_i = f(x_i)
    end if
    if pbest_i is better than gbest then
      gbest = pbest_i
      s = p_i
    end if
  end for
  /* Update Loop */
  Rank all pbest_i according to their quality
  Compute w =< w_1, ..., w_k > using Equation 5
  Compute all p_l using Equation 6
  for i = 1 to k do
    for j = 1 to n do
      v_ij = χ(v_ij + φ_1 U_1(0,1)(p_ij − x_ij) + φ_2 U_2(0,1)(s_ij − x_ij))
      x_ij^candidate = x_ij + v_ij
      Select a Gaussian function from the kernel according to p_l, name it g_l^i.
      Compute σ_l^i using Equation 7
      prob_move = σ_l^i √(2π) g_l^i(x_ij^candidate) /* σ_l^i √(2π) unscales the function */
      if U_3(0,1) < prob_move then
        x_ij = x_ij^candidate /* The particle moves normally */
      else
        x_ij = sample(g_l^i) /* New position is a sample from the chosen function */
      end if
    end for
  end for
  t = t + 1
end while
```

All the benchmark functions we used have the global optimum at or very near the origin, i.e., at the center of the search domain and hence a symmetric uniform initialization would induce a possible bias [15]. To avoid this problem, all functions were shifted to a random location within the search range. This

Table 2. Parameters used by the algorithms

Algorithm	Parameter	Value
Canonical PSO	φ_1	2.05
	φ_2	2.05
	χ	0.729
EDPSO	φ_1	2.05
	φ_2	2.05
	χ	0.729
	q	0.1
	ξ	0.85

approach has been used before and does not confine the swarm to a small region of the search space as is usually done with asymmetrical initializations [16].

Table 2 shows the parameter settings for the algorithms used in our experiments.

5 Results

The benefits of estimating the probability distribution of good regions in the search space and to guide the swarm to search them are reflected (in general) in the quality of the solutions achieved, as well as in the number of function evaluations needed to achieve a solution of certain quality. Table 3 shows the average fitness value (of the best particle in the swarm) after the maximum number of allowed function evaluations.

Table 3. Average fitness value after the maximum number of allowed function evaluations over 30 runs

Algorithm	Dimension	Sphere	Rosenbrock	Rastrigin	Griewank	Ackley
Canonical PSO	30	0.0	37.48	73.52	0.023	13.35
	40	0.0	55.06	133.15	0.037	18.78
	50	0.0	102.4	203.8	0.1	18.3
EDPSO	30	0.0	22.3	25.6	0.0012	0.000019
	40	0.0	37.3	33.43	0.00098	0.00004
	50	0.0	48.12	56.18	0.0029	0.7

Table 4. Average number of function evaluations needed to achieve the solution qualities defined in Table 1 and the probability of achieving them

Algorithm	Dimension	Sphere	Rosenbrock	Rastrigin	Griewank	Ackley
Canonical PSO	30	13049,1.0	20969,0.86	7880,0.9	11907,0.93	13980,0.06
	40	19365,1.0	38442,0.83	13296,0.16	17563,0.93	–
	50	27451,1.0	61124,0.66	–	24584,0.76	–
EDPSO	30	5988,1.0	20921,0.96	18549,1.0	5520,1.0	5656,1.0
	40	8717,1.0	24896,0.9	28045,1.0	7866,1.0	8437,1.0
	50	11971,1.0	50442,0.86	41659,1.0	10741,1.0	20284,0.96

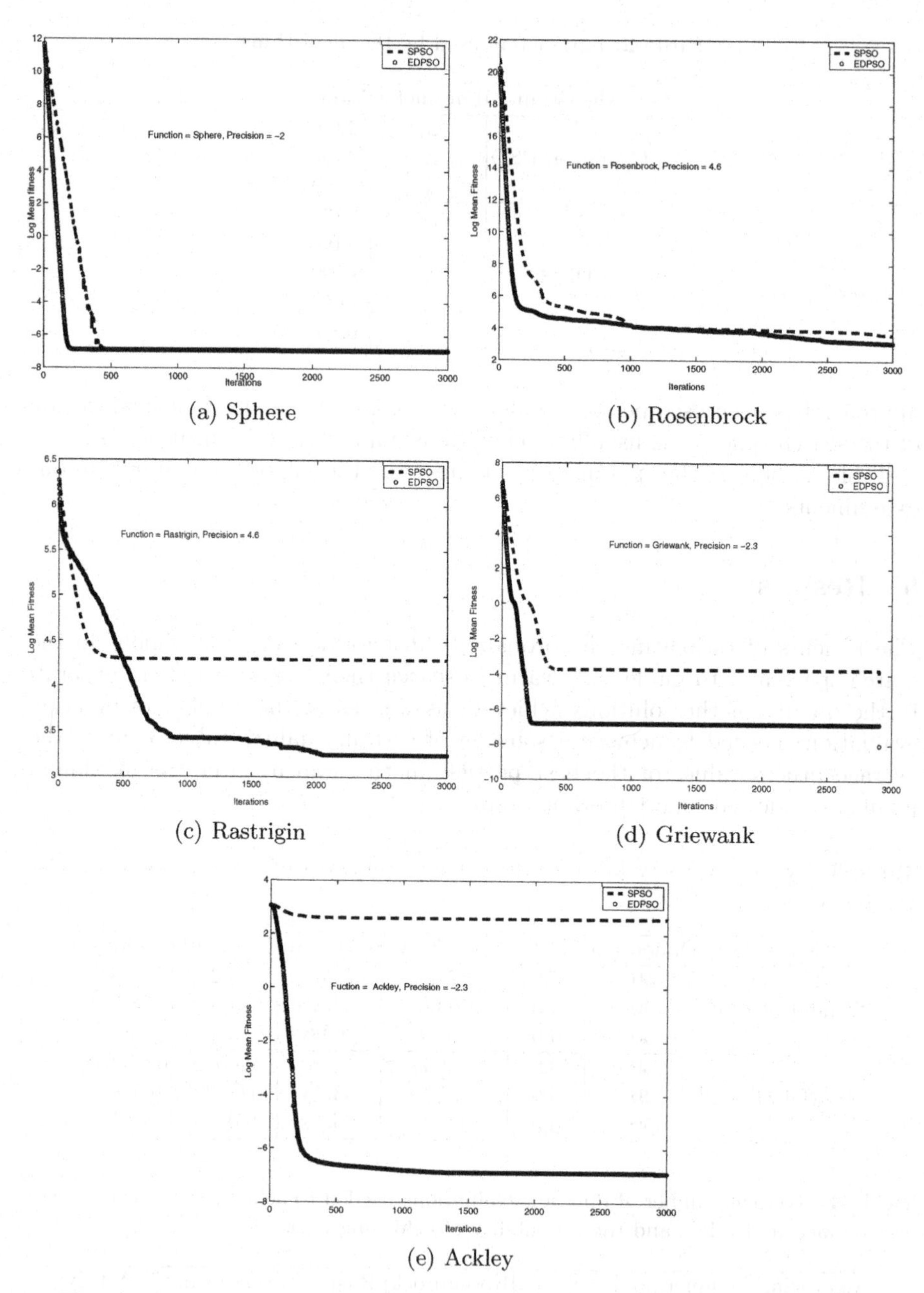

(a) Sphere

(b) Rosenbrock

(c) Rastrigin

(d) Griewank

(e) Ackley

Fig. 2. Solution quality over time. Lines represent the average solution value.

In all benchmark functions, except in the case of the Sphere function, a tendency can be immediately recognized: EDPSO can find better solution qualities after the same number of function evaluations. This is particularly true in the case of the Rastrigin and Ackley functions.

Regarding the issues of speed and reliability, Table 4 shows the average number of function evaluations needed to achieve the solution qualities defined in Table 1 and the probability of achieving them, defined as the success rate (a successful run is one that achieves the specified goal). The average was computed over the successful runs only and rounded off to the nearest integer number greater than the actual number.

From Table 4, it can be seen that EDPSO is faster than the Canonical PSO in getting to the desired objective function value in all functions except in Rastrigin. Entries marked with "–" specify cases in which the goal was not reached in any run. From our experiments, it can be observed that EDPSO shows a significant improvement in terms of the number of function evaluations it needs to get to a certain solution quality. It should also be noted that the behavior of the Canonical PSO is not robust as we go into higher dimensions. In contrast, EDPSO is quite consistent. The Rastrigin case is better explained after examining Figure 2 which shows how the solution quality improves over time for the benchmark problems in 30 dimensions.

The data in Tables 3 and 4 give only a partial view of the behavior of the algorithms. Specifically, they do not show how the solution quality evolves over time. Knowing this is particularly useful to identify which algorithm is best suited for real-time applications in which there are hard time limits or for applications in which we are interested in the solution quality only. In Figure 2(c) it can be seen how the goal defined in Table 1 was reached first by the Canonical PSO but it can also be seen how it stagnates and cannot find better solutions after some more iterations.

6 Conclusions

In this paper we have introduced an Estimation of Distribution Particle Swarm Optimization (EDPSO) algorithm. It is in fact a modular extension that can be used in any other PSO variant that uses a position update rule based on previously found solutions. In effect, it is a learning mechanism that helps the particle swarm explore potentially good regions of the search space. It benefits from the information gathered during the optimization process that is encoded in the array of *pbests*. The end result is a PSO variant that not only finds better solutions than the Canonical PSO, but also does it with fewer function evaluations. There are some cases, however in which speed is sacrificed for the sake of finding better solutions.

EDPSO is not a pure Estimation of Distribution Algorithm (EDA). It explores the search space in the same way as the Canonical PSO but becomes an EDA whenever particles are pushed further away from good regions (so learnt by the whole swarm). It remains a research issue the problem of handling interactions between variables and the correct parameterization of the probability distributions. The results reported here are encouraging enough to continue looking for ways to allow the particle swarm learn from its past experience.

Acknowledgments. Mudassar Iqbal acknowledges support from the Computing Laboratory, University of Kent and the EPSRC grant GR/T11265/01 (eXtended Particle Swarms). Marco A. Montes de Oca acknowledges support from the Programme Alβan, the European Union Programme of High Level Scholarships for Latin America, scholarship No. E05D054889MX. His work was also supported by the ANTS project, an *Action de Recherche Concertée* funded by the Scientific Research Directorate of the French Community of Belgium.

References

1. Kennedy, J., Eberhart, R.: Particle swarm optimization. In: Proceedings of the IEEE International Conference on Neural Networks, Piscataway, NJ, IEEE Press (1995) 1942–1948
2. Eberhart, R., Kennedy, J.: A new optimizer using particle swarm theory. In: Proceedings of the 6th International Symposium on Micro Machine and Human Science, Piscataway, NJ, IEEE Press (1995) 39–43
3. Clerc, M., Kennedy, J.: The particle swarm–explosion, stability, and convergence in a multidimensional complex space. IEEE Transactions on Evolutionary Computation **6**(1) (2002) 58–73
4. Eberhart, R., Shi, Y.: Comparing inertia weights and constriction factors in particle swarm optimization. In: Proceedings of the 2000 IEEE Congress on Evolutionary Computation. (2000) 84–88
5. Trelea, I.C.: The particle swarm optimization algorithm: Convergence analysis and parameter selection. Information Processing Letters **85**(6) (2003) 317–325
6. Dorigo, M., Stützle, T.: Ant Colony Optimization. The MIT Press (2004)
7. Pelikan, M., Goldberg, D.E., Lobo, F.: A survey of optimization by building and using probabilistic models. Computational Optimization and Applications **21**(1) (2002) 5–20
8. Kern, S., Müller, S.D., Hansen, N., Büche, D., Ocenasek, J., Koumoutsakos, P.: Learning probability distributions in continuous evolutionary algorithms–a comparative review. Natural Computing **3**(1) (2004) 77–112
9. Larrañaga, P., Lozano, J.: Estimation of Distribution Algorithms: A New Tool for Evolutionary Computation. Genetic Algorithms and Evolutionary Computation, Vol. 2. Springer (2001)
10. Kennedy, J., Eberhart, R., Shi, Y.: Swarm Intelligence. Morgan Kaufmann Publishers, San Francisco, CA, USA (2001)
11. Bonabeau, E., Dorigo, M., Theraulaz, G.: Swarm Intelligence: From Natural to Artificial Systems. Santa Fe Institute Studies on the Sciences of Complexity. Oxford University Press, USA (1999)
12. Socha, K.: ACO for Continuous and Mixed-Variable Optimization. In Dorigo, M., Birattari, M., Blum, C., eds.: Proceedings of ANTS 2004 – Fourth International Workshop on Ant Colony Optimization and Swarm Intelligence. Volume 3172 of LNCS., Springer-Verlag, Berlin, Germany (2004) 25–36
13. Socha, K., Dorigo, M.: Ant colony optimization for continuous domains. Technical Report TR/IRIDIA/2005-037, IRIDIA, Université Libre de Bruxelles (2005)
14. Janson, S., Middendorf, M.: A hierarchical particle swarm optimizer and its adaptive variant. IEEE Transactions on Systems, Man and Cybernetics–Part B **35**(6) (2005) 1272–1282

15. Monson, C.K., Seppi, K.D.: Exposing origin-seeking bias in PSO. In et al., H.G.B., ed.: Proceedings of the Genetic and Evolutionary Computation Conference (GECCO), New York, NY, ACM Press (2005) 241–248
16. Suganthan, P.N., Hansen, N., Liang, J.J., Deb, K., Chen, Y.P., Auger, A., Tiwari, S.: Problem definitions and evaluation criteria for the CEC 2005 special session on real-parameter optimization. Technical Report 2005005, Nanyang Technological University, Singapore and IIT Kanpur, India (2005)

Ant-Based Approach to the Knowledge Fusion Problem

David Martens[1], Manu De Backer[1], Raf Haesen[1],
Bart Baesens[1,2], Christophe Mues[1,2], and Jan Vanthienen[1]

[1] Department of Decision Sciences & Information Management, K.U. Leuven, Belgium
{David.Martens, Manu.Debacker, Raf.Haesen}@econ.kuleuven.be,
{Bart.Baesens, Christophe.Mues, Jan.Vanthienen}@econ.kuleuven.be
[2] University of Southampton, School of Management, United Kingdom
{Bart, C.Mues}@soton.ac.uk

Abstract. Data mining involves the automated process of finding patterns in data and has been a research topic for decades. Although very powerful data mining techniques exist to extract classification models from data, the techniques often infer counter-intuitive patterns or lack patterns that are logical for domain experts. The problem of consolidating the knowledge extracted from the data with the knowledge representing the experience of domain experts, is called the knowledge fusion problem. Providing a proper solution for this problem is a key success factor for any data mining application. In this paper, we explain how the AntMiner+ classification technique can be extended to incorporate such domain knowledge. By changing the environment and influencing the heuristic values, we can respectively limit and direct the search of the ants to those regions of the solution space that the expert believes to be logical and intuitive.

1 Introduction

Over the past decades we have witnessed a true explosion of data. Although much information is available in this data, it is typically hidden. Data mining entails the overall process of extracting knowledge from this data. Different types of data mining are discussed in the literature [1], such as regression, classification and clustering. The task of interest here is classification, which is the task of assigning a datapoint to a predefined class or group according to its predictive characteristics. The classification problem and accompanying data mining techniques are relevant in a wide variety of domains such as credit scoring (predicting whether a client will default on his/her loan or not) and medical diagnosis (e.g. classifying a breast mass as either benign or malignant). The result of a classification technique is a model which makes it possible to classify future data points, based on a set of specific characteristics in an automated way.

Although many powerful classification algorithms have been developed, they generally rely solely on modeling repeated patterns or correlations which occur in the data. However, it may well occur that observations, that are very evident

M. Dorigo et al. (Eds.): ANTS 2006, LNCS 4150, pp. 84–95, 2006.

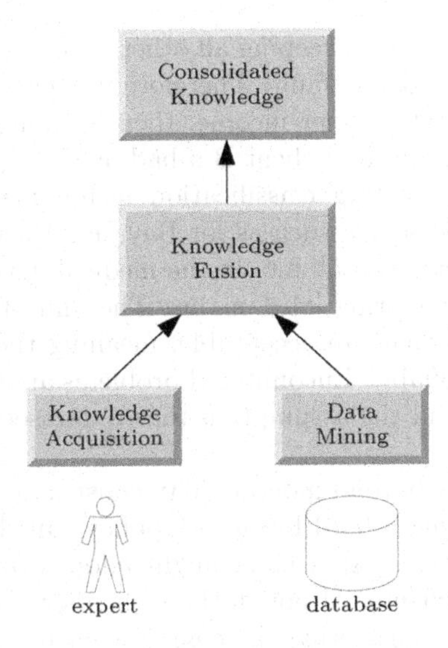

Fig. 1. The knowledge fusion process

to classify by the domain expert, do not appear frequently enough in the data in order to be appropriately modeled by a data mining algorithm. Hence, the intervention and interpretation of the domain expert still remains crucial. A data mining approach that takes into account the knowledge representing the experience of domain experts is therefore much preferred and of great focus in current data mining research.

2 The Knowledge Fusion Problem

The main performance criterion for data mining models has been accuracy; however accuracy is often not enough. Domains where the models need to be validated and justifiable, for instance credit scoring and medical diagnosis, require that the model is in line with existing domain knowledge. The academically challenging problem of consolidating the automatically generated data mining knowledge with the knowledge reflecting experts' domain expertise, constitutes the knowledge fusion problem (see Fig. 1). The final goal of the knowledge fusion problem is to provide models that are both accurate, comprehensible and justifiable, and thus acceptable for implementation.

The most frequently encountered and researched aspect of knowledge fusion is the monotonicity constraint. This constraint demands that an increase in a certain input(s) cannot lead to a decrease in the output. More formally (similarly to [2]), given a dataset $D = \{x^i, y^i\}_{i=1}^n$, with $x^i = (x_1^i, x_2^i, \ldots, x_m^i) \in X = X_1 \times X_2 \times \ldots X_m$, and a partial ordering $\leq$ defined over this input space X. Over the space Y of class values y^i, a linear ordering $\leq$ is defined. Then the classifier $f : x^i \mapsto f(x^i) \in Y$ is monotone if Eq. 1 holds.

$$x^i \leq x^j \Rightarrow f(x^i) \leq f(x^j),\ \forall i,j \quad (\text{or } f(x^i) \geq f(x^j), \forall i,j) \qquad (1)$$

For instance, increasing income, keeping all other variables equal, should yield a decreasing probability of loan default. Therefore if client A has the same characteristics as client B, but a lower income, then it cannot be that client A is classified as a good customer and client B a bad one.

Several adaptions to existing classification techniques have been put forward to deal with monotonicity, such as for Bayesian learning [3], classification trees [2,4] and neural networks [5]; a.o. in the medical diagnosis [6], house price prediction [7] and credit scoring [8] domains. The aim of all these approaches is to generate classifiers that are acceptable, meaning they are both accurate, comprehensible and justifiable. Encountered problems are how to deal with non-monotonic, noisy data and the assumption that a dataset has to be monotonic in all its variables.

Knowledge fusion goes beyond monotonicity constraints, as preferred policies cannot be declared in such a hard format. A policy can show a preference of a certain value for a variable; e.g. a bank might adopt a policy that focuses on young, highly-educated clients. Although these clients will tend to have a rather low income and savings, considering the complete customer lifetime value they can be very profitable to the bank. Incorporating such policies are of great importance as well, and to the best of our knowledge, not incorporated in data mining techniques. As we will discuss in the next sections, we can incorporate existing domain knowledge in an elegant manner using the AntMiner+ classification technique.

3 AntMiner+

AntMiner+ is a recently proposed classification technique based on a $\mathcal{MAX}$-$\mathcal{MIN}$ ant system [9,10]. The first application of ant systems for classification was reported in [11], where the authors introduced the AntMiner algorithm for the discovery of classification rules. The aim of AntMiner+ is to extract simple **if** *rule antecedent* **then** *rule consequent* rules from data, where the rule antecedent is a conjunction of terms. A directed acyclic construction graph is created that acts as the environment of the ants. All ants begin in the *Start* vertex and walk through their environment to the *Stop* vertex, gradually constructing a rule. To allow for rules where not all variables are involved, hence shorter rules, an extra *dummy vertex* is added to each variable whose value is undetermined, meaning it can take any of the values available. Although only categorical variables are allowed, we make a distinction between nominal and ordinal variables. Each nominal variable has one vertex group (with the inclusion of the mentioned dummy vertex), but for the ordinal variables however, we build two vertex groups to allow for intervals to be chosen by the ants. The first vertex group corresponds to the lower bound of the interval and should thus be interpreted as $< V_{i+1} \geq Value_k >$, the second vertex group determines the upper bound, giving $< V_{i+2} \leq Value_l >$ (of course, the choice of the upper bound is constrained by the lower bound). This allows to have less, shorter and actually better rules. To extract a ruleset that is complete, such that all future data

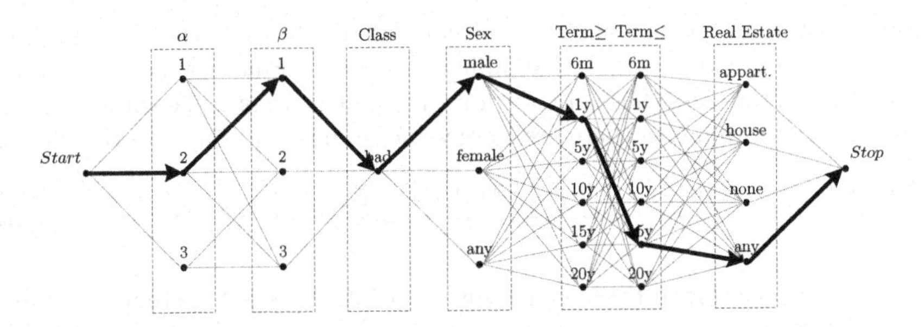

Fig. 2. Example of path described by an ant for the AntMiner+ construction graph

points can be classified, the majority class is not included in the vertex group of the class variable, and will be the predicted class for the final **else** clause.

Only the ant that describes the best rule will have the pheromone (τ) of its followed trail increased. Evaporation decreases the pheromone of all other edges. Supplementary modifications of the pheromone levels may be needed since the $\mathcal{MAX}$-$\mathcal{MIN}$ approach additionally requires the pheromone levels to lie within a given interval [12]. Convergence occurs when all the edges of one path have a pheromone level τ_{max} and all others edges have pheromone level τ_{min}. Next, the rule corresponding to the path with τ_{max} will be extracted and training data covered by this rule removed from the training set. This iterative process will be repeated until early stopping occurs. A credit scoring example of the construction graph described so far is shown in Fig. 2, where we have three variables: sex of the applicant, term of the loan and information concerning real estate property of the applicant. An ant that follows the path indicated in bold from Start to Stop describes the rule:

if *Sex=male* **and** *Term≥1y* **and** *Term≤5y* **then** *customer=bad*

The edge probability $P_{(v_{i,j},v_{i+1,k})}$ is the probability that an ant which is in vertex $v_{i,j}$ (vertex for which variable V_i is equal to its jth value) will go to vertex $v_{i+1,k}$. This probability is dependent on the heuristic value $\eta_{v_{i+1,k}}$ (Eq. 3) and the pheromone value $\tau_{(v_{i,j},v_{i+1,k})}$, as defined by Eq. 2. The α and β are weight parameters that indicate the relative importance of the pheromone and heuristic value. Although these are typically set by trial and error, the parameters are included in the construction graph, allowing them to be dynamically chosen and thus be automatically optimized to the dataset at hand.

$$P_{(v_{i,j},v_{i+1,k})} = \frac{[\tau_{(v_{i,j},v_{i+1,k})}]^{\alpha}.[\eta_{v_{i+1,k}}]^{\beta}}{\sum_{l=1}^{p_{i+1}}[\tau_{(v_{i,j},v_{i+1,l})}]^{\alpha}.[\eta_{v_{i+1,l}}]^{\beta}} \quad (2)$$

The heuristic value gives for each vertex in the construction graph a notion of its quality and importance in the problem domain. For the classification task, we define the importance of a vertex, with a certain value for a variable, by the number of training cases that are covered (described) by this value. A formal definition for this heuristic value for the vertex $v_{i,k}$ is given by Eq. 3, with T_{ik} a shortened notation for the term $V_i = Value_k$, and the $|c|$ operator returning the number of uncovered training data fulfilling condition c. Since the heuristic value is dependent on the class chosen by the ant (denoted as $class_{ant}$), each vertex has as many heuristic values as there are class values minus one (the majority class).

$$\eta_{v_{i,k}} = \frac{|T_{ik} \;\&\; CLASS = class_{ant}|}{|T_{ik}|} \tag{3}$$

Updating the pheromone trail of the environment of an ant system is accomplished in two phases, being evaporation and reinforcement (Eq. 4). In a $\mathcal{MAX}$-$\mathcal{MIN}$ ant system, reinforcement of the pheromone trail is only applied to the best ant's path.

$$\tau_{(v_{i,j},v_{i+1,k})}(t+1) = \rho \cdot \tau_{(v_{i,j},v_{i+1,k})}(t) + Q^{+}_{best} \tag{4}$$

Clearly, the reinforcement of the best ant's path, Q^{+}_{best}, should be proportional to the quality of the path, which we define as the sum of the *confidence* and the *coverage* of the corresponding rule. Confidence measures the fraction of the number of uncovered data points correctly classified by a rule compared to the total number of not yet covered data points covered by that rule. The coverage gives an indication of the overall importance of the specific rule by measuring the number of correctly classified (not yet covered) data points over the total number of uncovered data points. A more formal definition is provided by Eq. 5, with $rule_{ant}$ the rule antecedent (if part) comprising of a conjunction of terms corresponding to the path chosen by the ant, $rule^{c}_{ant}$ the conjunction of $rule_{ant}$ with the class chosen by the ant, and Cov a binary variable expressing whether a data point is already covered by one of the extracted rules ($Cov = 1$) or not ($Cov = 0$).

$$Q^{+} = \underbrace{\frac{|rule^{c}_{ant}|}{|rule_{ant}|}}_{\textbf{confidence}} + \underbrace{\frac{|rule^{c}_{ant}|}{|Cov = 0|}}_{\textbf{coverage}} \tag{5}$$

Previous experiments show that AntMiner+ is competitive with state-of-the-art classification techniques, such as C4.5, logistic regression and support vector machines [9,10]. AntMiner+, as C4.5 and other rule-based classifiers, has the supplementary benefit of providing comprehensible rules that are interpretable to the experts that use the classification model in practice. However, inconsistencies with previous knowledge can still be present, yet hidden in the rules. As we will show in the next section, decision tables can assist in easily detecting such anomalies.

4 Visualising and Validating the Models

4.1 Decision Tables

Decision tables are a tabular representation used to describe and analyze decision situations [13] and consists of four quadrants. The vertical line divides the table

1. Owns property?	2. Years client	3. Savings amount	1. Applicant=good	2. Applicant=bad
yes	≤ 3	low	–	×
		high	×	–
	> 3	low	×	–
		high	×	–
no	≤ 3	low	–	×
		high	×	–
	> 3	low	–	×
		high	×	–

(a) Expanded decision table

1. Owns property?	2. Years client	3. Savings amount	1. Applicant=good	2. Applicant=bad
yes	≤ 3	low	–	×
		high	×	–
	> 3	–	×	–
no	–	low	–	×
		high	×	–

(b) Contracted decision table

Fig. 3. Minimizing the number of columns of a lexicographically ordered DT

into a condition (left) and an action part (right), while the horizontal line separates subjects (above) from entries (below). The condition subjects are the problem criteria (the variables) that are relevant to the decision-making process. The action subjects describe the possible outcomes of the decision-making process; i.e., the classes of the classification problem. Each condition entry describes a relevant subset of values (called a state) for a given condition subject (variable), or contains a dash symbol ('–') if its value is irrelevant within the context of that row. Every row in the entry part of the decision table thus comprises a classification rule. For example, in Fig. 3a, the final row tells us to classify the applicant as good if owns property = no, years client > 3 and savings amount = high. Decision tables can be contracted by combining logically adjacent (groups of) rows that lead to the same action configuration, as shown in 3b.

Decision tables can easily be checked for potential anomalies, such as inconsistency with monotonicity constraints: by placing the assumingly monotone variable in the last column, adjacent rows are found with data entries that are equal in all variables except the last one. It can then be easily seen whether or not the class variable changes in the expected manner. For example, based on Fig. 3, we can see that it is indeed reflected in the model that a high savings amount can only have a positive effect on the applicants assessment, if any.

4.2 Using Decision Tables to Validate AntMiner+ Rulesets

An example ruleset that was extracted by AntMiner+ on the german credit scoring dataset [14] is provided in Table 1. When the rules are transformed into a decision table, shown by Table 2, an anomaly is revealed. Consider a client that requests a loan with a maturity of less than 15 months and low savings and checking amounts; when this client has no credits taken in the past, the client will be rejected. On the other hand, when he or she has a critical account, the loan will be granted. This decision policy is counter-intuitive and unwanted.

Table 1. Example credit scoring ruleset

if (Checking Account < 100€ **and** Duration > 15 m **and**
  Credit History = no credits taken **and** Savings Account < 500€)
then class = bad
else if (Purpose = new car/repairs/education/others **and**
  Credit History = no credits taken/all credits paid back duly at this bank **and**
  Savings Account < 500€)
then class = bad
else if (Checking Account < 0€ **and**
  Purpose = furniture/domestic appliances/business **and**
  Credit History = no credits taken/all credits paid back duly at this bank **and**
  Savings Account < 250€)
then class = bad
else if (Checking Account < 0€ **and** Duration > 15 m **and**
  Credit History = critical account **and** Savings Account < 250€)
then class = bad
else class = good

Table 2. Discrepancies in the classification model, visualized by decision table

Purpose	Duration	Checking Account	Savings Account	Credit History	Bad	Good
furniture/business	≤15m	<0€	<250€	no credits taken/all credits paid back duly	×	–
				critical account	–	×
			≥250€	–	–	×
		≥0€	–	–	–	×
	>15m	<0€	<250€	–	×	–
			≥250 and <500€	no credits taken/all credits paid back duly	×	–
				all credits at this bank paid back duly or critical account	–	×
			≥500€	–	–	×
		≥0 and <100€	<500€	no credits taken/all credits paid back duly	×	–
				all credits at this bank paid back duly or critical account	–	×
			≥500€	–	–	×
		≥100€	–	–	–	×
car/retraining or others	≤ 15m	–	<500€	no credits taken/all credits paid back duly	×	–
				critical account	–	×
			≥500€	–	–	×
	>15m	<0€	<250€	–	×	–
			≥250 and <500€	no credits taken/all credits paid back duly	×	–
				critical account	–	×
			≥500€	–	–	×
		≥0€	<500€	no credits taken/all credits paid back duly	×	–
				critical account	–	×
			≥500€	–	–	×

5 Incorporating Domain Knowledge in AntMiner+

We suggest two approaches to incorporate domain knowledge in AntMiner+. The first approach implements so-called hard constraints, which are constraints that must not be violated according to the expert. This is done by emitting some vertex groups all together, thereby limiting the solution space. Alternatively, one can also consider soft constraints, which are constraints that are preferred but not mandatory. This only biases our search toward certain regions in the solution space, but does not limit the search. To do so, we will manipulate the heuristic values of the vertexes in the AntMiner+ environment. It is up to the domain

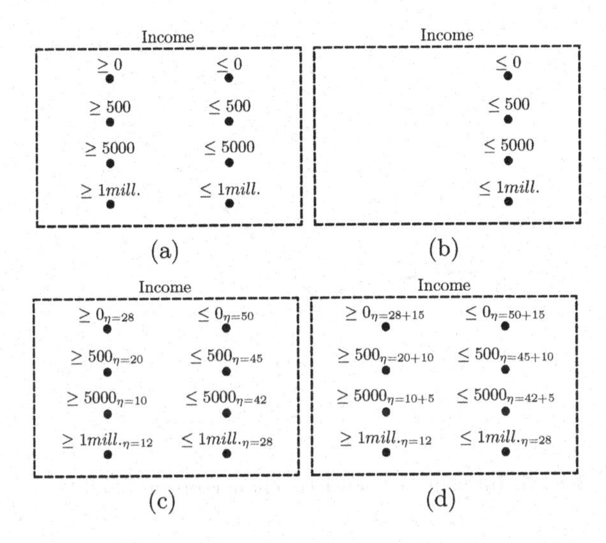

Fig. 4. Vertex groups for variable *income* without (a) and with (b) environment adjustment to reflect hard monotonicity constraint; and without (c) and with (d) heuristic adjustments to reflect soft constraint

expert to decide for which variables a constraint should be incorporated, and whether it is hard or soft.

5.1 Hard Constraints to Incorporate Domain Knowledge

The most straightforward manner of avoiding anomalies in the solutions is to close those parts of the solution space, effectively hard-coding the constraints. Our AntMiner+ environment allows to elegantly do so, by emitting vertexes or vertex groups. A typical constraint that can be fulfilled in this manner is the monotonicity constraint. Referring back to our credit scoring example, we can make sure that increasing income cannot lead to a customer changing from good to bad by removing the vertex group corresponding to *income* $\geq$ (see Fig. 4(a,b)): since the ants look only for rules to classify bad customers (only the final else clause will classify a customer as good), the term with *income* can only be in the form *income* $\leq X$. Using hard constraints guarantees that the resulting classifier is monotone, even when the dataset is not. This is not true for some of the previous approaches that incorporate monotonicity constraints [2].

5.2 The Use of Heuristics to Incorporate Domain Knowledge

The unique approach of AntMiner+ for the classification task allows for an easy extension to incorporate soft constraints. As described in Section 3, the heuristic values provide a problem dependent preference measure for each vertex. The higher the heuristic, the more probable that it will be chosen by the ant. Therefore, if we suspect that a variable's value is monotonically related to the predicted class, we can adjust the heuristic values to incorporate this preference,

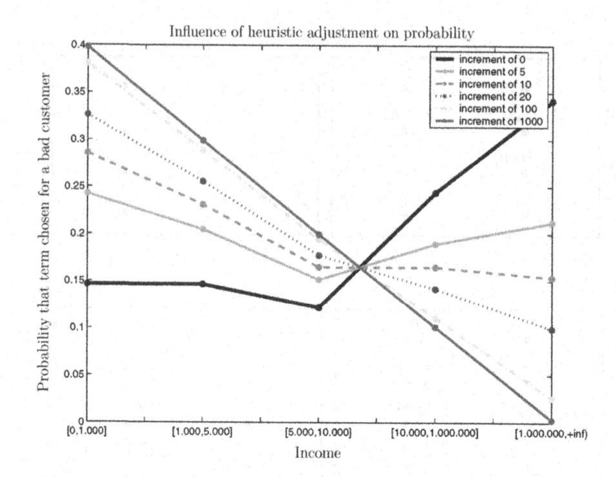

Fig. 5. Influence of heuristic change on probability

as done in Fig. 4(c,d). Note that the difference with hard constraints is that the soft constraints are only preferred, not demanded. Another example of a soft constraint might be the preference of young, highly-educated clients for banks which demonstrates that we are not limited to ordinal variables. Soft constraints are ideal to incorporate a policy, where we increase the heuristics of the preferred value(s). Fig. 5 gives an example of the influence of the change in heuristic value on the edge probability (with α and β set to 1, and pheromone values equal to 1 for all edges to keep the calculations simple). The example concerns the variable income. For increasing income, we would expect a lowering probability that the edge is chosen: lower incomes should be more related to bad customers than high incomes. However, this (rather obvious) intuition is not present in the data. By increasing the heuristic values in a linear manner, we can see that the probability increasingly meets our expectations.

5.3 Experiments

To empirically assess the performance of the suggested approach, we applied AntMiner+ to several publicly available datasets from the UCI data repository [14]. The data is split up into a training set (4/9), validation set (2/9) to implement the stopping criterion and test set (3/9) to obtain an unbiased performance indicator. To eliminate any chance of having unusually good or bad training and test sets, 10 runs are conducted where the order of observations is first randomized before the training, validation and test set are chosen. For each randomization AntMiner+ is run with and without hard monotonicity constraints (with 1000 ants and evaporation factor of 0.85 [9]).

The *auto* dataset concerns the car fuel consumption prediction in gallons-per-mile (higher or lower than 28), based on 8 car properties. Domain knowledge states that higher weight and displacement will lead to higher fuel consumption, where more recent models and cars manufactured in Japan are presumed to be

Table 3. Classification model fulfilling all monotonicity constraints, visualized by decision table

Duration	Purpose	Checking Account	Savings Account	Credit History	Bad	Good
≤ 15m	–	–	–	–	–	×
> 15m	car(old)/others	–	–	–	–	×
	furniture/business	<0€	<250€ or unknown/no savings	–	×	–
			≥250€	–	–	×
		≥0 and <100€ or no checking account	<50€	all credits paid back duly or all credits at this bank paid back duly	–	×
				existing credits paid back duly till now or critical account	×	–
			≥50€ or unknown/no savings	–	–	×
		≥100€	–	–	–	×
	radio/television	<0€	–	all credits paid back duly or all credits at this bank paid back duly	–	×
				existing credits paid back duly till now or critical account	×	–
		≥0€ or no checking account	–	–	–	×
	car(new)/retraining	–	–	–	–	×

more fuel-efficient. For the *breast cancer ljubljana* dataset we need to predict whether breast cancer will recur, where we expect that an increase in tumor-size, number of nodes involved and the degree of malignancy will lead to a higher probability of recurrence. The problem of the *pima* dataset is to predict whether a person shows signs of diabetes. Increasing age, number of pregnancies, body mass index and pedigree risk would suggest a higher chance of being diabetic. In the already discussed *german* credit scoring dataset, the classification problem is to predict clients as either good or bad (defaulted). Expert knowledge suggests bad customers will have less amount on checking and savings accounts, and have had more problems with their credit history. Finally, the *haberman* dataset concerns the prediction of the survival status of a patient that has undergone breast cancer surgery. Medical knowledge suggests a lower survival rate for patients that are older, have more detected positive axillary nodes and whose operation was less recent. As Table 3 shows, the rules extracted for the german credit scoring dataset are compliant with all monotonicity constraints, making the ruleset a justifiable credit scoring model.

Table 4 shows the results of our experiments. The performances given are the average accuracy, standard deviation of the accuracies, average number of extracted rules (#R) and average number of terms per rule (#T/R). The best average test set performance over the 10 runs is denoted in bold face for each dataset. By definition, the rulesets that are extracted by AntMiner+ with hard constraints no longer show counter-intuitive patterns. A first observation from Table 4 is that less and shorter rules are extracted when incorporating the hard constraints, which is of course beneficial to the comprehensibility of the classifier. Although the overall accuracy increases slightly (but not significantly), the change depends on the existence of monotonicity in the dataset: if the required constraints are not fulfilled by the dataset, a ruleset that does comply will

Table 4. Average out-of-sample AntMiner+ performance

		No constraints				Hard constraints			
		Acc		#R	#T/R	Acc		#R	#T/R
	instances	Avg	σ			Avg	σ		
auto	392	86.3	2.9	3.5	6.4	**87.6**	2.2	**1.9**	**3.4**
bcl	277	79.4	3.2	3.0	7.3	**80.0**	2.9	**2.2**	**5.6**
pima	768	68.9	3.6	5.1	6.2	**69.2**	2.3	**4.6**	**5.3**
ger	1000	**72.3**	1.3	4.3	4.4	71.5	1.3	**4.2**	**3.6**
hab	306	**75.0**	3.4	**3.2**	3.8	74.2	3.3	3.6	**2.5**

naturally experience an accuracy decrease. Note that similar accuracy gains and losses are observed in [3].

6 Conclusion

As knowledge fusion concerns the acceptability of data mining models, it is of great relevance for any successful data mining application. Although the generated models may be very accurate, a lack of comprehensibility or justifiability will result in a reluctance or rejection to put the models into practice. Previous research was focused on hard monotonicity constraints, but this fails to properly address the complete knowledge fusion problem. The unique approach of AntMiner+, with its use of an environment to represent the solution space, heuristics to guide the search and the distinction between nominal and ordinal variables allow us to extend the technique to deal with both hard and soft constraints to respectively limit and direct the search of the ants. Incorporating hard monotonicity constraints in AntMiner+ provides monotone classifiers, even when the dataset is not. The soft constraints are very important as well, and are suitable to incorporate a specific policy in the resulting model.

Our experiments show that, whereas the rules extracted without constraints did not satisfy domain expert's intuition, by inserting domain knowledge we are able to improve both the comprehensibility (less and shorter rules) and justifiability (compliant with monotonicity constraints) of the AntMiner+ rulesets with a similar level of predictive accuracy.

References

1. Witten, I., Frank, E.: Data Mining: Practical machine learning tools and techniques. Morgan Kaufmann (2005)
2. Feelders, A., Pardoel, M.: Pruning for monotone classification trees. In: Advanced in intelligent data analysis V. LNCS 2810, Springer (2003) 1–12
3. Altendorf, E., Restificar, E., Dietterich, T.: Learning from sparse data by exploiting monotonicity constraints. In: Proceedings of the 21st Conference on Uncertainty in Artificial Intelligence, Edinburgh, Scotland (2005) 18–26
4. Ben-David, A.: Monotonicity maintenance in information-theoretic machine learning algorithms. Machine Learning **19**(1) (1995) 29–43

5. Velikova, M., Daniels, H., Feelders, A.: Solving partially monotone problems with neural networks. In: Proceedings of the International Conference on Neural Networks, Vienna, Austria (2006)
6. Pazzani, M., Mani, S., Shankle, W.: Acceptance by medical experts of rules generated by machine learning. Methods of Information in Medicine **40**(5) (2001) 380–385
7. Velikova, M., Daniels, H.: Decision trees for monotone price models. Computational Management Science **1**(3–4) (2004) 231–244
8. Daniels, H., Velikova, M.: Derivation of monotone decision models from non-monotone data. Discussion Paper 30, Tilburg University, Center for Economic Research (2003)
9. Martens, D., De Backer, M., Haesen, R., Baesens, B.: A $\mathcal{MAX}$-$\mathcal{MIN}$ ant system working towards comprehensible classifiers. (Under Review)
10. Martens, D., De Backer, M., Haesen, R., Baesens, B., Holvoet, T.: Ants constructing rule-based classifiers. In: Swarm Intelligence and Data Mining. Studies in Computational Intelligence. Springer (2006)
11. Parpinelli, R.S., Lopes, H.S., Freitas, A.A.: Data mining with an ant colony optimization algorithm. IEEE Transactions on Evolutionary Computation **6**(4) (2002) 321–332
12. Stützle, T., Hoos, H.H.: $\mathcal{MAX}$-$\mathcal{MIN}$ ant system. Future Generation Computer Systems **16**(8) (2000) 889–914
13. Vanthienen, J., Mues, C., Aerts, A.: An illustration of verification and validation in the modelling phase of KBS development. Data and Knowledge Engineering **27**(3) (1998) 337–352
14. Hettich, S., Bay, S.D.: The uci kdd archive [http://kdd.ics.uci.edu] (1996)

Beam-ACO Applied to Assembly Line Balancing*

Christian Blum[1], Joaquín Bautista[2], and Jordi Pereira[3]

[1] ALBCOM, Dept. Llenguatges i Sistemes Informàtics
Universitat Politècnica de Catalunya, Barcelona, Spain
cblum@lsi.upc.edu
[2] ETSEIB, Nissan Chair
Universitat Politècnica de Catalunya, Barcelona, Spain
joaquin.bautista@upc.es
[3] ETSEIB, Dept. d'Organització d'Empreses
Universitat Politècnica de Catalunya, Barcelona, Spain
jorge.pereira@upc.edu

Abstract. Assembly line balancing concerns the design of assembly lines for the manufacturing of products. In this paper we consider the time and space constrained simple assembly line balancing problem with the objective of minimizing the number of necessary work stations. This problem is denoted by TSALBP-1 in the literature. For tackling this problem we propose a Beam-ACO approach, which is an algorithm that results from hybridizing ant colony optimization with beam search. The experimental results show that our algorithm is a state-of-the-art metaheuristic for this problem.

1 Introduction

The simple assembly line balancing problem (SALBP) [1] concerns the manufacturing of products via assembly lines. An assembly line is a sequence of work stations that are connected by a transport system moving the product to be manufactured along the line. The product is manufactured by performing a given set of tasks $T = \{1, \ldots, n\}$. Each task $j \in T$ has a processing time $t_j > 0$. A solution to a SALBP instance is obtained by the assignment of all tasks to work stations subject to precedence constraints. The assembly line moves in constant speed, which leads to a maximum time c (called cyle time) in which the tasks assigned to a work station must be performed. An assignment of tasks to work stations is only valid if the the sum of the processing times of the tasks assigned to a work station does not exceed the cycle time. Among different optimization objectives, the minimization of the number of necessary work stations is quite

* This work was supported by grants TIN-2005-08818-C04-01 and DPI2004-03475 of the Spanish government, and by the "Juan de la Cierva" program of the Spanish Ministry of Science and Technology of which Christian Blum is a post-doctoral research fellow. Moreover, we acknowledge Nissan Spain and the UPC Nissan Chair for partially funding this work as well as for providing real data.

M. Dorigo et al. (Eds.): ANTS 2006, LNCS 4150, pp. 96–107, 2006.

popular. This particular problem is denoted by SALBP-1 in the literature. It can be seen as a bin packing problem with additional side constraints (see [2]).

Approaches for solving SALBP-1 include constructive heuristics based on priority rules (see [3]), complete techniques such as branch & bound approaches (see [4,5,6,7]), and several metaheuristics such as tabu search [8,9], simulated annealing [10], evolutionary computation [11], and ant colony optmization [12,13]. The current situation for SALBP-1 is quite unusual: Complete techniques such as SALOME [6] still appear to be at least as successful as metaheuristics. Nonetheless, the interest in well-working metaheuristics is high. This is because the existing complete techniques are limited to academic formulations of the problem. Even slight differences between a real problem and the academic SALBP-1 make existing complete techniques unapplicable. In this work we consider a generalization of SALBP-1, called TSALBP-1 (the time and space constrained simple assembly line balancing problem with the objective of minimizing the number of necessary work stations). This generalization was proposed by Bautista and Pereira in [13]. The generalization consists in adding space constraints to the processing time constraints, and was motived by a real assembly line balancing problem at the Nissan plant in Barcelona, Spain.

Motivation. Many existing metaheuristic techniques for SALBP-1 do not employ a direct solution approach. They rather solve the SALBP-1 as follows: Given an initial solution with m work stations, a metaheuristic is applied to find a solution with a fixed number of $m-1$ work stations with a cycle time c' as low as possible (that is, the cycle time is considered variable). If a solution can be found with $c' \leq c$ (where c is a parameter of the SALBP-1 instance), the found solution is a valid solution for SALBP-1 with $m - 1$ work stations. In the next step, the number of work stations is reduced by one, and the metaheuristic is applied again. The existing ACO approach for SALBP-1 (see [13]) works in this way.

In this work we wanted to study if a clever ACO hybrid can be directly applied to solve the SALBP-1 (and its generalization, the TSALBP-1) with the goal of improving over the indirect approaches. In order to achieve this, we tackle the TSALBP-1 with an hybrid ant colony optimization (ACO) [14] algorithm called Beam-ACO [15], which is obtained by hybridizing ACO with beam search.[1] The main idea of Beam-ACO is to allow the extension of partial solutions in several different ways. An accurate and computationally inexpensive lower bound is used to limit the number of partial solutions that are visited by the algorithm. In Beam-ACO, artificial ants perform a probabilistic beam search in which the extension of partial solutions is done in the ACO fashion rather than deterministically.

Paper outline. In Section 2 we present a technical definition of the TSALBP-1 problem. In Section 3 we outline Beam-ACO. Finally, in Section 4 we present the computational results, and in Section 5 we offer conclusions and an outlook on future work.

[1] Beam search is a classical tree search method that was introduced in the context of scheduling [16].

2 TSALBP-1

An instance (T, G, c, a) of the TSALBP-1 problem consists of four components. $T = \{1, \ldots, n\}$ is a set of n tasks that must be processed by a line of work stations. Each task $j \in T$ has a processing time $t_j > 0$, and a space requirement $a_j > 0$ (both values may be integer or real values). Furthermore, given is a precedence graph $G = (T, A)$, which is a directed graph without cycles whose nodes are the tasks. An arc $l_{i,j} \in A$ indicates that i must be processed before j. Given a task $j \in T$, we denote by $\text{Pre}_j \subset T$ the set of tasks that must be processed before j. Finally, c is the processing time limit of a work station (called the cycle time), and a is the available space of a work station. Note that all work stations are equal with respect to c and a.

A solution is obtained by assigning each task to exactly one work station. In this work we represent a solution s as an ordered set $\langle S_1, \ldots, S_m \rangle$ of $m \leq n$ work stations S_k. Each work station $S_k \subseteq T$ is a set of tasks. A solution s is valid if the following conditions are fullfilled:

1. $\bigcup_{k=1}^{m} S_k = \{1, \ldots, n\}$ and $\bigcap_{k=1}^{m} S_k = \emptyset$. These conditions ensure that each task is assigned to exactly one work station.
2. $\sum_{j \in S_k} t_j \leq c$, for $k = 1, \ldots, m$. This ensures that no work station has too much load.
3. $\sum_{j \in S_k} a_j \leq a$, for $k = 1, \ldots, m$. Herewith is ensured that the space limits of the work stations are not exceeded.
4. For each $j \in S_k$ it is required that $\bigcup_{l=1}^{k} S_l$ contains Pre_j, which ensures that the precedence constraints between the tasks are respected.

All our algorithms exclusively generate valid solutions. Finally, note that each SALBP-1 instance can be transformed into a TSALBP-1 instance by setting $t_j := a_j \; \forall \, j \in T$, and by setting $a := c$.

Objective function. The objective function of TSALBP-1 is the number of work stations of a solution, which must be minimized. Given a solution s, this objective function is denoted by $c_1(s) = |s|$. However, this objective function contains large plateaus, that is, many different solutions will have the same objective function value. Therefore, we introduce a second criteria in order to distiguish between solutions with the same number of work stations. The second criteria concerns the remaining time and space in the last work station $S_m \in s$. We use the following notations:

$$t^{\text{rem}}(s) := c - \sum_{j \in S_m} t_j \qquad \text{and} \qquad a^{\text{rem}}(s) := a - \sum_{j \in S_m} a_j \tag{1}$$

Using these notations, the second criteria is defined as $c_2(s) = \frac{t^{\text{rem}}(s)}{c} + \frac{a^{\text{rem}}(s)}{a}$. Having c_1 and c_2 we can indirectly define a new objective function $f(\cdot)$ as follows. Given $s \neq s'$,

$$f(s) < f(s') \Leftrightarrow c_1(s) < c_1(s') \quad \textbf{OR} \quad c_1(s) = c_1(s') \text{ and } c_2(s) < c_2(s') \; . \tag{2}$$

This means that—in the case of equality concerning the first criteria—a solution s is regarded as better than a solution s' if and only if more space and time is remaining in the last work station of solution s. Note that despite the fact that Beam-ACO uses the objective function $f(\cdot)$, we will present the results only in terms of the original objective function.

Reverse problem instances [1]. Given a problem instance (T, G, c, a), the corresponding reverse problem instance (T, G^r, c, a) is obtained by inverting all the arcs in the precedence graph G. Each solution $s^r = \langle S_1, \ldots, S_m \rangle$ to the reverse problem instance (T, G^r, c, a) can be converted into a solution s to the original problem instance (T, G, c, a) by inverting the ordered list of tasks, that is, $s = \langle S_m, \ldots, S_1 \rangle$. It is known from the literature (see, for example, [1]) that the reverse problem instance may be easier to solve than the original one, or vice versa.

3 The Algorithm

Our Beam-ACO approach—shown in Algorithm 1—works as follows. First, the heuristic information for the computation of the transition probabilities is determined in function DetermineHeuristicInformation(). Then, the pheromone values are initialized. At each algorithm iteration, a_{o} ants construct solutions to the original problem instance, and a_{r} ants construct solutions to the reverse problem instance. The solutions to the reverse problem instance are subsequently converted to solutions to the original problem instance. During each solution construction a lower bound is used for detecting situations in which the resulting final solution must be worse than the best solution found by the algorithm so far. In this case the corresponding solution construction is aborted. Finally, the pheromone values are updated in function UpdatePheromoneTrail($\mathcal{T}$,$*$). The pheromone values are re-initilized in case of algorithm convergence. In Algorithm 1 we use the following notations: $\mathcal{T} = \{\tau_{j,k}\}_{j,k=1,\ldots,n}$ is the set of pheromone values. A pheromone value $\tau_{j,k}$ represents the desirability of assigning task j to work station k. Furthermore, s_{ib} is the best solution constructed at an iteration, and s_{bsf} is the best solution found since the start of the algorithm. The functions of our algorithm are outlined in more detail below.

DetermineHeuristicInformation(): The heuristic information is obtained by optimizing the parameters of a so-called parametrized greedy heuristic [17]. The result is a static heuristic value $\eta_j^{\text{aco}} \in (0, 1]$ for each $j \in T$. A by-product of this function is the greedy solution that is obtained with the respective heuristic values. We allowed 0.5 seconds for the application of this function. Even though the derivation of the heuristic information is one of the innovative parts of our algorithm we can—due to space limitations—not give a detailed description. We refer the interested reader to [18] instead.

ConstructSolution($\mathcal{T}$) (see Algorithm 2): Solutions are constructed by filling work stations successively one after the other. Hereby, an ant fills each work station in k_{ext} different ways (see lines 8-11 of Algorithm 2), of which the best one is

Algorithm 1. Beam-ACO for TSALBP-1

```
1: input: An instance (T, G, c, a) of TSALBP-1
2: s_bsf ← DetermineHeuristicInformation()
3: forall τ_{j,k} ∈ 𝒯 do τ_{j,k} := 0.5 end forall
4: cf := 0
5: while termination conditions not satisfied do
6:    I := ∅
7:    for i = 1 to a_o do
8:       s_i ← ConstructSolution(𝒯) {see Algorithm 2}
9:       if s_i ≠ NULL then I := I ∪ {s_i}
10:   end for
11:   for i = 1 to a_r do
12:      s_i^r ← ConstructReverseSolution(𝒯)
13:      if s_i^r ≠ NULL then
14:         Obtain a solution s_i to the original instance from s_i^r
15:         I := I ∪ {s_i}
16:      end if
17:   end for
18:   if I = ∅ then
19:      UpdatePheromoneTrail(𝒯,s_bsf)
20:   else
21:      s_ib := min{f(s_i) | s_i ∈ I}
22:      UpdatePheromoneTrail(𝒯,s_ib)
23:      if f(s_ib) < f(s_bsf) then s_bsf := s_ib
24:   end if
25:   cf ← ComputeConvergenceFactor(𝒯)
26:   if cf < 0.05 then
27:      forall τ_{j,k} ∈ 𝒯 do τ_{j,k} := 0.5 end forall
28:   end if
29: end while
30: output: s_bsf
```

selected. In the context of beam search, k_{ext} is the number of extensions that may be obtained from a partial solution. For all our experiments we have used the setting of $k_{\text{ext}} = 50$.

Let $T^{\text{av}} \subseteq T'$ be the set of tasks such that $\text{Pre}_j \cap T' = \emptyset$ for all $j \in T^{\text{av}}$, that is, T^{av} is the set of available tasks. Moreover, let $T^{\text{sat}} \subseteq T^{\text{av}}$ be the set of available tasks such that $c_{\text{rem}} - t_j = 0$ or $a_{\text{rem}} - a_j = 0$; henceforth called the set of saturating available tasks. A work station is filled by applying the function FillWorkStation(T, k) which is shown in Algorithm 3. Function ChooseTask(T') of Algorithm 3 is implemented as follows. First, we flip a coin in order to decide if the construction step is performed deterministically, or probabilistically. In case of a deterministic construction step, the set of tasks from which to choose an operation, denoted by T^{c}, is determined as follows: If the set of available saturating tasks T^{sat} is non-empty, we set $T^{\text{c}} := T^{\text{sat}}$, otherwise $T^{\text{c}} := T^{\text{av}}$. Then, from T^{c} is chosen the task that maximizes

$$\mathbf{p}_j = \frac{\left(\sum_{i=1}^{k} \tau_{j,i}\right) \cdot \eta_j^{\text{aco}}}{\sum_{l \in T^{\text{c}}} \left(\sum_{i=1}^{k} \tau_{l,i}\right) \cdot \eta_l^{\text{aco}}} . \tag{3}$$

This formula uses the summation rule introduced by Merkle and Middendorf for scheduling problems [19]. In case of a probabilistic construction step, T^{c} is set to T^{av}, and a task is chosen by roulette-wheel-selection with respect to the probabilities shown in Equation 3.

After filling a work station, a lower bound LB(·) is applied to the current partial solution s extended by the filled work station (see line 10 of Algorithm 2). The value of the lower bound indicates the minimum number of work stations needed by a solution that contains the current partial solution extended by the filled work station. In this work we used a very simple lower bound: Given a partial solution s, let T^{rem} be the set of tasks that are not yet assigned to work stations. Then:

$$\text{LB}(s) = \max\left\{ \left\lceil \frac{\sum_{j \in T^{\text{rem}}} t_j}{c} \right\rceil, \left\lceil \frac{\sum_{j \in T^{\text{rem}}} a_j}{a} \right\rceil \right\} \tag{4}$$

Only if the lower bound value is not worse than the number of work stations of the best solution found so far, the extension of the current partial solution is considered (see line 10 of Algorithm 2). Finally, from all the possible extensions of a partial solution, the one with the lowest lower bound value is chosen (see line 13 of Algorithm 2). Note that this corresponds to a beam search algorithm with a beam width equal to 1. We decided for this setting, because in this work we only wanted to test the potential of a beam search approach. However, we want to make the reader aware of the fact that a proper setting of the beam width might improve the results of the algorithm even further.

Finally, function ConstructReverseSolution($\mathcal{T}$) works in the same way as function ConstructSolution($\mathcal{T}$), just that it constructs a solution for the reverse problem instance. The same pheromone values are used, just in a slightly different way. For example, the pheromone value that expresses the desirability to assign task j to the first work station of the reverse problem instance is pheromone value $\tau_{j,|s_{\text{bsf}}|}$, instead of $\tau_{j,1}$, and so on.

ComputeConvergenceFactor($\mathcal{T}$): Given the current pheromone values, this function computes a value cf to indicate the state of convergence of the algorithm:

$$cf = 2 \cdot \left(\frac{\sum_{j=1}^{n} \sum_{k=1}^{|s_{\text{bsf}}|} \min\{\tau_{\max} - \tau_{j,k}, \tau_{j,k} - \tau_{\min}\}}{n \cdot |s_{\text{bsf}}| \cdot (\tau_{\max} - \tau_{\min})} \right) \tag{5}$$

Algorithm 2. Function ConstructSolution($\mathcal{T}$) of Algorithm 1

1: **input:** The set of pheromone values $\mathcal{T}$, and s_{bsf}
2: $T = \{1, \ldots, n\}$
3: $k := 0$
4: $s := \langle\rangle$
5: **while** $T \neq \emptyset$ **do**
6: $\quad k := k + 1$
7: $\quad I := \emptyset$
8: $\quad$ **for** $i = 1, \ldots, k_{\text{ext}}$ **do**
9: $\quad\quad S_k^i$ =FillWorkStation(T, k) {see Algorithm 3}
10: $\quad\quad$ **if** $\text{LB}(s \cup S_k^i) \leq |s_{\text{bsf}}|$ **then** $I = I \cup S_k^i$
11: $\quad$ **end for**
12: $\quad$ **if** $I \neq \emptyset$ **then**
13: $\quad\quad S_k^* := \text{argmax}\{\text{LB}(s \cup S_k^i) \mid S_k^i \in I\}$
14: $\quad\quad T := T \setminus S_k^*$
15: $\quad\quad$ Add S_k^* to s, that is, $s := \langle S_1, \ldots, S_{k-1}, S_k^* \rangle$
16: $\quad$ **else**
17: $\quad\quad$ **output:** NULL
18: $\quad$ **end if**
19: **end while**
20: **output:** Solution s

When the pheromone values are initialized, *cf* is 1; on the other side, when all pheromone values are either equal to $\tau_{\max}$ or to $\tau_{\min}$, *cf* is 0. We have set $\tau_{\max}$ to 0.99, and $\tau_{\min}$ to 0.01. Note that the use of these bounds and their value setting is motivated by the implementation of $\mathcal{MAX}$-$\mathcal{MIN}$ AS algorithms implemented in the hyper-cube framework (see, for example, [20]).

UpdatePheromoneTrail($\mathcal{T}$,$*$): This function either uses solution s_{ib} or solution s_{bsf} for updating the pheromone values. s_{bsf} is only used in case no iteration best solution exists, due to solution construction abortions. Let us denote the updating solution by s_{upd}. Then, for $j = 1, \ldots, n$ and $k = 1, \ldots, |s_{\text{upd}}|$ the corresponding pheromone value $\tau_{j,k}$ is updated as follows:

$$\tau_{j,k} = \min\left\{\max\left\{\tau_{\min}, \tau_{j,k} + \rho \cdot (\delta_{j,k} - \tau_{j,k})\right\}, \tau_{\max}\right\} \ , \tag{6}$$

where $\rho \in (0, 1]$ is a learning rate (which we have set to 0.1 for all the experiments). Moreover, $\delta_{j,k}$ is 1, if task j is assigned to work station k in solution s_{upd}, and 0 otherwise.

This concludes the description of our algorithm. The experimental results are outlined in the following section.

4 Computational Results

We implemented the Beam-ACO algorithm in ANSI C++ using GCC 3.2.2 for compiling the software. Our experimental results were obtained on a PC with

Algorithm 3. Function $\mathsf{FillWorkStation}(T, k)$ of Algorithm 2

1: **input:** A set T of tasks, and the index k of the work station to be filled
2: $T' := T$
3: $S_k := \emptyset$
4: $c_{\text{rem}} := c$
5: $a_{\text{rem}} := a$
6: **while** $T' \neq \emptyset$ **and** $\exists\, i \in T'$ **s.t.** $c_{\text{rem}} - t_i \geq 0$ and $a_{\text{rem}} - a_i \geq 0$ **do**
7: $\quad j \leftarrow \mathsf{ChooseTask}(T')$
8: $\quad T' := T' \setminus \{j\}$
9: $\quad S_k := S_k \cup \{j\}$
10: $\quad c_{\text{rem}} := c_{\text{rem}} - t_j$
11: $\quad a_{\text{rem}} := a_{\text{rem}} - a_j$
12: **end while**
13: **output:** Filled work station S_k

Table 1. Results obtained by Beam-ACO in comparison to the results of solution techniques from the literature. The second table row provides the number of SALBP-1 instances solved to optimality (out of 269). The third table row contains the average computation times (in seconds) for finding the best solution of each run. Note that the computation time comparison is not really useful, because the computers that were used are quite different. We refer the interested reader to the corresponding publications.

	SALOME [6]	**PrioTabu** [8]	**EurTabu** [8]	**ANTS** [13]	**HGA** [11]	**Beam-ACO**
solved	227	200	214	227	214	245
avg. time	98.6	101.8	62.6	13.84	n. g.	1.92

Intel Pentium 4 processor (3.06 GHz) and 1 Gb of memory, running Debian Linux. We performed three series of computational tests, which are outlined in the following.

4.1 Results for SALBP-1 Instances

First we applied Beam-ACO to all 269 SALBP-1 instances from the benchmark obtainable from `http://www.assembly-line-balancing.de`. The results of Beam-ACO are presented in a summarized form and compared to other approaches in Table 1. Beam-ACO was applied 10 times for 120 CPU time seconds to each problem instance. We can note that Beam-ACO solves more problem instances to optimality than any other available technique. In particular, we can note that Beam-ACO solves more instances than ANTS, which is an ACO algorithm that utilizes the indirect resolution approach as outlined in the introduction.

Exemplary we show the results of Beam-ACO for the 26 difficult instances based on the precedence graph called SCHOLL in Table 2. The results show that Beam-ACO can solve more instances to optimaliy (namely, 9) than the two most recent metaheuristic approaches. The computation times show that,

Table 2. Results obtained by Beam-ACO in comparison to the results of two of the best techniques available for SALBP-1: ANTS is a standard ACO approach proposed in [13], and TABU is a recent tabu search approach proposed in [9]. The comparison is performed on the 26 difficult instances based on the precedence graph called SCHOLL. The instances differ in the cycle time, which is indicated in the first table column. The second column (headed by **bks**) contains the best known solution, and the third and fourth column contain the values of the best solutions found by ANTS, respectively TABU. Finally, the last 3 table columns provide the results of Beam-ACO, concerning the best solution found in 10 runs (**best**), the average and standard deviation of the results (**average (std)**), and the times including the standard deviation at which the best solutions were found (**average time (std)**).

c	**bks**	**ANTS**	**TABU**	**Beam-ACO**		
				best	**average (std)**	**average time (std)**
1394	50	52	51	51	51.00 (0.00)	0.82 (0.87)
1422	50	51	**50**	**50**	50.00 (0.00)	0.29 (0.39)
1452	48	50	49	49	49.00 (0.00)	0.78 (0.99)
1483	47	49	48	48	48.00 (0.00)	1.47 (1.35)
1515	46	48	47	47	47.00 (0.00)	0.39 (0.77)
1548	46	**46**	**46**	**46**	46.00 (0.00)	0.67 (0.26)
1584	44	46	45	45	45.00 (0.00)	1.48 (0.92)
1620	44	**44**	**44**	**44**	44.00 (0.00)	1.27 (0.82)
1659	42	44	43	43	43.00 (0.00)	0.52 (0.57)
1699	42	**42**	**42**	**42**	42.00 (0.00)	2.71 (0.29)
1742	40	41	41	41	41.00 (0.00)	0.47 (0.33)
1787	39	40	40	40	40.00 (0.00)	0.59 (0.27)
1834	38	39	39	39	39.00 (0.00)	0.34 (0.27)
1883	37	38	38	38	38.00 (0.00)	0.022 (0.024)
1935	36	37	37	37	37.00 (0.00)	0.21 (0.15)
1991	35	37	36	**35**	35.90 (0.32)	0.33 (0.97)
2049	34	35	35	35	35.00 (0.00)	0.013 (0.0057)
2111	33	34	34	34	34.00 (0.00)	0.010 (0.0042)
2177	32	33	33	**32**	32.90 (0.32)	9.28 (29.31)
2247	31	32	32	32	32.00 (0.00)	0.0090 (0.0054)
2322	30	31	31	31	31.00 (0.00)	0.012 (0.0090)
2402	29	30	30	30	30.00 (0.00)	0.010 (0.0051)
2488	28	29	29	29	29.00 (0.00)	0.011 (0.0047)
2580	27	28	28	**27**	27.80 (0.42)	14.02 (31.99)
2680	26	27	27	**26**	26.10 (0.32)	47.14 (36.68)
2787	25	26	26	**25**	25.20 (0.42)	50.19 (30.97)

in case the optimal solution can be found, this is usally the case after 30 to 40 seconds. In case the optimal solution is not found, the gap is never greater than 1, and the computation times are very low. This means that Beam-ACO finds very easily near-optimal solutions.

Table 3. Results obtained by Beam-ACO in comparison to the results of ANTS [13], which is so far the only available technique for TSALBP-1. The comparison is performed on the 26 instances based on the precedence graph called SCHOLL. The instances differ in the cycle time (which is at the same time the space limit). Cycle time, respectively space limit, are indicated in the first table column. The second column (headed by **bks**) contains the values of the best known solution. The arrow indicates that Beam-ACO was the first algorithm to generate this best known solution value. The third column contains the values of the best solutions found by ANTS. Finally, the last 3 table columns provide the results of Beam-ACO, concerning the best solution found in 10 runs (**best**), the average and standard deviation of the results (**average (std)**), and the times including the standard deviation at which the best solutions were found(**average time (std)**).

c, a	**bks**	**ANTS**	**Beam-ACO**		
			best	**average (std)**	**average time (std)**
1394	→ 59	60	**59**	60.00 (0.47)	30.50 (35.27)
1422	58	**58**	59	59.00 (0.00)	9.21 (10.07)
1452	→ 57	58	**57**	57.40 (0.52)	13.82 (36.23)
1483	→ 55	56	**55**	56.20 (0.63)	34.23 (33.35)
1515	54	**54**	**54**	54.40 (0.52)	51.43 (40.23)
1548	53	**53**	**53**	53.10 (0.32)	3.22 (5.79)
1584	→ 51	53	**51**	51.70 (0.48)	26.61 (38.82)
1620	→ 49	50	**49**	49.70 (0.48)	15.79 (23.57)
1659	→ 48	49	**48**	48.30 (0.48)	40.80 (34.67)
1699	→ 46	47	**46**	46.90 (0.32)	14.33 (14.77)
1742	→ 45	46	**45**	45.00 (0.00)	38.65 (31.60)
1787	→ 44	45	**44**	44.00 (0.00)	20.90 (21.57)
1834	43	**43**	**43**	43.00 (0.00)	13.13 (7.59)
1883	42	**42**	**42**	42.00 (0.00)	9.04 (4.05)
1935	41	**41**	**41**	41.00 (0.00)	9.36 (5.87)
1991	40	**40**	**40**	40.00 (0.00)	6.24 (3.45)
2049	→ 38	39	**38**	38.00 (0.00)	12.64 (10.05)
2111	37	**37**	**37**	37.00 (0.00)	2.02 (4.28)
2177	36	**36**	**36**	36.00 (0.00)	2.06 (3.70)
2247	→ 34	35	**34**	34.80 (0.42)	10.37 (19.81)
2322	→ 33	34	**33**	33.40 (0.52)	33.66 (36.84)
2402	→ 32	33	**32**	32.70 (0.48)	27.83 (45.99)
2488	→ 31	32	**31**	31.00 (0.00)	42.11 (22.34)
2580	30	**30**	**30**	30.00 (0.00)	4.20 (3.00)
2680	29	**29**	**29**	29.00 (0.00)	3.83 (3.92)
2787	28	**28**	**28**	28.00 (0.00)	4.41 (4.15)

4.2 Results for the TSALBP-1 Instances

We also applied our algorithm to the 269 TSALBP-1 instances that were generated by Bautista and Pereira (see [13]) from the 269 SALBP-1 instances.[2] We

[2] This was done by setting $a_j := t_{n-j+1}$ for all $j \in T$, and $a := c$.

exemplary show the results concerning the 26 instances based on the precedence graph called SCHOLL in Table 3. The results are compared to the results of the standard ACO approach ANTS by Bautista and Pereira (see [13]), which is the only available technique for TSALBP-1. The results show that in 14 out of 26 cases, Beam-ACO improves the best known solution. In other 11 cases, Beam-ACO matches the best known solution values. Only in one case Beam-ACO does not match the performance of ANTS. This case is characterized by a relatively small cycle time. In general, we noticed the tendency that Beam-ACO—when applied to TSALBP-1 instances—is better when the cycle time is bigger. This might be caused by the fact that the quality of the lower bound is higher for bigger cycle times. Exchanging our simple lower bound for a more sophisticated lower bound might help to improve the performance of Beam-ACO when smaller cycle times are concerned.

4.3 Results for the Nissan TSALBP-1 Instance

Finally we applied Beam-ACO to the real-life instance provided by the Nissan plant in Barcelona, Spain. This instance consists of 140 tasks, a cycle time of 180 seconds, and a space limit of 4. This real-life instance is easily solvable: Our Beam-ACO approach finds in a fraction of a second an optimal solution with 21 work stations. Also when disregarding space constraints (that is, regarding it as a SALBP-1 isntance), it is easily solvable. Our Beam-ACO approach finds an optimal solution with 17 work stations in a fraction of a second.

5 Conclusions and Outlook to the Future

In this work we have proposed a hybrid ant colony optimization approach called Beam-ACO for the TSALBP-1 problem. Beam-ACO is obtained by hybridizing ant colony optimization with beam search. Our approach differs from existing ant colony optimization approaches for TSALBP-1 (and from most existing metaheuristics) by the fact that it solves the problem in a direct way. The results show that Beam-ACO is a state-of-the-art metaheuristics, for example, for the well-studied SALBP-1 problem, which is a specific case of TSALBP-1.

In the future we plan to study the influence of the different algorithmic components on the performance of Beam-ACO. It would be interesting to know, for example, which influence the heuristic information (obtained by parametrizing the priority rule) exactly has. Furthermore, we plan to extend the experimental evaluation of the algorithm to beam widths greater than one. We expect that this will further improve the algorithms' performance.

References

1. Scholl, A., Becker, C.: State-of-the-art exact and heuristic solution procedures for simple assembly line balancing. European Journal of Operational Research **168**(3) (2006) 666–693
2. Baker, K.R.: Introduction to sequencing and scheduling. Wiley, New York (1974)

3. Talbot, F.B., Patterson, J.H., Gehrlein, J.H.: A comparative evaluation of heuristic line balancing techniques. Management Science **32** (1986) 430–454
4. Johnson, R.V.: Optimally balancing large assembly lines with "FABLE". Management Science **34** (1988) 240–253
5. Hoffmann, T.R.: EUREKA: A hybrid system for assembly line balancing. Management Science **38** (1992) 39–47
6. Scholl, A., Klein, R.: SALOME: A bidirectional branch and bound procedure for assembly line balancing. INFORMS Journal on Computing **9** (1997) 319–334
7. Sprecher, A.: Dynamic search tree decomposition for balancing assembly lines by parallel search. International Journal of Production Research **41** (2003) 1423–1430
8. Scholl, A., Voss, S.: Simple assembly line balancing—Heuristic approaches. Journal of Heuristics **2** (1996) 217–244
9. Lapierre, D.L., Ruiz, A., Soriano, P.: Balancing assembly lines with tabu search. European Journal of Operational Research **168** (2006) 826–837
10. Vilarinho, P.M., Simaria, A.: A two-stage heuristic method for balancing mixed-model assembly lines with parallel workstations. International Journal of Production Research (2002) 1405–1420
11. Gonçalves, J.F., de Almeida, J.R.: A hybrid genetic algorithm for assembly line balancing. Journal of Heuristics **8** (2002) 629–642
12. Bautista, J., Pereira, J.: Ant algorithms for assembly line balancing. In Dorigo, M., Di Caro, G., Sampels, M., eds.: Ant Algorithms – Proceedings of ANTS 2002 – Third International Workshop. Volume 2463 of Lecture Notes in Computer Science., Springer Verlag, Berlin, Germany (2002) 65–75
13. Bautista, J., Pereira, J.: Ant algorithms for a time and space constrained assembly line balancing problem. European Journal of Operational Research (2006) In press.
14. Dorigo, M., Stützle, T.: Ant Colony Optimization. MIT Press, Cambridge, MA (2004)
15. Blum, C.: Beam-ACO—Hybridizing ant colony optimization with beam search: An application to open shop scheduling. Computers & Operations Research **32**(6) (2005) 1565–1591
16. Ow, P.S., Morton, T.E.: Filtered beam search in scheduling. International Journal of Production Research **26** (1988) 297–307
17. Fügenschuh, A.: Parametrized greedy heuristics in theory and practice. In Blesa, M.J., Blum, C., Roli, A., Sampels, M., eds.: Proceedings of HM 2005 – 2nd International Workshop on Hybrid Metaheuristics. Volume 3636 of Lecture Notes in Computer Science., Springer Verlag, Berlin, Germany (2005) 21–31
18. Blum, C., Bautista, J., Pereira, J.: Beam-ACO applied to assembly line balancing. Technical Report LSI-06-22-R, LSI, Universitat Politècnica de Catalunya, Barcelona, Spain (2006)
19. Merkle, D., Middendorf, M.: An ant algorithm with a new pheromone evaluation rule for total tardiness problems. In: Proceedings of the EvoWorkshops 2000. Volume 1803 of Lecture Notes in Computer Science., Springer Verlag, Berlin, Germany (2000) 287–296
20. Blum, C., Dorigo, M.: The hyper-cube framework for ant colony optimization. IEEE Transactions on Systems, Man, and Cybernetics – Part B **34**(2) (2004) 1161–1172

Boundary Search for Constrained Numerical Optimization Problems in ACO Algorithms

Guillermo Leguizamón[1] and Carlos A. Coello Coello[2]

[1] LIDIC - Universidad Nacional de San Luis, San Luis, Argentina
legui@unsl.edu.ar
[2] Evolutionary Computation Group (EVOCINV) at CINVESTAV-IPN
Electrical Engineering Department, Computer Science Section, México D.F., México
ccoello@cs.cinvestav.mx

Abstract. This paper presents a novel boundary approach which is included as a constraint-handling technique in an ant colony algorithm. The necessity of approaching the boundary between the feasible and infeasible search space for many constrained optimization problems is a paramount challenge for every constraint-handling technique. Our proposed technique precisely focuses the search on the boundary region and can be either used alone or in combination with other constraint-handling techniques depending on the type and number of problem constraints. For validation purposes, an ant algorithm is adopted as our search engine. We compare our proposed approach with respect to constraint-handling techniques that are representative of the state-of-the-art in constrained evolutionary optimization using a set of standard test functions.

1 Introduction

One of the first ACO extensions to operate on continuous spaces can be found in Bilchev et al. [1] in which the whole search space is discretized in order to represent a finite number of search directions. This approach was validated using a small set of constrained problems. Since then, several other researchers have proposed schemes to apply the ACO algorithm to continuous search spaces (see for example [2,3,4]). However, all of these approaches only deal with unconstrained optimization problems.

In this paper we introduce a boundary approach for solving nonlinear constrained problems, which is presented as a possible extension of ACO algorithms in continuous search spaces. It is worth noting, however, that our proposal can be coupled to other metaheuristics (e.g., particle swarm optimization or an evolutionary algorithm), and it is expected to be highly competitive in problems with active constraints. Our ACO approach is mainly based on the work of Bilchev et al. [1]. The reason for not adopting one of the more recent ACO extensions for continuous search spaces is that, as indicated before, none of them deals with constrained optimization problems.

M. Dorigo et al. (Eds.): ANTS 2006, LNCS 4150, pp. 108–119, 2006.

2 Constraint-Handling and Boundary Search

Michalewicz et al. [5] is one of the first papers on boundary search through the use of evolutionary algorithms. The efficiency of this approach was shown by using two constrained optimization problems: Keane's function (also known as $G02$) [6] and another function with one equality constraint (also known as $G03$). For solving $G02$ the authors proposed two genetic operators: the *geometrical* crossover and a special mutation operator. Both operators generate offspring lying on the boundary between the feasible and infeasible search space. Similarly for $G03$, Michalewicz et al. [5] proposed the *spherical* crossover which only generates points on the surface of the sphere given as the only constraint. Schoenauer et al. [7] proposed several evolutionary operators capable of exploring a general surface of dimension $n - 1$ (n is the number of variables). The design of these operators depends on the surface representation: curves-based, plane-based, and parametric representation. Wu et al. [8] proposed a GA for the optimization of a water distribution system, which is a highly constrained optimization problem. The proposed approach co-evolves and self-adapts two penalty factors in order to guide and preserve the search towards the boundary of the feasible search space. The reduction of the search space is one of the most relevant characteristics of the boundary search approach since the exploration considers only the boundary of the feasible search space. However, many of the test cases considered so far by other researchers only include problems with one or two constraints (e.g., $G02$ and $G03$[1], respectively). In these cases, it is possible to define *ad hoc* genetic operators that fit perfectly the boundary of the feasible region. However, this sort of approach is impractical in an arbitrary problem with many constraints, and it is therefore necessary to define a more general approach for boundary search which can be as robust as possible to deal with different types of constraints. This was precisely the motivation for the research reported in this paper.

3 An Alternative Boundary Search Approach

We are interested in solving the general nonlinear programming problem whose aim is to find $\mathbf{x}$ so as to optimize: $f(\mathbf{x}) \quad \mathbf{x} = (x_1, x_2, ..., x_n) \in \mathbb{R}^n$ where $\mathbf{x} \in \mathcal{F} \subset \mathcal{S}$. The set $\mathcal{S} \in R$ defines the search space and sets $\mathcal{F} \subseteq \mathcal{S}$ and $\mathcal{U} = \mathcal{S} - \mathcal{F}$ define the *feasible* and *infeasible* search spaces, respectively. The search space $\mathcal{S}$ is defined as an n-dimensional rectangle in $\mathbb{R}^n$ (domains of variables defined by their lower and upper bounds): $l(i) \leq x_i \leq u(i)$ for $1 \leq i \leq n$ whereas the feasible set $\mathcal{F}$ is defined by the intersection of $\mathcal{S}$ and a set of additional $m \geq 0$ constraints: $g_i \leq 0$, for $j = 1, \ldots, q$ and $h_j = 0$ for $j = q+1, \ldots, m$. At any point $\mathbf{x} \in \mathcal{F}$, the constraints g_k that satisfy $g_k(\mathbf{x}) = 0$ are called the active constraints at $\mathbf{x}$. Equality constraints h_j are active at all points of $\mathcal{F}$.

[1] Keane's function can be considered as having one constraint since one of them is ignored focusing the search only on the active constraint.

3.1 The Boundary Operators

We propose here a general boundary operator which is based on the notion that each point $\mathbf{b}$ of the boundary region can be represented by means of two different points $\mathbf{x}$ and $\mathbf{y}$ where $\mathbf{x}$ is some feasible point and $\mathbf{y}$ is some infeasible one, i.e., $(\mathbf{x}, \mathbf{y})$ can represent one point lying on the boundary by applying a "binary search" on the straight line connecting the points $\mathbf{x}$ and $\mathbf{y}$. Figure 1 shows a hypothetical search space including the feasible and infeasible (shadowed area) regions. We can identify four points lying on the boundary $\mathbf{b}_1$, $\mathbf{b}_2$, $\mathbf{b}_3$, and $\mathbf{b}_4$ which are respectively obtained from $(\mathbf{x}_1, \mathbf{y}_1)$, $(\mathbf{x}_2, \mathbf{y}_2)$, $(\mathbf{x}_3, \mathbf{y}_3)$, and $(\mathbf{x}_4, \mathbf{y}_4)$.

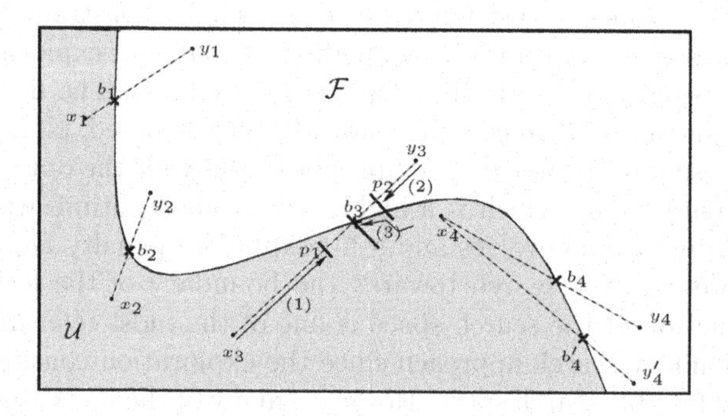

Fig. 1. Given one feasible and one infeasible point, the respective point lying on the boundary can be easily reached by using a simple binary search

The binary search applied to each pair of points $(\mathbf{x}, \mathbf{y})$ is achieved following the steps described in function BS (see Figure 2). For example, a possible application of this process can be seen in Figure 1 where we adopt the pair of points $(\mathbf{x}_3, \mathbf{y}_3)$ from which we obtain the point $\mathbf{b}_3$, which lies on the boundary. The first step (labeled (1)) indicates that the first mid point found is infeasible. Consequently, the left side of the straight line $(\mathbf{x}_3)$ is moved to point $\mathbf{p}_1$. In the next step ((2)) we consider the points $\mathbf{p}_1$ and $\mathbf{y}_3$ as extreme points for which the mid point is the feasible point $\mathbf{p}_2$. Thus, the new feasible point or right extreme of the line is now the point $\mathbf{p}_2$. Finally, the last point generated is $\mathbf{b}_3$ which can be either lying on or close to the boundary. Condition (dist_to_boundary($\mathbf{m}$) $> \xi$) defines a threshold to stop the process of approaching the boundary. It is worth noticing that parameters $\mathbf{x}$ and $\mathbf{y}$ are local to BS, i.e., function BS behaves as a decoder of the pair of feasible and infeasible points passed as parameters. Therefore, the number of "mid_points_between" $\mathbf{x}$ and $\mathbf{y}$ before approaching the boundary within a distance less that ξ is given by $log_2(r)$ where $r = (dist(\mathbf{x}, \mathbf{y})))/\xi$. Thus, the closer to the boundary, the larger $log_2(r)$.

So far, we have shown how a point lying on the boundary can be represented through a pair of points. Now we need to consider the exploration of the search space. For example, from the perspective of evolutionary algorithms, the candidate operators are crossover and mutation. However, for the ACO approach we only adopt a mutation-like operator. Here, a pair of points is considered (one

```
function BS(x,y: real vector): int
begin
  do
    m = mid_point_between(x, y);
    if ( Is_on_Boundary(m) ) return m; /* m is a point lying on the boundary */
    if ( Feasible(m) ) x = m; else y = m;
  while ( dist_to_boundary(m)> ξ );
  return mid_point_between(x, y); /* The closest point to the boundary */
end
```

Fig. 2. Given one feasible and one infeasible point, function BS returns either a point on the boundary or one which is close enough to the boundary according to a parameter ξ

feasible and one infeasible). Alternatively, any of these two points could be modified. For example, we can consider the pair of points $(\mathbf{x}_4, \mathbf{y}_4)$ in Figure 1 which represents point $\mathbf{b}_4$ on the boundary. In this case, the feasible point $\mathbf{y}_4$ can be modified giving as a result a point $\mathbf{y}'_4$ in the feasible search space. After this process, the new point lying on the boundary is obtained by decoding $(\mathbf{x}_4, \mathbf{y}'_4)$, which gives us $\mathbf{b}'_4$.

3.2 The Proposed Method

The simplest case to apply the boundary approach is when the problem has only one constraint which could be either an equality or an inequality constraint. For the last case, it is important to remember that we are assuming active constraints at the global optimum to proceed with this method where the search is always performed on the boundary of the space defined by any of the constraints.

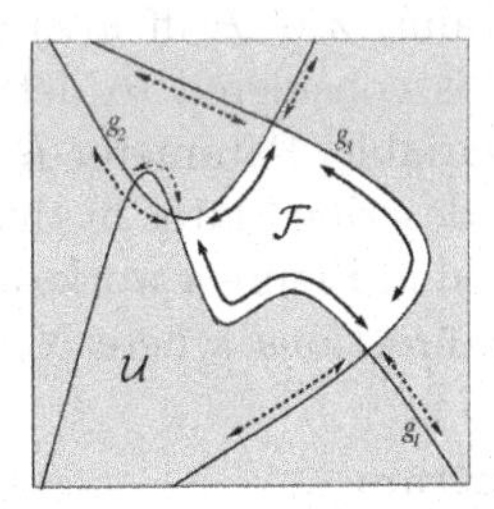

Fig. 3. Feasible search space defined by 3 inequality constraints

For facing the typical situation in which we have more than one constraint, it is necessary to define an appropriate policy to explore the boundary as efficiently as possible. One possibility is to explore in turn the boundary of each constraint. The selection of the constraints to search for can be determined using different methods. If the problem includes at least one equality constraint, such equality constraints are the most appropriate candidates to be selected first. In order to show the robustness of our method in the absence of information about the active constraints of a problem, we will show in our experimental study (see

Section 5) a more general approach to apply the boundary operators. As an illustrative example, Figure 3 shows a hypothetical search space determined by three inequality constraints. Let's suppose that the search proceeds starting on constraint g_1 by using the boundary operator O_{g_1} (filled line in Figure 3). The application of this operator will eventually produce points violating constraints g_2 and g_3 (dotted line in Figure 3). One of the simplest methods to deal with this situation is the application of a penalty function for the infeasible solutions. In addition, if g_1 is active at the global optimum, the method will focus the search on the boundary in order to restrict the explored regions of the whole search space. Note however, that other (more sophisticated) constraint-handling techniques can also be adopted.

4 Boundary Approach in ACO Algorithms

A possible design to apply the ACO approach in continuous search problems is by discretizing the continuous search space in some way. In this work we use a discrete structure to represent a set of different points spread on the search space. These points are called *directions*, following Bilchev et al.'s proposal. The discrete structure can be seen as a set $\{d_1, d_2, ..., d_k\}$, where k is a parameter for the number of directions. Each direction d_i is represented as a real n-dimensional vector. A general outline of the ACO algorithm is shown in Figure 4. It is worth remarking that the original proposal [1] for ACO in continuous domains is used to proceed with the local exploration after a genetic algorithm has finished with the global search. However, the algorithm proposed here, is in charge of performing the entire search process. More precisely, our ACO algorithm starts with a set of k pairs of points $(\mathbf{x}, \mathbf{y})$ randomly generated with $\mathbf{x} \in \mathcal{F} \iff \mathbf{y} \in \mathcal{U}$ (when considering an equality constraint, $\mathbf{z} \in \mathcal{F}$ iff $h(\mathbf{z}) \leq 0$; otherwise, $\mathbf{z} \in \mathcal{U}$) . In addition, a value $0 \leq R \leq 1$ is considered to define the extent of the search interval with respect to each variable. Parameter R starting at value 1 will vary down to 0 in each iteration as described later in this section.

The ACO algorithm displayed in Figure 4 works as follows: `initialize A(t)` "distributes" N_a ants in the k directions, where $N_a > k$ in order to allocate one

ACO algorithm
begin
  t = 0; initialize $A(t)$; evaluate $A(t)$;
  while (stop condition not met) **do**
  begin
    t = t + 1
    update_trail; reallocate_ants $A(t)$;
    evaluate $A(t)$;
  end
end

Fig. 4. General outline of the ACO algorithm for continuous problems adopted in this paper

or more ants to the same direction. Each ant randomly generates one possible solution, i.e. a pair $(\mathbf{x}, \mathbf{y})$ with $x \in \mathcal{F}$ and $y \in \mathcal{U}$; `evaluate A(t)` obtains the objective value for the new points generated; *update_trail* is in charge of accumulating pheromone trial in each direction proportionally to the quality of the objective function values found in the corresponding direction, i.e., $\tau_d = (1 - \rho) \cdot \tau_d + \Delta\tau_d$ where $\Delta\tau_d$ is a value proportional to the best objective value in direction d and $0 \leq \rho \leq 1$ is the pheromone trail evaporation rate; *reallocate_ants $A(t)$* redistributes the population of ants in the k directions, proportionally to the accumulated pheromone trail values. Thus, the ants in direction $d \in \{1, \ldots, k\}$ are in charge of searching in the neighborhood of the respective boundary feasible point in direction d. The changes on the values of ratio R to control the extent of the search interval for each dimension can be implemented as $\Delta_R(t) = R(1 - r^{(1-t/T)})$ where r is a random number in the range $[0..1]$; T is the maximum number of iterations. Consequently, the value $\Delta_R(t)$ falls in the range $[0..R]$ and gets closer to 0 as the elapsed number of iterations t increases. Therefore, the ACO algorithm can be seen as a trajectory approach which simultaneously searches in different directions and exploits the past experience to guide the search towards the most promising regions according to the quality of the results. Furthermore, the accumulated pheromone trail will decrease in the direction that produces low-quality solutions due to the effects of the evaporation process focusing the ants' attention on more promising regions of the feasible search space. In order to avoid premature convergence of the algorithm, a potentially useful direction can remain as an alternative search region by bounding with lower and upper values the amount of pheromone trial in each direction following the principle of the $\mathcal{MM}$AS algorithm [9]. The main characteristics of this method include two abstraction levels:

1. *individual search*: involves the strategy followed by each ant to search in its neighborhood (in our case, a mutation-like operator).
2. *cooperation*: involves information exchange among the ants in order to guide the search to certain regions of the search space. This information is represented by the pheromone trial structure (τ) where τ_j represents the accumulation of pheromone trail on direction j. The distribution of the ants in the different directions is achieved by the formula: $P_d(t) = \frac{\tau_d(t)}{\sum_{h=1}^{k} \tau_h(t)}$

5 Analysis of Results

The application of our approach (called ACO$_\mathcal{B}$) requires minimum changes when applied to the different test cases considered: the objective function, the number of variables, the range of each variable, and the constraints. However, the policy to determine on which constraint the search should focus needs to be considered when the problems have more than one constraint: a) we can focus the search on all the constraints, but considering one constraint in turn by controlling the change through a particular condition (S$_{all}$), b) similar to the previous alternative but considering only the active constraints (S$_{act}$), or c) just considering one constraint during the whole run (S$_c$ where $c \in \{1, \ldots, m\}$). These three ways of exploring the search space are presented first in our experimental study in order to analyze the performance of the $ACO_{\mathcal{B}}$ on each of the considered problems. In our experiments, the condition to produce a change on the search from one to another constraint is given by an elapsed number of iterations and it is represented by the parameter t_c. In addition, for problems with more than one constraint, we incorporate a penalty function of the form:

$$\phi(x, \mu) = f(x) + \mu(t)(\sum_{j=1}^{q} \max\{0, g_j(x)\} + \sum_{j=q+1}^{m} |h_j(x)|) \quad (1)$$

where $\mu(t)$ is a dynamic penalty factor which could change as t, the elapsed iteration, increases with $\mu(0) \leq \mu(1) \leq \mu(2) \ldots \leq \mu(T)$. Alternatively, the penalty factor can be fixed throughout the run, i.e., $\mu(t) = \mu_0$ for all $1 \leq t \leq T$. Regardless of the penalty function adopted, it is worth remarking that each solution is always lying on the boundary of the feasible space corresponding to the constraint under consideration. Note that this approach was adopted due to its simplicity, since our interest was to assess the advantages of our proposed approach. However, other constraint-handling techniques are evidently possible. The parameter values used in the experimental study are the following: 50 ants (population size), 20 directions (number of points), maximum number of iterations 30000, the evaporation rate $\rho = 0.5$, $t_c > 0$ is the number of iterations the $ACO_{\mathcal{B}}$ focuses on one constraint in turn ($t_c = 200$). When $t_c = 0$, the $ACO_{\mathcal{B}}$ focuses on only one constraint throughout the whole run. The penalty factor $\mu(t)$ was experimentally determined for each particular problem and is shown in the corresponding tables of results. The $ACO_{\mathcal{B}}$ was executed 30 times with different seeds for each parameter combination. The problems studied include a set of well-known test cases traditionally adopted in the specialized literature: $G01$ to $G13$ [10].

5.1 Study of the Application of $ACO_{\mathcal{B}}$

We have divided the presentation of the results into two groups according to the following criteria: the first group, is displayed in Tables 1 and 2. Table 1 includes two special cases since they where the first problems on which the boundary approach was applied (problems $G02$ and $G03$). In addition, these problems have one and two constraints respectively. However, the second constraint of problem $G02$ is not considered since it is not active at the best known value. The columns in this table show the setting for the number of variables, the best value found (BF), Mean, Standard Deviation (Std), Worst, number of feasible solutions out of 30 runs (#F), and the mean number of evaluations, expressed in thousands, to get the best value found (M(#E)). On the other hand, Table 2 shows two problems both of which include one equality constraint (problems $G11$ and $G25$). Accordingly, no penalty values (μ) need to be applied for this first group of problems. In the remaining tables, the column "No. of variables" is replaced by "Cnst", indicating the criteria adopted to proceed with the boundary search, i.e., S_c ($c \in \{1, \ldots, m\}$), S_{act}, or S_{all}. In addition, the best known or global optimum value for each problem is shown in parenthesis.

Table 1. Results for problems $G02$ (Keane's function) and $G03$

No. of Variables (n)	BF	Mean	Std	Worst	#F	M(#E)
			Problem G02			
20	0.8036190867	0.8025656939	0.0032	0.7930839658	30	29
50	0.8352618814	0.8339309692	0.0021	0.8259508014	30	35
100	0.8456841707	0.8446936011	0.0007	0.8423509002	30	46
			Problem G03			
20	1.0	1.0	0.0	1.0	30	140
50	1.0	1.0	0.0	1.0	30	389

Table 2. For problems G11 and G25 it is unnecessary to use a penalty factor

Cnst.	BF	Mean	Std	Worst	#F	M(#E)
			Problem G11 (0.75)			
S_1	0.75	0.75	0.0	0.75	30	70
			Problem G25 (16.73889)			
S_1	-16.73889	-16.73889	0.0	-16.73889	30	10

We tested $G02$ setting the number of variables as $n = 20, 50$, and 100. ACO$_{\mathcal{B}}$ succeeded in finding the best known value for $n = 20$ [11]. In addition, it was able to find a better quality result than the best objective reported in [7] where $n = 50$ and $f(\mathbf{x}^*) = 0.831937$. For $n = 100$, we found 0.8456841707 as the best value in our experimental study. Also, it is worth remarking that all the solutions found were feasible for all n and very similar among themselves as can be observed in the columns Mean, Std, and Worst. With respect to problem $G03$, we considered $n = 20$ and $n = 50$ variables. ACO$_{\mathcal{B}}$ found the optimum feasible solution for both cases in all runs. Similarly to $G03$, the remaining problems of this group ($G11$ and $G25$), our approach reached the optimum in all cases.

The second group of the test cases is conformed by some problems having more than one constraint which have been frequently used in the specialized literature: $G01$, $G04$, $G05$, $G06$, $G07$, $G09$, $G10$, and $G13$. Also we include problem $G24$ [12] in this subgroup. Only for $G10$, we adopted a dynamic penalty ($\mu(t) = 1.05 \times \mu(t-1)$ for $t = 0, 1, \cdots t_{max}$, with $\mu(0) = 200000$). The static penalty factors adopted for the remaining problems are (i.e., for $t = 0, 1, \cdots T$): $G01$, $\mu(t) = 1000$; $G04$, $\mu(t) = 800000$; $G05$, $\mu(t) = 10$; $G06$, $\mu(t) = 10000$; $G07$, $\mu(t) = 20000$; $G09$, $\mu(t) = 2000$, $G13$, $\mu(t) = 0.2$; and $G24$, $\mu(t) = 1000$. The results for this group of problems are displayed in Tables 3 and 4. It must be noticed that these problems include different numbers and complexities of the equality and inequality constraints which are active at the best known or optimum solution. As indicated in column "Cnst.", each row shows the results when ACO$_{\mathcal{B}}$ is applied one of the following criteria: search exclusively on constraint j (S_j, $j = 1, \ldots, m$), on all the active constraints in turn (S_{act}), and over all the constraints in turn (S_{all}). For example, problem $G01$ has 6 active constraints. Accordingly, ACO$_{\mathcal{B}}$ performs optimally when searching on those active constraints. Similarly, the algorithm succeeded in finding the optimal solution when using both S_{act} and S_{all}. However, its performance slightly decays when searching on the non active

Table 3. Results for problems $G01$, $G04$, $G05$, $G06$, and $G07$. Column $M(\#E)$ is not showed for these problems due to space constraints.

Cnst.	BF	Mean	Std	Worst	#F	Cnst.	BF	Mean	Std	Worst	#F
	Problem G01 (-15.00)						Problem G07 (24.306)				
S_1	-15.00	-14.99	0.001	-14.996	30	S_1	24.37	29.59	4.83	42.97	30
S_2	-15.00	-14.96	0.012	-14.995	30	S_2	24.51	35.10	23.02	121.56	30
S_3	-15.00	-14.99	0.001	-14.965	30	S_3	24.56	28.31	5.54	50.83	30
S_4	-14.27	-13.54	.38	-13.18	29	S_4	24.79	54.17	70.46	380.03	30
S_5	-13.84	-13.48	0.32	-13.04	25	S_5	24.52	34.52	16.39	77.19	30
S_6	-14.22	-13.39	0.47	-13.00	26	S_6	24.79	31.12	6.46	48.40	30
S_7	-15.00	-14.78	0.2	-14.65	26	S_7	33.08	38.86	4.01	46.53	30
S_8	-15.00	-14.74	0.49	-14.46	27	S_8	41.03	46.86	20.92	127.06	30
S_9	-15.00	-14.67	0.76	-13.08	30						
S_{act}	-15.00	-15.00	0	-15.00	30	S_{act}	24.37	24.64	0.15	24.92	30
S_{all}	-15.00	-15.00	0	-15.00	30	S_{all}	24.38	24.76	0.16	25.22	30
	Problem G04 (-30655.539)						Problem G05 (5126.49)				
S_1	-30665.539	-30665.357	0.04	-30665.157	30	S_1	-	-	-	-	-
S_2	-	-	-	-	-	S_2	-	-	-	-	-
S_3	-	-	-	-	-	S_3	5126.50	5133.29	9.284	5147.81	6
S_4	-	-	-	-	-	S_4	5126.51	5134.70	11.219	5164.91	11
S_5	-	-	-	-	-	S_5	5126.68	5130.55	3.656	5136.08	11
S_6	-30655.539	-30665.302	0.01	-30665.290	30						
S_{act}	-30655.539	-30655.539	0.001	-30655.539	30	S_{act}	5126.50	5138.37	8.20	5132.14	6
S_{all}	-	-	-	-	-	S_{all}	5126.50	5143.77	10.60	5163.56	5
	Problem G06 (6961.81)										
S_1	-6961.79	-6961.71	0.075	-6169.54	11						
S_2	-6961.81	-6961.72	0.097	-6961.34	25						
S_{act}	-6961.81	-6961.74	0.070	-6961.71	25						

constraints as could be expected. This situation is more dramatic for problem $G04$ which has two active constraints. In this case, ACO$_{\mathcal{B}}$ only finds the optimum solution when searching on the respective active constraints and S_{act}. Although strategy S_{all} fails in finding any feasible solution, this strategy worked well for all the other problems considered. A similar situation can be seen for problem $G05$ which has three equality constraints. Accordingly, ACO$_{\mathcal{B}}$ finds a high quality solution for this problem (very near to the optimal one) when searching on the equality constraints, S_{act}, and S_{all}. On the other hand, problem $G06$ has two inequality constraints which are active at the optimum. ACO$_{\mathcal{B}}$ performs optimally for this problem by following any of the three applicable strategies: S_1, S_2, and S_{act}. The last problem in Table 3 has six active constraints and ACO$_{\mathcal{B}}$ performs similarly to $G01$ since the best results were obtained when searching on the active constraints or by using S_{act} or S_{all}.

Problem $G09$ has two active constraints for which ACO$_{\mathcal{B}}$ found the optimum value. However, searching on the non active constraints can give results far from the expected value (see S_2 and S_3). $G10$ constitutes one of the most difficult test cases not only for our approach, but also for any other constraint-handling technique. ACO$_{\mathcal{B}}$ found feasible solutions with all the search strategies except for S_5 and S_6. Note the small number of feasible solutions found for this problem, as well as the large standard deviation value produced (with respect to the deviations of the other problems). Another interesting problem is $G13$ whose feasible search space is defined by three nonlinear equality constraints. For this problem ACO$_{\mathcal{B}}$ found the optimal solution following either of the four applicable search

Table 4. Results for problems $G09$, $G10$, $G13$, and $G24$

Cnst.	BF	Mean	Std	Worst	#F	M(#E)
	Problem G09 (680.63)					
S_1	680.63	680.66	0.10	681,29	30	80
S_2	1664.00	1890.01	119.92	1982.72	5	108
S_3	840.00	880.82	15.06	890.56	29	22
S_4	680.63	680.96	0.96	681.95	29	43
S_{act}	680.63	680.67	0.026	680.72	30	7
S_{all}	680.65	680.75	0.056	680.89	30	19
	Problem G10 (7049.331)					
S_1	7101.50	7346.61	202.15	7682.20	9	147
S_2	7063.02	8169.68	1866.32	10325.00	3	131
S_3	7057.27	7406.51	148.60	7518.91	9	148
S_4	7095.27	7349.83	360.00	7604.39	2	128
S_5	-	-	-	-	-	-
S_6	-	-	-	-	-	-
S_{act}	7052.30	7199.01	175.01	7943.15	30	42
S_{all}	7068.04	7141.87	52.27	7239.54	30	9.8
	Problem G13 (0.053950)					
S_1	0.053950	0.054908	0.00054	0.055386	6	29
S_2	0.053950	0.054372	0.00044	0.054968	4	7
S_3	0.053950	0.054637	0.00017	0.054394	6	7
S_{act}	0.053950	0.054736	0.001	0.058462	15	19
	Problem G24 (-5.508013)					
S_1	-5.508013	-5.508013	0.0	-5.508013	30	5
S_2	-5.508013	-5.508013	0.0	-5.508013	30	24
S_{act}	-5.508013	-5.508013	0.0	-5.508013	30	21

strategies. Finally, it can be seen that $ACO_{\mathcal{B}}$ performs optimally on problem $G24$ which has two active inequality constraints where the optimal solution was found for all strategies in each run (see #F).

5.2 Comparison with a State-of-the-Art Algorithm

In this section we compare the best quality results from $ACO_{\mathcal{B}}$ (we use S_{act} as the most efficient search criteria) with respect to the results of a constraint-handling technique representative of the state-of-the-art in the area: Stochastic Ranking (SR) [10]. Table 5 shows for each problem considered, the optimum, and the corresponding Best value found (BF), average (Mean), and Worst values respectively from $ACO_{\mathcal{B}}$ and SR (reported in [10][2]). The performance of $ACO_{\mathcal{B}}$ is comparable in many ways with respect to SR.

From the perspective of the best values (BF) found $ACO_{\mathcal{B}}$ reaches similar values as SR in all the problems considered. For $G02$, $ACO_{\mathcal{B}}$ reached the best

[2] Except for problems $G24$ and $G25$ for which SR was run by the authors using Thomas Runarsson's code.

known value reported in [5] by using an *ad hoc* boundary operator. On the opposite side, for $G10$, ACO$_{\mathcal{B}}$ did not obtain the optimal solution. However, the results achieved in all cases are highly competitive.

Table 5. Comparison of ACO$_{\mathcal{B}}$ with respect to a constraint-handling technique representative of the state-of-the-art in the area: stochastic ranking (SR)

		BF		Mean		Worst	
Prob.	Opt[3]	ACO$_{\mathcal{B}}$	SR	ACO$_{\mathcal{B}}$	SR	ACO$_{\mathcal{B}}$	SR
G01	-15.000	-15.000	-15.000	-15.000	-15.000	-15.000	-15.000
G02	0.803619	0.803619	0.803515	0.802656	0.781975	0.793083	0.726288
G03	1.000	1.000	1.000	1.000	1.000	1.000	1.000
G04	-30665.539	-30665.539	-30665.539	-30665.539	-30665.539	-30666.539	-30665.539
G05	5126.498	5126.50	5126.497	5138.37	5128.881	5132.14	5142.472
G06	-6961.814	-6961.81	-6981.814	-6961.74	-6875.940	-6961.71	-6350.262
G07	24.306	24.37	24.307	24.64	24.374	24.92	24.642
G09	680.630	680.63	680.63	680.67	680.56	680.72	680.763
G10	7049.331	7052.30	7054.316	7199.01	7559.192	7943.15	8835.655
G11	0.75	0.75	0.75	0.75	0.75	0.75	0.75
G13	0.053950	0.053950	0.053957	0.054908	0.057006	0.055386	0.216915
G24	-5.508013	-5.508013	-5.508013	-5.508013	-5.508013	-5.508013	-5.508013
G25	-16.73819	-16.73819	-16.73819	-16.73819 -	16.73819	-16.73819	-16.73819

6 Conclusions and Future Work

In this paper we presented an alterative approach to reach the boundary between the feasible and infeasible search space which could be useful when facing problems with active constraints. For the initial testing of this method we have used an ant colony algorithm as a search engine (ACO$_{\mathcal{B}}$) and a penalty function as a complementary mechanism for problems with more than one constraint. The overall performance of ACO$_{\mathcal{B}}$ was satisfactory for all of the problems considered. The comparison with a state-of-the-art algorithm shows the potential of this method as an alternative or complementary approach for constrained optimization problems. In fact, for some problems, ACO$_{\mathcal{B}}$ was able to improve the respective best known solutions (e.g., $G02$ (with $n = 50$ variables)). It is clear that further improvements should be considered. For example, it is desirable to implement a self-adaptation mechanisms and to try different search engines (e.g., differential evolution and evolution strategies). We also plan to study the performance of our approach in several additional test functions.

References

1. Bilchev, G., Parmee, I.C.: The ant colony metaphor for searching continuous design spaces. Lecture Notes in Computer Science **993** (1995) 25–39
2. Ling, C., Jie, S., Ling, Q., Hongjian, C.: A method for solving optimization problems in continuous space using ant colony algorithm. In Dorigo, M., Di Caro, G., Sampels, M., eds.: Proceedings of the Third International Workshop, (ANTS'2002). Volume 2463 of Lecture Notes in Computer Science. Springer Verlag, (Brussels, Belgium) 288–289

3. Lei, W., Qidi, W.: Further example study on ant system algorithm based continuous space optimization. In: Proceedings of the 4th Congress on Intelligent and Automation, Shangai, P.R. China (2002) 2541–2545
4. Pourtakdoust, S.H., Nobahari, H.: An extension of ant colony systems to continuos optimization problems. In Dorigo, M., Birattari, M., Blum, C., Gambardella, L.M., Mondada, F., Stützle, T., eds.: Ant Colony Optimization and Swarm Intelligence, 4th International Workshop, ANTS 2004. (Springer-Verlag) 294–301
5. Michalewicz, Z., Nazhiyath, G., Michalewicz, M.: A note on usefulness of geometrical crossover for numerical optimization problems. In et al., L.J.F., ed.: Proceedings of the Fifth Annual Conference on Evolutionary Programming, Cambridge, MA, MIT Press (1996) 305–311
6. Keane, A.: Genetic algorithms digest, v8n16 (1994)
7. Schoenauer, M., Michalewicz, Z.: Evolutionary computation at the edge of feasibility. In Voigt, H.M., Ebeling, W., Rechenberg, I., Schwefel, H.P., eds.: Parallel Problem Solving from Nature – PPSN IV, Berlin, Springer (1996) 245–254
8. Wu, Z., Simpson, A.: A self-adaptive boundary search genetic algorithm and its application to water distribution systems. Journal of Hidraulic Research **40**(2) (2002) 191–203
9. Corne, D., Dorigo, M., Glover, F., eds.: New Ideas in Optimization. McGraw-Hill International (1999)
10. Runarsson, T.P., Yao, X.: Stochastic ranking for constrained evolutionary optimization. IEEE Transactions on Evolutionary Computation **4**(3) (2000) 284–294
11. Hamida, S.B., Schoenauer, M.: ASCHEA: New Results Using Adaptive Segregational Constraint Handling. In: Proceedings of the Congress on Evolutionary Computation 2002 (CEC'2002). Volume 1., Piscataway, New Jersey, IEEE Service Center (2002) 884–889
12. Liang, J., Runarsson, T.P., Mezura-Montes, E., Clerc, M., Suganthan, P.N., Coello Coello, C., Deb, K.: Problem definitions and evaluation criteria for the cec 2006 special session on constrained real-parameter optimization. Technical report, School of Electrical and Electronic Engineering Nanyang Technological University, Singapore, http://www.ntu.edu.sg/home5/lian00 12/cec2006/technical_report.pdf (2006)

Chain Based Path Formation in Swarms of Robots

Shervin Nouyan and Marco Dorigo

IRIDIA, CoDE, Université Libre de Bruxelles, Brussels, Belgium
{snouyan, mdorigo}@ulb.ac.be

Abstract. In this paper we analyse a previously introduced swarm intelligence control mechanism used for solving problems of robot path formation. We determine the impact of two probabilistic control parameters. In particular, the problem we consider consists in forming a path between two objects which an individual robot cannot perceive simultaneously.

Our experiments were conducted in simulation. We compare four different robot group sizes with up to 20 robots, and vary the difficulty of the task by considering five different distances between the objects which have to be connected by a path.

Our results show that the two investigated parameters have a strong impact on the behaviour of the overall system and that the optimal set of parameters is a function of group size and task difficulty. Additionally, we show that our system scales well with the number of robots.

1 Introduction

Environment exploration, navigation, and path formation are a prerequisite for the accomplishment of a wide range of tasks in the robotics domain. As environment exploration is a very general task, there are many different approaches to it. Often, researchers equip robots with an explicit, map-like representation of their environment [1,2]. Such a representation may be given a priori, mainly leaving the robot with the non-trivial task of localizing itself, or the map may be constructed by the robot itself while moving through the environment. Such strategies have proven efficient particularly for static environments when using a single robot. However, problems can arise when an environment changes dynamically, and in particular when multiple robots are considered. There are strategies [3] to approach this situation. However, complex navigation strategies do not naturally scale with respect to the number of robots, and require careful engineering of the controller in order to deal with the difficulties related to dynamic environments and multiple robots.

In swarm robotics, the goal is to emphasize the cooperation and the collectivity of a robot group. Rather than equipping an individual robot with a control mechanism that enables it to solve a complex task on its own, individual robots are usually controlled by simple strategies, and complex behaviours are achieved at the colony level by exploiting the interactions among the robots, as well as

M. Dorigo et al. (Eds.): ANTS 2006, LNCS 4150, pp. 120–131, 2006.

between the robots and the environment. When designing swarm robotics control algorithms, complex strategies are in general avoided, and instead principles such as locality of sensing and communication, homogeneity and distributedness, are followed. The main benefits that one can hope for when pursuing a swarm robotic approach are scalability with respect to the number of robots used, fault tolerance in case of individual failure, and robustness with respect to noisy sensory data.

In swarm robotics, inspiration is often taken from social insects, such as ants, bees or termites. For example, if we consider environment exploration, when foraging for food, ants of many species lay trails of pheromone, a chemical substance that attracts other ants. Deneubourg *et al.* [4] showed that laying pheromone trails is a good strategy for finding the shortest path between a nest and a food source. Similarly, the concept of robot chains relies on the idea of locally manipulating the environment in order to attract other individuals and to form a global path. However, due to their lack of a substance such as pheromone, the robots constituting a chain serve as trail markers themselves.

The concept of robot chains stems from Goss and Deneubourg [5]. In their approach, every robot in a chain emits a signal indicating its position in the chain. A similar system was implemented by Drogoul and Ferber [6]. Both works have been carried out in simulation. Werger and Matarić [7] used real robots to form a chain in a prey retrieval task. Neighbouring robots within a chain sense each other by means of physical contact: one robot in the chain has to regularly touch the next one in order to maintain the chain.

One of the main differences of our approach with respect to the previously mentioned approaches to robot chains is that we rely on the concept of chains with cyclic directional patterns in order to give the chains a directionality. In a previous work [8] we have shown how such chains of real robots can be used for forming a path, and how such a path is used by other robots to transport a heavy object. In this work we concentrate on the path formation and omit the transport. We have conducted a series of experiments in simulation using different robot group sizes and varying the difficulty of the task. Our goal is to determine the capabilities of our robot chains. We measure the speed of the environment exploration, and the scalability with respect to the number of robots. Furthermore, we study the impact of two parameters specifying the controller: the probability for the robots to aggregate to, and to disaggregate from, a chain. We show that these two parameters have a significant effect both on the overall behaviour of the robot group in terms of the number of formed chains and their length, and on the success rate with which they find the prey.

The remainder of this paper is organised as follows. In Section 2 we give a description of the considered problem and a short outline of our approach. In Sections 3 and 4 we give a brief overview of the simulator and the control algorithm we used. In Section 5 we present the experimental results. Finally, in Section 6 we draw some conclusions and discuss possible future works.

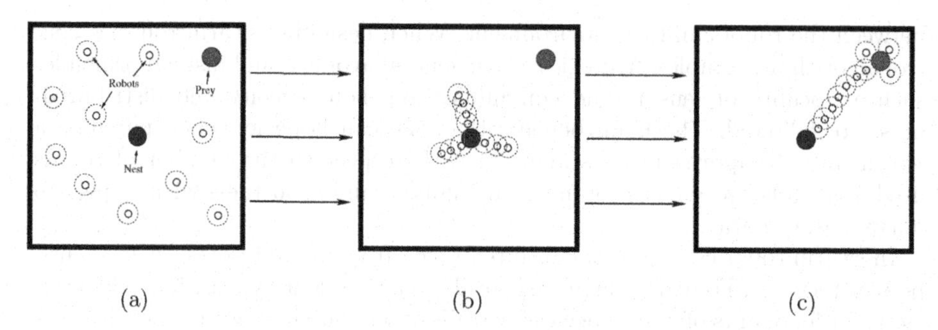

(a) (b) (c)

Fig. 1. (a) Initial situation. Robots are indicated by the small white circles. Their limited sensing range is indicated by dashed circles. The task is to form a path between the nest and the prey. (b) The robots search for the nest and once they find it they start self-organizing into randomly oriented chains. (c) When a chain perceives the prey a path is formed.

2 The Problem

The task that we have chosen as test-bed to analyse our control algorithm is illustrated in Figure 1: a group of robots has to form a path between two objects—denoted as nest and prey. The robots have no a priori knowledge about the dimensions, or the position of any object within the environment, and a robot's perception range is small when compared to the distance between the nest and the prey. The difficulty of the task can be varied by changing the distance between nest and prey.

Initially, as displayed in Figure 1a, all robots are placed at random positions. They search the nest, and once they perceive it, they start self-organizing into chains (Figure 1b), where robots act as trail markers and attract other robots. Neighbouring robots within a chain have to be able to sense each other in order to assure the connectivity of the chain. As the robots have no knowledge about the position of the prey, the chains are oriented in random directions. Due to a self-organized process where robots disaggregate from chains and start new ones into possibly new, unexplored directions, the environment is continuously explored until eventually the prey is perceived by a chain. As shown in Figure 1c, a path is then formed, and can for instance be used by other robots to navigate between nest and prey, or to transport the prey to the nest.

3 The S-bot and Its Simulator

All our experiments have been conducted in simulation. Our simulation platform, called twodee, is a multi-robot simulator based on a custom high-level dynamics engine optimized for the use with the *s-bot*, a robot on which we have previously implemented and tested our controller [8]. Figure 2 shows the physical implementation of an *s-bot*. It has a diameter of 12 cm and weighs approximately 700 g. In the following, we briefly overview the actuators and sensors that we

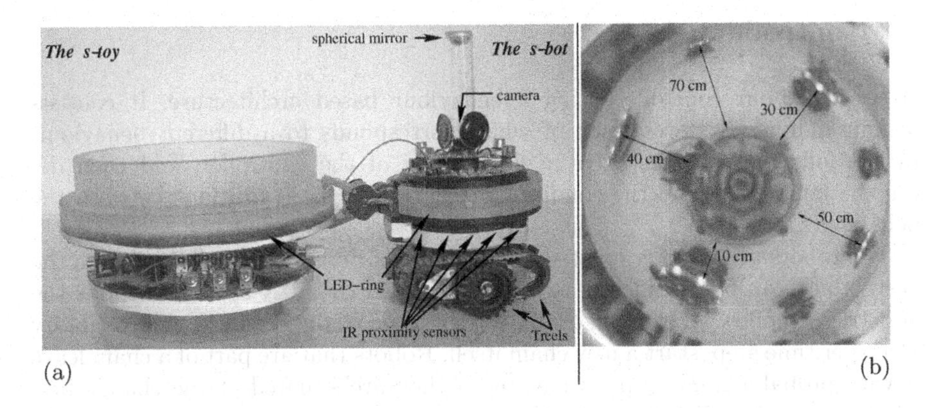

Fig. 2. The hardware. (a) The *s-toy* and the *s-bot*. (b) An image taken with the omni-directional camera of the *s-bot*. It shows other *s-bots* and an *s-toy* activating their red LEDs at various distances.

use in this study. For a more comprehensive description of the *s-bot*'s hardware see [9], and for the twodee simulator see [10].

The robot's traction system consists in a combination of tracks and two external wheels, called *treels*©. For the purpose of communication, the *s-bot* has been equipped with eight RGB LEDs distributed around the robot.

There are 15 infra-red proximity sensors distributed around the turret. They are used to avoid crashing into other objects. We have recorded samples of the proximity sensor activation for various angles and distances towards other objects. These samples have been integrated into twodee to allow a realistic simulation of the proximity sensors.

A VGA camera is directed towards a spherical mirror on top of the *s-bot*, in this way providing an omni-directional view. The camera is used to perceive the nest, the prey, and other *s-bots* emitting a colour with their LED ring. A snapshot taken from an *s-bot*'s camera is shown in Figure 2b. Due to differences among the robots' cameras, there are some variations in the perceptual range. The software we use to detect coloured objects allows a recognition of the red coloured prey up to a distance of $70 - 90$ cm, and of the three colours blue, green and yellow, up to $35 - 60$ cm (depending on which robot is used). Due to the spherical shape of the mirror, the distance to close objects can be approximated with good precision. For objects further away than 30 cm it becomes very difficult to deduce the distance from the camera image. The differences among the perception of different colours and among the robots are taken into account in simulation. Initially, each robot is given a different set of perceptual ranges for the four colours. Each value is chosen randomly from the ranges mentioned above.

Next to the *s-bot*, Figure 2a shows the *s-toy*, an object which we use either as nest or as prey (depending on its colour). It has a diameter of 20 cm and, like the *s-bot*, it is equipped with an RGB LED-ring. In our simulations, the nest and the prey are represented by coloured cylinders of the size of an *s-toy*, and are both immobile.

4 Controller

We realized our controller using a behaviour based architecture. It consists of four individual states, each of which corresponds to a different behaviour. In the following, we first give a global view of the controller, and then detail the behaviours and the conditions that trigger the transitions between the behaviours.

The robots are initially located at random positions. They have to search the nest, which can be considered as the root of each chain. A robot that finds the nest tries to follow an existing chain. If there is no chain, it will, with probability $P_{e \to c}$ per time step, start a new chain itself. Robots that are part of a chain leave it with probability $P_{c \to e}$ per time step if they are situated at the chain's tail. The process of probabilistically aggregating to, disaggregating from, a chain is fundamental for the exploration of the environment as it allows the formation of new chains in unexplored areas. The task is successfully finished when a chain encounters the prey and thereby establishes a path between nest and prey. The members of this chain do not decompose any more and are used by the other robots to reach the prey.

As mentioned in the introduction, our concept of chain relies on cyclic directional patterns. As displayed in Figure 3a, each robot emits one out of three signals depending on its position in the chain. By taking into account the sequence of the signals, a robot can determine the direction towards the nest. The main advantage of using a periodically repeating sequence of three signals is that each signal can be realized by the activation of a dedicated colour with the LED ring. Previous approaches to directed robot chains require the members of a chain to broadcast as many different signals as there are robots in a chain. This leads to increasing complexity of communication for chains of growing length. Therefore, we expect our approach to lead to better scalability with respect to the number of robots.

Behaviours. The behaviours are realized following the motor schema paradigm [11]. One behaviour is executed exclusively at a given *control time step.*[1] For each behaviour, a set of motor schemas are active in parallel. Each motor schema outputs a vector denoting the desired direction of motion. The vectors of active motor schemas are added and translated into motor activation at the beginning of each control time step. Common to all behaviours, and therefore permanently active, is a motor schema for collision avoidance. It simply returns vectors which are directed into the opposite direction of each proximity sensors activation. In the following, the four behaviours are detailed:

- **Search**: in order to search the nest, the robot performs a random walk which consists in straight motion and turning on the spot when an obstacle is encountered. No LEDs are activated.

[1] On the real *s-bot*, a control time step has a length of approximately 120 *ms*. We adopted the same value in simulation.

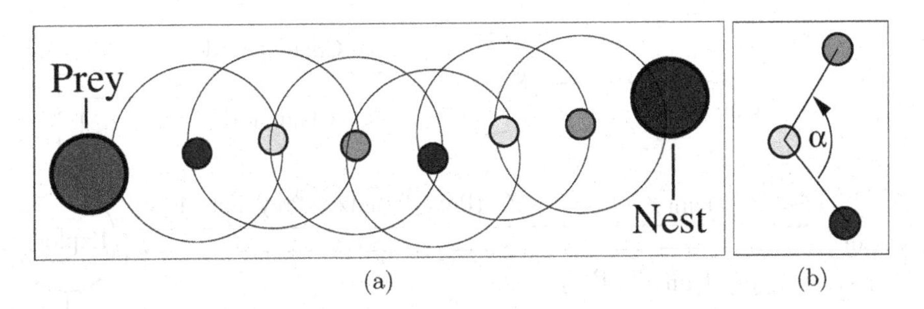

Fig. 3. (a) A chain with a cyclic directional pattern. The small circles represent robots that have formed a chain that connects a nest with a prey. Three colours are sufficient to give a directionality to the chain. The large circles surrounding the robots indicate their sensing range. (b) Alignment of a chain member. If the angle α is less than 120°, the central chain member aligns with respect to its closest neighbours.

- **Explore**: an explorer moves along a chain towards its tail. In case a robot becomes an explorer by leaving a chain, it moves back to the nest from where it can then start to follow a different chain. No LEDs are activated.
- **Chain**: a chain member activates an appropriate colour, which is defined by the previous neighbour. To avoid loops in chains and to improve the length of the chains, we implemented an alignment behaviour, that is, the robot aligns with its two closest neighbours in the chain in case the angle between them is smaller than 120° (see Figure 3b). Furthermore, a chain member adjusts his distance with respect to its previous neighbour to roughly 30 cm in order to avoid breaking up the chain, and to increase the chain length.
- **Finished**: a path has been established and the robot stays in the vicinity of the prey.

Behaviour Transitions. The set of rules governing the transition from one behaviour to another is illustrated in Figure 4, and detailed in the following:

- **Search → Explore**: if a chain member is perceived. Note that the nest is perceived as a chain member, and that a robot searching for the nest does not react when it perceives the prey.
- **Explore → Search**: if no chain member is perceived.
- **Explore → Chain**: (i) if the tail of a chain is reached (i.e., only one chain member is perceived), the robot joins the chain with probability $P_{e \to c}$ per time step, or (ii) if the prey is detected at a distance larger than 30 cm.
- **Explore → Finished**: if the prey is detected at a distance of less than 30 cm.
- **Chain → Search**: if the previous neighbour in the chain is no longer detected.
- **Chain → Explore**: if a chain member is situated at the tail of a chain, it leaves the chain with probability $P_{c \to e}$ per time step.

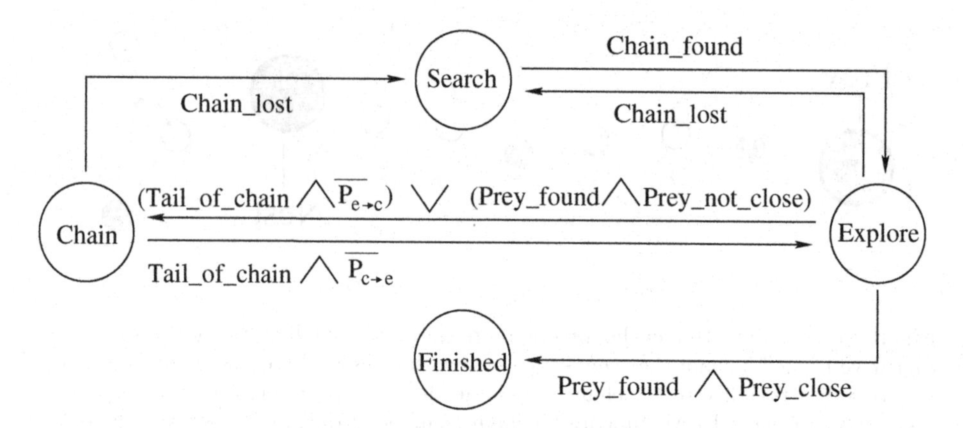

Fig. 4. State diagram of the control. Each circle represents a state (i.e., a behaviour). Edges are labelled with the corresponding conditions that trigger a state transition. The initial state is the search state. $\overline{P_{e\to c}}$ (and $\overline{P_{c\to e}}$ respectively) is a boolean variable which is set to *true*, if $R \leq P_{e\to c}$ ($R \leq P_{c\to e}$), and to *false* otherwise, where R is a stochastic variable sampled from the uniform distribution in $[0, 1]$, and $P_{e\to c}$ ($P_{c\to e}$) is the probability per time step to aggregate to (disaggregate from) a chain.

5 Experiments

The main objectives of our experiments are to determine the impact of the two probabilistic control parameters on the chain formation process, and to find the optimal parameter combination for a given task. In the following, we explain the experimental procedure and detail the results.

5.1 Setup

A group of N simulated robots is placed within a bounded arena of size $5\,\mathrm{m}\times 5\,\mathrm{m}$. The nest is placed in the centre of the arena, and the prey is put at distance D (in m). The initial position and orientation of the robots are chosen randomly, and defined by an initial seed. We investigated all setups (N, D), with $N \in \{5, 10, 15, 20\}$, and $D \in \{0.6, 1.2, 1.8, 2.4, 3.0\}$. For $D = 0.6$ the task is rather trivial, as the prey can be perceived from the proximity of the nest and only one robot is required to form a path. An additional two robots are required for each distance increase, meaning that it makes sense to test group size $N = 5$ only up to distances $D \leq 1.8$.

The probabilities per control time step to aggregate to a chain, $P_{e\to c}$, and to disaggregate from a chain, $P_{c\to e}$, are the parameters that we intend to optimize. For both parameters we have chosen to examine the same logarithmic range of values defined by $2^{-x}, x \in \{0, 1, 2, 3, \ldots, 10\}$.

For each combination of the setups (N, D), and of the parameter settings $(P_{e\to c}, P_{c\to e})$, we conducted 100 trials with different initial seeds. A trial is considered to be successful if a chain establishes a connection to the prey which is

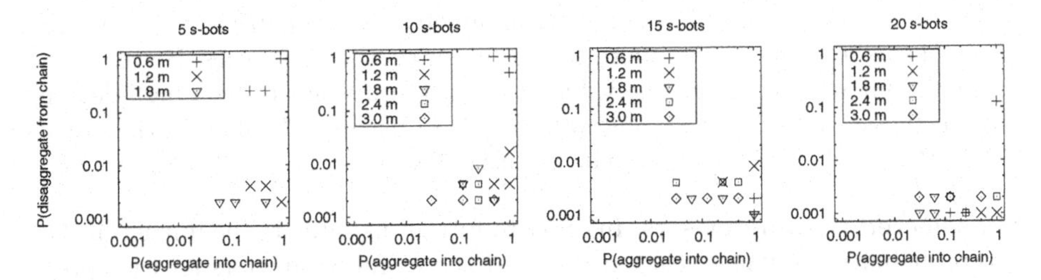

Fig. 5. The three most successful parameter combinations are displayed for all prey distances ordered by robot group size

kept for at least 100 seconds. The time by which a trial is successfully completed is denoted completion time. The limit to accomplish this is set to 10,000 seconds.

5.2 Results

Let us first describe the impact of the two probabilistic parameters $P_{e \to c}$ and $P_{c \to e}$ on the overall behaviour of the robot group. In general, values for $P_{e \to c}$ close to 0 result in a rather patient behaviour; in most cases a single chain is formed slowly. For $P_{e \to c}$ close to 1, several chains are formed fast and in parallel. The second parameter, $P_{c \to e}$, determines the stability of the formed chains, directly influencing their lifetime and the frequency of chain disbandment. High values of $P_{c \to e}$ lead to an impatient behaviour where robots joining a chain more or less immediately disaggregate from it.

Overall success. For all except one of the considered setups (N, D) there is at least one parameter set that reaches a success rate of more than 90%. The only exception is $(N, D) = (10, 3.0)$, where the highest success rate is 77%, still a reasonable value when considering that for this setup nine out of the ten robots have to form a chain in the right direction in order to form a path. Adding more robots increases the success rate to 91% for $(N, D) = (15, 3.0)$, and to 94% for $(N, D) = (20, 3.0)$. In all other setups the maximum success rate is at least 97%. Figure 5 summarizes the most successful parameters. Ordered by group size, the four plots show the three best performing parameter combinations for each prey distance. If different parameters achieve the same maximum success rate, the one with the lowest median completion time is chosen.

For the two smallest prey distances, there is a wide range of parameter values that reach a 100% success rate. It appears that for these rather simple tasks, combinations of higher values for both probabilities are more successful. They lead to a fast creation of short chains with often just a single chain member, and a fast disaggregation of the chain in case it has not already encountered the prey. However, this can be considered as an advantage only for short prey distances because a high probability to disaggregate from a chain makes it very unlikely to form long chains.

For setups with distances $d \geq 1.8$ the most successful parameter combinations employ low values of $P_{c \to e} \leq 2^{-8}$. The corresponding value of $P_{e \to c}$ is always higher than the one of $P_{c \to e}$. For both parameters, there appears to be a tendency of smaller values being more successful for growing distances. However, in particular for what concerns the value of $P_{e \to c}$, there seems to be a high degree of robustness, that is, for the same value of $P_{c \to e}$, usually all values in the range $2^{-8} \leq P_{e \to c} \leq 2^{-2}$ achieve a similar performance.

Six selected parameter sets. In order to further investigate the impact of the two probabilities, we have selected the most successful parameters for each prey distance when using 10 robots.[2] Additionally, after some initial analysis, we have selected two parameter combinations to allow for a better understanding of the overall effect of the two probabilities. For each of the selected parameter sets we run 100 trials for 10,000 seconds in an environment without a prey, and measure the exploration rate, which we define as the percentage of the explored area within the arena, and the length of the longest chain. Figure 6 shows the results for these two measures at nine different temporal instants. For those parameters which are the most successful in a given setup, the respective setup is indicated under the probability values.

Looking at the two plots in Figure 6, one would intuitively separate the six parameters into two groups. The two parameter sets on the left perform quite poorly, reaching a median exploration rate of less than 25% at the end of the trial. Comparably high values for both probabilities are employed, and as we stated earlier, this may lead to an initial speedup for exploring the direct vicinity of the nest on the one hand, but on the other hand the robot chains remain very short, often consisting of a single robot.

Differently, the other four parameter sets perform quite well. After 10,000 seconds they all reach exploration rates of more than 85%. The main reason for the better long term performance is the lower probability to leave a chain, resulting in a higher fraction of robots aggregated into chains, and therefore longer chains. The differences among these four parameter sets are less obvious. The two right ones with lower values for the probabilities reach approximately 30 cm longer distances, which is equivalent to one additional chain member. And even if their exploration rate is initially lower than for the other parameters, in the end it is slightly higher.

Scalability. Let us now look more closely at the performance of the most successful parameter sets. Figure 7a shows the shortest completion times reached for all setups. The results are ordered by robot group size, and we can see that the completion time increases more than linearly with growing prey distance. This is not surprising, as the area to explore grows quadratically with respect to the prey distance.

In Figure 7b the normalized completion time, defined as the product of completion time and robot group size, is displayed. This measure indicates the efficiency of the system as it represents the added amount of time spent by all

[2] Note that the combination $(P_{e \to c}, P_{c \to e}) = (0.125, 0.004)$ is the most successful one for both setups $(N, D) = (10, 1.8)$ and $(N, D) = (10, 2.4)$.

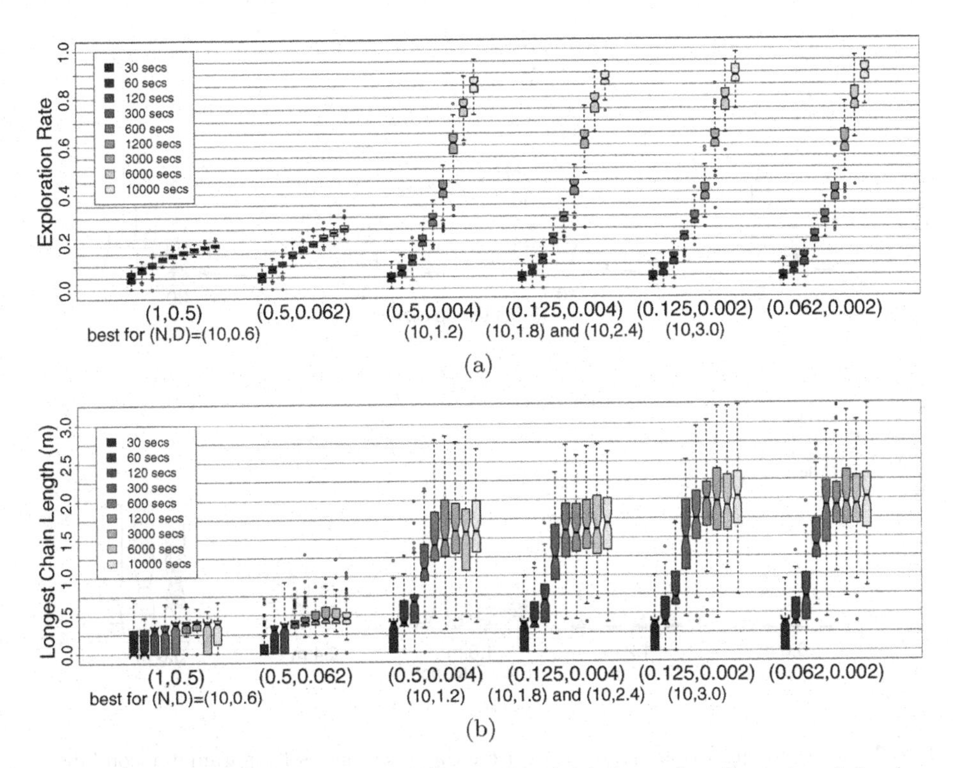

Fig. 6. For six selected parameter sets $(P_{e\to c}, P_{c\to e})$ (a) the exploration rate—defined as the percentage of the explored area within the arena—and (b) the length of the longest chain are displayed. The parameters were selected according to their success in the setups with $N = 10$ robots. The setup for which a parameter combination is most succesful is indicated below the probability values. Note that the combination $(P_{e\to c}, P_{c\to e}) = (0.125, 0.004)$ is the most successful one for both setups $(N, D) = (10, 1.8)$ and $(N, D) = (10, 2.4)$. Additionally, two parameter sets were selected by hand in order to allow for a better understanding of the overall effect of the probability values.

robots until completion of a trial. The results are ordered by prey distance, and show that our system scales quite well with respect to the number of robots.

6 Conclusions and Future Work

We have presented an experimental study of a system that employs robot chain formation for forming a path between two objects that are too distant from each other for a single robot to be able to perceive them both at the same time. Our control system is completely distributed and homogeneous, and makes use of local information and communication only. Our concept of robot chain relies on cyclic directional patterns in order to give the chains a directionality.

Our results reveal the impact of the two probabilistic parameters which determine the rate at which a robot aggregates into, and disaggregates from, a

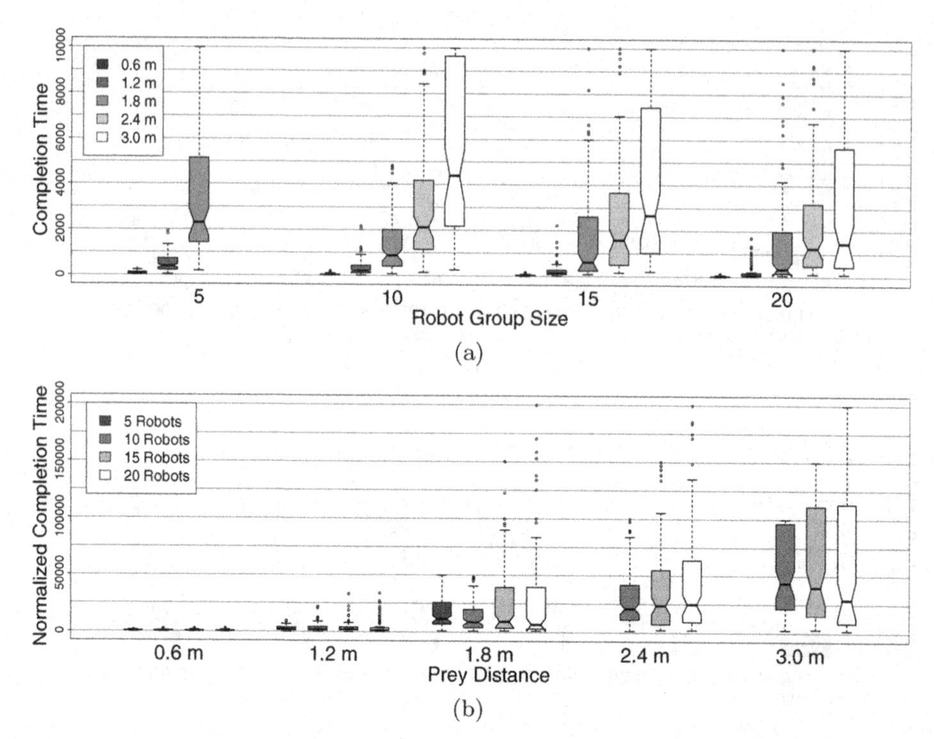

Fig. 7. (a) The completion time is shown for the most successful parameter combinations of all setups, ordered by the robot group size. (b) The normalized completion time is shown and ordered by the prey distance. It is an indicator of efficiency, and is calculated as the product of completion time and robot group size.

chain. We have shown that for simple tasks where a required path is short, high values for the two probabilities result in a faster success. On the contrary, for growing difficulty of the task, smaller values, in particular for the probability to disaggregate, should be employed in order to allow the chains to grow longer.

Furthermore, we have shown that our system scales quite well with respect to the number of robots. However, for growing distances of the prey, it seems to take at least a quadratically growing amount of time to establish a connection. In the future, we will extend our controller to improve the performance in particular for larger prey distances. A simple idea that seems promising is to start chains not only from the nest, but also from the prey.

Finally, we would like to investigate more complex environments. The problem of using robot chains the way we currently implemented them is their linear shape. For this purpose we are interested in studying control algorithms that allow swarm of robots to spread in the environment in a more uniform way and form arbitrary shapes.

Acknowledgments. This work was supported by the "ANTS" project, an "Action de Recherche Concertée" funded by the Scientific Research Directorate of the French Community of Belgium, and by the "SWARM-BOTS Project", funded by the Future and Emerging Technologies programme (IST-FET) of the European Commission, under grant IST-2000-31010. The information provided is the sole responsibility of the authors and does not reflect the Community's opinion. The Community is not responsible for any use that might be made of data appearing in this publication. Marco Dorigo acknowledges support from the Belgian FNRS, of which he is a Research Director.

References

1. Filliat, D., Meyer, J.A.: Map-based navigation in mobile robots - I. A review of localization strategies. J. of Cognitive Systems Research **4** (2003) 243–282
2. Meyer, J.A., Filliat, D.: Map-based navigation in mobile robots - II. A review of map-learning and path-planning strategies. J. of Cognitive Systems Research **4** (2003) 283–317
3. Howard, A.: Multi-robot mapping using manifold representations. In: Proc. of the 2004 IEEE Int. Conf. on Robotics and Automation, IEEE Computer Society Press, Los Alamitos, CA (2004) 4198–4203
4. Deneubourg, J.L., Aron, S., Goss, S., Pasteels, J.M.: The self-organizing exploratory pattern of the argentine ant. J. Insect Behavior **3** (1990) 159–168
5. Goss, S., Deneubourg, J.L.: Harvesting by a group of robots. In: Proc. of the 1st European Conf. on Artificial Life, MIT Press, Cambridge, MA (1992) 195–204
6. Drogoul, A., Ferber, J.: From Tom Thumb to the dockers: Some experiments with foraging robots. In: From Animals to Animats 2. Proc. of the 2nd Int. Conf. on Simulation of Adaptive Behavior (SAB92), MIT Press, Cambridge, MA (1992) 451–459
7. Werger, B., Matarić, M.: Robotic food chains: Externalization of state and program for minimal-agent foraging. In: From Animals to Animats 4, Proc. of the 4th Int. Conf. on Simulation of Adaptive Behavior (SAB96), MIT Press, Cambridge, MA (1996) 625–634
8. Nouyan, S., Groß, R., Bonani, M., Mondada, F., Dorigo, M.: Group transport along a robot chain in a self-organised robot colony. In: Proc. of the 9th Int. Conf. on Intelligent Autonomous Systems, IOS Press, Amsterdam, The Netherlands (2006) 433–442
9. Mondada, F., Gambardella, L.M., Floreano, D., Nolfi, S., Deneubourg, J.L., Dorigo, M.: The cooperation of swarm-bots: Physical interactions in collective robotics. IEEE Robotics & Automation Magazine **12**(2) (2005) 21–28
10. Christensen, A.L.: Efficient neuro-evolution of hole-avoidance and phototaxis for a swarm-bot. Technical Report TR/IRIDIA/2005-14, Université Libre de Bruxelles, Belgium (2005) DEA Thesis.
11. Arkin, R.: Behavior-Based Robotics. MIT Press, Cambridge, MA (1998)

Communication, Leadership, Publicity and Group Formation in Particle Swarms*

Riccardo Poli[1], William B. Langdon[1], Paul Marrow[2], Jim Kennedy[3], Maurice Clerc[4], Dan Bratton[5], and Nick Holden[6]

[1] Department of Computer Science, University of Essex, UK
[2] BT Pervasive ICT Research Centre, Adastral Park, Ipswich, UK
[3] US Bureau of Labor Statistics, Washington DC, USA
[4] Independent Consultant, Groisy, France
[5] Department of Computing, Goldsmiths College, University of London, UK
[6] Computing Laboratory, University of Kent, UK

Abstract. We look at how the structure of social networks and the nature of social interactions affect the behaviour of Particle Swarms Optimisers. To this end, we propose a general model of communication and consensus which focuses on the effects of social interactions putting the details of the dynamics and the optimum seeking behaviour of PSOs into the background.

1 Introduction

In the standard Particle Swarms Optimiser, there are three features that bias the particle to look in a better place: 1) the particle remembers its best position, 2) it identifies its best neighbour, and 3) it knows that neighbour's best position so far. However, not all three forms are needed for good performance. For example, in Mendes' Fully Informed Particle Swarm (FIPS) model [1,2,3], the first two are absent. However, the topology of the social network of a PSO is considered a critical factor in determining performance.

Researchers have investigated how different topologies for the social network affect performance [1–8]. [4]. For example, it has been reported that with unimodal problems a *gbest* topology provides better performance, while the *lbest* PSO topology performs well on multimodal functions. Also, we know that with an appropriate topology, FIPS performs significantly better than the standard best-neighbour PSO on an array of test functions.

Albeit several lessons have been learnt in previous research on PSO social networks, the focus has mostly been on PSO performance rather than behaviour. However, it is clear behaviour is what matters, since performance on any particular fitness function is the result of coupling the features of that function with the natural behaviour of that particular PSO. If the mode of exploration fits the features of a problem, we expect good performance and vice versa.

Although PSOs are inspired by natural swarms, shoals, flocks, etc., the social network in a PSO has some important differences from its natural counterparts.

* Work supported by EPSRC XPS grant GR/T11234/01.

M. Dorigo et al. (Eds.): ANTS 2006, LNCS 4150, pp. 132–143, 2006.

In particular, a particle is fully aware of what happens in its neighbourhood. E.g., if at one iteration a particle in a neighbourhood achieves an improvement in its personal best, by next iteration all the other particles in that neighbourhood will be influenced by this change of state. It is as if the new local or global leader broadcasted its new state to the whole neighbourhood in one go. In nature this is not possible. Very often one can acquire information only by one-to-one interactions with another individual in the social network. So, the propagation of information is a stochastic diffusive process, rather than a deterministic broadcast.

Another feature of a PSO, which is not very common in nature, is that the best individual in the population, as long as it remains the swarm best, is unaffected by the others. In a shoal of fish, an individual may know about the presence of food or of a predator and act accordingly independently of other individuals. This individual will tend to act as a leader and will be followed by other fish. However, if, for whatever reasons, the rest of the shoal does not follow, it stops acting as a leader and rejoins the shoal.

Most researchers consider the social network as the communication medium through which information is exchanged. An important question is what exactly we mean by that. Perhaps a way to understand this is to investigate the properties of social networks as systems for reaching consensus. Ultimately the social network in a PSO is expected to get all of the particles to "agree" on where the swarm will search. However, the process and stages through which a global consensus is reached may be very different in PSOs using different communication topologies. For example, initially, until some form of consensus emerges, a population will act as a set of independent individuals. In systems where only localised interactions are possible (e.g., in a *lbest*-type of PSO) consensus must emerge firstly on a local basis and then progressively at larger and larger scales. However, when groups of individuals reach some local consensus, the group may start showing emergent properties. To understand the behaviour of the system as a whole it then becomes important to understand in which ways local groups (as opposed to single individuals) interact. This is particularly important in systems and landscapes where different parts of the population can reach different consensual decisions. If two such domains of agreement can come into contact, can they both persist? If not, will the consensus reached in one domain invade the other? Or, will the interaction produce a domain of a third type?

Another unnatural feature of PSOs, is that, once provided with a fitness function, they are closed systems. I.e. no external signals affect a PSO. In natural populations and many other systems there are always external influences. For example, in a social network of human customers or voters, individuals will influence each other and over time local or global consensus may emerge. However, the process of consensus formation and its final outcomes may be influenced by external factors, such as advertisements or broadcast party messages. These factors have been considered in agent-based models of customer relationship management where interactions between the social network of agents (customers) influence the response to marketing and other business activities (e.g., see [5]).

Naturally, it would be possible to modify PSOs to incorporate these three features: non-deterministic communication, democratic (as opposed to dictatorial) leaders and external influences. The question would then be, in what way would the search *behaviour* of a PSO be affected?

Rather than implementing PSO variants and testing them as optimisers for some set of functions (which we may do in future studies), we prefer to abstract away from the particular details of the dynamics of a PSO and to model social networks as consensus machines. In doing so we introduce approximations. So, in a sense, our model will not represent exactly any particular PSO. However, in return for these approximations we will gain a much deeper understanding of the social dynamics in particle swarms for different choices of interaction mechanisms and external influences.

The paper is organised as follows. In Section 2 we describe how we can abstract PSOs with very simple models which put many detailed aspects of the particle dynamics into the background. In Section 3 we consider the elitist communication strategies used in current PSOs and model them within a more general framework, which allows us to evaluate other, less-elitist strategies. These models are executable and so we were able to test them. Section 4 describes the parameters and fitness functions used in our experiments, while Section 5 reports on some key results. These are discussed in Section 6.

2 Abstracting PSO States and State Transitions

Let us consider a standard PSO. For as long as a particle's best and a neighbourhood best do not change, the particle will swarm around the point where the forces are zero. That is, around

$$x_i^* = \frac{x_{s_i} R_{\max_1} + x_{p_i} R_{\max_2}}{R_{\max_1} + R_{\max_2}}$$

where x_{s_i} is the ith component of the best point visited by the neighbours of the current particle, x_{p_i} is the ith component of its personal best, and $R_{\max_1}$ and $R_{\max_2}$ are the upper bounds of the intervals from which random numbers are drawn. This means that the points sampled by the PSO will be distributed about x_i^* until a new personal best or a new neighbourhood best are found. Because of this, on average a new personal best will tend to be found somewhere in between x_s and the old personal best. If the personal best gets closer to x_s, the new sampling distribution will progress even more towards x_s.

Naturally, when the distance between x^* and x_s reduces, the variance of the sampling distribution also reduces. This is effectively a step-size adjustment, which may be very important for good optimisation performance. The effect of this adjustment is that the probability of sampling the region "between" x_s and x_p remains high despite this region shrinking. If we took the centre of the sampling distribution x^* (rather than x_p) to represent the state of a particle, and approximate the whole swarm as a set of stochastic oscillators, each centred at its own x^*. We could then approximately model the dynamics of x^* by imagining that, on average, when a particle interact socially with another particle in its neighbourhood a sort of stochastic interpolation process between the particles' states takes place.[1]

A natural modelling choice for PSOs is to consider states as continuous variables which change with an interpolation rule such as

$$x^*_{new} = \alpha x^*_s + (1 - \alpha) x^*_{old} \tag{1}$$

where x^*_{old} is the current state of a particle, x^*_s is the state of the particle with which interaction takes place (e.g., the neighbourhood best), and α is a random variable uniformly distributed in $[0, 1]$. These choices seem particularly suitable for smooth monotonic landscapes and for PSOs where velocities are limited (e.g., when ψ_1 and ψ_2 are small or friction is high). In highly multimodal landscapes each particle's best and, consequently x^*, may change less continuously.

To explore what happens at this other extreme of the behaviour, we idealise the multi-modal case and imagine that only a finite number, n, of different states x^* are possible. State transitions can then be represented using a Markov chain, where the entries of state transition matrix represent the probability of x^* moving from one state to another. The transition probabilities would have to be such as to effectively implement a stochastic interpolation process between x^* and x_s like the one in Equation 1.

If n is a power of two, a simple realisation of this is to represent states in binary form. So, a generic state x could be represented as $x_1 x_2 \cdots x_\ell$ (with $\ell = \log_2 n$). The interpolation process between states could then be implemented by the replacement of one or more random bits in x^* with corresponding bits from x_s. That is, for discrete states we could use the update rule: $x^*_{new_i} = x^*_{old_i}$ with probability β and $x^*_{new_i} = x^*_{s_i}$ with probability $1 - \beta$, where β is a constant.

Irrespective of the model used, we can see that state update rules have the form $x^*_{new} = \chi(x^*_s, x^*_{old})$ where χ is a stochastic function.

In a standard PSO, and also in many natural and artificial systems, an individual's choice as to whom to interact with depends on quality, desirability, or, more generally, on whatever is an appropriate measure of success for an individual. We will just call this *fitness.* Without any information about the fitness function it would be impossible to say much about the behaviour of a system. So, even in our effort to hide the finer details of the dynamics of PSOs, we will need to keep a notion of fitness. In particular, we will imagine that each state, x, has an associated fitness value f_x. Note that this may be distinct from the value taken by the fitness function, f, in x (seen as a point in the search space), $f(x)$. For example, for standard PSOs, we could take this to be the value of fitness associated with the current particle best, i.e. $f_x = f(x_p)$, or the expected fitness value for a sampling distribution centred at x^*. In a maximisation problem, with any reasonable definition of f_x, in a PSO we would expect to see $f_{x_s} \geq f_{x^*_{new}} \geq f_{x^*_{old}}$.

[1] This is an approximation which is expected to be reasonably accurate if we look at states at a suitably coarse time scale, and, of course, as long as x_s remains constant.

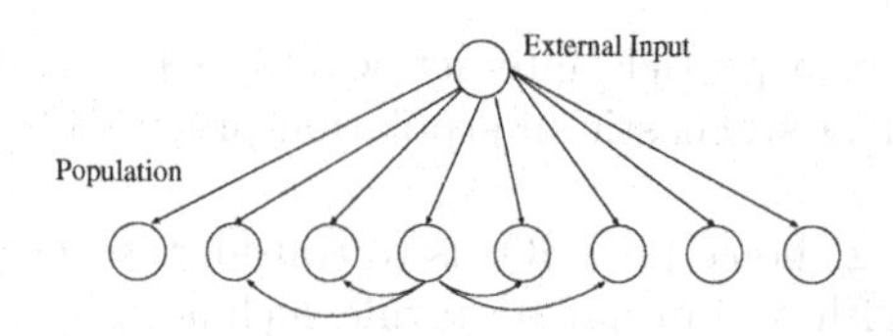

Fig. 1. Extended *lbest* social network topology

If we use integer states then we can represent the state-fitness function f_x as a table. For example, if the search space was one-dimensional then the table would look like the following

x	0	1	$\cdots$	$n-1$
f_x	f_0	f_1	$\cdots$	f_{n-1}

In order to keep things simple, for our real-valued state representation we will define f_x explicitly only over a discrete lattice of points in the search space and then use linear interpolation to construct fitness values elsewhere. This allows us to still represent f_x as a table, albeit at the cost of a reduced generality.[2]

3 Particle Communication as It Is and as It Could Be

We are interested in looking at current forms of inter-particle communication in PSOs as part of a bigger family of social interactions. To allow the study of the effects of external sources of information, noise or publicity we extended the classical *lbest* ring topology as shown in Figure 1. We imagine that the external input is simply another individual which is in the neighbourhood of every particle and which has a pre-defined and constant state x_{ext}. Naturally, we don't necessarily want the external input to act at all times and for all individuals at the same time. So we model the effect of the external input using a probabilistic update rule: $x^*_{new} = \chi(x_s, x^*_{old})$ with probability γ, and $x^*_{new} = \chi(x_{ext}, x^*_{old})$ with probability $1-\gamma$, where γ is a constant controlling the intensity of the action of the external input. When $\gamma = 0$ the system is again a closed system.

As we mentioned in Section 1, we want to study the effects of reducing the ability of individuals to perceive the state of their neighbourhood. This implies that it is possible for them to decide to have social interactions with individuals other than the neighbourhood best. We do this by defining a stochastic social-selection function σ which, given the states x_1, x_2, etc. of the individuals in the neighbourhood returns one of such states. For a standard PSO we have $\sigma(x_1, x_2, \cdots) = \text{argmax}_{x \in \{x_1, x_2, \cdots\}} f(x)$. Naturally, this form of selection function guarantees that, if an individual is the neighbourhood best, the state of that

[2] PSOs and other rank based optimisers behave identically on any pair of isomorphic landscapes f and g such that for every pair of points x and y, $f(x) > f(y)$ if and only if $g(x) > g(y)$. So, fine details on the shapes of landscapes are often not very important.

individual will always be returned by σ. Since the state update functions χ defined above are such that $\chi(x, x) = x$, in these circumstances

$$x^*_{new} = \chi(\sigma(\cdots, x^*_{old}, \cdots), x^*_{old}) = \chi(x^*_{old}, x^*_{old}) = x^*_{old}$$

which guarantees that neighbourhood-best individuals are not affected by state updates.

To explore what happens at the other extreme of the spectrum, while still favouring social interactions with above average fitness individuals, we introduce another version of σ where the value returned is the result of a binary tournament. That is, we draw (with repetition) two random states from the set of neighbouring states $\{x_1, x_2, \cdots\}$. Out of these two, we then return the state with higher fitness. That is

$$\sigma(x_1, x_2, \cdots) = \begin{cases} x' & \text{if } f(x') > f(x'') \\ x'' & \text{otherwise} \end{cases} \tag{2}$$

where x' and x'' are two randomly (and uniformly) chosen states from the set $\{x_1, x_2, \cdots\}$. With this social-selection rule the neighbourhood best individual can still be required to interact with some individuals other than itself. This means that its state may change and it may no longer remain a leader forever. Unlike the standard PSO, this leadership change can happen even if an individual with a better state has not been found in the neighbourhood.

4 Experimental Setup

The models described in the previous section are executable models. Once a fitness table f_x, the size of the population and the structure of the social network are defined, it is possible to iterate the state update and inter-particle interaction rules for as many iterations (generations) as desired. The objective of running these models, however, is not to see how well the population can locate global optima. Rather, we want to see the different qualitative behaviours the different types of social interactions and states can provide.

For this study we decided to keep the search space one-dimensional. Both for the integer state representation and the real-valued representation we used numbers in the range [0,7]. For the integer representation $n = 8$, so we used $\ell = 3$ bits. We used populations of $P = 80$ individuals (we chose a relatively large population because this slows down all transients, making it possible to better see the phenomena we are interested in). Runs lasted for 200 generations. In the initial generation, binary individuals were given random states. (Since there are only 8 possible states, with a population of 80, in virtually all runs, all states had non-zero frequency). In the case of the real-valued representation,to ensure a wide spread, we randomly initialised individuals by drawing states *without replacement* from the set $\{\frac{7\times i}{79}\}_{i=0}^{79} = \{0, 7/79, 14/79, \ldots, 7\}$. With our choice of fitness functions (see below), these initialisation strategies virtually guarantee that the global optimum is always represented in the initial population. This puts

Table 1. Fitness functions used in this study

	Fitness values								
Function	f_0	f_1	f_2	f_3	f_4	f_5	f_6	f_7	Description
flat	1	1	1	1	1	1	1	1	a base line case that reveals the natural biases associated with the social interactions
linear	1	2	3	4	5	6	7	8	a function where different states are associated with different fitness, and where individuals that are neighbours *in state space* have similar fitness
twopeaks	8	4	4	4	4	4	4	8	a function with two symmetric peaks at opposite extremes of the state space embedded in a flat landscape that is useful to study the stability of consensus domains
nondeceptive	8	2	2	2	2	2	2	7	a function almost identical to *twopeak*, but where one peak is actually a local optimum, which provides a comparative basis for other functions
deceptive (binary)	8	2	2	4	2	4	4	7	a function with the same optima as *nondeceptive*, but where the neighbours of the optima in state space have different fitness, which allows us to show why the consensus reached by the population may be counterintuitive
deceptive (float)	8	2	2	2	4	4	4	7	see binary deceptive fitness function
trap	8	1	2	3	4	5	6	7	a variant of *linear* where the global optimum is in the area of the search space with the lowest fitness and all gradient information points towards the local optimum, thereby making the problem very deceptive

the search for optima in the background, allowing us to focus on the dynamics of information flows and on consensus/group formation and stability.

We considered *lbest*-type (ring) topologies for the social network. However, we tested values for the neighbourhood radius r in the set $\{1, 2, 3, 4, 5, 10, 20, 40\}$, thereby going from the most extreme form of local interaction to a *gbest*-type of topology (with $P = 80$, an *lbest* topology with $r = 40$ is a fully connected network).

We tested applying the external input at probabilities of $\gamma \in \{0, 0.01, 0.03, 0.05, 0.1, 0.2, 0.5, 1\}$ (per individual). When $\gamma > 0$, in both binary and real-valued state representations, we tested two different values (2 and 7) for x_{ext}. We tested both the deterministic social-selection strategy (typical of PSOs) and the probabilistic one based on tournaments. In the binary representation, social interaction occur by randomly exchanging bits with a probability of $\beta = 0.5$ (per bit).

For all settings we tested the six fitness functions shown in Table 1.

Combined together the experimental settings described above produce over 3,000 conditions. In each condition, we gathered statistics over 200 independent runs of the model, for a total of over 600,000 runs. Due to space limitations, we are able to report only a tiny subset of our experimental results here. However, a fuller set of results is provided online in [6].

5 Results

For each of the settings described in the previous section, we did both single runs and multiple runs to assemble statistically reliable state histograms. The single runs are represented by 3–D plots of the state of each individual in the population in each generation. (For space limitations here we do not show any such runs, but in [6] we plots 144 of them.) The state histograms represent the average behaviour of the system over 200 independent runs. These are also represented by 3–D plots, but this time they represent how the proportion of individuals in each particular state changed over time (so the topological organisation of the individuals is not represented). In these plots the x axis represent the generation number, the y axis the state and the z axis the proportion of individuals in that state at that generation. For binary states only 8 different states are possible and so we collected statistics for all of them. For maximum comparability, for the continuous representation, we divided the state space into 8 bins, centred at 0, 1, 2, 3, 4, 5, 6 and 7. States were associated to the nearest bin. (As a consequence, in the initial generation bins 0 and 7 have only half the number of samples of other bins.)

As shown in Figure 2, with a real-valued representation, deterministic communication and in the absence of exogenous inputs, the model behaves (as expected) like a PSO. The only difference between the small neighbourhood ($r = 1$) case, shown in Figure 2, and other cases is the speed at which a steady state is reached. Note: there are dynamics in the system even on a flat landscape. In particular, the interactions produce an implicit bias towards the average state value (3.5) (Figure 2 top left). The model produces a distribution which is very similar to the bell-shaped sampling distribution obtained in real PSOs when x_s and x_p are kept constant [7], which corroborates the approach. Note that this bias can be masked by the presence of fitness gradients, but cannot be removed.

In addition, we can see that local optima have an effect, with initially a part of the population (except for a fully connected swarm, $r = 40$) forming a consensus towards exploring them (Figure 2 middle right, bottom left and right). This slows down convergence to the global optimum. This effect, however, does not simply depend on local optima, but is also markedly influenced by the fitness of neighbouring individuals. This is a form of *deception*. The deception is only partial since, with real-valued states and deterministic communication, eventually all individuals settle for the highest fitness state. If we change the social-selection strategy to tournaments (cf. Equation 2), however, we see that the deceptive pressure can become predominant and force the population towards a suboptimal state, as shown in Figure 3. Even with bigger neighbourhoods the swarm will be deceived in these conditions. The only difference is that with more global communication the algorithm settles for some neighbour of the local optimum, due to the implicit bias of the interaction mechanism discussed above (see Figure 4).

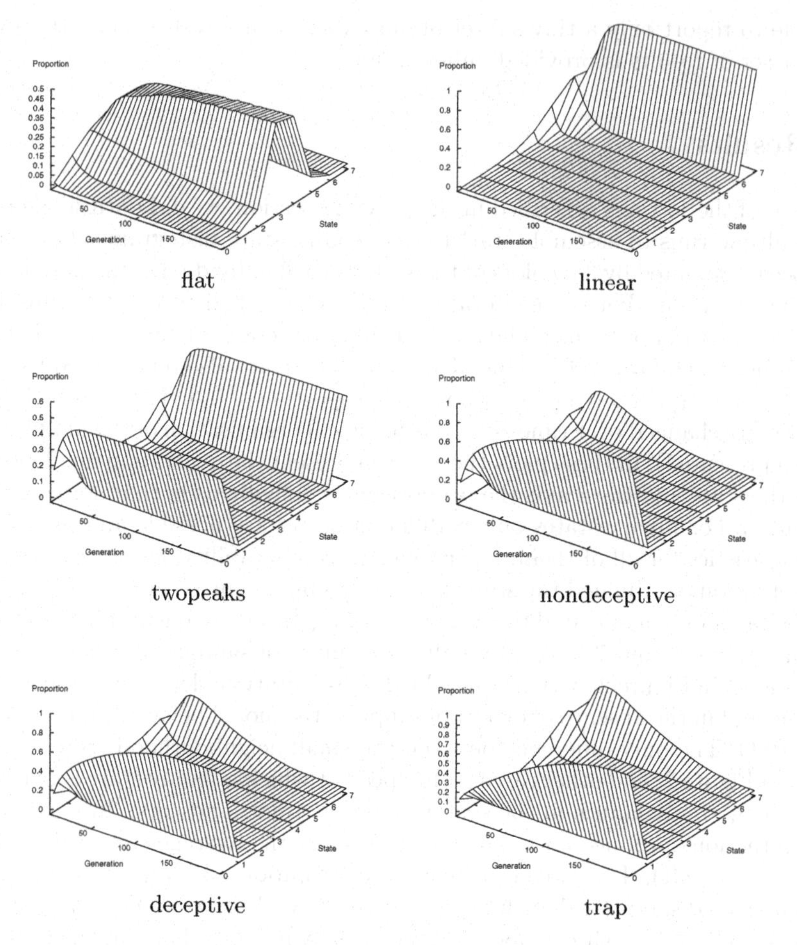

Fig. 2. State evolution histograms for the real-valued model with deterministic communication, $r = 1$ and no external input, for the 6 fitness functions used in our study

Interestingly, a form of deception has been recently reported in real PSOs [8,9], which again corroborates our model.

The injection of an external input $x_{ext} = 2$ with a very low probability ($\gamma = 0.03$) has a clear impact on behaviour. Except for *flat*, for the functions we considered $x = 2$ is not a desirable state. So, this input can be considered as publicity for a suboptimal product (state). We can see the effects of exogenous inputs by considering the flat fitness case. What happens in this case, is that the population initially moves towards the natural fixed-point state (3.5), but later on, with the reiterated introduction of spurious states, eventually converges to state 2. This bias cannot be completely eliminated in non-flat landscapes, and it can cause, for example, the breaking of symmetries, as shown in Figure 5 left, or even the convergence to a low-fitness state, as shown in Figure 5 right.

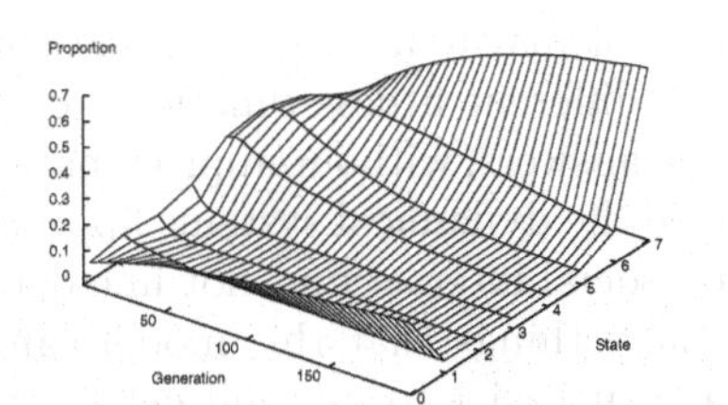

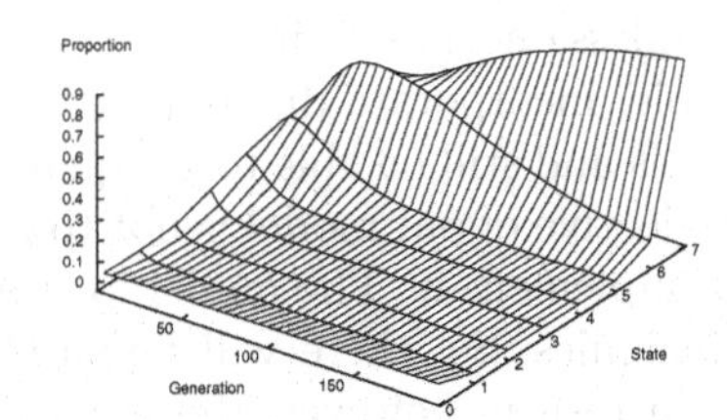

Fig. 3. State evolution histograms for the real-valued model with *probabilistic* communication, $r = 1$ and no external input, for the *deceptive* (left) and *trap* (right) functions

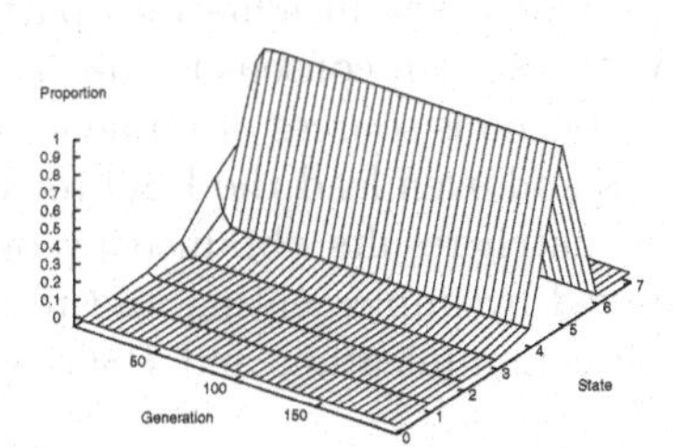

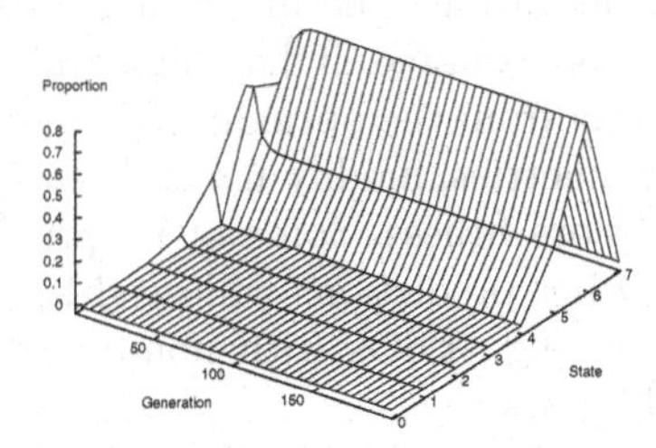

Fig. 4. As Figure 3 but for $r = 40$

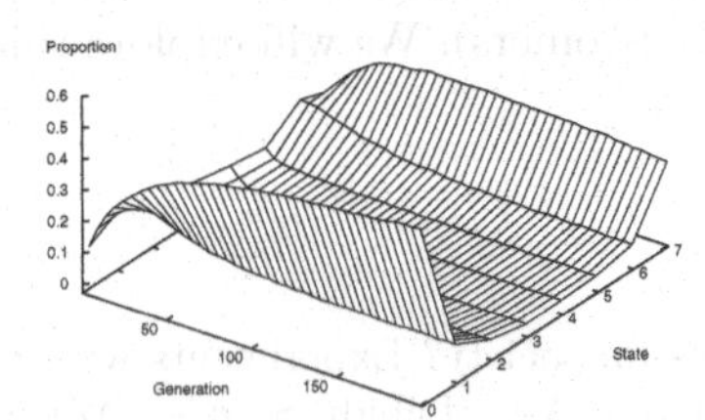

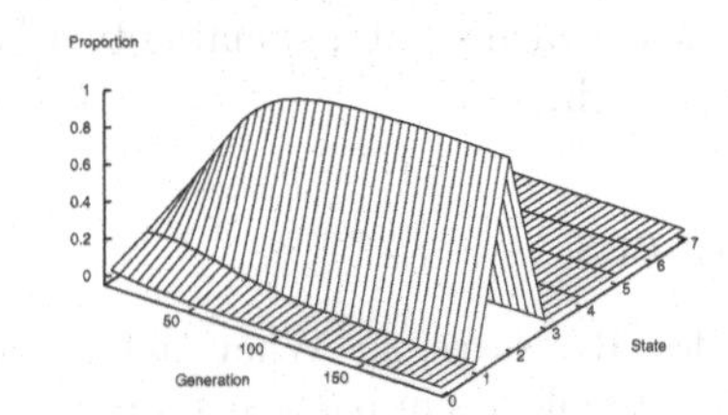

Fig. 5. State evolution histograms for the real-valued model with deterministic (left) and probabilistic (right) communication ($r = 1$, $x_{ext} = 2$, $\gamma = 0.1$, and $f_x = twopeaks$)

6 Discussion

We have proposed a simplified but general model of communication and consensus dynamics in PSOs. The model was specifically designed to look at how the structure of the social network and the nature of the social interactions affect the behaviour of these systems. So, we made an effort to conceal as much as possible of the implementation details, of the dynamics and of the optimum seeking behaviour of PSOs.

Models are useful tools to understand systems, but, except for very simple systems, no model can tell everything there is to know. That is, every model will make it easy to answer certain questions, and hard or impossible to answer different questions. That is why it is important to build models starting from different points of view. For PSOs, nothing can replace dynamical system models of PSOs [10]. However, these models become immensely complex and are difficult to study unless one makes simplifications. It is not easy to imagine how one could, for example, incorporate different topologies, different fitness functions, etc. in such models and still be able to get qualitative answers, without approximations.

The objective of our model is not to replace the dynamical systems approach. It is to complement it. The model focuses on social interactions, not fine details of the dynamics. So, there will be surely many things that this model cannot capture. In return, however, the model makes it easier to ask and get answers to other questions.

With all models, they are only as good as the predictions and understanding they can produce. These need to be checked against empirical data for further corroboration. The model has already provided new insights into particle swarms. For example, it has highlighted how the swarm consensus can be deceived away from the optimal state and how exogenous sources of influence can break symmetries, modify natural search biases and even lead the PSO to completely ignore fitness. Some of its predictions match results obtained by other means. Other predictions will need to be checked by implementing and testing new PSOs, e.g., PSOs with non-deterministic communication and PSOs with external inputs.

The model includes the forms of communication currently implemented in PSOs, but it is significantly more general. As a result, we believe that this model and our results may be applicable to natural and artificial systems other than just particle swarms (e.g., social networks of customers). We will explore this in future research.

References

1. Mendes, R., Neves, J.: What makes a successful society? Experiments with population topologies in particle swarms. In Bazzan, A.L., Labidi, S., eds.: Advances in Artificial Intelligence. XVII Brazilian Symposium on Artificial Intelligence - SBIA'04. Volume 3171 of Lecture Notes in Computer Science, Springer (2004) 346–355
2. Mendes, R., Kennedy, J., Neves, J.: The fully informed particle swarm: Simpler, maybe better. IEEE Transactions of Evolutionary Computation **8**(3) (2004) 204–210
3. Mendes, R., Kennedy, J., Neves, J.: Avoiding the pitfalls of local optima: How topologies can save the day. In: Proceedings of the 12th Conference Intelligent Systems Application to Power Systems (ISAP2003), Lemnos, Greece, IEEE Computer Society (2003)
4. van den Bergh, F.: An Analysis of Particle Swarm Optimizers. PhD thesis, Department of Computer Science, University of Pretoria, Pretoria, South Africa (2001)
5. Baxter, N., Collings, D., Adjali, I.: Agent-based modelling – intelligent customer relationship management. BT Technology Journal **21**(2) (2003) 126–132
6. Poli, R., Langdon, W.B., Marrow, P., Kennedy, J., Clerc, M., Bratton, D., Holden, N.: Communication, leadership, publicity and group formation in particle swarms. Technical Report CSM-453, Department of Computer Science, University of Essex (2006)
7. Kennedy, J.: Bare bones particle swarms. In: Proceedings of the IEEE Swarm Intelligence Symposium (SIS) 2003, Indianapolis, Indiana (2003) 80–87

8. Langdon, W.B., Poli, R., Holland, O., Krink, T.: Understanding particle swarm optimisation by evolving problem landscapes. In Gambardella, L.M., Arabshahi, P., Martinoli, A., eds.: Proceedings SIS 2005 IEEE Swarm Intelligence, Pasadena, California, USA, IEEE (2005) 30–37
9. Langdon, W.B., Poli, R.: Evolving problems to learn about particle swarm and other optimisers. In Corne, D., Michalewicz, Z., Dorigo, M., Eiben, G., Fogel, D., Fonseca, C., Greenwood, G., Chen, T.K., Raidl, G., Zalzala, A., Lucas, S., Paechter, B., Willies, J., Guervos, J.J.M., Eberbach, E., McKay, B., Channon, A., Tiwari, A., Volkert, L.G., Ashlock, D., Schoenauer, M., eds.: Proceedings of the 2005 IEEE Congress on Evolutionary Computation. Volume 1., Edinburgh, UK, IEEE Press (2005) 81–88
10. Clerc, M., Kennedy, J.: The particle swarm-explosion, stability, and convergence in a multidimensional complex space. IEEE Transactions on Evolutionary Computation **6**(1) (2002) 58–73

Covering a Continuous Domain by Distributed, Limited Robots

Eliyahu Osherovich, Alfred M. Bruckstein, and Vladimir Yanovski

Technion - Israel Institute of Technology, Computer Science Department, Haifa, Israel
{oeli, freddy, volodyan}@cs.technion.ac.il

Abstract. We present an algorithm for covering continuous domains by primitive robots whose only ability is to mark visited places with pheromone and to sense the level of the pheromone in their neighborhood. These pheromone marks can be sensed by all robots and thus provide a way for indirect communication between the robots. Apart from this, the robots have no means to communicate. Additionally they are memoryless, have no global information such as the domain map, own position, coverage percentage, etc. Despite the robots' simplicity, we show that they are able to cover efficiently any connected domains, including non-planar ones.

1 Introduction

We say that a domain is *covered* by a robot if each and every point of the domain was swept by the robot's effector. In fact, every time we want to build an automatic machine suitable for applications such as floor cleaning, snow removal, lawn mowing, painting, mine-field de-mining, unknown terrain exploration and so forth, we face the problem of complete covering of corresponding domains by our machine.

A particular solution of the covering problems depends, of course, on the capabilities of our robots and various environmental constraints. Hence a vast number of algorithms can be, and actually have been, developed to accommodate the numerous constraints of the covering problem.

In this paper we adopt the model used in [1], which assumes that our robots are anonymous, i.e., any two robots are the same, memoryless, i.e., they have no ability to "remember" anything from the past and have no means of direct communication. This model was originally inspired by ants and other insects that use chemicals called *pheromones* that are left on the ground and used for some kind of indirect communication and coordination tasks. Ant colonies, despite primitivism of single ants, demonstrate surprisingly good results in global problem solving and pattern formation [2,3,4,5,6]. Consequently, some ideas borrowed from these insects are becoming increasingly popular in ant-robotics and distributed systems [5,6,7,8,9,10]. Such robots are usually capable of performing quite complex distributed tasks while providing the benefits of being small, cheap, easy to produce and easy to maintain.

M. Dorigo et al. (Eds.): ANTS 2006, LNCS 4150, pp. 144–155, 2006.

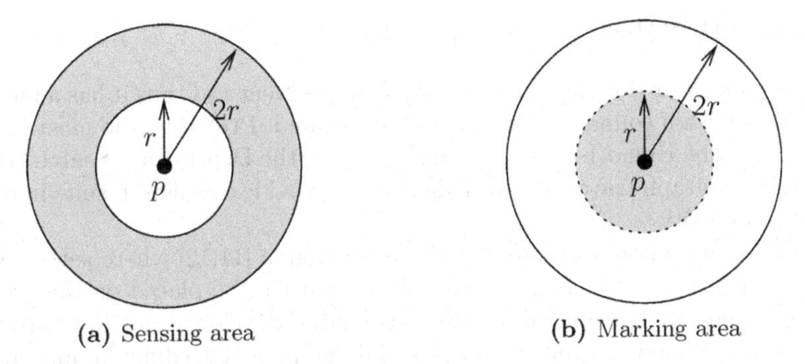

(a) Sensing area (b) Marking area

Fig. 1. Robot's sensing and marking areas

2 Agent Model

Mathematical formulation of the problem is as follows. The domain will be denoted by Ω. At the moment we consider only flat two-dimensional domains; further extensions will be given in Section 6.4. Given any two points $a, b \in \Omega$ we denote the *distance* between a and b as $\|a - b\|$. Again, we assume, initially, that the distance is the common Euclidean distance in two-dimensional space; extensions to other distance measures will be given in Section 6.4. The robot is able to sense the pheromone level at its current position p and in a closed ring of radii r and $2r$ around p denoted by $R(r, 2r, p)$. Additionally, our robot is able to set an arbitrary pheromone level in an open disk of radius r around its current location p denoted by $D(r, p)$, We assume that our time steps are discrete and denote by $\sigma(a, t)$ the pheromone level of point $a \in \Omega$ at time instance t.

3 The Mark-Ant-Walk (MAW) Algorithm

Initially, no point is marked with the pheromone and thus all σ values are assumed to be equal to zero: $\sigma(a, 0) = 0$; $\forall a \in \Omega$. A starting point is chosen (randomly) for the robot and then the MAW step rule is applied repeatedly. There is no explicit stopping condition for this algorithm; nevertheless, one can use the upper bound, provided later in this paper, on the cover time in order to stop robots after a sufficient time period that guarantees complete covering.

Table 1. MAW step rule

Mark-Ant-Walk step rule (current time is t and agent location is p)	
(A)	$x :=$ a point from $R(r, 2r, p)$ with *minimal* value of $\sigma(x, t)$ */* In case of a tie - make an arbitrary decision */*
(B)	If $\sigma(p) \leq \sigma(x)$: $\forall u \in D(r, p)$ $\sigma(u) = \sigma(x) + 1$ */* we mark open disk of radius r around current location */*
(C)	$t := t + 1$
(D)	move to x

4 Related Work

Covering of discrete domains (graphs) is an old problem and thus it has a number of solutions with a sound mathematical background. Probably, the most known examples are the Breadth-First Search (BFS) and the Depth-First Search (DFS) algorithms for graph traversal. Both algorithms provide excellent results in terms of time complexity.

A step toward an odor-oriented model was taken in [11,12] where *pebbles* were used to assist the search. Pebbles are tokens that can be placed on the ground and later removed. The idea of pebbles was further developed in [13] where they were used for unknown graph exploration and mapping. Two different algorithms that fit our paradigm entirely, i.e. fully distributed autonomous agents that mark the ground with pheromones, were suggested for efficient and robust graph covering. One, called the Edge-Ant-Walk, marks the graph edges [14]. Another one, called the Vertex-Ant-Walk, leaves marks on graph vertices instead [1,15]. Both algorithms provided significant improvement over DFS in robustness terms along with quite efficient cover time.

Random walks are defined for both discrete and continuous domains and provide unrivaled robustness and scalability; however, they cannot guarantee complete coverage, providing only expected time. We would like to concentrate on solutions that can guarantee complete coverage after a limited time period.

One possible approach is to introduce an *artificial potential field* in order to accomplish the robot motion planning task (e.g. [16,17]). This approach can easily be adopted by our robots where the potential is represented by the odor level. However, it assumes that the potential field is constructed prior to the start of robot motion and thus requires a global knowledge of the domain boundaries and obstacles, which is unavailable in our model. Some authors used trails that mark the path travelled by the agent so far and performed some kind of peeling/milling. This approach often fails with non-convex domains and thus the whole domain may be approximated as a union of convex non-overlapping cells [18,19,20], ,however, this approach, in fact, takes us back to a graph whose vertices are associated with the cells and edges between vertices that are defined according to the corresponding inter-cell connectivity. Another representative of trail-based algorithms is the Mark-And-Cover (MAC) algorithm [21], which is actually an adaptation of the DFS to continuous domains. This algorithm provides efficient and effective coverage with excellent provable cover time. Additionally, the agent model used in the paper fit our paradigm entirely. Nevertheless, the problem of the MAC algorithm, and probably all trail-based algorithms, is their sensitivity to noise and agents failure. Moreover, trails of one agent may hamper performance of another agent. Another shortcoming of these algorithms is seen in the situation when the domain is required to be covered repeatedly, e.g., in surveillance tasks or in the scenario described in [22] where autonomous agents are used to de-mine minefields using imperfect sensors, i.e. the probability of a mine detection is less than 1. Our algorithm guarantees that the whole domain is covered repeatedly time after time. Furthermore, the time between two successive visits at any point is bounded (see Section 6.1).

5 MAW - Formal Proof of Correctness and Upper Time Bound

Let us show that a single robot governed by the MAW rule covers any connected bounded domain in a finite number of steps. The outline of the proof is as follows.

First, we prove that at any time instance, any two points that are close enough, i.e., their distance from each other is less than or equal to r, must have pheromone levels that differ by one at most. We call this *the proximity principle*. It has also been used in several other research studies, e.g., [1,15,14].

Second, we look at the diameter d of the domain that is defined as the length of the longest geodesic line embedded in the domain, i.e., $d = \sup_{a,b \in \Omega} \|a - b\|$. Assuming that d is finite, we easily conclude with the aid of the proximity principle that at any time t for any two points $a, b \in \Omega$, the difference between the pheromone levels of these two points is limited by $\lceil d/r \rceil$. This, in turn, means that once the value of $\lceil d/r \rceil + 1$ is reached at any point, no unmarked point remains and thus the whole domain has been covered. Finally, we show that we eventually reach value of $\lceil d/r \rceil + 1$. A formal proof is given below.

Lemma 1
The difference between marker values of close points is bounded.

$$\forall t; \forall a, b \in \Omega : \textit{if } \|a - b\| \le r \textit{ then } |\sigma(a, t) - \sigma(b, t)| \le 1$$

PROOF: We shall prove the lemma by mathematical induction on the step number. The lemma is clearly true at $t = 0$. Assuming it is also true at time $t = n$, we shall show it remains true at time $t = n + 1$. Let us look at two points $a, b \in \Omega$, such that $\|a - b\| \le r$. In the trivial case neither a nor b changes its marker value at the $(n + 1)$th step; therefore, the lemma holds according to the induction hypothesis. If both a and b change their values, then $\sigma(a, t + 1) = \sigma(b, t + 1)$ since the algorithm assigns the same values to all the points it changes. Hence the only interesting case is when only one point (say a) changes its marker value Assuming the current agent's location is p_t we conclude that $a \in D(r, p_t)$, otherwise it could not change its marker value. And therefore, $\|a - p_t\| < r$. b, however, does not change its marker value and thus $\|b - p_t\| \ge r$. Combining these constraints we get $r \le \|b - p_t\| \le 2r$ or, equivalently, $b \in R(r, 2r, p_t)$. Now let us recall how the new marker value of a is determined. First, we look for the minimal marker value among all points in $R(r, 2r, p_t)$. Assume that this value is attained at some point $x \in R(r, 2r, p_t)$. The new marker value of a is then set if and only if $\sigma(p_t, t) \le \sigma(x, t)$:

$$\sigma(a, t + 1) = \sigma(x, t) + 1. \tag{1}$$

Since both points x and b belong to $R(r, 2r, p_t)$, we have

$$\sigma(b, t) \ge \sigma(x, t), \tag{2}$$

because of the way the point x was chosen. Now, on the one hand, we have:

$$\begin{cases} |\sigma(a, t) - \sigma(b, t)| \le 1 \\ \sigma(b, t) \ge \sigma(x, t) \end{cases} \Rightarrow \sigma(a, t) \ge \sigma(x, t) - 1; \tag{3}$$

and on the other hand:

$$\begin{cases} |\sigma(a,t) - \sigma(p_t,t)| \leq 1 \\ \sigma(p_t,t) \geq \sigma(x,t) \end{cases} \Rightarrow \sigma(a,t) \leq \sigma(x,t) + 1. \tag{4}$$

Combining inequalities (3) and (4), we get

$$|\sigma(a,t) - \sigma(x,t)| \leq 1. \tag{5}$$

Using the system of inequalities (3), we conclude that

$$0 \leq \sigma(b,t) - \sigma(x,t) \leq 2. \tag{6}$$

Combining the above inequality with the fact that $\sigma(a,t+1) = \sigma(x,t)+1$ and $\sigma(b,t+1) = \sigma(b,t)$, we get the desired result: $|\sigma(a,t+1) - \sigma(b,t+1)| \leq 1$. Thus the lemma is proven. ∎

Lemma 2
The difference between marker values of any two points is bounded at all times.

$$\forall t; \forall a, b \in \Omega : |\sigma(a,t) - \sigma(b,t)| \leq \left\lceil \frac{d}{r} \right\rceil$$

where d - diameter of Ω.

PROOF: Follows immediately from Lemma 1. ∎

Our next step will be to show that the maximal marker value tends to ∞ as t goes to ∞. First, we prove that marker values can only grow and never decrease.

Lemma 3
Marker values of any point form a non-decreasing series; that is

$$\forall t; \forall u \in \Omega : \sigma(u,t+1) \geq \sigma(u,t).$$

PROOF: Let us assume the contrary, i.e., there exists a point $u \in \Omega$ and time instance t such that the pheromone level of u decreases during the t-th step: $\sigma(u,t+1) < \sigma(u,t)$. Let us now look at point p_t – the location of the agent at time t. Obviously $u \in D(r, p_t)$ (otherwise it could not change its value), hence $\|u - p_t\| < r$. Assume that the minimal marker value among all points in $R(r, 2r, p_t)$ was attained at some point x. We know also that $\sigma(p_t,t) \leq \sigma(x,t)$; otherwise, the robot does not change the pheromone values. Thus we have

$$\begin{cases} \sigma(x,t) + 1 = \sigma(u,t+1) < \sigma(u,t) \\ \sigma(p_t,t) \leq \sigma(x,t) \\ \|u - p_t\| < r \end{cases} \tag{7}$$

This implies

$$\begin{cases} |\sigma(u,t) - \sigma(p_t,t)| > 1 \\ \|u - p_t\| < r \end{cases} \tag{8}$$

which contradicts Lemma 1. ∎

At this point we are ready to prove the main result of this work.

Theorem 1
The domain Ω will be covered within a finite number of steps.

PROOF: Imagine that the domain Ω is tessellated into n cells so that every such cell can be inscribed into a circle of diameter less than r. Let us examine the following sum:

$$S_t = \sum_{i=1}^{n} m_t^i - \sigma(p_t, t), \tag{9}$$

where m_t^i is the minimal marker value over the ith cell at time t and $\sigma(p_t, t)$ is the marker value at the agent's location p_t at time instance t. With the aid of Lemma 3 one can easily verify that

$$S_{t+1} > S_t. \tag{10}$$

Given that $S_0 = 0$, we easily conclude that

$$S_t \geq t \Rightarrow \sum_{i=1}^{n} m_t^i \geq t \quad \forall t, \tag{11}$$

which leads us to the conclusion that after $n\lceil \frac{d}{r} \rceil + 1$ steps, at least one of the m_{nd+1}^i values will be greater than $\lceil \frac{d}{r} \rceil$ and thus the whole domain will be covered. ∎

In order to find an approximation to n , we can tile the domain with regular hexagons of side length $r/2$. In order to guarantee full coverage by the hexagons we look at the "augmented" domain $\bar{\Omega}$, which results from Ω that has undergone morphological dilation with a disk of radius r. Using a development similar to the one shown in [21], we get the following bound on the area of $\bar{\Omega}$

$$A_{\bar{\Omega}} \leq A_\Omega + rP_\Omega + \pi r^2, \tag{12}$$

where A_Ω and P_Ω are the area and the perimeter of Ω, respectively. Thus we have

$$n \leq \frac{A_\Omega + rP_\Omega + \pi r^2}{\frac{3\sqrt{3}}{8} r^2}, \tag{13}$$

where $\frac{3\sqrt{3}}{8} r^2$ represents the area of a hexagon of side length $r/2$.

6 Extensions

6.1 Repetitive Coverage

In some scenarios we might be interested in repetitive coverage of the domain, e.g., the aforementioned scenario of minefield de-mining with imperfect sensors [22] or tasks such as surveillance and patrolling. In all cases we would like to bound the time between two successive visits.

Lemma 4
For any two time instances t_1 and t_2, if only $t_2 > t_1$ then the following inequality must hold:

$$S_{t_2} - S_{t_1} \geq t_2 - t_1.$$

PROOF: The proof is very simple. We can always write $t_2 = t_1 + n$ for some natural n and prove the lemma by mathematical induction. For $n = 1$ the lemma holds due to Equation (10). Assuming that the lemma holds for some n, we can easily conclude that the lemma holds for $n + 1$ as well. ∎

Theorem 2
For any point $a \in \Omega$, the time period between two successive visits of the robot is bounded by $2n\left(\lceil \frac{d}{r} \rceil + 1\right)$.

PROOF: If we show that after a sufficient time period the pheromone level changes at all locations in the domain Ω, we can obviously be sure that all points were re-visited by the robot during this time period. Let us look at time instance t_s when the robot covers our point of interest a. We denote by $\sigma_{max}(t_s)$ the maximal pheromone level over Ω at that time. If we show that at some time instance t_e the minimal pheromone level denoted by $\sigma_{min}(t_e)$ becomes greater than the maximal value that was at time t_s: $\sigma_{min}(t_e) > \sigma_{max}(t_s)$, then we can easily conclude that during the time period $t_e - t_s$ the pheromone level changed at all points and thus all points (including a) were re-covered by the robot. Let us examine S_{t_s} and S_{t_e} as defined in the Equation (9). On the one hand:

$$S_{t_s} = \sum_i^n m_{t_s}^i - \sigma(p_{t_i}, t_i) \geq \sum_i^n m_{t_s}^i \geq \sum_i^n \sigma_{min}(t_s) = n\,\sigma_{min}(t_s) \tag{14}$$

According to Lemma 2

$$\sigma_{min}(t_s) \geq \sigma_{max}(t_s) - \left\lceil \frac{d}{r} \right\rceil. \tag{15}$$

Combining Equations (14) and (15) we get

$$S_{t_s} \geq n\left(\sigma_{max}(t_s) - \left\lceil \frac{d}{r} \right\rceil\right). \tag{16}$$

On the other hand we want to know the time instance t_e that guarantees that $\sigma_{min}(t_e) \geq \sigma_{max}(t_s) + 1$. Instead of estimating t_e directly from $\sigma_{min}(t_e)$, we shall look for t_e that guarantees the existence of σ value greater than or equal to $\sigma_{max}(t_s)+1+\lceil \frac{d}{r}+1 \rceil$, which guarantees by Lemma 2 that $\sigma_{min}(t_e) \geq \sigma_{max}(t_s)+1$. Now, in the same way as the proof of Theorem 1, we can say that once $S_{t_e} \geq n(\sigma_{max}(t_s) + 1 + \lceil \frac{d}{r} \rceil + 1)$, we have $\sigma_{min}(t_e) \geq \sigma_{max}(t_s) + 1$. Thus we have

$$S_{t_e} - S_{t_s} \leq n\left(\sigma_{max}(t_s) + 1 + \left\lceil \frac{d}{r} \right\rceil + 1\right) - \left(\sigma_{max}(t_s) - \left\lceil \frac{d}{r} \right\rceil\right) = 2n\left(\left\lceil \frac{d}{r} \right\rceil + 1\right). \tag{17}$$

According to Lemma 4 we have

$$t_e - t_s \leq S_e - S_s \leq 2n \left(\left\lceil \frac{d}{r} \right\rceil + 1 \right), \tag{18}$$

which completes the proof. ∎

6.2 Noise Immunity

Until now we always assumed that there is no noise in the input, i.e., the robot starts with a domain that does not contain any pheromone marks. Unfortunately, in the real life such a clear environment is not always available hence, we shall consider situation when the initial pheromone level is not zero. Unlike trail-based algorithms that cannot cope with noise our algorithm, can easily overcome this problem as demonstrated by the experiments in Section 7.3.

6.3 Multiple Robots

As a natural extension we would like to analyze how the MAW algorithm can be applied to multi-robot environments. First of all, we must address problems such as collisions both between the robots themselves (if we deal with physical robots and not programs) and between different pheromone levels when two (or more) robots try to mark the same point in the domain.

At the moment we assume that the clock phases of all robots are slightly different so that no two robots are active at the same time. Thus each robot sees other robots as regular stationary obstacles and acts accordingly. This approach also resolves the problem of different pheromone levels that might be assigned to the same point by different robots, since only one robot is active at given time.

Let us find the upper bound for complete coverage provided we have k robots. Using the same notation as in Equation (9) we have:

$$S_t = \sum_{i=1}^{n} m_t^i - \sum_{j=1}^{k} \sigma(p_t^j, t), \tag{19}$$

where p_t^j denotes the location of the j-th robot at time t. Using exactly the same reasoning as before, we again obtain:

$$S_t \geq t, \tag{20}$$

which leads us to the same upper bound we got for a single robot. Hence adding more robots does not necessarily guarantees better coverage time. However our simulations (see Section 7) demonstrate that there is a substantial improvement when we use more robots.

6.4 Using Other Metrics

Until now we always used the usual notion of the distance, nevertheless, it is easy to verify that all the proofs remain valid if we change the Euclidean (L_2) distance to another one. For example, we used L_∞ in our simulations. Since corresponding effector shape is a square in this case.

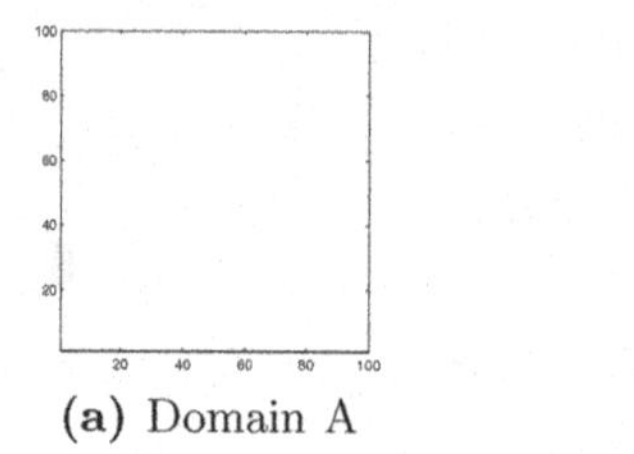

(a) Domain A

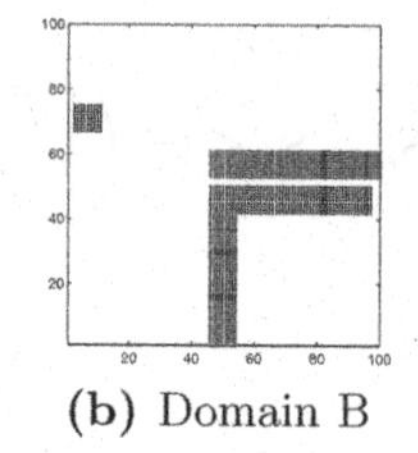

(b) Domain B

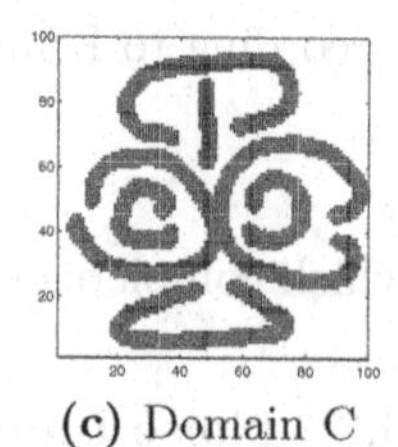

(c) Domain C

Fig. 2. Simulation domains

7 Simulations and Experiments

7.1 General Notes

Experiments were conducted on the domains shown in Figure 2. All domains are of size 100×100 pixels and marking radius in all experiments was set to 3, i.e., each step robot marks a square of 5×5 pixels. Figure 3 demonstrates some stages of covering Domain B by ten robots.

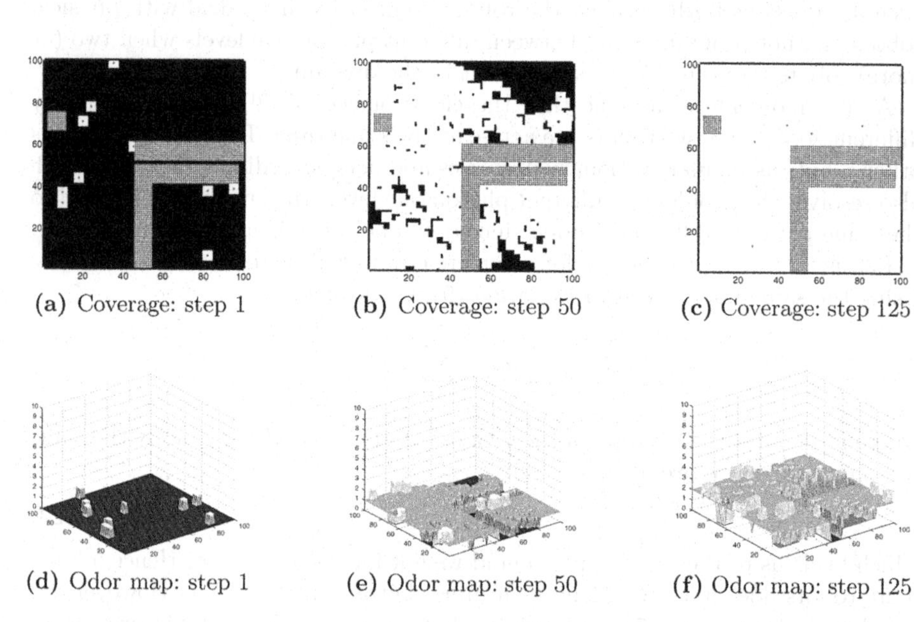

(a) Coverage: step 1 (b) Coverage: step 50 (c) Coverage: step 125

(d) Odor map: step 1 (e) Odor map: step 50 (f) Odor map: step 125

Fig. 3. MAW progress on Domain B

7.2 Comparing MAW to Other Algorithms

In this experiment we studied performance of three different algorithm: MAW, MAC [21], and Random Walk. All algorithms used the same square effector

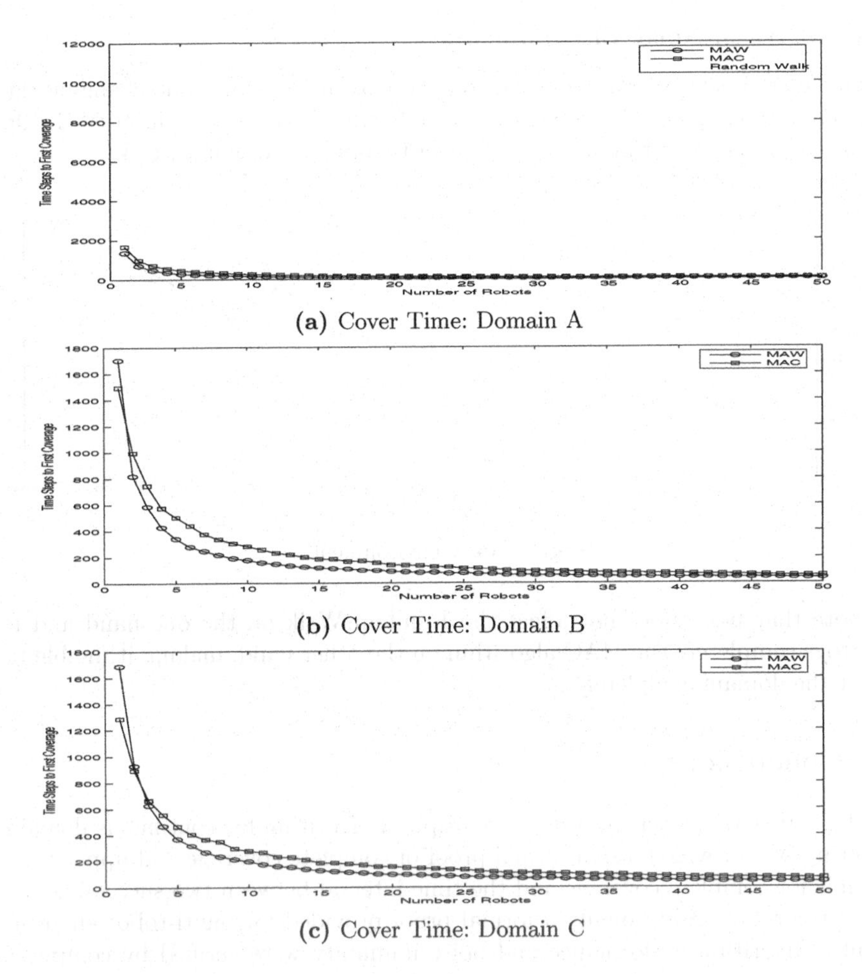

(a) Cover Time: Domain A

(b) Cover Time: Domain B

(c) Cover Time: Domain C

Fig. 4. Cover Time

of size 5×5 pixels; additionally, the steps of the Random Walk algorithm were restricted to be in interval $[r, 2r]$ just like the steps in the MAW algorithm.

In all experiments the robots were modeled as points and multiple robots were allowed to occupy the same location. We always measured the number of time steps until the robots covered the domain for the first time, averaged over 100 runs.

As we can see the MAW algorithm is a clear winner when we use three or more robots. For fewer robots the MAC algorithm performs better on complex domains. Note that the MAW algorithm in general performs better than the theoretical upper bound we got in Section 5. Cover time of the Random Walk was omitted from Figures 4b and 4c because the values were so big that the difference between the MAC and the MAW algorithms became invisible on this scale.

7.3 MAW in Noisy Environments

In this experiment we ran one robot on the Domain A, each time changing the amount of noisy pixels. Noise values are uniformly distributed in interval [1, 10]. Figure 5 shows cover time as a function of the amount of noisy pixels.

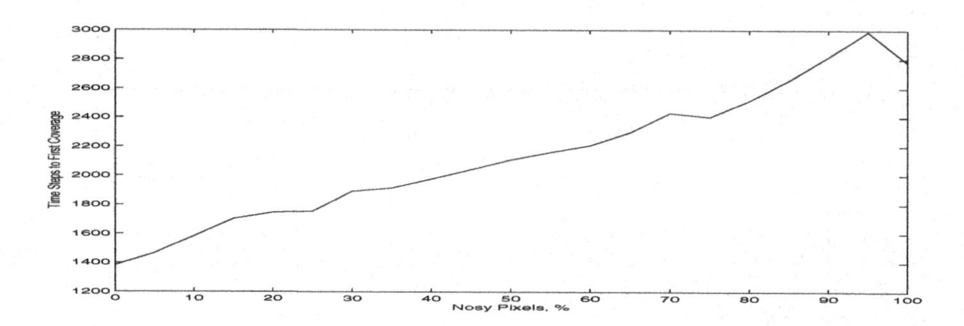

Fig. 5. Noisy environment

Note that noise does not affect the Random Walk on the one hand and it destroys completely the MAC algorithm on the other hand, making it unable to cover the domain completely.

8 Conclusions

In this paper we presented a new ant-inspired algorithm for continuous domain covering. We provided also a formal proof of complete coverage and upper time bounds for complete coverage and the time interval between two successive visits of the robot. Additionally a formal proof provided for multi-robot environments. Algorithm performance and noise immunity were verified by computer simulations.

References

1. Yanovski, V., Wagner, I.A., Bruckstein, A.M.: Vertex-ant-walk - A robust method for efficient exploration of faulty graphs. Annals of Mathematics and Artificial Intelligence **31**(1-4) (2001) 99–112
2. Bruckstein, A.M.: Why the ant trails look so straight and nice. The Mathematical Intelligencer **15**(2) (1993) 58–62
3. Hölldobler, B., Wilson, E.O.: The Ants! Harvard University Press (1990)
4. Schöne, H.: Spatial orientation : the spatial control of behavior in animals and man. Princeton University Press, Princeton, N.J. (1984)
5. Dorigo, M., Maniezzo, V., Colorni, A.: Ant system: Optimization by a colony of cooperating agents. IEEE Trans. on Systems, Man, and Cybernetics–Part B **26**(1) (1996) 29–41
6. Dorigo, M., Di Caro, G., Gambardella, L.M.: Ant algorithms for discrete optimization. Artificial Life **5**(2) (1999) 137–172

7. Wagner, I.A., Bruckstein, A.M.: From ants to a(ge)nts: A special issue on ant-robotics. Annals of Mathematics and Artificial Intelligence **31**(1-4) (2001) 1–5
8. Bonabeau, E., Théraulaz, G.: Swarm smarts. Scientific American **282**(3) (2000) 72–79
9. Russell, R.A.: Ant trails - an example for robots to follow? In: ICRA. (1999) 2698
10. Koenig, S., Liu, Y.: Terrain coverage with ant robots: a simulation study. In: AGENTS '01: Proceedings of the fifth international conference on Autonomous agents, New York, NY, USA, ACM Press (2001) 600–607
11. Blum, M., Sakoda, W.: On the capability of finite automata in 2 and 3 dimensional space. In: Ann. Symp. on Foundations in Computer Science. (1977) 147–161
12. Blum, M., Kozen, D.: On the power of the compass. In: Proc. 19th Ann. Symp. on Foundations in Computer Science. (1978) 132–142
13. Bender, M.A., Fernández, A., Ron, D., Sahai, A., Vadhan, S.: The power of a pebble: exploring and mapping directed graphs. In: STOC '98: Proceedings of the Thirtieth Annual ACM Symposium on Theory of Computing, New York, NY, USA, ACM Press (1998) 269–278
14. Wagner, I.A., Lindenbaum, M., Bruckstein, A.M.: Smell as a computational resource — A lesson we can learn from the ant. In: Proceedings of the 4th Israel Symposium on Theory of Computing and Systems, ISTCS'96 (Jerusalem, Israel, June 10-12, 1996), Los Alamitos-Washington-Brussels-Tokyo, IEEE Computer Society Press (1996) 219–230
15. Wagner, I.A., Lindenbaum, M., Bruckstein, A.M.: Efficiently searching a graph by a smell-oriented vertex process. Annals of Mathematics and Artificial Intelligence **24**(1-4) (1998) 211–223
16. Khatib, O.: Real-time obstacle avoidance for manipulators and mobile robots. The International Journal of Robotics Research **5**(1) (1986) 90–98
17. Zelinsky, A., Byrne, J.C., Jarvis, R.A.: Planning paths of complete coverage of an unstructured environment by a mobile robot. In: International Conference on Advanced Robotics (ICAR). (1993)
18. Choset, H., Pignon, P.: Coverage path planning: The boustrophedon decomposition. In: International Conference on Field and Service Robotics. (1997)
19. Butler, Z.J.: Distributed coverage of rectilinear environments. PhD thesis, Carnegie Mellon University (2000)
20. Acar, E.U., Choset, H., Zhang, Y., Schervish, M.J.: Path planning for robotic demining: Robust sensor-based coverage of unstructured environments and probabilistic methods. I. J. Robotic Res **22**(7-8) (2003) 441–466
21. Wagner, Lindenbaum, Bruckstein: MAC vs. PC: Determinism and randomness as complementary approaches to robotic exploration of continuous unknown domains. ROBRES: The International Journal of Robotics Research **19** (2000)
22. Gage, D.W.: Randomized search strategies with imperfect sensors. In Chun, W.H., Wolfe, W.J., eds.: Proc. SPIE. Volume 2058. (1994) 270–279

Incremental Local Search in Ant Colony Optimization: Why It Fails for the Quadratic Assignment Problem

Prasanna Balaprakash, Mauro Birattari, Thomas Stützle, and Marco Dorigo

IRIDIA, CoDE, Université Libre de Bruxelles, Brussels, Belgium
{pbalapra, mbiro, stuetzle, mdorigo}@ulb.ac.be

Abstract. Ant colony optimization algorithms are currently among the best performing algorithms for the quadratic assignment problem. These algorithms contain two main search procedures: solution construction by artificial ants and local search to improve the solutions constructed by the ants. Incremental local search is an approach that consists in re-optimizing partial solutions by a local search algorithm at regular intervals while constructing a complete solution. In this paper, we investigate the impact of adopting incremental local search in ant colony optimization to solve the quadratic assignment problem. Notwithstanding the promising results of incremental local search reported in the literature in a different context, the computational results of our new ACO algorithm are rather negative. We provide an empirical analysis that explains this failure.

1 Introduction

Ant colony optimization (ACO) is a recent metaheuristic technique that is inspired by the pheromone trail laying and following behavior of some ant species [1]. In ACO algorithms, artificial ants are stochastic solution construction procedures that generate solutions using artificial pheromones and heuristic information; the ants' solutions are then used to modify the artificial pheromone trails. This mechanism shifts the stochastic solution construction procedure towards the construction of solutions similar to the better ones seen previously in the algorithm. The definition of the ACO metaheuristic includes also the possibility of using local search [1]: Once ants complete their solution construction phase, local search algorithms can be used to refine their solutions before using them for the pheromone update. Various experimental researches have shown that the combination of solution construction by ants and local search procedures is a promising approach [1].

There exist a large number of possible choices when using local search in ACO algorithms. We refer the reader to [1,2] for a recent review of these techniques. The primary goal of this paper is to investigate the opportunity of adopting incremental local search in ACO, that is, to improve via a local search algorithm the ants' partial solutions at regular intervals during the solution construction

M. Dorigo et al. (Eds.): ANTS 2006, LNCS 4150, pp. 156–166, 2006.

process. Previous works on incremental local search in non-ACO algorithms have reported promising results: Russel [3] introduced a method for re-optimizing partial solutions by means of an interchange procedure after every k steps of the solution construction. Gendreau et al. [4] introduced a generalized insertion heuristic to solve the traveling salesman problem and extended the same approach to a vehicle routing problem [5]. In a nutshell, generalized insertion can be described as an insertion procedure that uses a limited form of incremental local search. Caseau and Laburthe [6] introduced an approach that applies local search after each step of the solution construction process to solve a large constrained vehicle routing problem. This methodology is compared with a customary technique that constructs solutions by greedy insertion and uses local search at the end to improve solutions. The computational results showed that, in this particular context, incremental local search was not only faster, but also produced much better solutions. They conclude that incremental local search is able to perform some improvements during the construction process that full local search may not be able to perform once the solution is complete. In the context of constructive methods, Fleurent and Glover [7] proposed a strategy called proximate optimality principle that consists in re-optimizing partial solutions of a greedy randomized adaptive search procedure to solve the quadratic assignment problem. They suggest that imperfections introduced during the construction step of the procedure can be removed by applying local search on the partial solutions. Since the method we investigate here is very similar to this experimental study, we have also chosen the quadratic assignment problem for our analysis.

The main motivation behind our research is that *a priori* the idea of re-optimizing the partial solutions of the ants during the solution construction looks promising, since the use of local search in ACO algorithms has already proven to often lead to a strong improvement of performance, and incremental local search has been successfully applied in other settings where constructive methods were used. However, the results of our computational experiments are negative and, at least for the quadratic assignment problem (QAP), the inclusion of incremental local search actually worsens the performance. In this paper, we analyse the possible reasons for this effect by studying the convergence behavior of the ACO algorithm. In fact, our analysis also gives hints on conditions under which incremental local search may become useful in ACO algorithms. For instance, since the empirical analysis shows that the incremental local search introduces a strong exploration in the search process of the ACO algorithm studied here, one might try to use it to generate new solutions when the search stagnates.

The paper is organized as follows. Section 2 shows how to use incremental local search in ACO for solving the quadratic assignment problem. In Section 3, we report our computational results, which show that incremental local search in ACO obtains rather poor results. An analysis of why incremental local search in ACO is not effective for the quadratic assignment problem is presented in Section 4. Section 5 concludes the paper.

2 Incremental Local Search in Ant Colony Optimization for the Quadratic Assignment Problem

The QAP can be described in the following way: Consider a set of n facilities that have to be assigned to n locations. A matrix $\mathcal{A} = [a_{rs}]$ gives the distances between locations, where a_{rs} is the distance between locations r and s. A matrix $\mathcal{B} = [b_{tu}]$ characterizes the flows among facilities, where b_{tu} is the flow between facility t and facility u. An assignment can be represented by a permutation π of $\{1,\cdots,n\}$, where $\pi(r)$ is the facility that is assigned to location r. The problem is to find a permutation π^* that minimizes the sum of the products of the flows among facilities by the distance between their locations.

Among the various metaheuristics, ACO has been shown to be particularly successful on the QAP as best exemplified by the high performance reached by $\mathcal{MAX}-\mathcal{MIN}$ ant system ($\mathcal{MM}$AS-QAP) [8,9]. Hence, we have chosen $\mathcal{MM}$AS-QAP as a starting point for our analysis. $\mathcal{MM}$AS-QAP constructs solutions by assigning at each construction step a facility to some location. Pheromone trails τ_{ij} refer to the desirability of assigning facility j to location i and the usual probabilistic choice known from ant system is used; since $\mathcal{MM}$AS-QAP does not use any form of heuristic information, the probability p_{ij} of assigning facility j to location i is directly proportional to τ_{ij} for feasible assignments. The pheromone update is done by lowering the pheromone trails by a constant factor ρ and depositing pheromone on the individual solution components of either the best solution in the current iteration (*iteration-best*), the best solution found so far by the algorithm (*best-so-far*), or the best solution found since the last re-initialization (*restart-best*) of the pheromone trails. We refer the reader to [9] for a more detailed description of $\mathcal{MM}$AS-QAP.

$\mathcal{MM}$AS-QAP uses an iterative improvement algorithm based on the 2-exchange neighborhood, where the set of neighbors of a permutation π comprises all permutations that can be obtained by exchanging the location of two facilities. This iterative improvement algorithm is referred to as 2-opt. When using 2-opt in $\mathcal{MM}$AS-QAP, each ant constructs a feasible solution and improves it by this local search.

It is straightforward to include incremental local search in $\mathcal{MM}$AS-QAP. While in the original $\mathcal{MM}$AS-QAP the local search is applied only to complete solutions, in a version that uses incremental local search, the local search is performed on an ants' partial solution. For convenience, let us define some terminology: We denote for each ant the *number of local searches* applied to its (partial) solutions by i, where i, $1 < i \leq n$, is a user defined parameter. We call $\mathcal{MM}$AS-QAP with incremental local search as $\mathcal{MM}$AS-QAP(i). For example, $\mathcal{MM}$AS-QAP(2) refers to the $\mathcal{MM}$AS-QAP algorithm with an incremental local search in which for each of the m ants the (partial) solution is re-optimized **twice**; for $\mathcal{MM}$AS-QAP(3) three local searches are applied. For the sake of uniformity, we denote the original $\mathcal{MM}$AS-QAP algorithm in which a single local search is performed at the end of the solution construction by $\mathcal{MM}$AS-QAP(1). We use the convention that the local searches are applied after equal sized intervals in the solution construction. Let k be the *number of assignments*;

then, in $\mathcal{MM}$AS-QAP(2) local search is applied after $k = 1 \cdot \lfloor n/2 \rfloor$ assignments and on the complete solutions, while in $\mathcal{MM}$AS-QAP(3), local search is applied after $k = 1 \cdot \lfloor n/3 \rfloor$ and $k = 2 \cdot \lfloor n/3 \rfloor$ assignments are done, and again once the assignment is completed.

In the local search on partial solutions, the cost difference for exchanging two facilities s and r in the partial solution is obtained by the instance data restricted to the occupied locations and used facilities of the current partial solution. The current partial solution is replaced if the local search finds a better neighboring one, and the local search continues until there is no more improvement. From this locally optimized partial solution, the particular ant continues its solution construction process. We expect that the computation time for $\mathcal{MM}$AS-QAP(i) increases with i, since more local searches need to be applied. However, each local search applies to a smaller instance and the final local search on the complete solution may start from an already improved solution, hence, requiring less improvement steps; this counteracts the effect on the computation time incurred by the increased number of local searches. Thus, the rate of increase in the computation time per solution is expected to be less than the rate of increase of i.

3 Experiments

We studied the impact of incremental local search in $\mathcal{MM}$AS-QAP on ten instances from QAPLIB [10] ranging in size from $n = 60$ to $n = 150$. The tested instances fall into one of the following groups: (i) instances with the distance and flow matrix entries generated randomly according to a uniform distribution, (ii) instances whose distance matrix is defined as Manhattan distance between points on a grid, and (iii) randomly generated instances in which the matrix entries are similar to those of real-life QAP instances. We allowed 10 independent trials for each algorithm and the code was run on a dual AMD Opteron^{TM}244 1.75GHz processor, 2 GB RAM and 1 MB L2-Cache. The parameter values for $\mathcal{MM}$AS-QAP are set as proposed in [9] except that the value of ρ is set to 0.1 which results in slightly better performance than the setting $\rho = 0.8$ proposed in the literature. (We run additional experiments that verified that the conclusions drawn in the following do not depend on the parameter value for ρ.) For $\mathcal{MM}$AS-QAP(i), we vary the value of i from 1 to 10 and report the solution quality obtained as the percentage deviation from the best known solutions.

For each instance, we first run $\mathcal{MM}$AS-QAP(1) for 1000 iterations and measured the average time over 10 trails. This average time is then taken as the termination criterion for all algorithms to ensure that we compare the algorithms using a same computation time. Table 1 shows the average solution cost for all values of i as the percentage deviation from the best known solution.

From Table 1, we can observe that the average solution cost obtained by $\mathcal{MM}$AS-QAP(i), for $i \geq 2$ is worse than that of $\mathcal{MM}$AS-QAP(1) for all instances; the only exception is that $\mathcal{MM}$AS-QAP(2) is better than $\mathcal{MM}$AS-QAP(1) on instance `tai150b`. In Table 2, we give the average number of iterations that each of the algorithm variants was able to do in the computation time

Table 1. Experimental results of $\mathcal{MM}$AS-QAP(i) algorithms on several QAP instances; given is, for each instance, the average percentage deviation from the best known solution. All algorithms were stopped after the same computation time. Best results are in bold-face.

	$i=1$	$i=2$	$i=3$	$i=4$	$i=5$	$i=6$	$i=7$	$i=8$	$i=9$	$i=10$
`tai60b`	**0.0004**	0.0086	0.0335	0.0449	0.0610	0.0557	0.0428	0.0578	0.0694	0.0690
`tai80a`	**1.2304**	1.6836	2.0250	2.1809	2.3091	2.2281	2.2934	2.3209	2.3216	2.3115
`tai80b`	**0.0204**	0.0535	0.1294	0.0806	0.0932	0.1052	0.1110	0.1107	0.1176	0.1207
`sko81`	**0.0672**	0.1433	0.2556	0.2714	0.2786	0.3114	0.3215	0.3125	0.3415	0.3331
`sko90`	**0.1376**	0.2089	0.2382	0.2863	0.3469	0.3477	0.3837	0.3436	0.4042	0.4168
`sko100a`	**0.1222**	0.1888	0.2193	0.2826	0.3342	0.3517	0.3536	0.3530	0.3505	0.4477
`tai100a`	**0.3579**	0.8090	0.9702	1.2808	1.2440	1.1532	1.1214	1.0984	1.0902	1.1808
`tai100b`	**0.0452**	0.1009	0.1289	0.1655	0.2456	0.2432	0.2391	0.2540	0.2632	0.2723
`tho150`	**0.2115**	0.2909	0.3530	0.3527	0.4230	0.4336	0.4677	0.5053	0.5104	0.5344
`tai150b`	0.2757	**0.1935**	0.2858	0.2852	0.4685	0.5319	0.5184	0.6217	0.6654	0.6968

Table 2. Experimental results of $\mathcal{MM}$AS-QAP(i) algorithms on several QAP instances; given is, for each instance, and algorithm pair, the average number of iterations done in a same computation time

	$i=1$	$i=2$	$i=3$	$i=4$	$i=5$	$i=6$	$i=7$	$i=8$	$i=9$	$i=10$
`tai60b`	983.1	837.3	711.2	614.9	544.4	489.3	373.2	352.0	354.9	354.0
`tai80a`	987.7	788.9	450.2	584.3	517.2	355.1	338.8	390.9	338.9	337.9
`tai80b`	991.2	782.2	460.3	569.6	503.6	353.7	333.9	376.0	325.4	325.8
`sko81`	1082.1	696.4	768.8	482.5	444.3	419.4	399.8	361.3	417.7	323.6
`sko90`	1097.7	909.9	769.1	486.7	591.1	531.7	408.6	363.0	414.6	387.6
`sko100a`	1070.6	890.1	522.5	653.5	578.3	411.5	376.8	363.9	333.8	380.7
`tai100a`	909.1	757.0	426.6	561.5	501.2	349.3	318.6	309.5	279.2	327.9
`tai100b`	981.7	760.8	455.4	555.5	488.0	344.0	318.2	302.5	280.4	316.2
`tho150`	998.2	791.2	663.6	416.0	504.7	454.8	329.1	318.9	299.1	330.3
`tai150b`	979.1	793.1	679.6	465.4	439.0	393.4	286.4	272.3	257.7	284.5

that was determined as described above. As we had conjectured in the previous section, with increasing value of i, generally also the number of iterations done by $\mathcal{MM}$AS-QAP(i) decreases.

Taking into account this latter observation on the number of iterations run, we could tentatively attribute the reason for the worse performance of $\mathcal{MM}$AS-QAP(i) to the fact that it could generate a smaller number of complete solutions in the same time. Naturally, the question arises: what will happen if all the algorithms are allowed to generate the same number of complete solutions? To answer this question, we re-run all the $\mathcal{MM}$AS-QAP(i) allowing each to perform 500 iterations. These results are given in Table 3. As it can easily been seen, the average solution quality of $\mathcal{MM}$AS-QAP(1) for most instances is still better than that of $\mathcal{MM}$AS-QAP(i) for $i \geq 2$. There are only two exceptions: instances `tai80b` and `tai150b`. This means that, in general, the usage of

Table 3. Experimental results of $\mathcal{MM}$AS-QAP(i) algorithms on several QAP instances; given is, for each instance, the average percentage deviation from the best known solution. All algorithms were stopped after 500 iterations.

	$i = 1$	$i = 2$	$i = 3$	$i = 4$	$i = 5$	$i = 6$	$i = 7$	$i = 8$	$i = 9$	$i = 10$
tai60b	**0.0056**	0.0195	0.0435	0.0649	0.0564	0.0535	0.0583	0.0729	0.0546	0.0546
tai80a	**1.2951**	2.0692	2.0393	2.3263	2.3028	2.1805	2.1112	2.2939	2.2204	2.2204
tai80b	0.1287	0.1128	0.1300	**0.0734**	0.0907	0.0800	0.1495	0.1042	0.1021	0.1021
sko81	**0.1279**	0.1872	0.2463	0.2573	0.2630	0.3391	0.3224	0.3072	0.3828	0.3074
sko90	**0.1391**	0.2669	0.2747	0.2925	0.4189	0.3746	0.3730	0.3768	0.3370	0.4409
sko100a	**0.1736**	0.2115	0.2318	0.3160	0.3307	0.3867	0.3369	0.3849	0.3267	0.3757
tai100a	**0.4320**	0.8002	1.0031	1.2218	1.3383	1.1167	1.0470	1.0566	1.0507	1.1497
tai100b	**0.0748**	0.0903	0.1507	0.1397	0.2198	0.2111	0.2164	0.2397	0.2133	0.2555
tho150	**0.2347**	0.3456	0.3827	0.3633	0.4236	0.4384	0.4236	0.4736	0.4787	0.4615
tai150b	0.3318	**0.2598**	0.4174	0.3835	0.4508	0.4320	0.4678	0.5603	0.4773	0.6855

incremental local search is actually causing a deterioration of $\mathcal{MM}$AS-QAP's performance, although it is allowed much more computation time. Said in other words, incremental local search is not only computationally expensive but also interferes negatively with the solution process of the ACO algorithm–at least for the QAP.

One may stop here and simply report this as a negative result. However, we were wondering as to why there may be a negative influence of incremental local search into the ACO algorithm's search process. A possible answer to this question is given in the next section.

4 Analysis

In this section, we try to explain why the incremental local search produces a detrimental effect on $\mathcal{MM}$AS-QAP's performance. For motivating the main line of attack of this analysis, let us consider first what is known about the convergence behavior of $\mathcal{MM}$AS-QAP, a high-performing ACO algorithms. Essentially, the search process of $\mathcal{MM}$AS-QAP shows a transition from an exploration phase, which is characterized by iteration-best update (the best solution in the current iteration is allowed to update the pheromones) and relatively high branching factor, to an exploitation phase, where the search is directed towards a search space region whose center is defined by the best-so-far solution seen by the algorithm. Interestingly, while in the exploration phase good quality solutions can already be found, typically the highest quality solutions in a trial of $\mathcal{MM}$AS-QAP are found when the search is in its exploitation phase–characterized by a low branching factor and by the fact that the solutions generated by the ants are relatively close to the best-so-far solution. In fact, $\mathcal{MAX}$–$\mathcal{MIN}$ Ant System was designed with the explicit intention to allow a careful transition between the exploration and exploitation phases and to further avoid search stagnation in the exploitation phase [11,8].

We suspected that the local changes introduced by the incremental local search disturb the behavior of $\mathcal{MM}$AS-QAP in the exploitation phase. This suspicion is based on the fact that the local search on the partial solution does not take into account the pheromone trails and, hence, may lead to significant changes to the partial solution. As an effect of this, the final complete solution before the final local search phase may actually be rather far from the best-so-far solution and, thus, hinder the exploitation phase from being effective. This effect is certainly increased for an increased frequency of partial re-optimizations. Differently, if ants do not apply partial re-optimization, the solutions they construct are relatively close to the best-so-far solution.

Hence, we decided to examine more carefully the variability of the generated solutions by the various settings of i in $\mathcal{MM}$AS-QAP(i). The main idea of our analysis is to examine the distance of the complete solutions generated in the current iteration from the best-so-far solution before and after the final application of the local search algorithm on the complete solutions. This is done for all the 10 settings of i. Since $\mathcal{MM}$AS-QAP(1) has proven to be a state-of-the-art ACO algorithm for the QAP, we use the computed distance as a yardstick for the analysis. For our analysis, we compute the distance $d(\pi, \pi')$ between two solutions π and π' as the minimum number of applications of 2–exchange moves that are required to convert one solution into the other one. This distance measure reflects that deviations in the individual assignments from the best-so-far solution can be undone by exchanging facilities between locations. Note that this distance measure can easily be computed using a linear time algorithm [12].

To make our analysis simpler, we consider a variant of $\mathcal{MM}$AS-QAP, denoted as *rb*$\mathcal{MM}$AS-QAP. In this variant, only the *restart-best* solution, the best solution found since the last re-initialization of the pheromones, is allowed to update the pheromones. (Hence, the *best-so-far* and *iteration-best* solutions are not taken into account in the pheromone deposit.) To justify the usage of *rb*$\mathcal{MM}$AS-QAP in the analysis, we first tested whether the versions *rb*$\mathcal{MM}$AS-QAP(i), $i = 1, \ldots, 10$ shows the same type of behavior as $\mathcal{MM}$AS-QAP(i), using as stopping criterion again 500 iterations. Table 4 shows that this is essentially the case, although *rb*$\mathcal{MM}$AS-QAP gives overall slightly worse results than $\mathcal{MM}$AS-QAP. Nevertheless, we can conclude that *rb*$\mathcal{MM}$AS-QAP captures the same trend as $\mathcal{MM}$AS-QAP and that it is safe to limit the analysis to *rb*$\mathcal{MM}$AS-QAP.

In our analysis, we compute at each iteration of *rb*$\mathcal{MM}$AS-QAP(i), $i = 1, \ldots, 10$, the distance between an ants' *complete* solution and the *restart-best* solution before and after the final local search on the complete solution has been applied. These distances are then averaged across the $m = 5$ ants. We denote these two measures as d^- (average distance *before* the final local search) and d^+ (average distance *after* the final local search), respectively. Figure 1 illustrates the observed results for the values of d^- and d^+ obtained by *rb*$\mathcal{MM}$AS-QAP(i), $i = 1, \ldots, 10$, over 500 iterations for the instance `sko100a`. The trend shown in these plots is representative for all the other instances we tested. (We used Tukey's (Running Median) Smoothing [13] for plotting the curves. If no

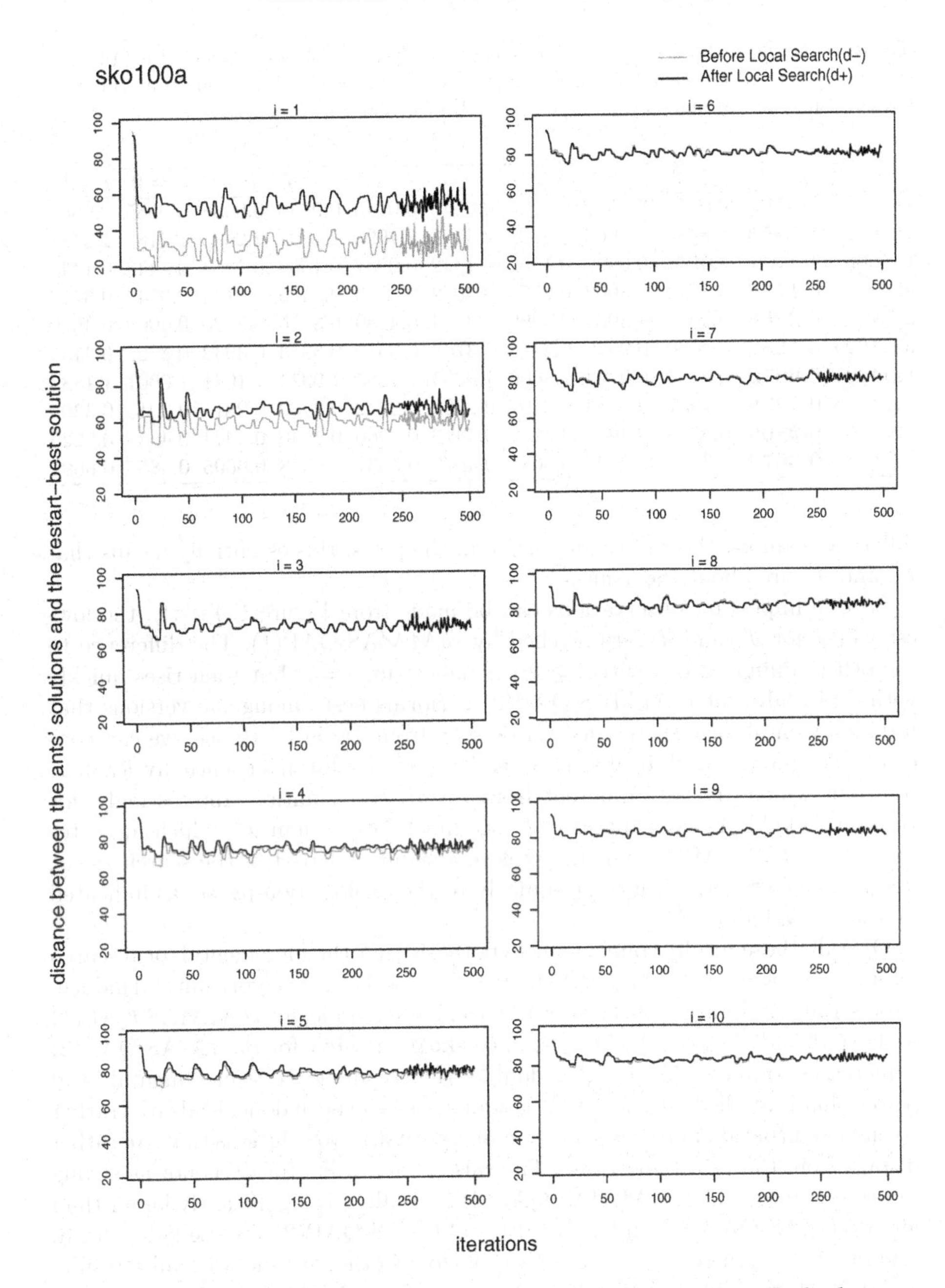

Fig. 1. Experimental results of *rbMMAS*-QAP on instance `sko100a`. Each plot represents the average distance between the ants' solutions and the *restart-best solution* before and after the final local search on a completed solution for *rbMMAS*-QAP(i), $i = 1, \ldots, 10$. The stopping criterion is set to 500 iterations.

Table 4. Experimental results of *rbMMAS*-QAP(i) algorithms on several QAP instances; given is, for each instance, the average percentage deviation from the best known solution. All algorithms were stopped after 500 iterations.

	$i=1$	$i=2$	$i=3$	$i=4$	$i=5$	$i=6$	$i=7$	$i=8$	$i=9$	$i=10$
tai60b	**0.0168**	0.0473	0.0562	0.0620	0.0540	0.0745	0.0643	0.0638	0.0585	0.0585
tai80a	**1.7958**	1.9787	2.1666	2.3629	2.4193	2.1998	2.1897	2.2693	2.2188	2.2188
tai80b	0.1708	**0.0952**	0.0999	0.1062	0.1342	0.0987	0.1069	0.1454	0.1448	0.1448
sko81	**0.1789**	0.2312	0.3063	0.2773	0.3072	0.3303	0.3459	0.3134	0.3749	0.3417
sko90	**0.2288**	0.2835	0.3000	0.3290	0.3664	0.4189	0.3875	0.3720	0.4062	0.4620
sko100a	**0.1982**	0.2830	0.2665	0.3414	0.3768	0.3519	0.3353	0.3942	0.3725	0.4135
tai100a	**0.6073**	0.9753	0.9763	1.3606	1.3204	1.1283	1.0974	1.1031	1.0001	1.1883
tai100b	**0.0938**	0.1231	0.1583	0.2297	0.2368	0.2242	0.2450	0.2811	0.3014	0.3303
tho150	**0.2800**	0.3405	0.4061	0.3990	0.4490	0.4966	0.4540	0.4924	0.4653	0.5229
tai150b	**0.3076**	0.3318	0.3907	0.3681	0.4677	0.5531	0.5978	0.6605	0.5857	0.6691

differences among the curves are visible in the plots, this essentially means that d^- and d^+ are about the same.)

Several important observations can be made from Figure 1. Firstly, the lowest values for d^- and d^+ are reached by *rbMMAS*-QAP(1). The difference to the other configurations with $i \geq 2$ is smallest for $i = 2$ but then rises quickly with i. (Recall that *rbMMAS*-QAP(2) performs best among the versions that use incremental local search, as can be seen from Table 1.) Hence, we can conclude that incremental local search on the partial solutions eventually leads to solutions which are very different from *restart-best* solution. Interestingly, for *rbMMAS*-QAP(1) the values for d^+ are much larger than d^-, which indicates that *rbMMAS*-QAP(1) can still explore a significant part of the search space, despite the fact that it converges quickly to the exploitation phase, as indicated by the low values of d^-.

Overall, these results confirm our hypothesis that the incremental local search interferes negatively with the exploitation phase of the ACO algorithm and induces a too strong exploration of the search space. For example, for *rbMMAS*-QAP(1) we have that d^- is around 30 for instance sko100a, while for *rbMMAS*-QAP(2) it increases to about 60—roughly double. Hence, already one incremental local search that is applied after $k = n/2$ assignments have been done, leads to a rather strong perturbation in the exploitation phase, that is, to solutions that are rather distant from the restart-best one. The raise in the values of d^+ is not as strong as for d^-; however, for *rbMMAS*-QAP(2) d^+ is already significantly larger than for *rbMMAS*-QAP(1), explaining also *rbMMAS*-QAP(2)'s worse behavior, in general. As said, these observations also hold for all other instances; detailed results are available from `http://iridia.ulb.ac.be/supp/IridiaSupp2006-002/`.

Finally, we run also experiments for the incremental local search when starting from random initial solutions, to check whether in such an environment the incremental local search can have some contribution. (We have chosen random initial solutions, since for the QAP no high-performing construction heuristics are available.) We run the random restart local search algorithm for the same

Table 5. Experimental results of the random-restart local search on several QAP instances; given is, for each instance, the average percentage deviation from the best known solution. All algorithms were stopped after the same computation time as the algorithms in Table 1. The best results are in bold-face.

	$i=1$	$i=2$	$i=3$	$i=4$	$i=5$	$i=6$	$i=7$	$i=8$	$i=9$	$i=10$
tai60b	**0.1461**	0.1833	0.1991	0.2210	0.1956	0.2202	0.2254	0.2298	0.1716	0.1716
tai80a	2.5402	2.5403	2.4038	2.5176	2.4891	**2.2537**	2.2732	2.2674	2.3156	2.3156
tai80b	0.6667	**0.3876**	0.6568	0.6274	0.6334	0.5372	0.6372	0.5625	0.6668	0.6668
sko81	0.7602	0.7085	0.7545	0.6219	0.6211	0.6041	0.6094	**0.5591**	0.6448	0.6006
sko90	0.8148	0.7860	0.7625	0.7026	0.7263	0.7559	0.7137	**0.6285**	0.6880	0.6734
sko100a	0.7448	0.7449	0.6419	0.6831	0.6356	0.6355	0.5878	0.6528	**0.5590**	0.6494
tai100a	1.4651	1.5655	1.2888	1.3980	1.4050	1.1792	1.1433	1.0761	**1.0518**	1.2299
tai100b	0.6754	0.6708	0.6458	0.5898	0.6056	**0.5688**	0.6294	0.5817	0.5809	0.5914
tho150	0.9266	0.8679	0.8999	0.8185	0.8746	0.8444	0.7996	**0.7739**	0.7825	0.8564
tai150b	1.1589	1.2146	1.2039	1.1491	1.234	1.1430	1.1819	1.1388	1.0972	**1.0522**

average computation time as needed for $\mathcal{MM}$AS-QAP(1) to perform 1000 iterations. Table 5 shows the average solution cost as the percentage deviation from the best known solution, obtained by this random restart local search with different numbers of local searches performed on the (partial) solutions. These results clearly show that for almost all instances, the usage of the incremental local search improves the performance over the version where only once a local search is run on a complete solutions. Hence, these results agree with the computational results reported in the literature [3,4,6,7] and indicate that the usage of incremental local search can be, in some situations, helpful. In fact, random restart has no means to exploit the possibility of learning and exploiting the most promising region of the search space. This suggests that the usefulness of the incremental local search depends strongly on the context where it is applied and the solution construction procedure.

5 Conclusions

Motivated by the promising results of incremental local search reported in the literature [3,4,6,7], we have investigated its behavior and performance in an ACO algorithm for solving the QAP. Our computational study has shown, however, rather poor results for this idea. Next, we have carried out an analysis that can explain this failure. In fact, we have shown that the incremental local search somehow destroys the behavior of the ACO algorithm in its exploitation phase by not allowing it to generate solutions that are rather close to the restart-best or global-best solutions.

Certainly, our results and explanation is limited to the QAP. However, we conjecture that the very same issue arises also in applications of ACO algorithms to other challenging combinatorial problems. More in general, our results also indicate that probably a more careful study of the behavior of ACO algorithms in the exploitation phase should be done to understand, which techniques may be more promising for improving the performance of ACO algorithms. Finally, our results indicate that incremental local search could be useful for increasing the exploration in convergence situations of ACO algorithms. Although this was not useful on the QAP, it may well be that the careful, occasional addition of incremental local searches in specific situation, could possibly result in improvements for ACO algorithms.

Acknowledgments. This research has been supported by COMP²SYS, a Marie Curie Early Stage Research Training Site funded by the European Community's Sixth Framework Programme under contract number MEST-CT-2004-505079, and by the ANTS project, an *Action de Recherche Concertée* funded by the Scientific Research Directorate of the French Community of Belgium. Thomas Stützle and Marco Dorigo acknowledge support from the Belgian FNRS of which they are a Research Associate and a Research Director, respectively. The information provided is the sole responsibility of the authors and does not reflect the opinion of the sponsors. The European Community is not responsible for any use that might be made of data appearing in this publication.

References

1. Dorigo, M., Stützle, T.: Ant Colony Optimization. MIT Press, Cambridge, MA (2004)
2. Stützle, T., Hoos, H.: Stochastic Local Search: Foundations and Applications. Morgan Kaufmann (2005)
3. Russell, R.: Hybrid Heuristics for the Vehicle Routing Problem with Time Windows. Transportation Science **29** (1995) 156–166
4. Gendreau, M., Hertz, A., Laporte, G.: New Insertion and Postoptimization Procedures for the Traveling Salesman Problem. Operations Research **40**(6) (1992)
5. Gendreau, M., Hertz, A., Laporte, G.: A Tabu Search Heuristic for the Vehicle Routing Problem. Management Science **40** (1994) 1276–1290
6. Caseau, Y., Laburthe, F.: Heuristics for Large Constrained Vehicle Routing Problems. Journal of Heuristics **5**(3) (1999) 281–303
7. Fleurent, C., Glover, F.: Improved Constructive Multistart Strategies for the Quadratic Assignment Problem Using Adaptive Memory. INFORMS Journal on Computing **11**(2) (1999) 198–204
8. Stützle, T., Hoos, H.: $\mathcal{MAX}$–$\mathcal{MIN}$ Ant System. Future Generation Computer Systems **16**(8) (2000) 889–914
9. Stützle, T., Dorigo, M.: ACO Algorithms for the Quadratic Assignment Problem. In Corne, D., Dorigo, M., Glover, F., eds.: New Ideas in Optimization. McGraw-Hill, London, UK (1999) 33–50
10. Burkard, R., Karisch, S., Rendl, F.: (`http://www.seas.upenn.edu/qaplib`)
11. Stützle, T., Hoos, H.H.: Improving the Ant System: A Detailed Report on the $\mathcal{MAX}$–$\mathcal{MIN}$ Ant System. Technical Report AIDA–96–12, FG Intellektik, FB Informatik, TU Darmstadt (1996)
12. Schiavinotto, T., Stützle, T.: Metrics on Permutations for Search Space Analysis. Computers & Operations Research (In press)
13. Cohen, P.R.: Empirical Methods for Artificial Intelligence. MIT Press, Cambridge, MA (1995)

Individual Discrimination Capability and Collective Choice in Social Insects

Jesus Millor, José Halloy, Jean-Marc Amé, and Jean-Louis Deneubourg

Service d'Ecologie Sociale, Université Libre de Bruxelles, Brussels, Belgium

Abstract. In an ant society individuals coming from different groups (lines, strains) bear it own chemical identity and those individuals present discrimination capabilities between different chemical profiles. However, at the collective level these groups may cooperate and act together. To understand this apparent contradiction we have to keep in mind that amplification is the main component of many collective phenomena in social and gregarious insects. We use a model of food recruitment where each group of foragers have its own blend of pheromone trail that is partly recognized by the others groups. We found that a low level of recognition between signals is sufficient to produce a collaborative pattern between groups and that beyond a critical value of recognition. The aggregation of all the groups around the same food source is observed. Such collective response is a generic property of social phenomena governed by amplification processes.

1 Introduction

1.1 The Individual Discrimination Capabilities

Many social insect are able to discriminate between nestmates and non-nestmates [1,2]. In honeybees the queen can mate with up to 20 drones that give rise to a colony with 20 patrilines or subfamilies [3]. Each sub-family has a hydrocarbon profile used by workers as sub-family discrimination [4]. In laboratory condition workers can discriminate between supersisters (same patriline) and half-sisters (other patriline). In gregarious insects like the cockroaches *Blattella germanica* L. the gregarious behaviour is mainly based on the cuticular hydrocarbons recognition characterising the strain odour and individuals prefer their own strain odour from those of another strain [5].

But in many collective behaviours it seems that workers behave independently of their sub-family origin. In bees swarming there is no differences between the sub-families composition of the primary and the after-swarm [6]. There are neither evidences for sub-family discrimination between bee dancers and followers in a colony of two sub-families [7]. In the polygynous ant *Lasius acervorum*, Heinze et al [8] have shown that colony fission didn't lead to the segregation of different matrilines. In the cockroaches *Blattella germanica* L., when tested group came from two different strains with different odours, they aggregated only on one site [9]. Individuals are facing up a mixing palette of odours coming

M. Dorigo et al. (Eds.): ANTS 2006, LNCS 4150, pp. 167–178, 2006.

from different lines or strains and in spite of individual kin discrimination capabilities, it seems that when the collective behaviour take place, there is a lack of sub-group recognition and segregation.

The lack of nepotism or segregation is in some cases, largely due to the mixing of individual signature. Everaerds et al [10] demonstrated that the mixing of the individuals of two cockroaches species increases with time and that this mixing is correlated with the apparition of a new individual cuticular signature. This new signature is a blend of the two specific signatures. However this process takes time and other mechanisms could be involved. There is a difference between the capacity of discrimination at the individual and the collective level. Many of these collective behaviours are based on amplification processes. These processes, through positive feed-backs mechanisms, are widespread in group-living organisms [11,12]. Amplification is an essential component of many collective phenomena observed in particular in social and gregarious arthropods e.g. aggregation [9], collective defence [13], recruitment to a food source or to a new nest [14].

Our hypothesis is that the collective level amplifies a low level of recognition between groups and is sufficient to lead to a collective response despite individual discrimination capabilities. To test this hypothesis we will use a model of food recruitment as a case study.

1.2 Hydrocarbons Profile and Communication in Ants

Insect body, and particularly ants body, is covered by a set of cuticular lipids including hydrocarbons (HCs). These components ensure a protection to the animal preventing against desiccation, due to their hydrophobicity, and against invasion of micro organisms. Several studies indicate that these cuticular hydrocarbons play a key role in social insect communication. The nest-mate recognition signal in ants seems to be constituted of an assemblage of cuticular hydrocarbons [1]. Each member of an ant colony has its reference model of the chemical signature by which it can discriminate between a member of the same colony or an alien. The acceptance or rejection of an individual depends on the degree of overlapping between its own chemical reference (sensory template) and the chemical signature of the encountered individual. Beside their role in nest-mate discrimination several lines of evidence indicate that cuticular compounds provide information about fertility status, caste/task recognition or kin recognition [1,4].

The HCs are biosynthesized in the abdomen close to the integuments probably in the oenocytes cells as in other insects [15]. They are secreted to the epicuticule or transported in the hydrophobic core of a lipoprotein: the high density lipophorin (HDLp) [15]. These long chain HCs are also found in a specific ant gland: the postpharyngeal gland (PPG). The PPG is an exocrine gland and allows the secretion or HCs exchanges through oral contact (trophallaxis) and grooming [16,15]. Generally two models are proposed to explain the elaboration of the colonial identity. The first one is the individualistic model where each member bears its determined odour and it's individually recognize by the other members of the colony. In ant species with large colonies, individual ant recognition becomes difficult and evolution favoured the emergence of a colonial odour where all individuals of the colony share their recognition cues to form a gestalt colony scent [17]. This gestalt model predicts that ants have to exchange HCs constantly each other to maintain themselves within the colonial odour. It was demonstrated that the PPG play a key role in this mechanism constituting a colonial odour reservoir. The HCs stored on the PPG are congruent with those find in the cuticule, this suggest that the PPG contains the recognition cues. An ant bearing alien PPG HCs becomes aggressive to nestmates, whereas an alien ant treated with PPG HCs from this colony is accepted [18]. The PPG gland, by facilitating the exchange of substances between members of the nest particularly by trophallaxis, promotes a fast distribution of the scent within the colony. In species that don't perform trophallaxis, grooming/allogrooming and other physical contacts allow the exchange of the body odour. The social interactions are crucial for the individual integration in the colony. A study of Dahbi et al [19] in the genus *Cataglyphis* identifies a total of 242 different hydrocarbons. The chemical repertoire in ant species is generally comprises between 20 and 60 substances with a minimum of 7 HC in the ant *Formica truncarum* until more than 60 HC in *Pachycondyla villosa* [20,15].

1.3 The Foraging Strategy in Social Insects

The foraging strategies of social insects, especially among ants and bees, show wide diversity in their organization. There is a series of factors which influences the choice of a particular foraging strategy, like the colony size (1), the size, proximity and predictability of food sources (2), the risks of predation and competition with other colonies (3), and variations in the degree of cooperation that occurs among foraging workers (4).

For the ant *Lasius niger*, for example, if a sugar solution is placed in the vicinity of the nest, after some time, a single forager will discover the sugar then through a recruitment process, a large number of foragers will build up rapidly at the food source. The increase of the number of individuals to this food source follows a logistic curve that reaches a plateau. The ants move along a pheromone trail that they create and they reinforce this trail with additional pheromone when they have ingested food and return to the nest and when they are following the pheromone trail on a subsequent journey back to the food source. The initial exponential growth phase is the consequence of this positive feedback process: each ant returning from the food source can stimulate many other nest mates to forage, which in turn stimulate others, and so on. When all the available foragers have been recruited no new ones are left to join the foraging system and the plateau phase is reached. This ant's recruitment system facilitates efficient collective decision making, allowing the colony to select the most profitable of an array of food sources in a heterogeneous environment.

But in the field foraging ants are in competition with ants coming from other colonies and can be in contact with trail pheromone laying by individuals from other colonies of the same specie or from other ant species. This pheromone can be recognized partly by the individuals of the considered colony.

The trail pheromone is generally a mixture of chemicals and different specific ratio from this chemicals produce different effects on the individuals [21]. Several compounds can be common to different pheromones produced by different species and a specific trail pheromone can be recognized and followed by individuals from other species. Moreover, in number of species it was reported the absence of species specificity in the chemical recruitment trails [22]. In many *Myrmica* ants species, trail fallowing is released by the same 3-ethyl-2,5-dimethylpyrazine compound. In several myrmicine genera, such as, *Tetramorium*, *Messor*, *Myrmica* or *Atta*, the pyrazines have bee identified as trail pheromones. Another example is giving by the two closely related ant species *Aphaenogaster cockerelli* and *A. albisetosus*. *A. cockerelli* follows only its own trail whereas *A. albisetosus* respond to the trail laying by both species. In addition, it should be said that in many other species, from the *Myrmica* genera for example, anonymous recruitment signals composed of a species-specific mixture of hydrocarbons, are used as home range markers and recognized by nest-mates [23].

The aim goal of this work is to study which will be the influence, on the collective food recruitment behaviours, of different and partially recognizable chemical signals product by individuals coming from different colonies or different species in presence of two equal food sources. We developed a theoretical model based on a differential system of equations, inspired from the Deneubourg model in order to analyze the influence of the level of recognition of a foreign trail pheromone scent on the collective foraging strategies developed by the scouts of two ant colonies that exploit the same set of food sources. To take in account the random aspect that characterize the recruitment dynamic, we also made a series of numerical Monte Carlo simulations based on the same mechanisms defined in the differential equations model. Recruitment is often seen as the archetype of amplification mechanism leading to collective decision. We could have chosen another type of behaviour like aggregation, collective homing or swarming and colony fission. The main idea is to see which collective behaviour will emerge when members of sub-groups with their own chemical signal may interact.

2 The Mean Field Model

The model, based upon empirical findings about the behaviour of the individual ants, describes the evolution of the concentration of trail pheromone and, as a consequence, the traffic of ants over each trail. It is an extension of a model that has already been applied to different types of choice experiments [11]. The ant departure from the nest to the food source is based on the probability ϕ (sec^{-1}) to leave the nest per time unit (the flow of departure). They leave the nest at a constant rate ingest food and promptly return to the nest laying a trail pheromone. One can quantify the ant decision at a choice point by equation (1) that depends on the values of the pheromone concentration c_i on each i trail [11]:

$$P_i = \frac{(k + c_i)^n}{\sum_{i=1}^{s}(k + c_i)^n} \qquad i = 1, ..., s \tag{1}$$

- n determines the level of nonlinearity in the response.
- k correspond to the intrinsic degree of attraction of a unmarked branch.

On its return journey from a food source, each ant lays a quantity of pheromone q. Since, in this case the time needed to visit the food sources and return is equal for both sources; we neglect the time delay between the choice and the ant's return. The evaporation rate of the trail pheromone will be proportional to the pheromone concentration and to a constant ν. This constant is the inverse of the mean-live time of the trail pheromone. Thus, pheromone concentration is directly proportional to the flux of individuals (ϕ) in the system. At each unit of time a quantity $q\phi P_i$ of pheromone is added to the trail i and νc_i is the rate of evaporation of this trail pheromone. We normalize the equation in relation to ν and k by replacing c_i and ϕ respectively by $C_i = c_i/k$ and $\Phi = q\phi/\nu$. These relationships can be expressed in the following system of s differential equations which describe the rate of change in concentration of pheromone on trails 1 to s.

$$\frac{dC_i}{dt} = \Phi P_i - C_i \qquad i = 1, ..., s \tag{2}$$

At the following, we will consider that we are in presence of g groups of foragers belonging g different strains. Each group have its own blend of pheromone (for example a mix of trail pheromone and footprint hydrocarbons) that may influence the decision of the individuals of the others groups. The probability to follow a trail increases with the concentration of the pheromone on each branch deposited by the individuals of the different groups (see Fig. 1). The trail pheromone of one group can be partially recognized by the other. The influence of individuals to the same group is more important than that of individuals belonging to the other group. To take this in account, we will include to equation (1) a parameter β_{jl} of inter-attraction between groups l and j.

$$P_{il} = \frac{(1 + \sum_{j=1}^{g} \beta_{jl} c_{il})^n}{\sum_{i=1}^{s}(1 + \beta_{jl} c_{il})^n} \qquad j, l = 1, ..., g \quad \text{and} \quad i = 1, ..., s \tag{3}$$

P_{il} is the probability for an individual of the group l to choose the trail i.

These probabilities will be increase with the concentration of the pheromone on each branch deposited by the individuals of the different groups adjusted by the parameter $0 \leq \beta_{jl} \leq 1$ that measures the level of recognition between both trails odour. $\beta_{jl} = 0$ corresponds to two independent groups and $\beta_{jl} = 1$ to an identical signal for the group j and l ($\beta_{ll} = 1$). We will suppose that the inter-attraction of group j on group l is the same that l on j and from then $\beta_{jl} = \beta_{lj}$ and all the parameters q, n, and ν are identical for the different groups-pheromones. The rate of change in concentration of pheromone on trail i for the group l can be expressed with taking into account the expression (2) and (3):

$$\frac{dC_{il}}{dt} = \Phi P_{il} - C_{il} \tag{4}$$

3 Results

At the stationary states, the flow of individuals of group l choosing the path i, is proportional to C_{il}. For two groups and two food sources, the system has four types of stationary states corresponding to different distributions of the individuals around the food sources (summarized in Fig. 1).

(1) The Symmetrical State: The pheromone concentrations are equal on both paths and for both groups: $C_{ij} = \Phi/2$ and the flows of both groups are equal on each trail. This symmetrical state exists for all the values of Φ and β. The stability analyse shows that this solution is only stable for low flux values:

$$\Phi < \frac{2}{1+\beta} \tag{5}$$

As $0 \leq \beta \leq 1$, this implies that for $\phi > 2$, $C_{ij} = \phi/2$ is always unstable.

(2) The Aggregative States: The activity of the two groups is focused on the same food source meaning both group have selected the same branch. The pheromone concentrations are equal for both groups on both branches. However, the selected branch presents a higher concentration than the non selected one. The selected branch is chosen randomly (i.e. with a probability of 0.5). This is summarised by the formula $C_{11} = C_{12} < C_{21} = C_{22}$ or $C_{11} = C_{12} > C_{21} = C_{22}$. These solutions are:

$$C_{11} = C_{12} = \frac{\Phi}{2} \pm \frac{1}{2}\sqrt{\Phi^2 - \frac{4}{(\beta+1}^2} \qquad C_{21} = C_{22} = \Phi - C_{11}$$

These stationary states exist if $\Phi > 2/(\beta+1)$ and under this condition is always stable for any value of Φ and β. In this case, the flows of both groups along one trail are equal.

(3) The Segregative States: The majority of the individuals of one strain focus their activity on a branch and the individuals of the second strain on the other one In this case, the segregative steady states are $C_{11} = C_{22}C_{21} = C_{12}$. In this case the pheromone concentrations are giving by:

$$C_{11} = C_{22} = \frac{\Phi}{2} \pm \frac{1}{2}\sqrt{\Phi^2 - \frac{4(\Phi\beta+1)^2}{(\beta-1}^2} \qquad C_{21} = C_{12} = \Phi - C_{11}$$

The segregative states exist if $\Phi > 2/(1-3\beta)$ and never appear for $\beta > 1/3$. They are stable under the condition:

$$\Phi > \frac{2(1+\beta)}{1-\beta^2-4\beta} \tag{6}$$

The stability of the segregative states depends on the level of recognition between the trails of the groups (β) but also on the flux of individuals. The lower the flux, the lower must be β to observe a stable split or segregation between both groups. Moreover, for $\beta > \sqrt{5}-2 \approx 0.236$, the segregative state are always unstable.

(4) The Mixed States: In this case $C_{11}C_{22}C_{21}C_{12}$, the pheromone concentrations and the traffic are different on both trails and for both groups. The four combinations of solutions are:

$$\begin{aligned}
C_{11} &= \frac{\Phi}{2} \pm \frac{1}{4}\sqrt{\frac{-A-2\sqrt{B}}{D}} & C_{21} &= \Phi - C_{11} \\
C_{12} &= \frac{\Phi}{2} \mp \frac{1}{4}\sqrt{\frac{-A+2\sqrt{B}}{D}} & C_{22} &= \Phi - C_{12} \\
C_{12} &= \frac{\Phi}{2} \pm \frac{1}{4}\sqrt{\frac{-A-2\sqrt{B}}{D}} & C_{22} &= \Phi - C_{11} \\
C_{11} &= \frac{\Phi}{2} \mp \frac{1}{4}\sqrt{\frac{-A+2\sqrt{B}}{D}} & C_{21} &= \Phi - C_{12}
\end{aligned} \tag{7}$$

$$\begin{aligned}
A &= 2(\beta-1)^2(2+\Phi+\Phi\beta)(\Phi\beta^3+2\beta^2+3\Phi\beta^2-\Phi\beta+\Phi-2) \\
B &= (\beta-1)^3(\beta+1)^3(2+\Phi+\Phi\beta)^2(\Phi\beta^2+2\beta+4\Phi\beta-\Phi+2) \\
&\quad *(\Phi\beta^2+2\beta+2\Phi\beta+\Phi-2) \\
D &= 1-2\beta^2+\beta^4
\end{aligned}$$

These four stationary states are always unstable and exist if $\Phi > \frac{2(1+\beta)}{(1-\beta^2-4\beta)}$. They never exist for $\beta > \sqrt{5}-2$.

The bifurcation diagrams show the value of the stationary states (the trail concentration for one group on one of the two branches) and their stability as a function of the individual flux Φ for a recognition parameter of β=0.22 (Fig. 2A) and as a function of the recognition factor β for a flux of Φ=10 (Fig. 2B). For $\Phi > 2$ the solution $C_{11} = C_{12} \geq C_{21} = C_{22}$ (aggregative state) exists and is always stable (Fig. 2). The segregative state ($C_{11} = C_{22} \neq C_{21} = C_{12}$) becomes stable only for large value of Φ (> 34) when β=0.22. For Φ=10 (Fig. 2B) the aggregative state is always stable and the segregative states are stable for small value of β. When $\beta < 0.185$, the segregative states become unstable.

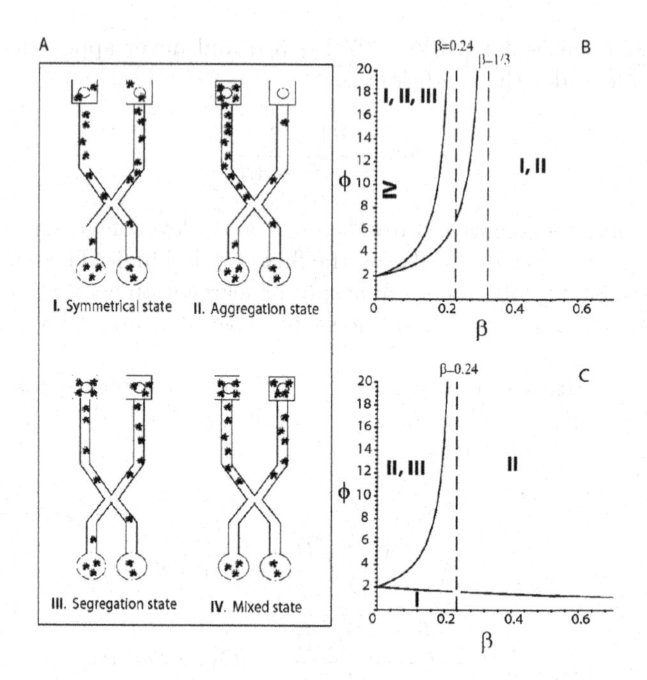

Fig. 1. A: Schema representing the different states of the model. Individuals from the two strains can choose between the two branches of the bridge leading to two identical food sources equidistant from the nests (I: the symmetrical state; II: the aggregative state; III: the segregative state and IV: the mixed state). Areas of the existence of the solutions (B) and the stability of the solutions (C) as a function of β and Φ.

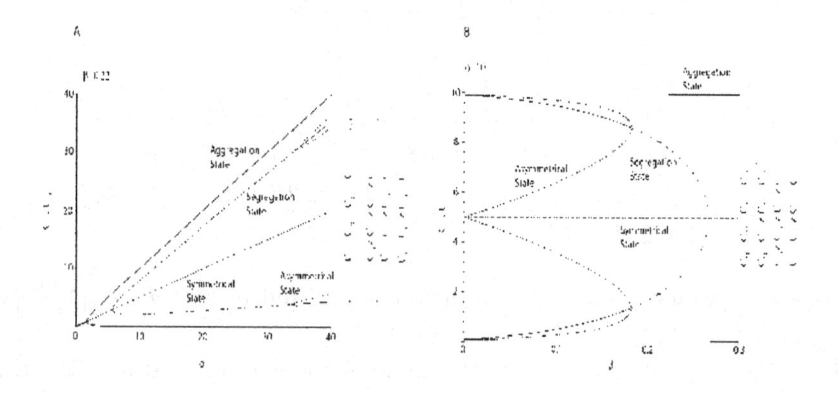

Fig. 2. Concentrations values of the trail odour for strain *1* on branches *1* of the bridge as a function of Φ for β=0.22 (Fig. 2A) and as a function of β for ϕ=10 (Fig. 2B). - - - - - - Unstable states

The space phase allows to illustrate the dynamics of the system. Figures 3 represent these spaces phase diagrams putting the values of the trail odour concentration on one branch for one group (C_{11}) in function of the trail odour concentration on the same branch for the other group (C_{12}). For low flux values of Φ=1 and low recognition β=0.1 (Fig. 3A) only the symmetrical state exists. When Φ is large and β small (β=0.1) all the states appear (Fig. 3B). The unstable mixed states surround the segregative one and will tend randomly to the aggregative or the segregative states that are stable for these parameters values. Figure 3B shows that the basin of attraction of the segregative states is small compared to those of the aggregative states. So despite the fact that both solutions are stable, the system is characterized by a higher probability to adopt one of the aggregative states. For Φ=10 and β=0.25 (Fig. 3C) the mixed state disappears and the system tends to the stable aggregative states. For $\beta > 0.33$ (Fig. 3D) just the symmetrical and aggregative states remain and the individuals from the two groups aggregate using the same path and visiting the same food source.

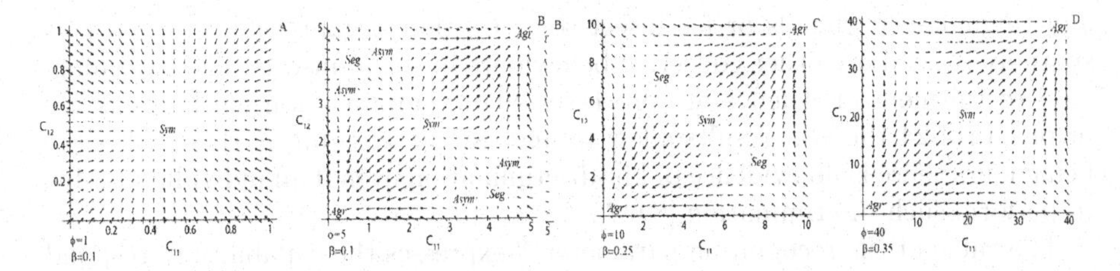

Fig. 3. Space phase representing the values of the trail odour concentration on one branch for one group (C_{11}) in function of the trail odour concentration on the same branch for the other group (C_{12}) for different values of Φ and β. Agr: aggregative state; Seg: segregative state, Sym: symmetrical state, Asym: mixed state.

4 Discussion

The choice function P_i (1) implies that the emission of a trail pheromone increases the probability to choose one branch and to lay an additional quantity of trail odour on it. With one unique strain, a bifurcation appears in the exploitation pattern of the food sources leading to the collective selection of only one path. In the present model the choice function P_{il} (3) assumes that the chemical signal of one group increases the probability of the other group to choose the same path. The pattern obtained by the recruitment amplification mechanism depends on the size of the groups in one hand and in the other hand on the level of discrimination between strain odours. For small population size, the individuals distribute themselves in a symmetrical way on the two food sources i.e. both branches are concomitantly exploited by both groups. As soon as the flow exceeds a certain threshold ($= 2/(1+\beta)$), the pattern switches to an aggregative or segregative state depending on the β value that measures the level of strain odours recognition. For low values of β (< 0.25), groups can segregate on the food sources because the odour signal from the other strain plays only a little role on the amplification process in comparison with its own trail odour. For these small values of β, aggregative states are also stable. The choice between segregative or aggregative solutions is random. However, the analysis of the model shows that the selection of the aggregative states is dominant due to the relative size of their attraction basins. As soon as β reached a critical value ($\beta > 0.24$), the two groups focus their activity on the same food source. To summarize, our results show that a good discrimination between groups (corresponding to a weak inter-attraction between them) is unable to keep segregative states between the groups when the individuals rely solely on the amplification mechanisms presented here. Moreover, the model shows that a collective behavioural plasticity occurs without the need of changing the individual behaviour or the communication system. The analysis for more than two groups and/or more than two sources (not summarise in this paper) predicts similar collective response and mainly the systematic aggregation between groups with a weak inter-attraction between them. This could be the case for many gregarious arthropods species that form aggregates which persist due to attraction modulated by chemical and thigmotactic signals or for insect societies that organise items or workers spatially [12]. Many other collective behaviours in social insects, like swarming, food recruitment or defence are processes where amplification mechanisms are involved [14,24,13]. Our results allow to understand why during these collective behaviours there is no evidence of kin-discrimination even if an individual is able to perform such discrimination [8,6,7].

The value of the recognition parameter β expresses the capability to respond to a signal of another individual belonging to a different group. If β =0, the individual is blind to this signal. We may formulate different hypothesis on the origin of this recognition factor. For social insects and many other arthropods, the communications are largely based on pheromones such as trail pheromones, cuticular hydrocarbons, Values of β may correspond to a certain overlapping between the chemical signatures of two groups. This chemical profile is under genetic control but is also determined by the environment and food consumption. So, genetic proximity or sharing common resource may increase the similitude between two signals and contribute to large value of β. In recruitment, the amplification can be mediated by generic signals common to all individuals (e.g. the trail pheromone) and specific signals such as the cuticular hydrocarbons (*Lasius sp*). In these situations, the value of parameter β is mediated by the relative contribution of the generic and specific signals.

In this model we considered that β was a constant and mutually equal for the two groups. In social insects there exist genetic determinants to explain behavioural differences among workers of a colony. However, the interactions between individuals may lead to a mixing of the different chemical signatures and to the apparition of a blended signature one that is also observed between different species. Our theoretical results suggest that a weak inter-attraction between groups is enough to increase the interactions between individuals from the different groups. These interactions will accelerate the apparition of a new common blend.

One of the main hypotheses of the model is that the resources are abundant and no crowding reaches significant level. We have shown that in such conditions, where competition is weak between individuals, the system adopts aggregative states. Preliminary analysis of a modified version of the model that includes crowding effects or limitation of the resource shows that the segregative states are favoured, but the aggregative states are still possible for a vast region in parameter space. A second important hypothesis of the model is that there is no agonistic behaviour between the individuals of the different groups. However, if such antagonistic behaviours are not too strong, qualitatively similar results are still found. Its main consequence is the shift of the β threshold leading to the aggregative state to higher value. A third simplification is the lack of modulation of the emission of the signals as a function of intensity of the other group's signal. Such modulation would favour the aggregative state. The last simplification is the lack or learning capabilities that seems involved in recognition. However as for the mixing of the chemical signatures and the modulation of the emission, learning must favour integration of different groups [25].

This simple model shows that collective choice can be shared by different groups without the need of specific signals or sophisticated behavioural.

References

1. Vander Meer, R.K., Morel, L.: Nestmate recognition in ants. In: Vander Meer, R.K., Breed, M., Winston, M., Espelie, K.E. (eds.): Pheromone Communication in Social Insects. Westview Press Boulder (1998) 79–103
2. Breed, M., Diaz, P.H., Lucero, K.D.: Olfactory information processing in honeybee, *Apis mellifera*, nestmate recognition. Anim. Behav. **68** (2004) 921–928
3. Estoup, A., Solignac, M., Cornuet, J.M.: Precise assessment of the number of patrillines and of genetic relatedness in the honeybee colonies. P. Roy. Soc. Lond. B **258** (1994) 1–7
4. Arnold, G., Quenet, B., Cornuet, J.M., Masson, C., De Schepper, B., Estoup, A., Gasqui, P.: Kin recognition in honeybees. Nature **379** (1996) 498
5. Rivault, C., Cloarec, A., Sreng, L.: Cuticular extracts inducing aggregation in the German cockroach, *Blattella germanica*. Anim. Behav. **55** (1998) 177–184
6. Kryger, P., Moritz, R.F.A.: Lack of kin recognition in swarming honeybees (*Apis mellifera*). Behav. Ecol. Sociobiol. **40** (1997) 271–276
7. Kirchner, W.H., Arnold, G.: Intracolonial kin discrimination in honey bees: do bees dance with their super-sisters. Anim. Behav. **61** (2001) 597–600
8. Heinze, J., Elsishans, C., Hölldobler, B.: No evidence for kin assortment during colony propagation in a polygynous ant. Naturwissenschaften **84** (1997) 249–250
9. Amé, J.M., Rivault, C., Deneubourg J.L.: Cockroach aggregation based on strain odour recognition. Anim. Behav. **68** (2004) 793–801
10. Everaerts, C., Farine, J.P., Brossut, R.: Changes of specific cuticular hydrocarbon profiles in the cockroaches *Nauphoeta cinera* and *Leucophaea maderae* reared in heteropsecific groups. Entomol. Exp. Appl. **85** (1997) 145–150
11. Deneubourg, J.L., Goss, S.: Collective patterns and decision making. Ethol. Ecol. Evol. **1** (1989) 295–311
12. Camazine, S., Deneubourg, J.L., Franks, N.R., Sneyd, J., Theraulaz, G., Bonabeau, E.: Self-organization in Biological Systems. Princeton University Press (2001)
13. Millor, J., Pham-Delegue, M., Deneubourg, J.L., Camazine, S.: Self-organized defensive behaviour in honeybees. P. Natl. Acad. Sci. USA **96** (1999) 12611–12615

14. Visscher, P.K., Camazine, S.: Collective decisions and cognition in bees. Nature **397** (1999) 400
15. Lucas, C., Pho, D.B., Fresneau, D., Jallon, J.M.: Hydrocarbon circulation and colonial signature in *Pachycondyla villosa*. J. Insect. Physiol. **50** (2004) 595–607
16. Lucas, C.: Etudes des bases chimiques et comportementales de la formation du "visa" colonial chez les Ponérines du genre *Pachycondyla*. Phd University of Paris XI (2002)
17. Crozier, R.H.: Genetic aspects of kin recognition: concepts, models, and synthesis. In: Fletcher, D.J.C., Michener, C.D. (eds.): Kin Recognition in Animals. John Wiley, New York (1987) 55–73
18. Lahav, S., Soroker, V., Hefetz, A., Vander Meer, R.K.: Direct behavioral evidence for hydrocarbons as ant recognition discriminators. Naturwissenschaften **86** (1999) 246–249
19. Dahbi, A., Lenoir, A., Tinaut, A., Taghizadeh, T., Francke, W., Hefetz, A.: Chemistry of the postpharyngeal gland secretion and its implication for the phylogeny of Iberian *Cataglyphis* species (Hymenoptera, Formicidae). Chemoecology **7** (1996) 163–171
20. Nielsen, J., Boomsma, J.J., Oldham, N.J., Petersen, H.C., Morgan, E.D.: Colony-level and season-specific variation in cuticular hydrocarbon profiles of individual workers in the ant *Formica truncorum*. Insect. Soc. **46** (1999) 58–65
21. Hölldobler, B.: The chemistry of social regulation: multicomponent signals in ant societies. P. Natl. Acad. Sci. USA **92** (1995) 19–22
22. Hölldobler, B., Wilson, E.O.: The Ants. Harvard University Press Cambridge Massachusetts (1990)
23. Attygalle, A.B., Morgan E.D.: Ant trail pheromones. Adv. Insect. Physiol. **18** (1985) 1–30
24. Camazine, S., Sneyd, S.: A model of collective nectar source selection by honeybees: self-organization through simple rules. J. Theor. Biol. **149** (1991) 547–571
25. Devigne, C., Renon, A.J., Detrain, C.: Out of sight but not out of mind: modulation of recruitment according to home range marking in ants. Anim. Behav. **67** (2004) 1023–1029

Iterated Ants: An Experimental Study for the Quadratic Assignment Problem

Wolfram Wiesemann[1] and Thomas Stützle[2]

[1] Fachbereich Wirtschaftswissenschaften
Technische Universität Darmstadt, Darmstadt, Germany
wolfram.wiesemann@gmail.com
[2] IRIDIA, CoDE, Université Libre de Bruxelles, Brussels, Belgium
stuetzle@ulb.ac.be

Abstract. Ant colony optimization (ACO) algorithms construct solutions each time starting from scratch, that is, from an *empty* solution. Differently from ACO algorithms, iterated greedy, another constructive stochastic local search method, starts the solution construction from *partial solutions*. In this paper, we examine the performance of a variation of $\mathcal{MAX}$–$\mathcal{MIN}$ Ant System, one of the most successful ACO algorithms, that exploits this idea central to iterated greedy algorithms. We consider the quadratic assignment problem as a case-study, since this problem was also tackled in a closely related research to ours, the one on the usage of *external memory* in ACO. The usage of external memory resulted in ACO variants, where partial solutions are used to seed the solution construction. Contrary to previously reported results on external memory usage, our computational results are more pessimistic in the sense that starting the solution construction from partial solutions does not necessarily lead to improved performance when compared to state-of-the-art ACO algorithms.

1 Introduction

Ant colony optimization (ACO) algorithms generate candidate solutions for an optimization problem by a construction mechanism where the choice of the solution component to be added at each construction step is probabilistically biased by (artificial) pheromone trails and heuristic information [1]. In usual ACO algorithms, each (artificial) ant starts from an initially *empty* solution and adds solution components to its current partial solution until a complete candidate solution is obtained.

In this article, we examine the possibility of changing this choice by starting the solution construction from partial solutions that are obtained by removing some solution components of an ant's current solution. This modification of ACO algorithms is in large part inspired by the iterated greedy (IG) method. IG algorithms iterate over construction algorithms in the following way. Given some initial solution s, first some solution components are removed, resulting in a partial candidate solution s_p. Starting from s_p a complete candidate solution s'

M. Dorigo et al. (Eds.): ANTS 2006, LNCS 4150, pp. 179–190, 2006.

is reconstructed by a greedy construction heuristic. An acceptance criterion then decides from which of the two solutions s and s' the next iteration continues. IG algorithms are at the core of high-performing algorithms for the set covering problem [2,3,4] and are state-of-the-art for the permutation flow-shop problem [5]. The idea underlying IG can easily be transferred to ACO. The extension followed here essentially says that some solution components of an ant's current solution are removed according to some rule and complete candidate solutions are reconstructed, following the usual construction rules of ACO algorithms. We integrate this process into the $\mathcal{MAX}$–$\mathcal{MIN}$ Ant System ($\mathcal{MM}$AS) algorithm [6], resulting into an iterated ants $\mathcal{MAX}$–$\mathcal{MIN}$ Ant System (ia$\mathcal{MM}$AS).

Another line of research relevant for this paper is the usage of external memory in ACO algorithms as proposed in the papers by Acan [7,8]. There, external memory involves the storage of partial solutions extracted from the best solutions obtained after each iteration of the ACO algorithm. Essentially, the partial solutions comprise a number of randomly chosen solution components of complete solutions. From these partial solutions, an ant chooses one according to a selection scheme that gives preference to better partial solutions. This idea was tested by Acan using a modified $\mathcal{MAX}$–$\mathcal{MIN}$ Ant System algorithm for the Traveling Salesman Problem [7] and the Quadratic Assignment Problem (QAP) [7,8] as a case study. Although in these papers the "iterated ants" variant of $\mathcal{MM}$AS was shown to reach better average solution qualities than a re-implementation of $\mathcal{MM}$AS, the overall computational results were rather poor (in fact, by far worse than previously published results with $\mathcal{MM}$AS, as can easily be seen when comparing the results across the papers [6] and [7,8]). The reason for this discrepancy may be that no well performing local search was included to improve the ants' solutions. Differently, in this article we extend a high-performing ACO algorithm by the iterated ants idea, namely the $\mathcal{MAX}$–$\mathcal{MIN}$ Ant System algorithm for the QAP by Stützle and Hoos [6], which is known to be a high performing ACO algorithm for the QAP [9]. This is done because we consider the ability to reach or surpass state-of-the-art performance a clear-cut and desirable benchmark for the significance of new features that are introduced in ACO algorithms.

The remainder of this article is structured as follows. In Section 2 we give the most important details of the $\mathcal{MM}$AS application for the QAP and the extension we used to implement the ia$\mathcal{MM}$AS algorithm. Section 3 gives the results of an extensive computational study and we conclude in Section 4.

2 QAP, $\mathcal{MAX}$–$\mathcal{MIN}$ Ant System and Iterated Ants

Quadratic Assignment Problem. The QAP is a widely studied $\mathcal{NP}$-hard problem [10] that models many real-life problems arising in the location of facilities like units in a hospital or the layout of keyboards [11,12]. In the QAP, one is given two $n \times n$ matrices A and B, where a_{ij} is the distance between locations i and j and b_{kl} is the flow (e.g. of material, patients etc.) between units k and l. The cost contribution of assigning units k and l to locations i and j, respectively, amounts to $a_{ij} \cdot b_{kl}$ and the goal is to minimize the cost arising from all interactions. A solution for the QAP can be represented by a permutation π, where $\pi(i)$ gives the unit assigned to location i. The objective function of the QAP is

$$f(\pi) = \sum_{i=1}^{n} \sum_{j=1}^{n} a_{ij} \cdot b_{\pi(i)\pi(j)}. \tag{1}$$

The QAP is one of the hardest optimization problems to solve to optimality. Currently, Stochastic local search (SLS) algorithms define the state-of-the-art approaches for finding near-optimal solutions in a reasonable amount of time [13]. Among the various available SLS methods, ACO has shown to be very successful on the QAP [9] and $\mathcal{MM}$AS to be among the top performing ACO algorithms for this problem [6].

$\mathcal{MAX}$–$\mathcal{MIN}$ Ant System for the QAP ($\mathcal{MM}$AS-QAP). In $\mathcal{MM}$AS-QAP, the artificial pheromone trails τ_{ij} give the desirability of assigning a unit j to location i. The solution construction of an ant in $\mathcal{MM}$AS-QAP is done by first ordering the locations randomly and then assigning in the logical construction step l a unit to the location at the lth position. The probability of assigning a still unassigned unit j to a location i in this step is given by

$$p_{ij} = \frac{\tau_{ij}}{\sum_{k \in \mathcal{N}(i)} \tau_{ik}}, \tag{2}$$

where $\mathcal{N}(i)$ is the set of still unassigned units, that is, those units that are still to be assigned to some location. Two remarks are noteworthy here: First, $\mathcal{MM}$AS-QAP does not use any heuristic information in the solution construction. Second, two variants of $\mathcal{MM}$AS-QAP were proposed – a first one makes at each construction step a probabilistic choice according to Equation 2 [6], whereas a second one uses the pseudo-random proportional action choice rule originally proposed for Ant Colony System [14]. Here we focus on the first version. The pheromone update in $\mathcal{MM}$AS-QAP is done by lowering the pheromone trails by a constant factor ρ and depositing pheromone on the individual assignments of either the best solution in the current iteration, the best solution found so far by the algorithm or the best solution found since the last re-initialization of the pheromone trails. A pheromone reinitialization is triggered if the branching factor is below a certain threshold and for a given number of iterations no improved candidate solution has been found. For details on $\mathcal{MM}$AS-QAP and its parameter settings we refer to [6] (the only difference to the earlier proposed parameter settings was the usage of a setting of $\rho = 0.1$ instead of the earlier $\rho = 0.8$ because of a slightly improved performance).

$\mathcal{MM}$AS-QAP uses local search for improving each candidate solution generated by the ants. Here we use an iterative improvement algorithm in the 2–exchange neighborhood, where two candidate solutions are neighbored if they differ in the assignment of exactly 2 units to locations. The local search algorithm uses a best-improvement pivoting rule.

Iterated Ants. In ACO algorithms, candidate solutions are generated by starting each solution construction from scratch. However, other constructive SLS

```
procedure Iterated_Greedy
    s0 := GenerateInitialSolution
    s := LocalSearch(s0)
    repeat
        sp := Destruct(s)
        s' := Construct(sp)
        s' := LocalSearch(s')        % optional
        s := AcceptanceCriterion(s, s')
    until termination condition met
end
```

Fig. 1. Outline of an IG algorithm. s_p is a partial candidate solution.

methods are known that use repeated solution constructions as well but start from partial solutions – this is the central idea of *iterated greedy* (IG). IG algorithms start with some complete candidate solution and then cycle through a main loop consisting of three procedures; first, a procedure *Destruct* removes from a complete solution s a number of solution components, resulting in a partial candidate solution s_p. Starting from s_p, a complete candidate solution s' is reconstructed by a procedure *Construct* and finally, a procedure *AcceptanceCriterion* decides whether the next iteration is continued from s or s'. In addition, it is straightforward to include a local search procedure that improves a candidate solution once it is completed by *Construct.* This results in the overall outline of an IG algorithm that is given in Figure 1.

Iterated ants applies the central idea underlying IG to ACO. This can be done in a rather straightforward way by considering each ant as implementing an IG algorithm. That is, an individual ant follows the steps of an IG algorithm and, hence, the solution construction in the ACO-IG hybrid algorithm starts from a partial solution that is obtained from deleting solution components from a complete candidate solution of an ant. The solution construction by the ants follows the same steps as usual in ACO algorithms, that is, solution components are added taking into account pheromones and possibly heuristic information.

In this article, we integrated this idea directly into the $\mathcal{MM}$AS algorithm and tested it for the QAP. We study the behavior of the algorithm considering various choices for the procedures *Destruct*, where we varied the choice of how solution components are removed from complete solutions, how many solution components are removed, and the type of pheromone update chosen in the algorithm. (Note that removing a solution component for the QAP refers to undoing an assignment of a unit to a location). For the solution components to be removed, we study three variants.

- `rand`: the solution components to be removed are chosen randomly according to a uniform distribution.
- `prob`: the probability of removing a solution component is proportional to the corresponding pheromone trail τ_{ij}, that is, the higher the pheromone trail, the more likely it is to remove a solution component.

- `iprob`: the probability of removing a solution component is inversely proportional to the associated pheromone trail τ_{ij}, that is, the lower the associated pheromone trail, the more likely it is to remove a solution component.

Regarding the number of solution components to be removed we consider two different possibilities.

- `fixed(k)`: Of each ant exactly k solution components are removed, where k is a parameter. In the experimental study, various values for k were tested.
- `variable`: In this case, the number of solution components to be removed is not fixed *a priori*, but is maintained variable similar to how the perturbation strength is modified in the simple variable neighborhood search: if for a current value k no improved solution is obtained for the ant, we set $k := k+1$; otherwise, k is set to some minimum value.

Finally, we also consider two possibilities for the pheromone update rule in $\mathcal{MM}$AS-QAP.

- $\texttt{gb}^+$: The pheromone update rule is the very same as the one used in the original $\mathcal{MM}$AS-QAP algorithm.
- $\texttt{gb}^-$: The best-so-far candidate solution and the best candidate solution since the pheromone re-initialization are not taken into account when depositing pheromone, that is, only the iteration-best ant deposits pheromone.

3 Computational Study

We tested iterated ants for the QAP on eight instances ranging in size from $n = 30$ to $n = 100$ from QAPLIB [15]. The eight instances are chosen such that they have different instance characteristics, representative for most of the available QAP instances that can be found at QAPLIB. The tested instances include (i) instances where both matrix entries are generated according to a uniform distribution, (ii) instances where the distance matrix corresponds to the Manhattan distances between locations on a grid, (iii) some instances derived from real-life applications of the QAP, and (iv) randomly generated instances that resemble the structure of real-life QAP instances.

All the experimental results, except where indicated differently, are measured across 25 independent trials of the algorithms and the code was run on a IBM X31 ThinkPad-Notebook with a Pentium M 1.5 GHz processor, 512 MB RAM and 1024 kB L2-Cache. For all instances we first ran the reference strategy, $\mathcal{MM}$AS-QAP for 500 iterations and measure the average time to finish the trials. This stopping time was then taken as the termination criterion for all variants; that is, all variants were given the same computation time. We compare the algorithms using the average percentage excess over the best-known solutions (proven optimal solutions are only available for two instances `kra30a` and `ste36a`) and for each comparison we use the non-parametric Wilcoxon test to check the statistical significance of the observed differences in performance. For each comparison, we give the corresponding p-values (we assume a test-wide

Table 1. Comparison of variant `var-rand-gb`$^{+}$ and `var-rand-gb`$^{-}$. The best results for each instance are indicated in bold-face.

	tai60a	**tai80a**	**sko81**	**kra30a**	**ste36a**	**tai60b**	**tai80b**	**tai100b**
`var-rand-gb`$^{+}$	**1.93**	**1.50**	**0.23**	0.36	**0.21**	0.23	0.66	0.38
`var-rand-gb`$^{-}$	2.03	1.54	0.23	**0.21**	0.27	**0.07**	**0.48**	**0.32**
p-value	0.2810	0.9062	0.4576	0.9938	0.9062	0.4676	0.1545	0.6994

Table 2. Comparison of variant `var-prob-gb`$^{+}$ and `var-prob-gb`$^{-}$. The best results for each instance are indicated in bold-face.

	tai60a	**tai80a**	**sko81**	**kra30a**	**ste36a**	**tai60b**	**tai80b**	**tai100b**
`var-prob-gb`$^{+}$	**1.99**	1.60	0.27	0.25	**0.33**	0.26	0.79	0.35
`var-prob-gb`$^{-}$	2.04	**1.58**	**0.26**	**0.19**	0.48	**0.11**	**0.54**	**0.31**
p-value	0.4676	0.9062	0.2810	0.9938	0.2810	0.2810	0.1545	0.1545

α-level of 0.05 when speaking of statistical significance). Note that in the case of multiple, say x comparisons, we assume to use Bonferroni corrections [16], that is we only speak of statistical significance if the p-values are lower than α/x. If the hypothesis test on equal performance between two algorithms is to be rejected, we will indicate this by marking the given p-values in boldface.

3.1 Study of ia$\mathcal{MM}$AS Parameters

In this section, we present the results on the parameter settings for the ia$\mathcal{MM}$AS-QAP algorithm, while the next will be concerned with a comparison with $\mathcal{MM}$AS-QAPand an analysis of the run-time behavior of ia$\mathcal{MM}$AS-QAP. In what follows, we refer to the variants by using abbreviations in dependence of the choice for the three strategies; for example, `var-rand-gb`$^{+}$ refers to the variant that uses a variable number of components to delete, deletes randomly chosen components, and uses the usual $\mathcal{MM}$AS pheromone update rule with the best-so-far update.

Pheromone Update. First, we tested the two possible variants for the pheromone update, namely `gb`$^{+}$ and `gb`$^{-}$. In this test, we used only the `variable` strategy for the number of solution components to be removed and we considered all three possible choices of {`rand`, `prob`, `iprob`}. The results in Tables 1 to 3 show that the differences between the two strategies `gb`$^{+}$ and `gb`$^{-}$ are, independent of the way solution components are removed, rather minor and most of the observed differences are statistically insignificant. Since `gb`$^{+}$ turned out to be superior for the few cases where statistically significant differences were observed, we continue to use this one for the remainder of the paper. (Note that additional tests confirmed that this conclusion is the same if the `fixed(k)` strategy were chosen.)

Strategies for Choosing Solution Components for Removals. As a next step we examined the influence of the type of deletions of solution components,

Table 3. Comparison of variant `var-iprob-gb`$^+$ and `var-iprob-gb`$^-$. The best results for each instance are indicated in bold-face.

	tai60a	tai80a	sko81	kra30a	ste36a	tai60b	tai80b	tai100b
`var-iprob-gb`$^+$	**1.77**	**1.38**	**0.18**	0.45	**0.12**	**0.17**	0.52	0.32
`var-iprob-gb`$^-$	2.00	1.51	0.20	**0.37**	0.17	0.18	**0.49**	**0.28**
p-value	**0.0002**	**0.0148**	0.4676	1.0000	0.9062	0.9938	0.6994	0.9062

Table 4. Comparison of the variants `var-rand-gb`$^+$, `var-prob-gb`$^+$ and `var-iprob-gb`$^+$

	tai60a	tai80a	sko81	kra30a	ste36a	tai60b	tai80b	tai100b
`var-rand-gb`$^+$ (1)	1.93	1.50	0.23	0.36	0.21	0.23	0.66	0.38
`var-prob-gb`$^+$ (2)	1.99	1.60	0.27	**0.25**	0.33	0.26	0.79	0.35
`var-iprob-gb`$^+$ (3)	**1.77**	**1.38**	**0.18**	0.45	**0.12**	**0.17**	**0.52**	**0.32**
p-value (1)/(2)	0.4676	0.1545	0.0366	0.9062	0.2810	**0.0158**	0.2810	0.2850
p-value (1)/(3)	**0.0063**	**0.0056**	0.2810	0.6994	0.2810	0.9062	0.1545	0.9938
p-value (2)/(3)	**0.0002**	≈ 0	**0.0002**	0.6994	0.0783	**0.0063**	**0.0023**	0.9062

which is one of {`rand`, `prob`, `iprob`}. The computational results in Table 4 indicate that the strategy `iprob` gives best overall results. On many instances the performance of `iprob` is statistically better than that of `prob` and for all instances the average solution quality obtained by `iprob` is better than that of `rand`. Note that `iprob` corresponds to an intensification of the search compared to the other two variants, since assignments that have associated a high pheromone value are more likely to remain in partial solutions.

Number of Removals. In a final step we examined the influence of the two strategies for removing solution components (`variable` vs. `fixed(k)`) and for the latter also the setting for the parameter k that determines how many solution components are removed. (The results of `var-iprob-gb`$^+$ can be taken from Table 4.) Ideally, we would see a pattern that suggests good settings for k. Yet, an inspection of Table 5 shows that there is no clear trend observable (best average performance is in boldface). One may tentatively conjecture that except for two instances with uniformly random distance and flow matrices (`tai60a` and `tai80a`) the number of solution components to be removed can be large, to be on the safe side (note that for four of the instances the lowest average deviation from the best known solutions is reached for the highest value for the parameter tested).

3.2 Comparison of $\mathcal{MM}$AS and ia$\mathcal{MM}$AS

As a next step we compare the performance of the original $\mathcal{MM}$AS-QAP with ia$\mathcal{MM}$AS-QAP. Before the comparison in terms of solution quality reached, we first consider the difference in the number of iterations applicable in an *a priori* fixed computation time for all variants.

Table 5. Comparison of various values for parameter k in variant `fixed(k)-iprob-gb`$^{+}$. The best results for each instance are indicated in bold-face.

	tai60a	**tai80a**	**sko81**	**kra30a**	**ste36a**	**tai60b**	**tai80b**	**tai100b**
k = 10	1.59	**1.24**	0.17	0.79	0.24	0.27	0.86	0.61
k = 20	1.68	1.38	0.11	**0.22**	0.31	0.11	**0.25**	0.11
k = 30	**1.43**	1.57	0.07	–	**0.06**	0.17	0.42	0.10
k = 40	1.53	1.37	0.11	–	–	0.25	0.58	0.16
k = 50	1.69	1.36	0.10	–	–	**0.00**	0.47	0.10
k = 60	–	1.32	**0.01**	–	–	–	0.31	0.10
k = 70	–	1.28	0.12	–	–	–	0.38	**0.07**

Table 6. Execution time for trials of 500 iterations for the reference strategy on the tested instances. Given are the 0.25, 0.5, and 0.75 quantiles of the measured times.

	tai60a	**tai80a**	**sko81**	**kra30a**	**ste36a**	**tai60b**	**tai80b**	**tai100b**
$q_{0.25}$	8.57	20.22	25.81	1.30	2.35	10.87	26.15	51.12
$q_{0.5}$	8.61	20.35	25.89	1.31	2.37	10.97	26.41	52.19
$q_{0.75}$	8.64	20.49	26.08	1.31	2.37	11.00	26.87	53.05

Speed. Table 6 gives the computation times $\mathcal{MM}$AS-QAP requires to complete 500 iterations, while Table 7 summarizes the number of iterations the three variants can apply until meeting the termination criterion (median time taken by $\mathcal{MM}$AS-QAP). The data show that the `variable` strategy and the setting `k=30` for most instances can do many more iterations than $\mathcal{MM}$AS-QAP in the same time. This is in part due to the more rapid solution construction. However, in the QAP the interpretation of this increased number of iterations must be slightly different, because about 98% of the computation time are taken by the local search. Hence, an explanation for the increased number of iterations is rather that the solutions generated by the iterated ants require less iterations of the iterative improvement local search, probably because a part of the solution is maintained from a previous local optimum. (Recall that in our algorithm we make the common use of local search in ACO and that removals of solution components start from locally optimal solutions.)

Comparison Based on Summary Statistics. Now we compare the performance of the reference strategy, $\mathcal{MM}$AS-QAP, to the best performing variant using the `variable` strategy and two parameter setting for the `fixed(k)` strategy – one keeping $k = 30$ constant and one choosing for each instance the settings that gave the best average solution quality. The comparison to the latter case is certainly unfair; however, it is interesting because it gives an impression of what would be the best-case performance if for each instance we would know *a priori* the appropriate parameter setting for the `fixed` strategy. The average solution quality reached by each of the four algorithms is given in Table 8. In Table 9 we give the p-values for the comparisons among the results, indicating those comparisons where the difference would be statistically significant. As it

Table 7. Number of iterations for variants `var-iprob-gb`$^+$ (`var`), `fixed(30)-iprob-gb`$^+$ (`k=30`) and `fixed(k`$_{opt}$`)-iprob-gb`$^+$ (`k = k`$_{opt}$). Given are again the corresponding 0.25, 0.5, and 0.75 quantiles.

		tai60a	**tai80a**	**sko81**	**kra30a**	**ste36a**	**tai60b**	**tai80b**	**tai100b**
var	$q_{0.25}$	864	888	1016	884	991	1006	1033	1030
	$q_{0.5}$	868	890	1019	913	1010	1029	1040	1033
	$q_{0.75}$	872	893	1027	925	1021	1053	1049	1040
k = 30	$q_{0.25}$	647	418	576	482	574	718	724	700
	$q_{0.5}$	678	420	719	485	578	723	739	729
	$q_{0.75}$	690	558	789	490	583	728	748	761
k = k$_{opt}$	$q_{0.25}$	647	841	600	638	574	563	616	583
	$q_{0.5}$	678	844	607	646	578	566	641	599
	$q_{0.75}$	690	845	611	650	583	570	733	608

Table 8. Average percentage deviation from best known solutions for $\mathcal{MM}$AS-QAP (`ref`), the best `variable` strategy, and two fixed settings. The lowest average percentage deviations on each instance are indicated in boldface.

	tai60a	**tai80a**	**sko81**	**kra30a**	**ste36a**	**tai60b**	**tai80b**	**tai100b**
ref	**0.13**	0.13	0.09	1.60	**0.00**	1.29	**0.06**	0.14
variable	0.45	0.18	0.12	1.77	0.17	1.38	0.52	0.32
k=30	–	**0.07**	**0.06**	**1.43**	0.17	1.57	0.42	0.10
k=k$_{opt}$	0.22	**0.07**	**0.06**	**1.43**	**0.00**	**1.24**	0.25	**0.07**

Table 9. p-values of the two-sided Wilcoxon tests comparing the algorithms' performance. We indicate in boldface the results that are statistically significant after α-corrections according to Bonferroni.

	tai60a	**tai80a**	**sko81**	**kra30a**	**ste36a**	**tai60b**	**tai80b**	**tai100b**
r/v	0.4676	0.0366	0.9938	**0.0063**	0.0158	0.4755	**≈ 0**	**0.0008**
r/k$_{30}$	–	**0.0008**	0.4676	0.1545	0.6994	**0.0019**	**0.0063**	0.1545
r/k$_{opt}$	0.9938	**0.0008**	0.4676	0.1545	0.9999	0.6994	0.0783	0.0158
v/k$_{30}$	–	**≈ 0**	0.2810	**0.0002**	0.2810	**0.0056**	0.0783	**0.0002**
v/k$_{opt}$	0.6994	**≈ 0**	0.2810	**0.0002**	0.0158	0.0783	**0.0023**	**≈ 0**
k$_{30}$/k$_{opt}$	–	1.0000	1.0000	1.0000	0.6994	**0.0006**	0.6994	0.6994

can be seen, the performance of iterated ants is roughly on par with the original $\mathcal{MM}$AS-QAP algorithm. Focusing on the variant with $k = 30$, we see that iterated ants perform statistically superior only on one instance, whereas they seem inferior on two others (other instances show no statistically significant differences). If the best setting of k would be known *a priori*, the results would be slightly more positive for the iterated ants. However, this is an ideal case and unlikely to be reachable in practice. Finally, the variant of iterated ants where the number of solution components to be removed is kept variable performs worse than $\mathcal{MM}$AS-QAP on all instances tested, although only on three the differences are statistically significant.

Comparison Based on Run-Time Distributions. In a final step we analyzed the run-time behavior of ia$\mathcal{MM}$AS-QAP and compared it with that of $\mathcal{MM}$AS-QAP. For doing so, we made use of the methodology based on measuring empirical run-time distributions [13]. The (qualified) run-time distribution (RTD) characterizes for a given algorithm and instance the development over time of the probability of reaching a candidate solution within a specific bound on the desired solution quality. As usual, we have chosen very high quality limits to be reached by the algorithms. In Figure 2 we show exemplary RTDs measured across 100 trials of the algorithms on several instances; 100 trials are chosen to make the RTDs reasonably stable. In the RTD plots we included two exponential distributions that give an indication of whether an algorithm shows stagnation behavior or not; one exponential distribution for the original $\mathcal{MM}$AS-QAP and one for the ia$\mathcal{MM}$AS-QAP algorithm that shows best performance. (For a detailed explanation of how to check for stagnation behavior see [13,17].) As it can be seen from these plots, the iterated ants variants are somewhat prone to stagnation behavior on instances `kra30a`, and `tai60b`, while no clear sign of

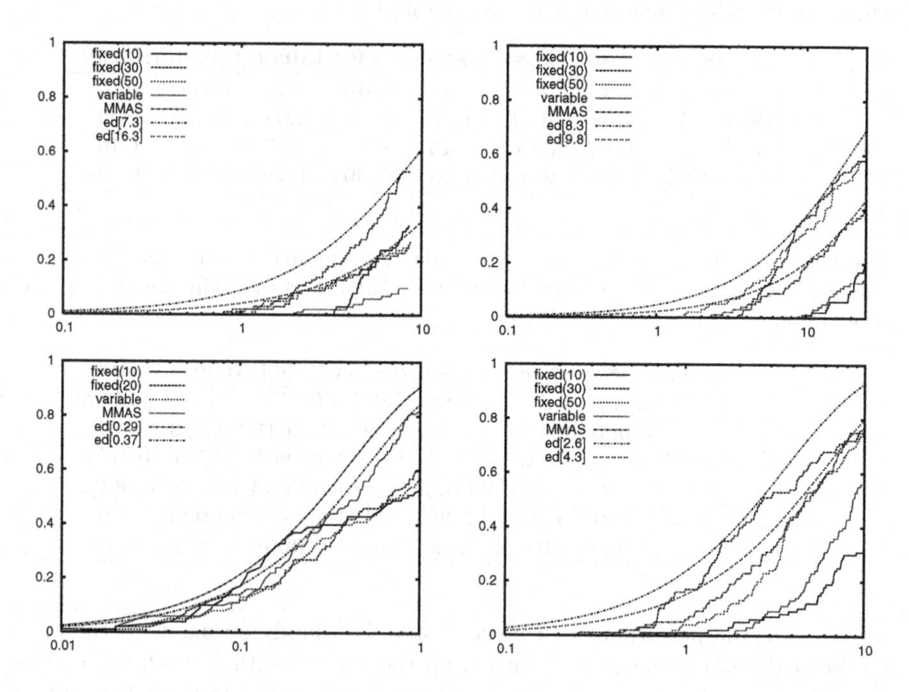

Fig. 2. RTDs measured with respect to high quality solutions for various of the QAP instances tested. From top left to bottom right the RTDs are given for instances `tai60a` (1.5%), `sko81` (0.1%), `kra30a` (optimum), `tai60b` (best–known solution value); in parentheses are given the target solution quality for the qualified RTDs. The exponential distributions in each plot indicate the idealized performance with an optimal restart strategy (indicated by $ed[d]$, where the value for d refers to a functional form of $ed(x) = 1 - 2^{-x/d}$).

stagnation is detected on the other instances. $\mathcal{MM}$AS-QAP, in contrast, apparently does not suffer from a significant stagnation behavior on any of the instances. For the algorithm variants on which stagnation behavior is observed, the associated exponential distribution indicates the behavior if an optimal restart-strategy or some effective diversification features would additionally be included into the algorithm. Hence, this analysis of the RTDs suggests that the ia$\mathcal{MM}$AS-QAP algorithm could for some instances strongly benefit from such strategies, contrary to $\mathcal{MM}$AS-QAP. As a result, by investing some further work, the ia$\mathcal{MM}$AS-QAP algorithm may become (more) competitive to $\mathcal{MM}$AS-QAP.

4 Discussion and Conclusions

In this paper, we examined the possibility of integrating the essential ideas of the iterated greedy method into ACO algorithms. The computational results of this adaptation show that the introduction of the idea of using partial solutions to seed the solution construction of ants does not necessarily improve the performance of a state-of-the-art ACO algorithm. This can be considered as a kind of negative result on the combination of two SLS methods into a higher performing algorithm, a combination that looked promising at first sight.

In fact, this negative result is a bit in contrast to the positive results reported in an earlier, similar approach that mainly differs in the way partial solutions are chosen by the ants [7,8]. However, these positive results were reported with respect to a version of $\mathcal{MM}$AS that performed rather poorly and for which the reported improved performance of the extended algorithm is still far away from the results reported here. Similarly, earlier research reported positive results for an "iterated ants" algorithm for the Unweighted Set Covering Problem, where standard local search algorithms appear not to be extremely high performing [18]. While the integration of the iterated ants concept into a state-of-the-art ACO algorithm failed to yield significant improvements, these two researches indicate that the exploitation of the iterated ants idea could be more promising if either the final solution quality reached by the ACO algorithm is still far from optimal or no effective local search for a problem exists.

Acknowledgments. Thomas Stützle acknowledges support of the Belgian FNRS, of which he is a research associate.

References

1. Dorigo, M., Stützle, T.: Ant Colony Optimization. MIT Press, USA (2004)
2. Brusco, M.J., Jacobs, L.W., Thompson, G.M.: A morphing procedure to supplement a simulated annealing heuristic for cost- and coverage-correlated set covering problems. Annals of Operations Research **86** (1999) 611–627
3. Jacobs, L.W., Brusco, M.J.: A local search heuristic for large set-covering problems. Naval Research Logistics **42**(7) (1995) 1129–1140

4. Marchiori, E., Steenbeek, A.: An evolutionary algorithm for large scale set covering problems with application to airline crew scheduling. In Cagnoni, S., et al., eds.: Real-World Applications of Evolutionary Computing, EvoWorkshops 2000. Volume 1803 of LNCS., Springer Verlag (2000) 367–381
5. Ruiz, R., Stützle, T.: A simple and effective iterated greedy algorithm for the permutation flowshop scheduling problem. European Journal of Operational Research (In press)
6. Stützle, T., Hoos, H.H.: $\mathcal{MAX}$–$\mathcal{MIN}$ Ant System. Future Generation Computer Systems **16**(8) (2000) 889–914
7. Acan, A.: An external memory implementation in ant colony optimization. In Dorigo, M., et al., eds.: ANTS'2004, Fourth Internatinal Workshop on Ant Algorithms and Swarm Intelligence. Volume 3172 of LNCS., Springer Verlag (2004) 73–84
8. Acan, A.: An external partial permutations memory for ant colony optimization. In Raidl, G., Gottlieb, J., eds.: Evolutionary Computation in Combinatorial Optimization. Volume 3448 of LNCS., springer-lncs (2005) 1–11
9. Stützle, T., Dorigo, M.: ACO algorithms for the quadratic assignment problem. In Corne, D., Dorigo, M., Glover, F., eds.: New Ideas in Optimization. McGraw Hill, UK (1999) 33–50
10. Sahni, S., Gonzalez, T.: P-complete approximation problems. Journal of the ACM **23**(3) (1976) 555–565
11. Burkard, R.E., Çela, E., Pardalos, P.M., Pitsoulis, L.S.: The quadratic assignment problem. In Pardalos, P.M., Du, D.Z., eds.: Handbook of Combinatorial Optimization. Volume 2. Kluwer Academic Publishers (1998) 241–338
12. Çela, E.: The Quadratic Assignment Problem: Theory and Algorithms. Kluwer Academic Publishers (1998)
13. Hoos, H.H., Stützle, T.: Stochastic Local Search—Foundations and Applications. Morgan Kaufmann Publishers (2004)
14. Stützle, T., Hoos, H.: $\mathcal{MAX}$–$\mathcal{MIN}$ Ant System and local search for combinatorial optimization problems. In Voss, S., et al., eds.: Meta-Heuristics: Advances and Trends in Local Search Paradigms for Optimization. Kluwer Academic Publishers (1999) 137–154
15. Hahn, P.: QAPLIB - a quadratic assignment problem library. `http://www.seas.upenn.edu/qaplib` (2006) Version visited last on 15 February 2006.
16. Sheskin, D.J.: Handbook of Parametric and Nonparametric Statistical Procedures. second edn. Chapman & Hall / CRC, Boca Raton, Florida, USA (2000)
17. Stützle, T., Hoos, H.H.: Analysing the run-time behaviour of iterated local search for the travelling salesman problem. In Hansen, P., Ribeiro, C.C., eds.: Essays and Surveys on Metaheuristics. Kluwer Academic Publishers (2001) 589–611
18. Lessing, L.: Ant colony optimization for the set covering problem. Master's thesis, Intellectics Group, Computer Science Department, Darmstadt University of Technology (2004)

Negotiation of Goal Direction for Cooperative Transport

Alexandre Campo, Shervin Nouyan, Mauro Birattari, Roderich Groß, and Marco Dorigo

IRIDIA, CoDE, Université Libre de Bruxelles, Brussels, Belgium
{acampo, snouyan, mbiro, rgross, mdorigo}@ulb.ac.be

Abstract. In this paper, we study the cooperative transport of a heavy object by a group of robots towards a goal. We investigate the case in which robots have partial and noisy knowledge of the goal direction and can not perceive the goal itself. The robots have to coordinate their motion to apply enough force on the object to move it. Furthermore, the robots should share knowledge in order to collectively improve their estimate of the goal direction and transport the object as fast and as accurately as possible towards the goal.

We propose a bio-inspired mechanism of negotiation of direction that is fully distributed. Four different strategies are implemented and their performances are compared on a group of four real robots, varying the goal direction and the level of noise. We identify a strategy that enables efficient coordination of motion of the robots. Moreover, this strategy lets the robots improve their knowledge of the goal direction. Despite significant noise in the robots' communication, we achieve effective cooperative transport towards the goal and observe that the negotiation of direction entails interesting properties of robustness.

1 Introduction

There are several advantages when using a group of robots instead of a single one. Ideally, the behaviour of a group of robots is more robust, as one robot can repair or replace another one in case of failure. Furthermore, a group of robots can overcome the limitations of a single robot and solve complex tasks than can not be solved by a single robot.

Within collective robotics, swarm robotics is a relatively new approach to the coordination of a system composed of a large number of autonomous robots. The coordination among the robots is achieved in a self-organised manner: the collective behaviour of the robots is the result of *local* interactions among robots, and between the robots and the environment. The concept of locality refers to a situation in which a robot alone can not perceive the whole system. Each single robot typically has limited sensing, acting and computing abilities. The strength of swarm robotics lies in the properties of robustness, adaptivity and scalability of the group [1].

Foraging has been outlined as a canonical problem by Cao *et al.* [2] among those studied in collective robotics and is an important topic in swarm robotics

M. Dorigo et al. (Eds.): ANTS 2006, LNCS 4150, pp. 191–202, 2006.

too. In foraging, a group of robots has to pick up objects that are scattered in the environment. The foraging task can be decomposed in an exploration subtask followed by a transport subtask. Foraging can be applied to a wide range of useful tasks. Examples of applications are toxic waste cleanup, search and rescue, demining and collection of terrain samples.

Central place foraging is a particular type of foraging problem in which robots must gather objects in a central place. Borrowing the terminology from biology, the central place is also called the *nest* and the objects are called *prey*. We focus on a specific case in which the transport of a prey requires the combined effort of several robots. This task is called cooperative transport. Several problems need to be solved to perform this task successfully. The coordination of the movement of the robots is one of them. This problem has been investigated by Groß *et al.* [3], in situations in which either all or some robots are able to perceive the nest.

In this paper we address the case in which all robots completely lose sight of the nest during the exploration subtask of foraging. We assume that the robots have partial knowledge of the goal direction. For instance, they may have perceived the nest earlier and kept track of its direction by means of odometry [4]. Odometry is achieved using internal, proprioceptive information [5] (*e.g.*, by measuring the rotation of the wheels of a robot). The information on the movement of a robot is integrated, thus the error made on localization increases with the distance covered. In our case, this leads to an erroneous indication about the direction of the nest. If several robots attempt to transport a heavy prey in different directions they may fail to move the prey at all. Therefore, we introduce a mechanism to let the robots negotiate the goal direction. In order to meet the general principles of swarm robotics [1], this system is fully distributed and makes use of local communication only.

The mechanism we introduce is strongly inspired by a natural mechanism that has been long studied by biologists. We rely on a particular property of models designed to explain and reproduce fish schools and bird flocks [6,7,8,9]. The models available in the literature are usually composed of three behaviours: an attraction behaviour that makes the individuals stick together, a repulsion behaviour that prevents collisions among individuals, and an orientiation behaviour that coordinates the individuals' motion. It is the last of these three behaviours that we transfer and implement in our robots. Informally, the orientation behaviour lets every individual advertise locally its own orientation and update it using the mean orientation of its neighbours.

We conduct experiments with a group of four real robots that have to transport a prey moving in a direction about which they have noisy knowledge. We assess quantitatively the performance of the negotiation mechanism implemented with respect to different levels of noise and different control strategies.

In Section 2 we detail the task, the hardware, the experimental setup and the different controllers. Section 3 is devoted to the presentation of the experimental results. Section 4 concludes the paper with a discussion of the results and some ideas for future work.

2 Methods

The Task. The task is the cooperative transport of a heavy prey towards a nest by a group of four real robots. The robots are physically connected to the prey using their grippers. The nest is out of sight and the robots have no means to perceive it. The initial knowledge of each individual about the goal direction is provided with a given amount of noise.

The mass of the prey is chosen such that a single robot can not transport it. At least three robots are necessary to move the prey. A high degree of coordination of the robots' motion is required to apply enough force to the prey to move it. If the robots lack coordination, that is, if they pull in different directions, they may not be able to move the prey at all.

The robots can share knowledge using visual communication in order to collectively improve their estimate of the goal direction and transport the prey as fast and as accurately as possible towards the goal.

Hardware. *The robots:* We use the *s-bot* (Figure 1(a)), a robot of 12 cm of diameter, designed and built within the context of the SWARM-BOTS project [10,11]. An *s-bot* moves using a combination of two wheels and two tracks, which we call "treels". This system notably allows the robot to efficiently turn on the spot. The robots can physically connect to a prey or to another *s-bot* using their grippers. They are supplied with a rotational base that lets them move in an arbitrary direction while maintaining the same physical connection pattern. The robots can send visual information by means of eight triplets of red, green and blue LEDs. The LEDs are positioned on a ring around the robot. An *s-bot* activating its LEDs can be perceived by another *s-bot* by means of an omnidirectional camera which provides a 360° view.

The prey: The mass of the prey is 1.5 kilograms. At least three robots are necessary to effectively pull the prey. This weight of the prey makes the transport by a group of robots very difficult if the robots are not well synchronised.

Experimental Setup. The experiments take place in an open space. Initially, four robots are connected to the prey in a regular arrangement, thus forming a cross pattern as shown in Figure 1(c). We test four levels of noise on the robots' initial estimate of the goal direction: *no noise* (*0*), *low noise* (L), *medium noise* (M) and *high noise* (H). In the case of no noise, the initial direction of the robots is the same and points towards the nest.

The initial imprecise knowledge of the robots about the direction of the nest is modeled by a random number drawn from a von Mises distribution, which is the equivalent of the Gaussian in circular statistics [12], and well suited for directional data. This distribution is characterised by two parameters μ and κ. The direction to the nest is indicated by μ, the mean of the distribution. The level of noise is indicated by κ. The smaller κ, the more the distribution resembles a uniform distribution in $[-\pi, \pi]$. When κ is large, the distribution resembles a Gaussian of mean μ and standard deviation σ, when $\kappa \to \infty$ the relationship $\sigma^2 = \sqrt{1/\kappa}$ holds. The three levels of noise L, M, H correspond to $\kappa = 3, 2, 1$, as displayed in Figure 2(a).

After each trial, the robots are randomly permuted, so that the possible differences among robots are averaged out and can be neglected in this study. We tested 4 possible goal directions of $0, 22.5, 45$ and $67.5°$. Any direction above $90°$ is redundant as the pattern of connected robots (a cross) is symmetrical on the two perpendicular axes, and the robots are permuted at each trial. Finally, we have tested 4 possible strategies for the robots to transport the prey towards the goal (see next section for more details). In total, we performed 256 replications: we tested 4 goal directions, 4 levels of noise and 4 distinct strategies for transport. Each combination of the aforementioned parameters was tested 4 times.

To extract the results, we used a camera placed above the initial position of the prey to record videos of each trial. The experiment is stopped either when the prey has been transported to a distance of 1 meter from its initial position or after 60 seconds (an average transport takes approximately 20 seconds). A trial can also be stopped if we judge that the robots are stuck in a situation that is potentially harmful to their hardware. Indeed, if the robots do not manage to coordinate their movements, they may pull in opposite directions and thus induce a high torque to their grippers. One gripper was broken during the experiments reported here, and we wished to avoid as much as possible any further damage. Any experiment stopped without the prey being transported for more than 1 meter of distance from the initial position is considered as a transport failure.

For each trial, we have extracted the position of the prey at each time step (5 pictures per seconds) using a simple tracking software. Using these data, we have categorized the trials in transport failure or success, and measured the duration of all trials. Furthermore, we measured the angular difference between the direction in which the prey has been moved and the goal direction, as shown in Figure 2(b). Later on, we also use the term deviation to refer to this angular difference.

Robot's Controller. *Vision software:* We employ a specific vision software that allows a robot to perceive the direction pointed by a neighbour in his visual range. The perception algorithm implemented in the software is probabilistic and approximates the directions communicated by the local neighbours using a triangular pattern shown by the LEDs (see Figure 1(b)). In order to assess the quality of the vision software, we have performed a series of basic tests. We have run in total 8000 times the vision software on 8 different pictures to obtain a distribution of direction estimates. Figure 3 summarises the pooled results of the tests for a communicated direction pointing towards direction 0. The tests show that it is possible to achieve a reliable estimate of the direction pointed by neighbouring robots. As the mechanism of negotiation of direction should be robust to noise, there is no need to improve the output of the vision software with any kind of signal filter. We directly feed the negotiation mechanism with a single estimate.

Negotiation mechanism: The negotiation mechanism is bio-inspired and implemented in a straightforward manner, following closely the rules used to model the orientation behaviour of fish schools or bird flocks [9]. Let n be the total number of robots. For each robot $i \in [1, n]$, let $\mathcal{N}_i(t)$ be the set of robots in the

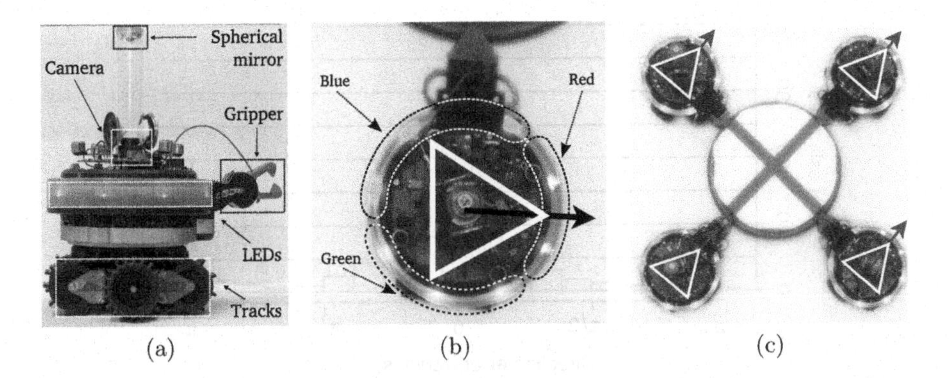

Fig. 1. (a) The *s-bot*. (b) An *s-bot* displaying a direction using a triangular LED pattern. (c) Star-like formation of four *s-bots* around the prey as used in the experiment.

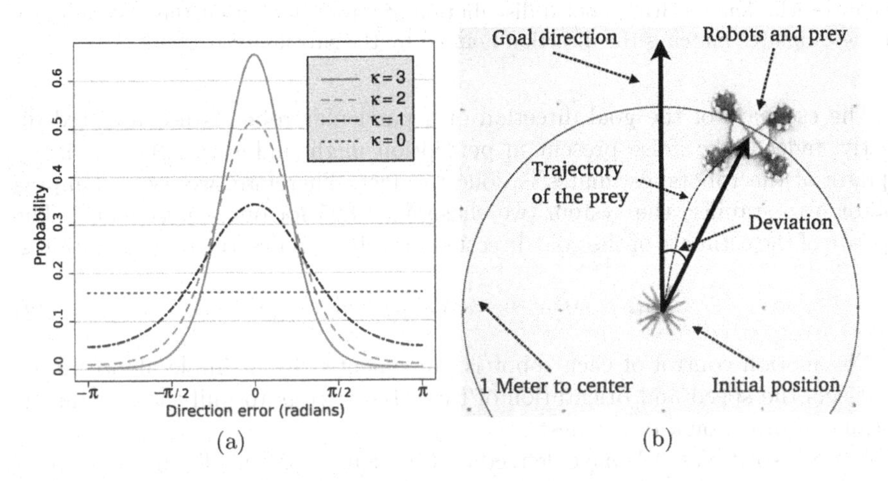

Fig. 2. (a) The effect of parameter κ on a von Mises distribution. (b) A snapshot describing the final situation of a successful transport. Note how the deviation of the transport direction from goal direction is measured.

visual range of robot i at time t. This defines the topology of the communication network. Let $d_i(t) \in [-\pi, \pi]$ be the goal direction estimated by robot i at time t. Let $D_j^i(t) = d_j(t) + \epsilon_j^i(t)$ with $j \in \mathcal{N}_i(t)$ the direction of robot j perceived by robot i assuming noise $\epsilon_j^i(t)$.

If robot i communicates and exchanges information with his neighbours, it will calculate what we call a desired direction $\overline{d}_i$ by using Equation 1 that basically computes a mean direction. To do so, we use the sum of unit vectors, which is a classical method in circular statistics [12]:

$$\overline{d}_i(t) = arctan\left(\frac{sin\left(d_i\left(t\right)\right) + \sum_{j\in\mathcal{N}(i)}\left(sin(D_j^i(t))\right)}{cos\left(d_i\left(t\right)\right) + \sum_{j\in\mathcal{N}(i)}\left(cos(D_j^i(t))\right)}\right). \quad (1)$$

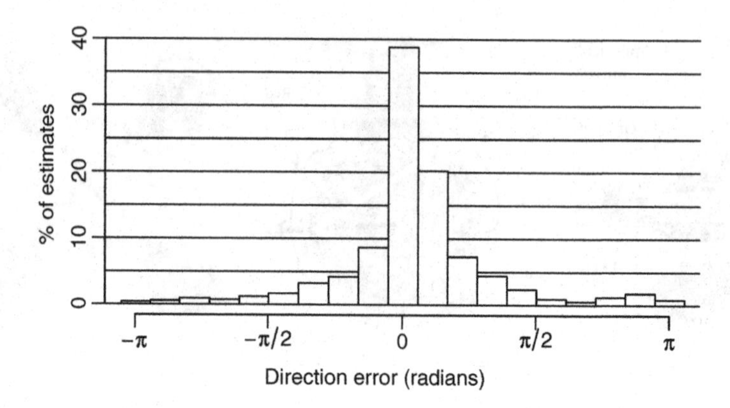

Fig. 3. We used four robots arranged in a circular pattern to display a common direction and a central robot to take pictures and estimate the direction pointed by the surrounding robots. For eight distinct directions, the central robot produced 1000 estimates each. The resulting pooled distribution of errors shows that the vast majority of the estimates matches the direction pointed by the surrounding robots.

The estimate of the goal direction of a particular robot is not updated directly. Indeed, the noise present in perception might induce oscillations if the update of the robots' estimates is done too fast. Therefore, we use a damping factor δ to stabilise the system (we chose $\delta = 0.05$ for our experiments). The update of the estimate of the goal direction for robot i is described by Equation 2:

$$d_i(t + \Delta t) = (1 - \delta) \cdot d_i(t) + \delta \cdot \overline{d}_i(t). \tag{2}$$

The motion control of each robot is implemented by a simple algorithm [3] that sets the speed and orientation of the robot's *treels* to pull the prey in the estimated direction d of the nest.

Control strategies: We have defined and implemented four distinct strategies. To refer to the strategies, we employ a notation in which **T** means transport, **N** means negotiation, and **:** marks the end of an optional and preliminary negotiation phase. If this preliminary phase takes place, it lasts 30 seconds. The second phase always involves transport and lasts 60 seconds.

- **Transport directly (T)**: a naive strategy that we use as a yardstick to show the improvement brought by the negotiation mechanism. The robots move along their initial direction. No communication and no update of the estimated direction is done.
- **Negotiate then transport (N:T)**: robots first negotiate their estimate of the direction of the goal for 30 seconds without moving. Afterwards, they all start moving without either communicating or updating their estimates.
- **Negotiate then transport and negotiate (N:NT)**: robots start by negotiating the direction of the goal for 30 seconds without moving. After this preliminary negotiation, they all start moving and at the same time they keep on negotiating together.

– **Negotiate and transport (NT)**: from the very beginning of the experiment, the robots start both moving and negotiating.

At the beginning of the experiments, robots have each a rough estimate of the direction of the goal, but they never perceive directly the goal. The three last strategies may appear identical to the reader, but in fact two important aspects, namely time and noise in communication should be considered. On one hand, the duration of the negotiation process affects the degree of synchronisation of the robots. On the other hand, visual communication is imperfect. When robots don't move, errors in visual communication are persistent and may have a strong impact on the outcome of the negotiation process. When robots move, they modify slightly their relative locations and this results in a reduction of the errors in visual communication.

3 Results

We report here the experimental results of the task of cooperative transport for all the strategies and levels of noise tested. We examine three different aspects of the system: the ability of the system to succeed in transporting the prey for a certain distance, the duration of transport and the accuracy in direction of transport. Data analysis was performed with the R software and the package *circular* [13].

Success in Transporting. We first study the ability of the robots to transport the prey. If the robots are not able to move the prey over a distance of at least 1 meter from the initial position within 60 seconds, we consider the trial as a transport failure. Figure 4 presents the performances in transport of the four strategies for the different levels of noise.

First, we observe that in absence of noise (level *0*), the robots manage very well to transport the prey without negotiating the direction. Therefore, negotiation is not necessary and it is desirable that strategies employing the negotiation mechanism do not perform worse. The strategy *N:T* yields only 75% of successful transports when there is no noise in the initial direction of the nest. When this strategy is employed, it is possible that negotiation is stopped while robots are not perfectly coordinated and no further correction can be done on the direction of the robots. The two other strategies *N:NT* and *NT* do not decrease the capability of the group of robots to transport the prey with respect to strategy *T*. We have observed that during motion, the formation of robots can alter slightly, mainly due to slippage of the grippers on the prey. Strategies using negotiation during transport allowed robots to quickly correct their direction and remain coordinated. Conversely, the strategies *T* and *N:T* were very sensitive to small errors.

When noise is present, the performance of the group of robots using strategy *T* decreases. For medium and high noise it is close to 10%. This result was expected as robots are not able to coordinate their motion at all and are initialised with different initial directions. We also notice that, although noise

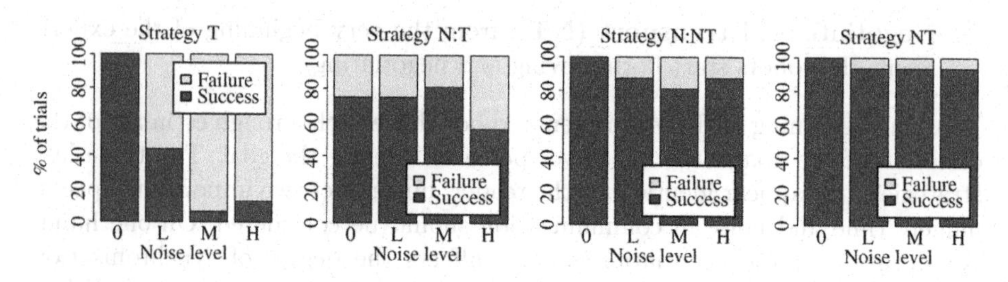

Fig. 4. The percentage of successful and failed transports grouped by strategy and by level of noise

has a non neglectable impact on the transport capability, the performances stay quite similar for different levels *L*, *M* and *H* of noise considering the strategies *N:T*, *N:NT* and *NT*. All strategies relying on the negotiation mechanism achieve better performances, and especially strategy *NT* is much less sensitive to noise than the others.

Duration of Transport. We focus now on the duration of the transport. For all the trials, we consider whether or not transport is successful. Figure 5 shows for all strategies and all levels of noise boxplots of the duration of the 16 transport tasks. Note that we do not take into account the preliminary negotiation period that lasts 30 seconds when strategies *N:T* or *N:NT* are employed.

Once again, the performance of strategy *T* in absence of noise is the best with respect to any other pair of strategy and level of noise. Only strategy *NT* reaches a comparable performance.

When the level of noise increases, the duration of transport of the strategy *T* increases too, in a quasi linear manner. Strategies that rely on the negotiation mechanism are much less sensitive to noise. The duration of transport using those strategies is very similar for the different levels of noise *L*, *M* and *H*, but strategy *N:T* has produced more failures. Because robots can not correct their coordination with this strategy, they easily rotate while transporting the prey. This constant error produces round or even circular trajectories and prevents the robots to quickly move the prey away from its initial position. Strategies can be clearly ranked: the slowest (*N:T*), the average (*N:NT*) and the fastest (*NT*).

Deviation From the Direction of the Nest. The last measure we study is the deviation of the direction of transport with respect to the direction of the nest. Again, we take into account all trials. The study of deviation from the direction of the nest confirms all previous observations (see Figure 6).

In absence of noise, the naive strategy *T* performs very well, and the only other strategy with a comparable result is strategy *NT*. When noise is introduced, the performance of strategy *T* decreases. The strategies that make use of negotiation perform better, and show only small differences for the different levels of noise tested. Among these strategies, the best is *NT*.

We have fitted von Mises distributions with the distributions of deviations in order to study strategy *NT* in further detail. The fit with a von Mises distribution

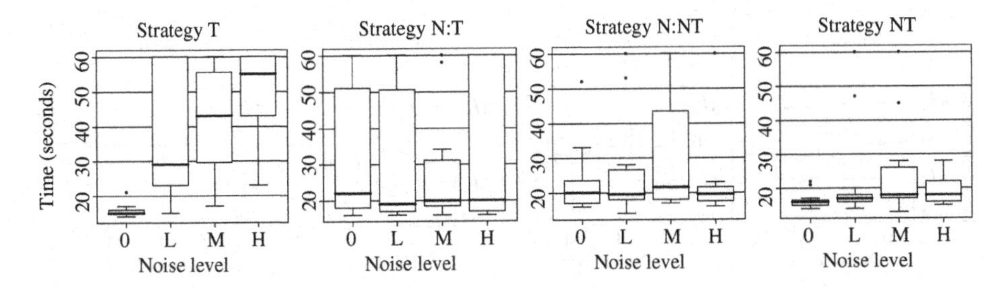

Fig. 5. Box-and-whisker plot [14] showing the duration of transport of the prey (in seconds), taking into account successful and failed transports. The distributions are grouped by strategy and by level of noise.

yields an estimate of κ, which corresponds to the error of the transport direction. The robots start with an initial knowledge affected by a noise that corresponds to an individual error of respectively 33.1°, 40.5° and 57.3°. After the application of the strategy *NT*, the final values of the error of transport direction for the levels of noise *L*, *M*, *H* are respectively 42.8° ± 76.2°, 42.3° ± 75.5° and 42.5° ± 75.7° (degrees ± standard error). These values are not significantly different. Moreover, it is observed that for the level of noise *H*, the strategy *NT* improves the robots' estimate of the direction of the nest.

4 Discussion

Achievements. We have compared different strategies to achieve efficiently the cooperative transport of a prey with partial knowledge of the direction of the nest. We performed systematic experiments to evaluate the characteristics of the different strategies under study for four distinct levels of noise. The comparison of the strategies has shown that negotiation during transport of a prey improves the coordination of motion. It has also been shown that negotiation without moving prior to transport (*N:NT*) performs worse than the straightforward strategy *NT* consisting in negotiating and transporting the prey at the same time.

It has been observed that the strategy *NT* is neutral: if negotiation is not mandatory to achieve efficient transport, making use of this strategy does not alter the transport performances with respect to the naive strategy *T*. Hence, it is not necessary to choose which strategy to employ depending on the level of noise affecting robots' knowledge of the direction of the nest. The strategy *NT* can be used at any time.

Besides the coordination of motion, our experimental results have also shown that the group of robots could improve their knowledge of the direction of the nest by means of visual negotiation. Strategy *NT* improves the robots' estimate of the direction of the nest and shows no discernible difference of the errors for the levels of noise *L*, *M*, *H*. The improvement of the accuracy of direction of transport with respect to the *s-bots* initial knowledge is most striking when the level of noise is high.

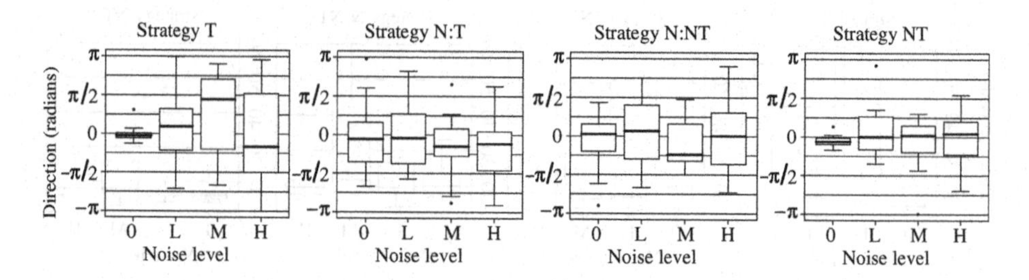

Fig. 6. Box-and-whisker plot [14] showing the average deviation (in radians) between direction of transport and direction of the nest. Both successful and failed transports are taken into account. The distributions are grouped by strategy and by level of noise.

Improvements. With respect to mechanisms of coordination of motion that use a traction sensor [3,15], our system appears to be more flexible, as visual communication is also available when the prey is not in motion, and it is not distracted if the prey moves in irregular steps. Additionally, visual communication leaves the door open to collective motion with or without transport or physical connections. The topology of the network of communications is also likely to be more flexible, allowing the robots to school in very diverse patterns.

The negotiation mechanism we have introduced is not only able to supply a group of robots with collective motion, but also to let each individual improve its own estimate of the goal direction by sharing knowledge with its neighbours. This mechanism may also be used to correct measures of odometry in multi-robot experiments, in a fully distributed fashion. This self-organised negotiation is likely to display properties of scalability besides the robustness shown in this paper.

Perspectives and Future Work. The difference in performance between strategies *N:NT* and *NT* is counter-intuitive, as the negotiation in the first strategy lasts in total longer and thus the robots are expected to achieve a better performance because they are granted more time to negotiate. However, the preliminary phase of negotiation without movement negatively affects the performance of the robots. This might be due to persistent errors in visual communication in absence of movement. It is also possible that transport is more efficient when robots align their tracks gradually, and not immediately as it happens with strategy *N:NT*. We plan to further investigate the exact reasons for this phenomenon.

We plan to integrate the cooperative transport in a more complex and challenging scenario of foraging, such as for instance the one used by Nouyan *et al.* [16]. This scenario would include an exploration phase preliminary to transport, in which robots lose sight of the nest before finding the prey. In this context, robots have a rough estimate of the direction of the nest by means of odometry. Improvement of this knowledge by means of negotiation is a critical feature of the scenario, necessary to let the robots transport the prey efficiently to the nest, even in presence of noisy communications and failed robots.

Acknowledgements. This research has been supported by COMP^{2}SYS, a Marie Curie Early Stage Research Training Site funded by the European Community's Sixth Framework Programme under contract number MEST-CT-2004-505079, and by the ANTS project, an *Action de Recherche Concertée* funded by the Scientific Research Directorate of the French Community of Belgium. Marco Dorigo acknowledges support from the Belgian FNRS of which he is a Research Director. The information provided is the sole responsibility of the authors and does not reflect the opinion of the sponsors. The European Community is not responsible for any use that might be made of data appearing in this publication.

The authors would like to thank Jean-Louis Deneubourg for fruitful discussions.

References

1. Dorigo, M., Şahin, E.: Swarm robotics – special issue editorial. Autonomous Robots **17**(2–3) (2004) 111–113
2. Cao, Y.U., Fukunaga, A.S., Kahng, Andrew, B.: Cooperative mobile robotics: Antecedents and directions. Autonomous Robots **4**(1) (1997) 7–27
3. Groß, R., Mondada, F., Dorigo, M.: Transport of an object by six pre-attached robots interacting via physical links. In: Proc. of the 2006 IEEE Int. Conf. on Robotics and Automation, IEEE Computer Society Press, Los Alamitos, CA (2006) 1317–1323
4. Dieter, F., Wolfram, B., Hannes, K., Sebastian, T.: A probabilistic approach to collaborative multi-robot localization. Autonomous Robots **8**(3) (2000) 325–344
5. Borenstein, J., Feng, L.: Measurement and correction of systematic odometry errors in mobile robots. IEEE Trans. on Robotics and Automation **12**(5) (1996) 869–880
6. Aoki, I.: A simulation study on the schooling mechanism in fish. Bulletin of the Japanese Society of Scientific Fisheries **48**(8) (1982) 1081–1088
7. Reynolds, C.W.: Flocks, herds, and schools: a distributed behavioral model. Computer Graphics **21**(4) (1987) 25–34
8. Huth, A., Wissel, C.: The simulation of the movement of fish schools. Journal of Theoretical Biology **156** (1992) 365–385
9. Couzin, I.D., Krause, J., James, R., Ruxton, G.D., Franks, N.R.: Collective memory and spatial sorting in animal groups. Journal of Theoretical Biology **218**(1) (2002) 1–11
10. Dorigo, M., Tuci, E., Groß, R., Trianni, V., Labella, T., Nouyan, S., Ampatzis, C., Deneubourg, J.L., Baldassarre, G., Nolfi, S., Mondada, F., Floreano, D., Gambardella, L.: The SWARM-BOTS project. In E. Şahin, Spears, W., eds.: Proceedings of the First International Workshop on Swarm Robotics at SAB 2004. Volume 3342 of LNCS., Springer Verlag, Berlin, Germany (2004) 31–44
11. Dorigo, M.: Swarm-bot: An experiment in swarm robotics. In Arabshahi, P., Martinoli, A., eds.: Proceedings of SIS 2005 – 2005 IEEE Swarm Intelligence Symposium, IEEE Press, Piscataway, NJ (2005) 192–200
12. Jammalamadaka, S.R., SenGupta, A.: Topics in Circular Statistics. World Scientific Press, Singapore (2001)
13. R Development Core Team: R: A Language and Environment for Statistical Computing. R Foundation for Statistical Computing, Vienna, Austria. (2005)

14. J. M. Chambers, W. S. Cleveland, B. Kleiner, P. A. Tukey: Graphical Methods for Data Analysis. The Wadsworth statistics / probability series. Wadsworth and Brooks/Cole, Pacific Grove, CA (1983)
15. Trianni, V., Nolfi, S., Dorigo, M.: Cooperative hole avoidance in a *swarm-bot.* Robotics and Autonomous Systems **54**(2) (2006) 97–103
16. Nouyan, S., Groß, R., Bonani, M., Mondada, F., Dorigo, M.: Group transport along a robot chain in a self-organised robot colony. In: Proc. of the 9th Int. Conf. on Intelligent Autonomous Systems, IOS Press, Amsterdam, The Netherlands (2006) 433–442

On $\mathcal{MAX}$ – $\mathcal{MIN}$ Ant System's Parameters

Paola Pellegrini, Daniela Favaretto, and Elena Moretti

Department of Applied Mathematics, Università Ca' Foscari, Venice, Italy
paolap@pellegrini.it, {favaret, emoretti}@unive.it

Abstract. The impact of the values of the most meaningful parameters on the behavior of $\mathcal{MAX}$–$\mathcal{MIN}$ Ant System is analyzed. Namely, we take into account the number of ants, the evaporation rate of the pheromone, and the exponent values of the pheromone trail and of the heuristic measure in the random proportional rule. We propose an analytic approach to examining their impact on the speed of convergence of the algorithm. Some computational experiments are reported to show the practical relevance of the theoretical results.

1 Introduction

The assignment of values to the parameters of ACO algorithms is analyzed for the first time in [1]. In the following, a growing number of papers have been produced for finding the *optimal* values, or more in general for identifying the influence of the parameters on the behavior of the algorithms. These studies can be divided into two groups: the ones that propose a method for finding suitable parameter settings, and the ones that propose experimental analysis from which a sort of general trend can be deduced. Among others, we can locate in the first group the works by Botee and Bonabeau [2], Pilat and White [3], and Zaitar and Hiyassat [4] who use genetic algorithms for setting the parameters of ACO algorithms, and Randall [5] who uses an ACO algorithm itself. In the second group we can include Gaertner and Clark [6], who try to find a correlation between the structure of a problem instance and the optimal values of the parameters, and Socha [7] and Solnon [8], who propose computational studies concerning some parameters.

Another branch of the literature has considered the problem of tuning the parameters of metaheuristics, more in general. Among others, these include: Adenso-Díaz and Laguna [9] and Coy et al. [10] whose approaches are based on the *response surface methodology*. Bartz-Beielstein and Markon [11] propose a method to determine relevant parameter settings, based on statistical design of experiments, classical regression analysis, tree based regression and DACE (design and analysis of computer experiments) models. Birattari et al. [12] propose a procedure based on the Friedman two-way analysis of variance by ranks. Finally, Battiti and Tecchioli [13] propose to tune the parameters while solving an instance, and Lau et al. [14] present a methodology called the *Visualizer for Metaheuristics Development Framework* (V-MDF).

Following this interest in the configurations of the parameters, the objective of this paper is to formalize the impact of the value chosen for the parameters of

M. Dorigo et al. (Eds.): ANTS 2006, LNCS 4150, pp. 203–214, 2006.

$\mathcal{MAX}$-$\mathcal{MIN}$ Ant System [15,16] on the speed of convergence to the best solution ants are able to find. In the following the term convergence will be used with this meaning.

Gaining understanding in this sense is important for two main reasons. First of all we want to stress the fact that it is not possible to define an *optimal* set of values for the parameters. While in general it is accepted that the values to assign depend from the problem and from the particular instances, it is not so infrequent to observe that some parameters are considered either *good* or *bad* (in terms of the average quality of the solution achieved) in absolute, without any reference to the computational time (t). The problem with this approach is that the *optimal* speed of convergence depends both on the instance and on the computational time available: If the solution is needed very fast, one might prefer a configuration of the parameters that reaches a local optimum, with respect to one that keeps exploring the search space and that probably in a longer time would reach a better local minimum. In this sense, analyzing the influence of each parameter on the searching behavior of the algorithm can be a way for emphasizing this element.

On the other hand, by gaining a deeper understanding of the dynamics underlying the algorithm, one might focus on a range of values of the parameters for the tuning phase. In this way, a finer choice would be possible.

For this analysis we consider a problem that can be represented on a graph $G = (N, A)$ with N set of nodes ($|N| = n$) and A set of arcs. For representing the time available we use an approximation: we suppose that the pheromone update is not time consuming. In other words, we consider the time in terms of number of solutions that can be built (T).

The paper is organized as follows. In Section 2 the relevant formulas characterizing $\mathcal{MAX}$-$\mathcal{MIN}$ Ant System are presented. In Sections 3, 4 and 5 the parameters of the algorithm are studied. Finally in Section 6 some computational results are presented. The well known traveling salesman problem is considered as case study.

2 $\mathcal{MAX}$-$\mathcal{MIN}$ Ant System

In $\mathcal{MAX}$-$\mathcal{MIN}$ Ant System the pheromone update is performed after the activity of each colony of ants according to $\tau_{ij} = (1-\rho)\tau_{ij} + \Delta\tau_{ij}^b$, where $\Delta\tau_{ij}^b = 1/C_b$ if arc (i, j) belongs to the best solution b, and $\Delta\tau_{ij}^b = 0$ otherwise. C_b is the cost associated with solution b, and solution b is either the iteration-best solution or the best-so-far solution. Intuitively, if the iteration-best solution is used, the level of exploration is greater. The schedule according to which the solution to be exploited is chosen, is described by Dorigo and Stützle [17].

Another element characterizing ACO algorithms is the random-proportional rule. In particular, ant k being in node i and not having visited the nodes belonging to the set $N_k \subset N$, randomly chooses node $j \in N_k$ to move to. Each node $j \in N_k$ has a probability of being chosen described in the random proportional rule: $p_{ij} = [\tau_{ij}]^\alpha [\eta_{ij}]^\beta / (\sum_{h \in N_k} [\tau_{ih}]^\alpha [\eta_{ih}]^\beta)$, where η_{ij} is a heuristic measure associated with arc (i,j) [17]. It is important to notice that this probability depends on the set of nodes not yet visited.

Finally, it is important to remember that the pheromone trail in $\mathcal{MAX}$–$\mathcal{MIN}$ Ant System is bounded between τ_{MAX} and τ_{min}. Following [17], we use the following values: $\tau_{MAX} = 1/(\rho C_{best-so-far})$, and $\tau_{min} = [\tau_{MAX}(1 - \sqrt[n]{0.05})]/[(\frac{n}{2} - 1)\sqrt[n]{0.05}]$. At the beginning of a run, the best solution corresponds to the one found by the nearest neighbor heuristic (NN).

3 Number of Ants m

Let us first of all analyze the effect of the number of ants on the behavior of the algorithm. One thing to notice is that, given a certain number of solutions T that can be built in the available run time, the number of ants m determines the number of iterations S that can be performed as $S = T/m$. A part from this element, the value of m affects the behavior of the algorithm for what is concerned the level of exploration. Given the pheromone update rule and the update schedule, the level of exploration mainly decided by the solutions used for the update. This is due to the fact that if only few solutions are used, after few iterations, only the arcs belonging to them will have a significant probability of being chosen. If the variation of the iteration best solution is small, then, the convergence will be fast. Let br_s be the iteration best solution at iteration s, and $BR = \{r_1, r_2, ..., r_{|BR|}\}$ the set of previous iteration best solutions. If a solution has been the iteration best more than once, obviously it will be inserted in BR only the first time. Let us analyze the probability of having as iteration best solution at iteration s a solution belonging to BR. Note that, given the solution construction procedure, we can easily associate to each feasible solution r a probability of being built ($\bar{p}_r$).

Let Ω be the set of all the possible solutions and $R_r = \{q \in \Omega : C_q \geq C_r\}$. The probability of having as iteration best solution at iteration s a solution r is: $p(br_s = r) = \bar{p}_r \left(\sum_{q \in R_r} \bar{p}_q\right)^{(m-1)}$. It is the product of the probability of having one ant constructing exactly r and all the other ants constructing solutions $q \in R_r$. In the following we consider all the solutions as having different costs, so that the ordering of the solutions is not ambiguous. Since $\sum_{q \in R_r} \bar{p}_q \leq 1$ (the case of equality is true only if r is the global optimum), $p(br_s = r)$ is decreasing in m. The meaning of this conclusion is that the higher the number of ants, the lower the probability of selecting as iteration best solution a specific one. This reasoning can be extended considering that at iteration s, $|BR|$ solutions $(r_i, i = 1, ..., |BR|)$ have already been selected. In particular, the probability of selecting as iteration best at iteration s a solution in BR is equal to:

$$p(br_s \in BR) = \sum_{r \in BR} \bar{p}_r \Big(\sum_{q \in R_r} \bar{p}_q \Big)^{(m-1)} . \tag{1}$$

This value is decreasing in m and increasing in $|BR|$. For the first property it is sufficient to observe that $\sum_{q \in R_r} \bar{p}_q < 1$ (we do not consider the global optimum). For the property related to $|BR|$, it is clear that the result of a sum of non negative terms is increasing in the number of addends, the probabilities $\bar{p}_r$ being equal. Obviously $|BR|$ is non decreasing in s.

The conclusion of this reasoning is that the higher the number of ants, the greater the exploration. On the other hand, the higher the number of iterations, the greater the exploitation of the cumulated knowledge. Moreover, given the available computational time, the greater the number of ants, the smaller the number of total iterations. It is clear, then, that there is a trade-off to be solved. Remark that this reasoning is independent from the probability of choice of any particular solution $\bar{p}$, if this probability is not null for all the solutions. This property is ensured by $\mathcal{MAX}$–$\mathcal{MIN}$ Ant System through the imposition of a positive lower bound of the pheromone.

4 Evaporation Rate ρ

The parameter ρ is present in the pheromone update rule. It fixes how much pheromone evaporates. The relevance of this parameter is related to the level of exploration of the search space performed: If ρ is high, the pheromone on the arcs belonging to solutions built a few iterations before will be roughly equal to the one on the arcs that have never been selected. In this way, the search will not be much biased toward the already visited areas.

On an arc (i,j) which has never been used, the pheromone at iteration $\bar{s}$ is equal to $(1-\rho)\tau_{ij} = (1-\rho)^{\bar{s}}\tau_0 = (1-\rho)^{\bar{s}}/(\rho C_{NN})$. Clearly this is a decreasing function of ρ. What we are interested in is the influence of the value of this parameter on the level of exploration. In particular, we want to know what are the conditions for having the minimum probability of choosing, after $\bar{s}$ iterations, an arc that has never been part of an iteration best solution, and is then supposed to be of *bad* quality. This objective is achieved by setting the pheromone in arc (i,j) equal to τ_{min}. As an approximation for this value we use τ_{min} at iteration 0, i.e. $[1/(\rho C_{NN})(1-\sqrt[n]{0.05})]/[(\frac{n}{2}-1)\sqrt[n]{0.05}]$. The investigation is then referred to the value of ρ such that $\tau_{ij} \leq \tau_{min}$:

$$(1-\rho)^{\bar{s}}\frac{1}{\rho C_{NN}} \leq \frac{\frac{1}{\rho C_{NN}}(1-\sqrt[n]{0.05})}{(\frac{n}{2}-1)\sqrt[n]{0.05}} \quad\Rightarrow\quad \rho \geq 1-\sqrt[\bar{s}]{\frac{(1-\sqrt[n]{0.05})}{(\frac{n}{2}-1)\sqrt[n]{0.05}}}. \qquad (2)$$

In this way, it is possible to fix a relation between ρ, the number of nodes of the graph (n) and the number of iterations ($\bar{s}$) after which an arc that has never been part of a solution used for the pheromone update has the minimum possible probability of being chosen.

Proposition 1. *If, given n, $\bar{s}$ is such that* $\rho \geq 1-\sqrt[\bar{s}]{\frac{(1-\sqrt[n]{0.05})}{(\frac{n}{2}-1)\sqrt[n]{0.05}}}$, *then* $\rho \geq 1-\sqrt[\bar{s}']{\frac{(1-\sqrt[n]{0.05})}{(\frac{n}{2}-1)\sqrt[n]{0.05}}}$, $\forall \bar{s}' \geq \bar{s}$.

Proof. $\rho \geq 1 - \sqrt[\bar{s}]{\frac{(1-\sqrt[n]{0.05})}{(\frac{n}{2}-1)\sqrt[n]{0.05}}}$, $0 < \rho < 1 \Rightarrow (1-\rho)^{\bar{s}} \geq (1-\rho)^{\bar{s}'}, \forall \bar{s}' \geq \bar{s} \Rightarrow$
$(1-\rho)^{\bar{s}'} \leq \frac{(1-\sqrt[n]{0.05})}{(\frac{n}{2}-1)\sqrt[n]{0.05}} \Rightarrow \rho \geq 1 - \sqrt[\bar{s}']{\frac{(1-\sqrt[n]{0.05})}{(\frac{n}{2}-1)\sqrt[n]{0.05}}}, \forall \bar{s}' \geq \bar{s}.$ □

Proposition 2. *If, given $\bar{s}$, n is such that $\rho \geq 1 - \sqrt[\bar{s}]{\frac{(1-\sqrt[n]{0.05})}{(\frac{n}{2}-1)\sqrt[n]{0.05}}}$, then*[1] *$\forall n'$ such that $3 \leq n' \leq n$, $\quad \rho \geq 1 - \sqrt[\bar{s}]{\frac{(1-\sqrt[n']{0.05})}{(\frac{n'}{2}-1)\sqrt[n']{0.05}}}$.*

Proof. $\rho \geq 1 - \sqrt[\bar{s}]{\frac{(1-\sqrt[n]{0.05})}{(\frac{n}{2}-1)\sqrt[n]{0.05}}} \Rightarrow \sqrt[n]{0.05}\left[(1-\rho)^{\bar{s}}\left(\frac{n}{2}-1\right)+1\right] \leq 1.$
$\sqrt[n]{0.05}$ is an increasing function of n, then

$$\sqrt[n']{0.05}\left[(1-\rho)^{\bar{s}}\left(\frac{n}{2}-1\right)+1\right] \leq \sqrt[n]{0.05}\left[(1-\rho)^{\bar{s}}\left(\frac{n}{2}-1\right)+1\right]. \quad (3)$$

Moreover, $\forall n' \leq n$

$$\sqrt[n']{0.05}\left[(1-\rho)^{\bar{s}}\left(\frac{n'}{2}-1\right)+1\right] \leq \sqrt[n']{0.05}\left[(1-\rho)^{\bar{s}}\left(\frac{n}{2}-1\right)+1\right], \quad (4)$$

from which the thesis is verified. □

Given propositions 1 and 2, it is quite easy to fix a lower bound for the value of ρ, both in a quite general case and in a specific one. For the first observation, one can compute the value of ρ which allows the algorithm to neglect the *bad* arcs (in terms of the average quality of the solutions they belong to) after a small number of iterations when dealing with a very big instance. To this aim, let $\bar{s} = 100$ and $n = 1000$, which implies $\rho \sim 0.1$. If one sets $\rho = 0.1$, after $\bar{s}' > \bar{s}$ iterations, for sure the algorithm will have neglected the *bad* arcs. In the same way, if n decreases, $\rho = 0.1$ will imply that after $\bar{s}$ iterations, the algorithm will have neglected the *bad* arcs.

In addition to this general purpose observation, if one needs to tackle instances of equal (or similar) size, one can fix a meaningful value for ρ after estimating $\bar{s}$. Clearly this estimate will depend on the available computational time. Figure 1 represents the trend followed by the value of this parameter when $\bar{s}$ and n vary. It is easy to see that the number of iterations is the leading force, at least until a certain threshold. Nonetheless, the number of nodes has a remarkable impact as well.

5 Exponent Values α and β

The last parameters we are going to consider for $\mathcal{MAX}$–$\mathcal{MIN}$ Ant System are α and β. They represent the exponent of the pheromone level and the heuristic

[1] If $n < 3$ the value of τ_{min} is not defined.

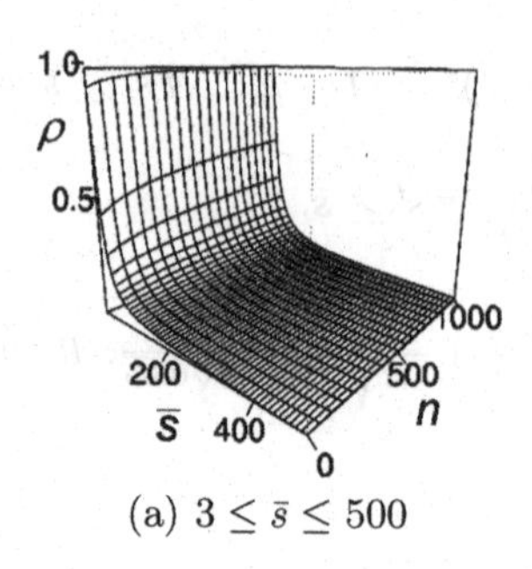

(a) $3 \leq \bar{s} \leq 500$

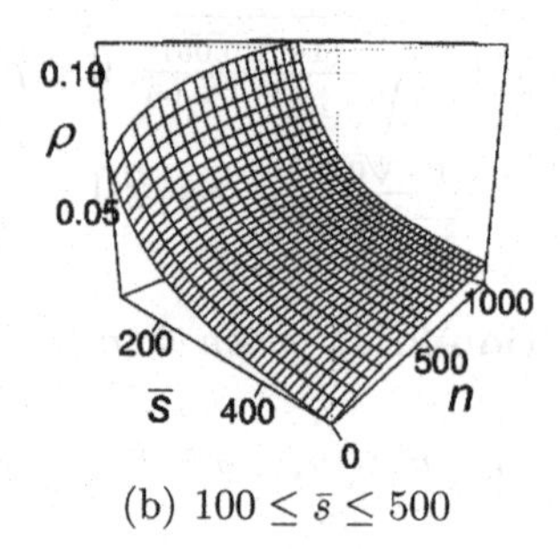

(b) $100 \leq \bar{s} \leq 500$

Fig. 1. Value of ρ necessary for having $\tau_{ij} = \tau_{min}$ on a never reinforced arc (i, j)

measure in the random proportional rule, respectively. Their main role consists in emphasizing the differences between arcs.

Instead of studying the trend of the probability of choosing the single arc, we consider the ratio between the probability of choosing two arcs (i, j) and (i, k). By analyzing this element it is possible not to consider the set of nodes still to visit. Let us write β as $c\alpha$, with $c \geq 0$. The ratio we want to study, then, is reported in formula (5).

$$\frac{p_{ij}}{p_{ik}} = \frac{[\tau_{ij}]^{\alpha}[\eta_{ij}]^{c\alpha}}{[\tau_{ik}]^{\alpha}[\eta_{ik}]^{c\alpha}} = \left[\frac{\tau_{ij}}{\tau_{ik}}\right]^{\alpha}\left[\frac{\eta_{ij}}{\eta_{ik}}\right]^{c\alpha} = \left[\frac{\tau_{ij}}{\tau_{ik}}\left(\frac{\eta_{ij}}{\eta_{ik}}\right)^{c}\right]^{\alpha} = f(\alpha, c). \tag{5}$$

Remark that being the pheromone limited by a positive lower bound, and being the length of the arcs a finite number, $\tau_{(\cdot)}$ and $\eta_{(\cdot)}$ are always strictly positive. Then, the sign of the first partial derivative with respect to α depends on $\ln\left[(\tau_{ij}/\tau_{ik})(\eta_{ij}/\eta_{ik})^{c}\right]$. This quantity is positive if and only if $(\tau_{ij}/\tau_{ik})(\eta_{ij}/\eta_{ik})^{c} > 1$. On the other hand, the sign of the first partial derivative with respect to c depends on $\ln\left[\eta_{ij}/\eta_{ik}\right]$, which is positive if and only if $\eta_{ij}/\eta_{ik} > 1$. A graphical representation of its trend is shown in Figure 2. The value of c determines both the magnitude of the variation, and the increase or decrease of the function. Then, let us have a look at function $g(c) = (\tau_{ij}/\tau_{ik})(\eta_{ij}/\eta_{ik})^{c}$. In particular we are interested in knowing in which cases the function is greater than 1.

$$g(c) > 1 \Rightarrow \ln\frac{\tau_{ij}}{\tau_{ik}} + c\ln\frac{\eta_{ij}}{\eta_{ik}} > 0 \Rightarrow c\begin{cases} > -\frac{\ln \tau_{ij}/\tau_{ik}}{\ln \eta_{ij}/\eta_{ik}} \text{ if } \eta_{ij} > \eta_{ik} \\ < -\frac{\ln \tau_{ij}/\tau_{ik}}{\ln \eta_{ij}/\eta_{ik}} \text{ if } \eta_{ij} < \eta_{ik} \end{cases} \tag{6}$$

Following the literature we consider only $\alpha, \beta \geq 0$. Let us first of all analyze the first inequality of (6). If $\tau_{ij} > \tau_{ik}$, then $\ln\left[\tau_{ij}/\tau_{ik}\right] > 0$ and the whole quantity on the right hand side of the inequality is negative, so there is no restriction on c for having $g(c)$ positive. If $\tau_{ij} < \tau_{ik}$, instead, there is a meaningful lower bound for c. A similar and opposite reasoning holds for the second inequality of (6): if $\tau_{ij} < \tau_{ik}$, then $\ln\left[\tau_{ij}/\tau_{ik}\right] < 0$ and the whole quantity on the right hand side of the inequality is negative, so there is no possible value of c such that $g(c)$ is positive. If $\tau_{ij} > \tau_{ik}$, instead, there is a meaningful upper bound for c.

Clearly the ratios between $\tau_{(\cdot)}$'s and between $\eta_{(\cdot)}$'s depend on the arcs we choose as (i, j) and (i, k). Moreover, as for what τ_{ij}/τ_{ik} is concerned, it depends

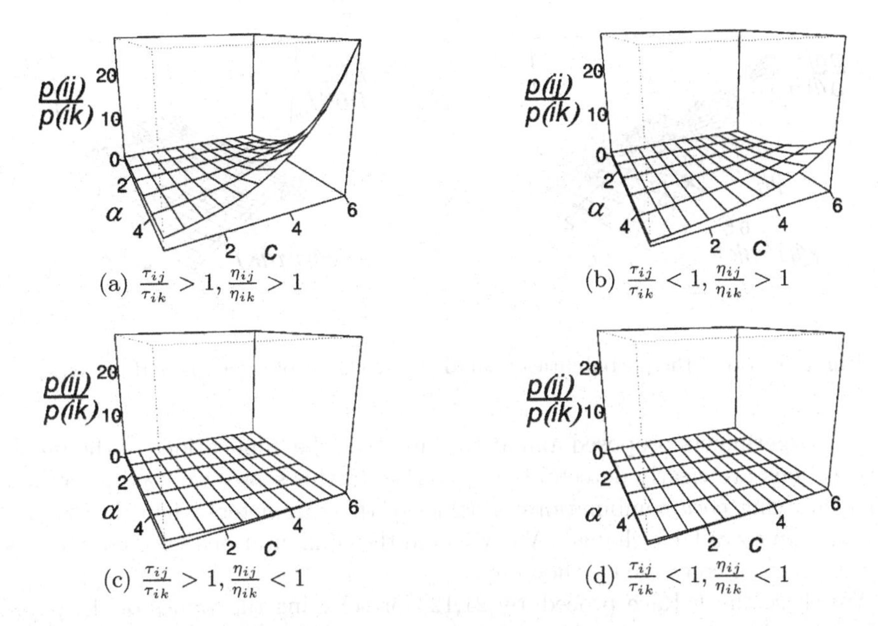

(a) $\frac{\tau_{ij}}{\tau_{ik}} > 1, \frac{\eta_{ij}}{\eta_{ik}} > 1$ (b) $\frac{\tau_{ij}}{\tau_{ik}} < 1, \frac{\eta_{ij}}{\eta_{ik}} > 1$

(c) $\frac{\tau_{ij}}{\tau_{ik}} > 1, \frac{\eta_{ij}}{\eta_{ik}} < 1$ (d) $\frac{\tau_{ij}}{\tau_{ik}} < 1, \frac{\eta_{ij}}{\eta_{ik}} < 1$

Fig. 2. Ratio of the probabilities related to the choice to arc (i, j) and arc (i, k)

on the behavior of the algorithm. Figure 3 represents the variation of $g(c)$ as a function of τ and η. In particular, we keep constant through the graphics the value of η_{ij}/η_{ik} and we vary the ratio between the pheromone levels. This schema follows the behavior of ACO algorithms in cases the heuristic measure is static. The values selected are $\eta_{ij}/\eta_{ik} = 1.1$ for Figure 3(a) and $\eta_{ij}/\eta_{ik} = 0.9$ for Figure 3(b). As it can be seen, there is an interval in which, even if the position with respect to 1 of τ_{ij}/τ_{ik} and $\eta_{ij}\eta_{ik}$ are opposite, for a while $g(c)$ keeps on following the sign of $1 - (\eta_{ij}/\eta_{ik})$. The greater c, the wider this interval. This observation is robust with respect to the value of η_{ij}/η_{ik}.

To sum up the reasoning on α and β, α amplifies the differences between the *good* and the *bad* arcs. The value of $c = \beta/\alpha$, instead, tells us how to distinguish the *good* from the *bad* arcs in case the heuristic information and the pheromone values lead to discordant orders. In particular, the higher the value of c, the more the order is driven by the heuristic information.

6 Experiments

The experimental analysis proposed is based on the traveling salesman problem (TSP). We consider the ACOTSP program implemented by Thomas Stützle as a companion software for [17]. The code has been released in the public domain and is available for free download on www.aco-metaheuristic.org/aco-code/. The TSP has been object of many studies, both practical and theoretical (see for example [18,19,20]). We consider this problem as a case study.

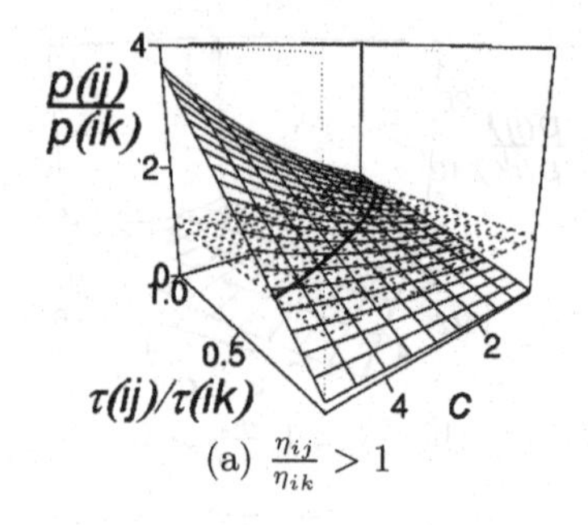

(a) $\frac{\eta_{ij}}{\eta_{ik}} > 1$

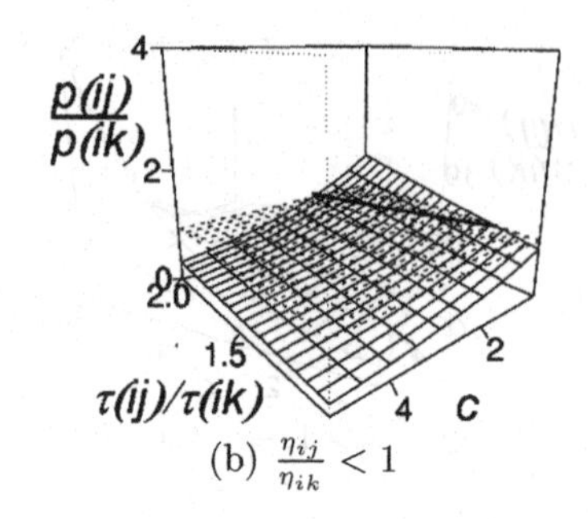

(b) $\frac{\eta_{ij}}{\eta_{ik}} < 1$

Fig. 3. Ratio of the probabilities related to the choice of arc (i, j) and arc (i, k)

The experiments proposed aim at showing that the implications of the previous sections are clearly detectable in practise. In this sense, we need a method for identifying *good* combinations of values of the parameters when the computational time available changes. We will read the configurations selected in terms of the speed of convergence they imply.

We chose the F-Race procedure [21,12] for selecting the values of the parameters. F-Race is a racing algorithm for choosing a combination of values (a candidate configuration) from a predefined range. A racing algorithm consists in generating a sequence of nested sets of candidate configurations to be considered at each step. The set characterizing a specific step is obtained by possibly discarding some configurations that appear to be suboptimal on the basis of the information available. This cumulated knowledge is represented by the behavior of the algorithm for which the tuning is performed, when using different candidates configurations. For each instance (each representing one step of the race) the ranking of the results obtained using the different configurations is computed and a statistical test is performed for deciding whether to discard some candidates from the following experiments. F-Race is based on the Friedman two-way analysis of variance by ranks [22].

The range of values considered for each parameter is the one that in our eyes one would test after the analysis of the literature. In particular the candidate configurations are 192. They are all those obtainable from combining the following values: $m \in \{50, 100, 200, 300\}$, $\rho \in \{0.02, 0.04, 0.06, 0.08\}$, $\alpha \in \{1, 2, 3\}$, $\beta \in \{2, 3, 4, 5\}$. Two sets of 220 instances are used. In one set each instance includes 300 customers. In the other one 600 customers are considered. The instances are generated through `portgen`, the instance generator adopted in the DIMACS TSP Challenge. In particular, the ones we consider here consist of two dimensional integer-coordinate cities grouped in clusters that are uniformly distributed in a square of size $10^6 \times 10^6$. They are available on the web page www.paola.pellegrini.it. On each set of instances, the F-Race is applied six times, varying the computational time available t in the set $\{5, 10, 30, 60, 90, 120\}$ seconds. The experiments are run on a processor AMD Athlon 1000 Mhz, 772 MB of memory, running GNU/Linux 2.4.20. No local search is applied, due to the fact that we want to investigate the relation between the values of the

Table 1. Configurations chosen by F-race with different computational time available

n	t	m	ρ	β	α	$\Rightarrow$	T	$S=T/m$	c
300	5	100	0.08	5	2		7000	70	2.5
300	10	100	0.08	4	2		14000	140	2
300	30	100	0.08	4	1		42000	420	4
300	60	100	0.08	3	1		84000	840	3
300	90	200	0.08	3	1		126000	630	3
300	120	200	0.08	3	1		168000	840	3
600	5	50	0.08	5	3		1700	34	1.66
600	10	50	0.08	5	3		3400	68	1.66
600	30	100	0.08	5	2		10200	102	2.5
600	60	200	0.08	4	2		20400	102	2
600	90	200	0.08	4	2		30600	153	2
600	120	200	0.08	4	2		40800	204	2

parameters and the speed of convergence, and we are not interested in the absolute quality of the solution. The candidate configuration chosen by F-Race for each set of instances/computational time are reported in Table 1. Beside the values of the parameters selected, the table reports the approximate number of tours that can be built in the available time (T), the total number of iterations performed (S), and the value of $c = \beta/\alpha$. The heuristic measure we consider is the typical one used for the TSP, i.e. the inverse of the length of arcs.

The trend followed by the values of the parameters are clear. They respect the expectations coming from the previous analysis. In particular it can be observed that the value of m increases with the increase of the computational time available. According to Section 3 this can be read as an increase of the level of exploration of the search space. The values of m are quite different through the cardinality of the set of nodes n. This is due to the fact that given the time available, the number of tours that can be constructed is noticeably different. Moreover, when considering different computational times which lead to the construction of a similar number of solutions, one can observe that the number of ants increases with n (see for example $n = 300, T = 42000 \Rightarrow m = 100$ and $n = 600, T = 20400 \Rightarrow m = 200$). This can be read as a greater need of exploration in case of a greater number of nodes. The explanation for this phenomenon can be found in the fact that the greater the number of nodes, in general the more complex the search space, and so the greater the risk of being entrapped in a local minimum.

The trend followed by the values of α mimics the prediction of Section 5. In fact, it is decreasing with the time available, reflecting the postposition of the need of convergence. For what concerns c, we can observe that, α being equal, its value is inversely correlated with the number of solutions that can be constructed. When α varies, the value of c changes in the opposite direction. When the value of α is high, the convergence is fast even if we do not consider the value of c. As a consequence, keeping c quite low is a way for smoothing the trend. On the other hand, speeding up the convergence is not the only role of c. As discussed

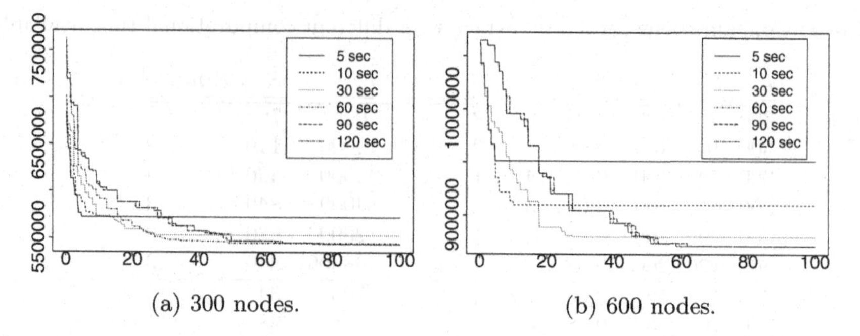

(a) 300 nodes.

(b) 600 nodes.

Fig. 4. Value of (C_b) found by the different configurations depending on t

in Section 5, it implies whether it is the heuristic information or the pheromone trail to state the distinction between the *good* from the *bad* arcs in case the respective indications are discordant. The higher the value of c, the more the decision is driven by the heuristic information. When we consider this element with the negative correlation between c and T, we can deduce that the earlier the algorithm needs to converge, the more it has to give importance to the heuristic measure, which has a more immediate link with the instance than the pheromone trail. In a similar way, let us consider the relation of c with the value of α. If the latter implies a very fast convergence, in general the algorithm will be reluctant to accept the indications of previous ants (and so of the pheromone trail) in case they are in contrast with the ones of the heuristic information. This is due to the fact that if they are misleading it will not be able to neglect them very soon.

Finally, it is not possible here to observe the trend followed by the value of ρ. The value selected, in fact, is always the same ($\rho = 0.08$), and it corresponds to the maximum value in the range set. Nonetheless, the choice has been motivated by the analysis of the literature and in particular of Dorigo and Stützle [17], where a value of 0.02 is proposed. By considering formula (2) it is possible to compute the number of iterations after which the pheromone on *bad* arcs becomes equal to τ_{min} with $\rho = 0.08$ and instances of 300 and 600 nodes. This quantity is equal to 116 and 132, respectively. These values are probably quite good when the total number of iterations is higher than 400-500, while it may be too low for shorter runs.

Figure 4 represents the trends followed by the values of the best solutions (C_b) found by $\mathcal{MAX}$–$\mathcal{MIN}$ Ant System with the configurations of parameters reported in Table 1 as function of the computational time (t). As it can be seen, the configurations selected by F-Race are the best performing up to the time available for the respective runs.

7 Conclusion

The relevance of the values of the parameters when dealing with metaheuristics is recognized in the literature. In this paper we analyze $\mathcal{MAX}$–$\mathcal{MIN}$ Ant

System: Theoretical aspects of the impact of the values of the parameters on its behavior are investigated. Some relations between the values chosen and the speed of convergence of the algorithm are proposed. Computational experiments are reported to show the practical reflections of the theoretical results.

Once fixed the constraints that one must satisfy, such as the characteristics of the instances to tackle and the computational time available, the comprehension of the impact of the parameters can give some indications about the range to use for the tuning phase. Further possible developments of this study can be the analysis of different problems and other ACO algorithms.

References

1. Colorni, A., Dorigo, M., Maniezzo, V.: An investigation of some properties of an "ant algorithm". In Männer, R., Manderick, B., eds.: PPSN, Brussels, Belgium, Elsevier (1992) 515–526
2. Botee, H., Bonabeau, E.: Evolving ant colony optimization. Advanced Complex Systems **1** (1985) 149–159
3. Pilat, M.L., White, T.: Using genetic algorithms to optimize acs-tsp. In Dorigo, M., Di Caro, G., Sampels, M., eds.: ANTS '02: Proceedings of the Third International Workshop on Ant Algorithms, London, UK, Springer-Verlag (2002) 282–287
4. Zaitar, R., Hiyassat, H.: Optimizing the ant colony optimization using standard genetic algorithm. In Hamza, M., ed.: Artificial Intelligence and Applications, Innsbruck, Austria, IASTED/ACTA Press (2005) 130–133
5. Randall, M.: Near parameter free ant colony optimisation. In Dorigo, M., Birattari, M., Blum, C., Gambardella, L.M., Mondada, F., Stützle, T., eds.: ANTS Workshop. Volume 3172 of Lecture Notes in Computer Science., Brussels, Belgium, Springer (2004) 374–381
6. Gaertner, D., Clark, K.: On optimal parameters for ant colony optimization algorithms. In Arabnia, H., Joshua, R., eds.: IC-AI 2005, Las Vegas, USA (2005) 83–89
7. Socha, K.: The influence of run-time limits on choosing ant system parameters. In Cantu-Paz, E., Livermore, L., Balakrishnan, K., Banzhaf, W., Bentley, P., Dasgupta, L.C.D., Jong, K.D., nad F. Herrera, J.F., Langdon, W., Lutton, E., Mazumder, P., Michielssen, E., Pedrycz, W., Roy, R., Rudnick, E., Soule, M.S.T., Spector, L., Verdegay, J., eds.: Proceedings of Genetic and Evolutionary Computation Conference, GECCO 2003. Volume 2611 of LNCS., Chicago, USA, Springer-Verlag (2003) 49–60
8. Solnon, C.: Boosting ACO with a preprocessing step. In Cagnoni, S., Gottlieb, J., Hart, E., Middendorf, M., Raidl, G., eds.: EvoWorkshop 2002. Volume 2279 of LNCS., Kinsale, Ireland, Springer-Verlag (2002) 163–172
9. Adenso-Díaz, B., Laguna, M.: Fine-tuning of algorithms using fractional experimental designs and local search. (to appear in *Operations Research*)
10. Coy, S., Golden, B., Runger, G., Wasil, E.: Using experimental design to find effective parameter settings for heuristics. Journal of Heuristics **7**(1) (2001) 77–97
11. Bartz-Beielstein, T., Markon, S.: Tuning search algorithms for real-world applications: A regression tree based approach. In Greenwood, G., ed.: Proc. 2004 Congress on Evolutionary Computation (CEC'04), Piscataway NJ, IEEE Press (2004) 1111–1118

12. Birattari, M., Stützle, T., Paquete, L., Varrentrapp, K.: A racing algorithm for configuring metaheuristics. In Langdon, W., ed.: GECCO 2002: Proceedings of the Genetic and Evolutionary Computation Conference, San Francisco, CA, USA, Morgan Kaufmann Publishers (2002) 11–18
13. Battiti, R., Tecchioli, G.: The reactive tabu search. ORSA Journal on Computing **6** (1994) 126–585
14. Lau, H., Wan, W., Halim, S.: Tuning tabu search strategies via visual diagnosis. In Doerner, K., Gendreau, M., Greistorfer, P., Gutjahr, W., Hartl, R., Reimann, M., eds.: Proceedings of Metaheuristics International Conference (MIC 2005), Vienna, Austria (2005) 630–636
15. Stützle, T., Hoos, H.: Improvements on the ant system: introducing the max-min ant system. In Albrecht, R., Smith, G., Steele, N., eds.: Proceedings of Artificial Neural Nets and Genetic Algorithms 1997, Norwich, U.K., Springer-Verlag (1998) 245–249
16. Stützle, T., Hoos, H.: Max-min ant system. Future Generation Computer Systems **16**(8) (2000) 889–914
17. Dorigo, M., Stützle, T.: Ant Colony Optimization. MIT Press, Cambridge (2004)
18. Bullnheimer, B., Hartl, R.F., Strauss, C.: A new rank based version of the ant system: A computational study. Central European Journal for Operations Research and Economics **7**(1) (1999) 25–38
19. Colorni, A., Dorigo, M., Maffioli, F., Maniezzo, V., Righini, G., Trubian, M.: Heuristics from nature for hard combinatorial problems. International Transactions in Operational Research **3**(1) (1996) 1–21
20. Stützle, T., Hoos, H.: The max-min ant system and local search for the traveling salesman problem. In Angeline, P., ed.: Proceedings of the IEEE International Conference on Evolutionary Computation, Indianapolis, USA, Springer-Verlag (1997) 308–313
21. Birattari, M.: The problem of tuning metaheuristics as seen from a machine learning perspective. PhD thesis, Université Libre de Bruxelles, Brussels, Belgium (2005)
22. Friedman, J.: Multivariate adaptive regression splines. The Annals of Statistics **19** (1991) 1–141

On the Invariance of Ant System

Mauro Birattari[1], Paola Pellegrini[1,2], and Marco Dorigo[1]

[1] IRIDIA, CoDE, Université Libre de Bruxelles, Brussels, Belgium
[2] Department of Applied Mathematics, Università Ca' Foscari, Venice, Italy
mbiro@ulb.ac.be, paolap@pellegrini.it, mdorigo@ulb.ac.be

Abstract. It is often believed that the performance of ant system, and in general of ant colony optimization algorithms, depends somehow on the scale of the problem instance at hand. The issue has been recently raised explicitly [1] and the *hyper-cube framework* has been proposed to eliminate this supposed dependency.

In this paper, we show that although the internal state of ant system—that is, the *pheromone* matrix—depends on the scale of the problem instance under analysis, this does not affect the external behavior of the algorithm. In other words, for an appropriate initialization of the pheromone, the sequence of solutions obtained by ant system does not depend on the scale of the instance.

As a second contribution, the paper introduces a straightforward variant of ant system in which also the pheromone matrix is independent of the scale of the problem instance under analysis.

1 Introduction

The *hyper-cube framework* [1] has been recently introduced with the aim of implementing ant colony optimization algorithms (ACO) [2] that are invariant with respect to a linear rescaling of problem instances. The need for the introduction of the hyper-cube framework has been explicitly motivated by the observation that

> *in standard ACO algorithms the pheromone values and therefore the performance of the algorithms, strongly depend on the scale of the problem.* [1]

In this paper, we formally show that this statement is only partially correct: Indeed, in standard ant colony optimization algorithms the pheromone trail and the heuristic values depend on the scale of the problem. Nonetheless, for an appropriate initialization of the pheromone, the sequence of solutions they find is independent of the scaling.

For definiteness, the paper focuses on ant system [3,4,5] for the traveling salesman problem. The theorems we enunciate in the paper are proved first for this specific algorithm and for this specific problem. The conditions under which these results extend to other problems are discussed in the following.

Although this paper shows that the main motivation for the introduction of the hyper-cube framework does not hold, the work of Blum and Dorigo [1] has the

M. Dorigo et al. (Eds.): ANTS 2006, LNCS 4150, pp. 215–223, 2006.

main merit of having explicitly attracted the attention of the research community on some important issues. Indeed, the fact that pheromone and heuristic values depend on the scale of the problem complicates the analysis of the algorithm and might cause numerical problems in the implementations. The hyper-cube framework is definitely a solution to this problem. Nonetheless, the hyper-cube version of ant system is effectively a new algorithm which shares with the original ant system the underlying ideas but that produces a different sequence of solutions. In other words, the hyper-cube ant system and the original ant system are not *functionally equivalent.* In this paper we propose *si*AS which is a trivial modification of ant system. Similar to the hyper-cube ant system, *si*AS has the property that the pheromone and the heuristic values do not depend on the scaling of the problem. Nevertheless, contrary to the hyper-cube ant system, *si*AS is *functionally equivalent* to the original ant system. This last property is particularly desirable: all theoretical and empirical studies previously performed on ant system immediately extend to *si*AS.

In this paper, we focus our attention on ant system. Nonetheless the same invariance property can be proved for other ACO algorithms. We refer the reader to [6] for an analysis of the invariance of $\mathcal{MAX}$–$\mathcal{MIN}$ ant system [7,8] and of ant colony system [9]. Moreover, in [6] the algorithms *si*$\mathcal{MM}$AS and *si*ACS are defined, which are *functionally equivalent* to $\mathcal{MAX}$–$\mathcal{MIN}$ ant system and ant colony system, respectively, and in which the pheromone and the heuristic values do not depend on the scaling of the problem.

The rest of the paper is organized as follows. Section 2 introduces some preliminary concepts. Section 3 defines ant system and formally proves its invariance. Section 4 introduces the *si*AS algorithm. Finally, Sect. 5 concludes the paper.

2 Preliminary Definitions

This section introduces a number of fundamental concepts that will be needed in the following.

Definition 1 (Linear transformation of a problem instance). *If I is an instance of a generic combinatorial optimization problem, $\bar{I} = fI$, $f > 0$, is a* linear transformation *of I if $\bar{I}$ is obtained by multiplying all costs in I by the coefficient f. In particular, it results that the cost $\bar{C}$ of a solution $\bar{T}$ of instance $\bar{I}$ is f times the cost C of the corresponding solution T of instance I.*

Definition 2 (Linear transformation of a traveling salesman instance). *With $\bar{I} = fI$, $f > 0$, we indicate that the instance $\bar{I}$ is a* linear transformation *of the instance I: The two instances have the same number of cities and the cost $\bar{c}_{ij}$ of traveling from city i to city j in $\bar{I}$ is f times the corresponding cost c_{ij} in instance I. Formally:*

$$\bar{c}_{ij} = fc_{ij},\ \textit{for all}\ \langle i, j\rangle. \tag{1}$$

Remark 1. The cost $\bar{C}$ of a solution $\bar{T}$ of instance $\bar{I}$ is f times the cost C of the corresponding solution T of instance I. Formally:

$$(\bar{I} = fI) \wedge (\bar{T} = T) \implies \bar{C} = fC. \quad (2)$$

Remark 2. In the following, if x is a generic quantity that refers to an instance I, then $\bar{x}$ is the corresponding quantity for what concerns instance $\bar{I}$, when $\bar{I}$ is a linear transformation of I.

Ant colony optimization algorithms are *stochastic*: Solutions are constructed incrementally on the basis of stochastic decisions that are biased by the pheromone and by some heuristic information. The following hypothesis will be used in the paper.

Hypothesis 1 (Pseudo-random number generator). When solving two instances I and $\bar{I}$, the stochastic decisions taken while constructing solutions are made on the basis of random experiments based on pseudo-random numbers produced by the same pseudo-random number generator. We assume that this generator is initialized in the same way (for example, with the same seed) when solving the two instances so that the two sequences of pseudo-random numbers that are generated are the same in the two cases.

Definition 3 (Invariance). *An algorithm A is* ***invariant*** *to linear transformations if the sequence of solutions S_I generated when solving an instance I and the sequence of solutions $S_{\bar{I}}$ generated when solving an instance $\bar{I}$ are the same, whenever $\bar{I}$ is a linear transformation of I.*

If A is a stochastic algorithm, it is said to be invariant if it is so under Hypothesis 1.

Definition 4 (Strong and weak invariance). *An algorithm A is said to be* ***strongly-invariant*** *if, beside generating the same solutions on any two linearly related instances I and $\bar{I}$, it also enjoys the property that the heuristic information and the pheromone at each iteration are the same when solving I and $\bar{I}$. Conversely, the algorithm A is* ***weakly-invariant*** *if it obtains the same solutions on linearly related instances but the heuristic information and the pheromone assume different values.*

If A is stochastic, it is said to be strongly-invariant (or weakly-invariant) if it is so under Hypothesis 1.

3 Ant System

Ant system is the original ant colony optimization algorithm proposed by Dorigo et al. [3,4,5]. The pseudo-code of the algorithm is given in Fig. 1. In our analysis, we refer to the application of ant system to the well-known traveling salesman problem, which consists in finding the Hamiltonian circuit of least cost on an edge-weighted graph.

Ant system:

Initialize pheromone trail
while (termination condition not met) **do**
  Construct solutions via the random proportional rule
  Update pheromone
end

Fig. 1. Pseudo-code of ant system

Definition 5 (Random proportional rule). *At the generic iteration h, suppose that ant k is in node i. Let $\mathcal{N}_i^k$ be the set of feasible nodes. The node $j \in \mathcal{N}_i^k$, to which ant k moves, is selected with probability:*

$$p_{ij,h}^k = \frac{[\tau_{ij,h}]^\alpha [\eta_{ij}]^\beta}{\sum_{l \in \mathcal{N}_i^k} [\tau_{il,h}]^\alpha [\eta_{il}]^\beta},$$

where α and β are parameters, $\tau_{ij,h}$ is the pheromone value associated with arc $\langle i,j \rangle$ at iteration h, and η_{ij} represents heuristic information *on the desirability of visiting node j after node i.*

Definition 6 (Heuristic information). *When solving the traveling salesman problem, the heuristic information η_{ij} is the inverse of the cost of traveling from city i to city j:*

$$\eta_{ij} = \frac{1}{c_{ij}}, \text{ for all } \langle i,j \rangle.$$

Definition 7 (Pheromone update rule). *At the generic iteration h, suppose that m ants have generated the solutions $T_h^1, T_h^2, \ldots, T_h^m$ of cost $C_h^1, C_h^2, \ldots, C_h^m$, respectively. The pheromone on each arc $\langle i,j \rangle$ is updated according to the following rule:*

$$\tau_{ij,h+1} = (1-\rho)\tau_{ij,h} + \sum_{k=1}^{m} \Delta_{ij,h}^k,$$

where ρ is a parameter called evaporation rate *and*

$$\Delta_{ij,h}^k = \begin{cases} 1/C_h^k, & \text{if } \langle i,j \rangle \in T_h^k; \\ 0, & \text{otherwise.} \end{cases} \tag{3}$$

Definition 8 (Ant system). *Ant system is an ant colony optimization algorithm in which solutions are constructed according to the random proportional rule given in Definition 5, and the pheromone is updated according to the rule given in Definition 7. The evaporation rate ρ, the number of ants m, and the exponents α and β are parameters of the algorithm.*

When ant system is used for solving the traveling salesman problem, it is customary to initialize the pheromone as follows.

Definition 9 (Nearest-neighbor pheromone initialization). *At the first iteration $h = 1$, the pheromone on all arcs is initialized to the value:*

$$\tau_{ij,1} = \frac{m}{C^{nn}}, \text{ for all } \langle i, j \rangle,$$

where m is the number of ants considered at each iteration, and C^{nn} is the cost of the solution T^{nn} obtained by the nearest-neighbor heuristic.

The following theorem holds true.

Lemma 1. *The random proportional rule is invariant to concurrent linear transformations of the pheromone and of the heuristic information. Formally:*

$$(\bar{\tau}_{ij,h} = g_1 \tau_{ij,h}) \wedge (\bar{\eta}_{ij} = g_2 \eta_{ij}), \text{ for all } \langle i, j \rangle \implies \bar{p}^k_{ij,h} = p^k_{ij,h}, \text{ for all } \langle i, j \rangle.$$

where $\bar{p}^k_{ij,h}$ is obtained on the basis of $\bar{\tau}_{ij,h}$ and $\bar{\eta}_{ij}$, according to Definition 5.

Proof. According to Definition 5:

$$\begin{aligned}
\bar{p}^k_{ij,h} &= \frac{[\bar{\tau}_{ij,h}]^\alpha [\bar{\eta}_{ij}]^\beta}{\sum_{l \in \mathcal{N}^k_i} [\bar{\tau}_{il,h}]^\alpha [\bar{\eta}_{il}]^\beta} = \frac{[g_1 \tau_{ij,h}]^\alpha [g_2 \eta_{ij}]^\beta}{\sum_{l \in \mathcal{N}^k_i} [g_1 \tau_{il,h}]^\alpha [g_2 \eta_{il}]^\beta} \\
&= \frac{[g_1]^\alpha [g_2]^\beta [\tau_{ij,h}]^\alpha [\eta_{ij}]^\beta}{\sum_{l \in \mathcal{N}^k_i} [g_1]^\alpha [g_2]^\beta [\tau_{il,h}]^\alpha [\eta_{il}]^\beta} = \frac{[g_1]^\alpha [g_2]^\beta [\tau_{ij,h}]^\alpha [\eta_{ij}]^\beta}{[g_1]^\alpha [g_2]^\beta \sum_{l \in \mathcal{N}^k_i} [\tau_{il,h}]^\alpha [\eta_{il}]^\beta} \\
&= \frac{[\tau_{ij,h}]^\alpha [\eta_{ij}]^\beta}{\sum_{l \in \mathcal{N}^k_i} [\tau_{il,h}]^\alpha [\eta_{il}]^\beta} = p^k_{ij,h}.
\end{aligned}$$

□

Theorem 1. *The ant system algorithm for the traveling salesman problem is weakly-invariant, provided that the pheromone is initialized as prescribed by Definition 9.*

Proof. Let us consider two generic instances I and $\bar{I}$ such that

$$\bar{I} = fI, \text{ with } f > 0.$$

The theorem is proved by induction: We show that if at the generic iteration h some set of conditions C holds, then the solutions generated for the two instances I and $\bar{I}$ are the same and the set of conditions C also holds for the following iteration $h+1$. The proof is concluded by showing that C holds for the very first iteration. With few minor modifications, this technique is adopted in the following for proving all theorems enunciated in the paper.

According to Definition 6, and taking into account (1), it results:

$$\bar{\eta}_{ij} = \frac{1}{f} \eta_{ij}, \text{ for all } \langle i, j \rangle.$$

According to Lemma 1, if at the generic iteration h, $\bar{\tau}_{ij,h} = \frac{1}{f}\tau_{ij,h}$, for all $\langle i, j\rangle$, then $\bar{p}^k_{ij,h} = p^k_{ij,h}$, for all $\langle i, j\rangle$. Under Hypothesis 1,

$$\bar{T}^k_h = T^k_h, \text{ for all } k = 1, \ldots, m,$$

and therefore, according to (2),

$$\bar{C}^k_h = fC^k_h, \text{ for all } k = 1, \ldots, m.$$

According to (3):

$$\begin{aligned}
\bar{\Delta}^k_{ij,h} &= \begin{cases} 1/\bar{C}^k_h, & \text{if } \langle i,j\rangle \in \bar{T}^k_h; \\ 0, & \text{otherwise;} \end{cases} = \begin{cases} 1/fC^k_h, & \text{if } \langle i,j\rangle \in \bar{T}^k_h = T^k_h; \\ 0/f, & \text{otherwise;} \end{cases} \\
&= \frac{1}{f}\begin{cases} 1/C^k_h, & \text{if } \langle i,j\rangle \in T^k_h; \\ 0, & \text{otherwise;} \end{cases} = \frac{1}{f}\Delta^k_{ij,h},
\end{aligned}$$

and therefore, for any arc $\langle i, j\rangle$:

$$\begin{aligned}
\bar{\tau}_{ij,h+1} &= (1-\rho)\bar{\tau}_{ij,h} + \sum_{k=1}^{m} \bar{\Delta}^k_{ij,h} = (1-\rho)\frac{1}{f}\tau_{ij,h} + \sum_{k=1}^{m} \frac{1}{f}\Delta^k_{ij,h} \\
&= (1-\rho)\frac{1}{f}\tau_{ij,h} + \frac{1}{f}\sum_{k=1}^{m} \Delta^k_{ij,h} = \frac{1}{f}\left((1-\rho)\tau_{ij,h} + \sum_{k=1}^{m} \Delta^k_{ij,h}\right) = \frac{1}{f}\tau_{ij,h+1}.
\end{aligned}$$

In order to provide a basis for the above defined induction and therefore to conclude the proof, it is sufficient to observe that at the first iteration $h = 1$, the pheromone is initialized as:

$$\bar{\tau}_{ij,1} = \frac{m}{\bar{C}^{nn}} = \frac{m}{fC^{nn}} = \frac{1}{f}\tau_{ij,1}, \text{ for all } \langle i, j\rangle.$$

□

Remark 3. Theorem 1 holds true for any way of initializing the pheromone, provided that for any two instances $\bar{I}$ and I such that $\bar{I} = fI$, $\bar{\tau}_{ij,1} = \frac{1}{f}\tau_{ij,1}$, for all $\langle i, j\rangle$.

Remark 4. Theorem 1 extends to the application of ant system to problems other than the traveling salesman problem, provided that the initialization of the pheromone is performed as prescribed in Remark 3 and for any two instances $\bar{I}$ and I such that $\bar{I} = fI$, with $f > 0$, there exists a coefficient $g > 0$ such that $[\bar{\eta}_{ij}]^\beta = [g\eta_{ij}]^\beta$, for all $\langle i, j\rangle$. In particular, it is worth pointing out here that one notable case in which this last condition is satisfied is when $\beta = 0$, that is, when no heuristic information is used.

4 Strongly-Invariant Ant System

A strongly invariant version of ant system (*si*AS) can be easily defined. For definiteness, we present here a version of *si*AS for the traveling salesman problem.

Definition 10 (Strongly-invariant heuristic information). *When solving the traveling salesman problem, the heuristic information* η_{ij} *is*

$$\eta_{ij} = \frac{C^{nn}}{nc_{ij}}, \text{ for all } \langle i, j \rangle. \tag{4}$$

where c_{ij} *is the cost of traveling from city* i *to city* j*,* n *is the number of cities, and* C^{nn} *is the cost of the solution* T^{nn} *obtained by the nearest-neighbor heuristic.*

Definition 11 (Strongly-invariant pheromone update rule). *The pheromone is updated using the same rule given in Definition 7, with the only difference that* $\Delta^k_{ij,h}$ *is given by:*

$$\Delta^k_{ij,h} = \begin{cases} C^{nn}/mC^k_h, & \text{if } \langle i, j \rangle \in T^k_h; \\ 0, & \text{otherwise;} \end{cases}$$

where C^{nn} *is the cost of the solution* T^{nn} *obtained by the nearest-neighbor heuristic and* m *is the number of ants generated at each iteration.*

Definition 12 (Strongly-invariant pheromone initialization). *At the first iteration* $h = 1$*, the pheromone on all arcs is initialized to the value:*

$$\tau_{ij,1} = 1, \text{ for all } \langle i, j \rangle.$$

Definition 13 (Strongly-invariant ant system). *The strongly-invariant ant system (siAS) is a variation of ant system. It shares with ant system the random proportional rule for the construction of solutions, but in siAS the heuristic values are set as in Definition 10, the pheromone is initialized according to Definition 12 and the update is performed according to Definition 11.*

Remark 5. In the definition of *si*AS given above, the nearest-neighbor heuristic has been adopted for generating a reference solution, the cost of which is then used for normalizing the cost of the solutions found by *si*AS. Any other algorithm could be used instead, provided that the solution it returns does not depend on the scale of the problem.

Remark 6. It is worth noting here that the presence of the term n in the denominator of the left hand side of (4) is not needed for obtaining an invariant heuristic information. It has been included for achieving another property. Indeed, η_{ij} as defined in (4) assumes values that do not depend on the size of the instance under analysis—that is, on the number n of cities. If this term were not present, since the numerator C^{nn} grows with n, η_{ij} would have been relatively larger in large instances and smaller in small ones.

Remark 7. Similarly, it should be noticed that by initializing the pheromone to $\tau_{ij,1} = 1/m$, for all $\langle i, j \rangle$, and by defining $\Delta^k_{ij,h}$ as:

$$\Delta^k_{ij,h} = \begin{cases} C^{nn}/C^k_h, & \text{if } \langle i, j \rangle \in T^k_h; \\ 0, & \text{otherwise;} \end{cases}$$

one would have obtained nonetheless an invariant algorithm. The advantage of the formulation given in Definitions 11 and 12 is that the magnitude of the pheromone deposited on the arcs does not depend on the number m of ants considered.

The strongly-invariant ant system is *functionally equivalent* to the original ant system, that is, the two algorithms produce the same sequence of solutions for any given instance, provided that the pheromone is properly initialized, their respective pseudo-random number generators are the same, and these generators are initialized with the same seed. Formal proofs of the functional equivalence of *si*AS and ant system and of the strong invariance of *si*AS, are given in [6].

5 Conclusions

We have formally proved that, contrary to what previously believed [1], ant system is invariant to the rescaling of problem instances. The same holds [6] for the two main other members of the ant colony optimization family of algorithms, namely, $\mathcal{MAX}$–$\mathcal{MIN}$ ant system and ant colony system.

Moreover, we have introduced *si*AS, which is a straightforward strongly-invariant version of ant system. In this respect, *si*AS is similar to the hyper-cube ant system [1] which is the first strongly-invariant version of ant system ever published in the literature. The main advantage of *si*AS over the hyper-cube ant system is that, while the latter is effectively a new algorithm, *si*AS is functionally equivalent to the original ant system. As a consequence, one can immediately extend to *si*AS all understanding previously acquired about ant system and all empirical results previously obtained. Following the strategy adopted in the definition of *si*AS, a strongly-invariant version of any ACO algorithm can be defined. In particular, *si*$\mathcal{MM}$AS and *si*ACS are introduced in [6]. These two algorithms are the strongly-invariant versions of $\mathcal{MAX}$–$\mathcal{MIN}$ ant system and ant colony system, respectively. Like *si*AS, also *si*$\mathcal{MM}$AS and *si*ACS are functionally equivalent to their original counterparts.

Acknowledgments. This work was supported by the ANTS project, an *Action de Recherche Concertée* funded by the Scientific Research Directorate of the French Community of Belgium. Marco Dorigo acknowledges support from the Belgian National Fund for Scientific Research (FNRS), of which he is a Research Director.

References

1. Blum, C., Dorigo, M.: The hyper-cube framework for ant colony optimization. IEEE Transactions on Systems, Man, and Cybernetics—Part B **34**(2) (2004) 1161–1172
2. Dorigo, M., Stützle, T.: Ant Colony Optimization. MIT Press, Cambridge, MA, USA (2004)
3. Dorigo, M., Maniezzo, V., Colorni, A.: The Ant System: An autocatalytic optimizing process. Technical Report 91-016 Revised, Dipartimento di Elettronica, Politecnico di Milano, Milano, Italy (1991)
4. Dorigo, M.: Ottimizzazione, apprendimento automatico, ed algoritmi basati su metafora naturale. PhD thesis, Politecnico di Milano, Milano, Italy (1992) In Italian.
5. Dorigo, M., Maniezzo, V., Colorni, A.: Ant System: Optimization by a colony of cooperating agents. IEEE Transactions on Systems, Man, and Cybernetics—Part B **26**(1) (1996) 29–41
6. Birattari, M., Pellegrini, P., Dorigo, M.: On the invariance of ant colony optimization. Technical Report TR/IRIDIA/2006-004, IRIDIA, Université Libre de Bruxelles, Brussels, Belgium (2006) Submitted for journal publication.
7. Stützle, T., Hoos, H.H.: The $\mathcal{MAX}$–$\mathcal{MIN}$ Ant System and local search for the traveling salesman problem. In Bäck, T., Michalewicz, Z., Yao, X., eds.: Proceedings of the 1997 IEEE International Conference on Evolutionary Computation (ICEC'97), Piscataway, NJ, USA, IEEE Press (1997) 309–314
8. Stützle, T., Hoos, H.H.: $\mathcal{MAX}$–$\mathcal{MIN}$ ant system. Future Generation Computer Systems **16**(8) (2000) 889–914
9. Dorigo, M., Gambardella, L.M.: Ant Colony System: A cooperative learning approach to the traveling salesman problem. IEEE Transactions on Evolutionary Computation **1**(1) (1997) 53–66

Parallel Ant Colony Optimization for the Traveling Salesman Problem

Max Manfrin, Mauro Birattari, Thomas Stützle, and Marco Dorigo

IRIDIA, CoDE, Université Libre de Bruxelles, Brussels, Belgium
{mmanfrin, mbiro, stuetzle, mdorigo}@iridia.ulb.ac.be

Abstract. There are two reasons for parallelizing a metaheuristic if one is interested in performance: (i) given a fixed time to search, the aim is to increase the quality of the solutions found in that time; (ii) given a fixed solution quality, the aim is to reduce the time needed to find a solution not worse than that quality. In this article, we study the impact of communication when we parallelize a high-performing ant colony optimization (ACO) algorithm for the traveling salesman problem using message passing libraries. In particular, we examine synchronous and asynchronous communications on different interconnection topologies. We find that the simplest way of parallelizing the ACO algorithms, based on parallel independent runs, is surprisingly effective; we give some reasons as to why this is the case.

1 Introduction

A system of n parallel processors is generally less efficient than a single n-times-faster processor, but the parallel system is often cheaper to build, especially if we consider clusters of PCs or workstations connected through fast local networks and software environments such as Message Passing Interface (MPI). This makes at the time of this research clusters one of the most affordable and adopted parallel architectures for developing parallel algorithms.

The availability of parallel architectures at low cost has widened the interest for the parallelization of algorithms and metaheuristics [1]. When developing parallel population-based metaheuristics such as parallel genetic algorithms and parallel ant colony optimization (ACO) algorithms, it is common to adopt the "island model" approach [2], in which the exchange of information plays a major role. Solutions, pheromone matrices, and parameters have been tested (see for example [3,4,5,6]) as the object of such an exchange. In [3] solutions and pheromone levels are exchanged, producing a rather high volume of communication which requires a significant part of the computational time. In [6] the communication of the whole pheromone matrix leads to a decrease in solution quality as well as worse runtime behavior, while the exchange of best-so-far and elite solutions produces the best results w.r.t. solution quality. In this paper, we study how different interconnection topologies affect the overall performance when we want to increase, given a fixed run time, the quality of the solutions found by a multi-colony parallel ACO algorithm to solve the traveling salesman problem (TSP).

M. Dorigo et al. (Eds.): ANTS 2006, LNCS 4150, pp. 224–234, 2006.

We use the TSP, an $\mathcal{NP}$-hard problem [7], as a case study, that also has been a central test bed in the development of the ACO field. For each interconnection topology, we implement both synchronous and asynchronous communication. In the first case, the sender waits for the receiver to exchange messages. In the second case, the sender forwards the message and continues, not waiting for the receiver. The communication strategy we adopt involves the exchange of the best-so-far solutions every r iterations, after an initial period of "solitary" search. The main advantage of using best-so-far solutions over pheromone matrices is that less data has to be exchanged: for the smallest instance that we consider, each pheromone matrix requires several megabytes of memory space, while a solution requires only some kilobytes.

For this study, we use $\mathcal{MAX}$–$\mathcal{MIN}$ Ant System ($\mathcal{MM}$AS) [8]—currently one of the best-performing ACO algorithms—as a basis for our parallel implementation. Our implementation of $\mathcal{MM}$AS is based on the publicly available ACOTSP code.[1]

Some research has been done on the parallelization of ACO algorithms, but, surprisingly enough, only few works used as a basis for the study of the effectiveness of the parallelization a high-performing ACO algorithm. An example is [9], where the effect of parallel independent runs was studied.

The article is structured as follows. Section 2 describes the details of our implementation of $\mathcal{MM}$AS, and describes the different interconnection topologies adopted. In Section 3, we report details about the experimental setup, and Section 4 contains the results of the computational experiments. Finally, in Section 5 we discuss the limitations of this work and summarize the main conclusions that can be drawn from the experimental results.

2 Parallel Implementation of $\mathcal{MAX}$–$\mathcal{MIN}$ Ant System

ACO is a metaheuristic introduced in 1991 by Dorigo and co-workers [10,11]. For an overview of the currently available ACO algorithms see [12]. In ACO, candidate solutions are generated by a set of stochastic procedures called artificial ants that use a parametrized probabilistic model which is updated using the previously seen solutions [13].

As said in Section 1, for this research, we use $\mathcal{MM}$AS as a basis for our parallel implementation. We extended the ACOTSP code by quadrant nearest neighbor lists. To have a version that is easily parallelizable, we removed the occasional pheromone re-initializations applied in the $\mathcal{MM}$AS described in [8], and we use only a best-so-far pheromone update. Our version also uses the 3-opt local search.

We aim at an unbiased comparison of the performance produced by communication among multiple colonies on five different interconnection topologies. In order to obtain a fair and meaningful analysis of the results, we have restricted the approaches to the use of a constant communication rate among colonies to exchange the best-so-far solutions. A colony *injects* in his current solution-pool a received best-so-far solution if and only if it is better than its current best-so-far

[1] `http://www.aco-metaheuristic.org/aco-code/public-software.html`

solution, otherwise it disregards it. In the following, we briefly and schematically describe the principles of the communication on each interconnection topology we considered. For each topology, with the exception of the Parallel independent runs, we have two versions: a first one, where the communication is synchronous, and a second one, where the communication is asynchronous. The topologies we studied are:

Fully-connected. In this parallel model, k colonies communicate with each other and cooperate to find good solutions. One colony acts as a master and collects the values of the best-so-far solutions found by the other $k-1$ colonies. The master then broadcasts to all colonies the identifier of the colony that owns the best solution among all k colonies so that everybody can get a copy of this solution. We consider a synchronous and an asynchronous implementation of this model identified by SFC and AFC, respectively, in the following.

Replace-worst. This parallel model is similar to the fully-connected, with the exception that the master identifies also the colony that owns the worst solution among the k colonies. Instead of broadcasting the identity of the best colony, the master sends only one message to the best colony, containing the identity of the worst colony, and the best colony sends its best-so-far solution only to the worst colony. We consider a synchronous and an asynchronous implementation of this model identified by SRW and ARW, respectively, in the following.

Hypercube. In this model, k colonies are connected according to the hypercube topology (see [14] for a detailed explanation of this topology). Practically, each colony is located on a vertex i of the hypercube and can communicate only with the colonies that are located in the vertices that are directly connected to i. Each colony sends to each of its neighbors its best-so-far solution. We consider a synchronous and an asynchronous implementation of this model respectively SH and AH in the following.

Ring. Here, k colonies are connected in such a way that they create a ring. We have implemented a unidirectional ring, so that colony i sends his best-so-far solution only to colony $[(i+1) \bmod k]$, and receives only the best-so-far solution from colony $[(i-1+k) \bmod k]$. We consider a synchronous and an asynchronous implementation of this model, called SR and AR in the following.

Parallel independent runs. In this model, k copies of the same sequential $\mathcal{MMAS}$ algorithm are simultaneously and independently executed using different random seeds. The final result is the best solution among all the k runs. Using parallel independent runs is appealing as basically no communication overhead is involved and nearly no additional implementation effort is necessary. In the following, we identify the implementation of this model with the acronym PIR.

These topologies allow us to consider decreasing communication volumes, moving from more global communication, as in fully-connected, to more local communication, as in ring, to basically no communication, as in parallel independent runs.

3 Experimental Setup

As said in Section 1, all algorithms are based on the ACOTSP software, which is coded in C. The parallel algorithms add, w.r.t. the sequential code, the communication capability, using MPI libraries. Experiments were performed on a homogeneous cluster of 4 computational nodes running GNU/Linux Debian 3.0 as Operating System and LAM/MPI 7.1.1 as communication libraries. Each computational node contains two AMD Opteron™ 244 CPUs, 2 GB of RAM, and one 1 Gbit Ethernet network card. The nodes are interconnected through a 48-ports Gbit switch.

The initial computational experiments are performed with $k = 8$ colonies of 25 ants each that exchange the best-so-far solution every 25 iterations, except for the first 100 iterations.

We consider 10 instances from TSPLIB [15] with a termination criterion based on run time, dependent on the size of the instance, as reported in Table 1. For each of the 10 instances, 10 runs were performed. In order to have a reference algorithm for comparison, we also test the equivalent sequential $\mathcal{MM}$AS algorithm. We considered two cases: in the first one (SEQ), it runs for the same overall wall-clock time as a parallel algorithm (8-times the wall-clock time of a parallel algorithm), while in the second one (SEQ2), it runs for the same wall-clock time as one CPU of the parallel algorithm. It is reasonable to request that a parallel algorithm performs at least not worse than SEQ2 within the computation time under consideration.

The parameters of $\mathcal{MM}$AS are chosen in order to guarantee robust performance over all the different sizes of instances; we use the same parameters as proposed in [8], except for the pheromone re-initializations and the best-so-far update, as indicated in Section 2.

To compare results across different instances, we normalize them with respect to the distance from the known optimal value. For a given instance, we denote as c_{MH} the value of the final solution of algorithm MH, and c_{opt} the value of the

Table 1. Instances with run time in seconds and average number of total iterations in a run done by the sequential algorithm SEQ2

instance	run time	SEQ2 average iterations
pr1002	900	11831
u1060	900	10733
pcb1173	900	10189
d1291	1200	11325
nrw1379	1200	8726
fl1577	1500	15938
vm1748	1500	6160
rl1889	1500	6199
d2103	1800	12413
pr2392	1800	8955

optimal solution; the normalized value is then defined as

$$\text{Normalized Value for MH} = \frac{c_{\text{MH}} - c_{\text{opt}}}{c_{\text{opt}}} \cdot 100. \quad (1)$$

This normalization method provides a measure of performance that is independent of the values of the different optimal solutions, allowing us to aggregate results form different instances.

4 Results

As stated in Section 2, we aim at an unbiased comparison of the performance produced by communication among multiple colonies on different interconnection topologies. The hypothesis is that the exchange of the best-so-far solution among different colonies speeds up the search for high quality solutions, having a positive impact on the performance of the algorithms. In order to test the effects of communication, we implement versions of $\mathcal{MM}$AS algorithm that differ only in the communication behavior. This setup allows us to use statistical techniques for verifying if differences in solutions quality found by the algorithms are statistically significant. Figure 1 contains the boxplot of the results[2] grouped by algorithm over all instances after the normalization described in Section 3. The boxplot indicates that, on average, all the parallel models, except SFC, seem able to do better than SEQ and SEQ2, but that the best performing approach is PIR. We check whether the differences in performance among the parallel models with exchange of information and PIR are statistically significant. The assumptions for a parametric method are not met, hence we rely on the *Wilcoxon rank sum* test [16] with *p-values* adjusted by Holm's method [17].

As can be seen from Table 2, the differences in performance of all the parallel models with information exchange from those of PIR are statistically significant; this confirms that PIR is the best performing approach under the tested conditions. We also check whether the differences in performance are statistically significant once we group the algorithms by interconnection topology, using again the Wilcoxon test with p-values adjusted by Holm's method and we report the results in Table 3. Differences in performance among interconnection topologies are not statistically significant.

Even though the boxplot indicates that parallel algorithms achieve, on average, better performance than the sequential ones, the impact of communication on performance seems negative. One reason might be that the run times are rather high, and $\mathcal{MM}$AS easily converges in those times. PIR can count on multiple independent search paths to explore the search space, reducing the effects of the "stagnation" behavior. In fact, the other parallel algorithms accelerate the convergence toward a same solution, due to the frequent exchange of information, as can be verified by the traces of the algorithms' outputs.

[2] We refer the reader interested in the raw data to the URL: `http://iridia.ulb.ac.be/supp/IridiaSupp2006-001/`

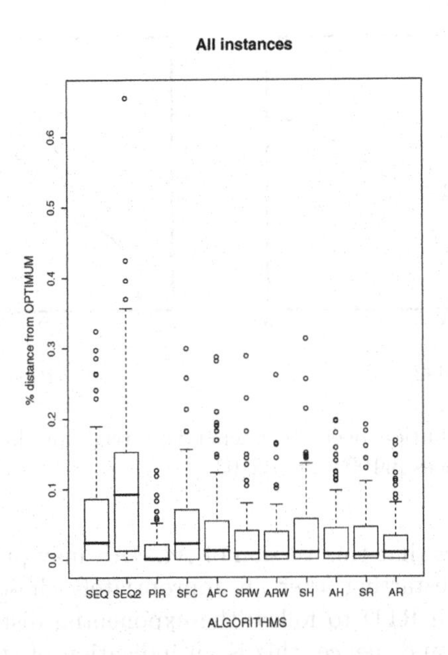

Fig. 1. Aggregate results over all instances. Boxplot of normalized results

Table 2. *p-values* for the null hypothesis "The distribution of the % distance from optimum of solutions for all instances is the same as PIR". The alternative hypothesis is that "The median of the PIR distribution is lower". The significance level with which we reject the null hypothesis is 0.05.

SFC	AFC	SRW	ARW	SH	AH	SR	AR
5.4e-4	0.01	0.02	0.02	1.2e-3	0.02	0.02	0.02

Table 3. *p-values* for the null hypothesis "The distributions of the % distance from optimum of solutions for all instances are the same". The significance level with which we reject the null hypothesis is 0.05.

	FC	RW	H
RW	0.55	-	-
H	1	1	-
R	0.55	1	1

Run time distributions. To examine the possibility of the "stagnation" behavior, we analyze the run-time distribution (RTD) of the sequential algorithm. A qualified run-time distribution measures the distribution of the time a stochastic local search algorithm requires to reach a specific target of solution quality, for example the optimal solution value. In Figure 2 we give plots of the measured RTDs for reaching the known optimal solution value for the two instances pr1002 and d2103. As explained in [18], the exponential distribution that is given

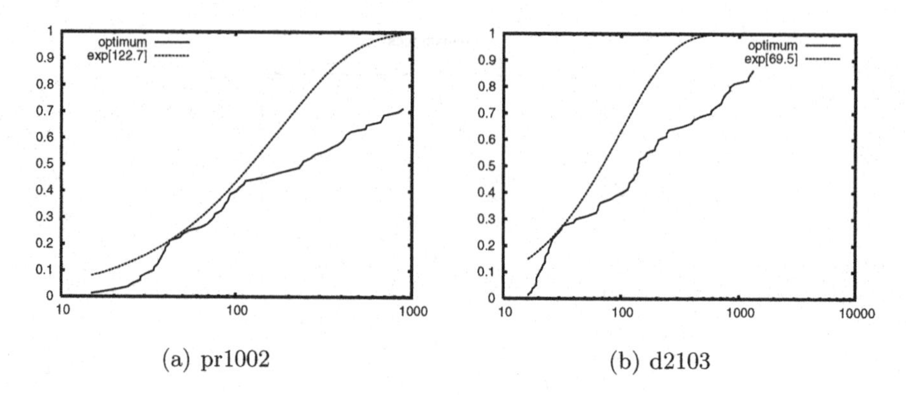

Fig. 2. Run-time distribution over 80 independent trials of the sequential $\mathcal{MM}$AS algorithm for the instances pr1002 and d2103

in these plots indicates that this version of $\mathcal{MM}$AS may profit from algorithm restarts (essentially, restarting after an appropriately chosen cutoff time, one can force the empirical RTD to follow the exponential distribution due to its statistical properties) and, hence, this is an indication of stagnation behavior. This explains to a large extent the good perfomance of parallel independent run given that PIR can count on multiple independent search paths to explore the search space, reducing the effects of the stagnation behavior.

Performance for reduced run-times. In order to check if our doubt on the "stagnation" behavior has some fundament, we re-analyze the results considering run times that are 1/4, 1/16, and 1/64 of the values reported in Table 1, showing the resulting boxplots in Figure 3. We observe that the more we reduce the run time, the smaller are the differences between the performance of the SEQ algorithm and the others, up to the reduced time of 1/64, for which SEQ performs on average better than all the parallel models (remember that SEQ has a run time that is 8-times the wall-clock time of a parallel algorithm).

Frequency of communication. As indicated, an apparent problem of our communication scheme is that communication is too frequent. To better understand the impact that the frequency of communication has on performance, we change the communication scheme to an exchange every $n/4$ iterations, except during the first $n/2$, where n is the size of the instance. We test this new communication scheme on the parallel models replace-worst (SRW2) and ring (SR2). Figure 4 shows the boxplots of the results. Once more, we rely on the Wilcoxon test with Holm's adjustment to verify whether the differences in performance are statistically significant. With the adoption of the new communication scheme, under the same experimental conditions, we are not able to reject the null hypothesis "The distributions of the % distance from optimum of solutions for all instances is the same as PIR" with a significance level of 0.05, given that the *p-values* relative to SRW2 and SR2 are both equal to 0.30. The reduced

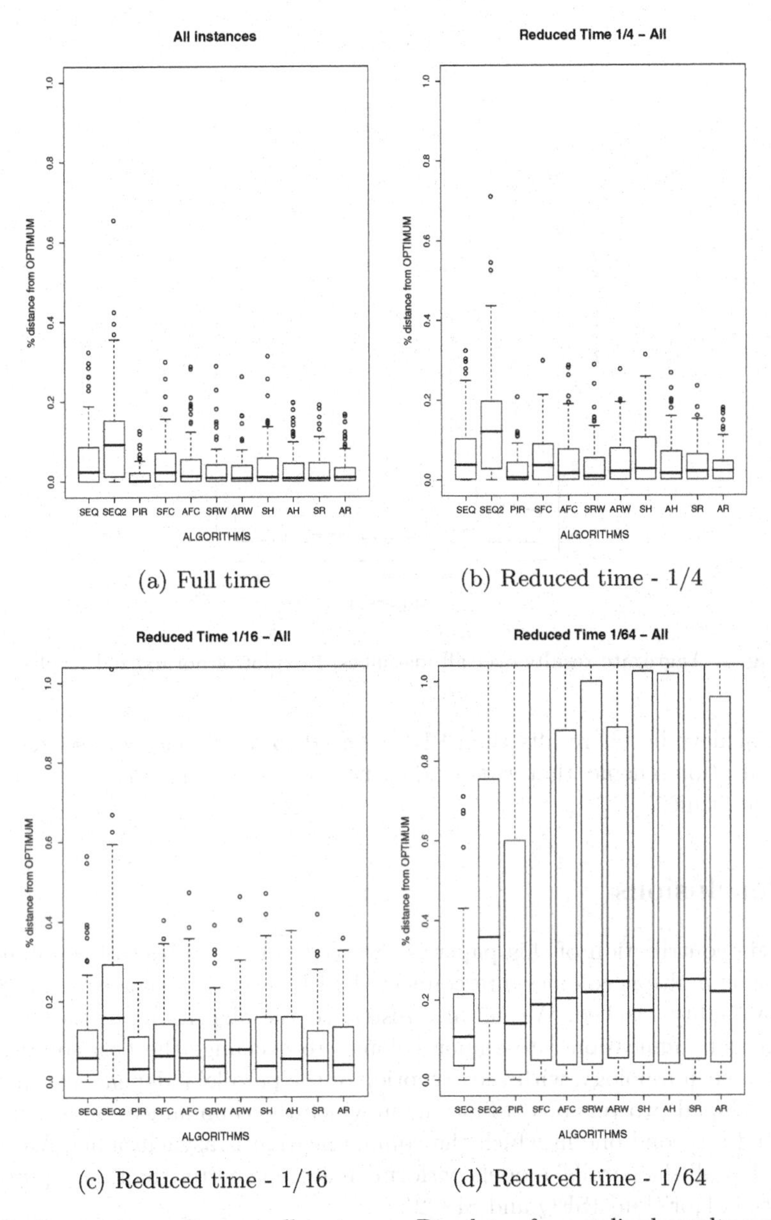

(a) Full time (b) Reduced time - 1/4

(c) Reduced time - 1/16 (d) Reduced time - 1/64

Fig. 3. Aggregate results over all instances. Boxplots of normalized results restricted to values in [0,1]. (a) full time is used; (b) run time reduced to 1/4 of full time; (c) run time reduced to 1/16 of full time; (d) run time reduced to 1/64 of full time.

frequency in communication has indeed a positive impact on the performance of the two parallel algorithms SRW2 and SR2, even though this is not sufficient to achieve better performance w.r.t parallel independent runs. We strongly believe

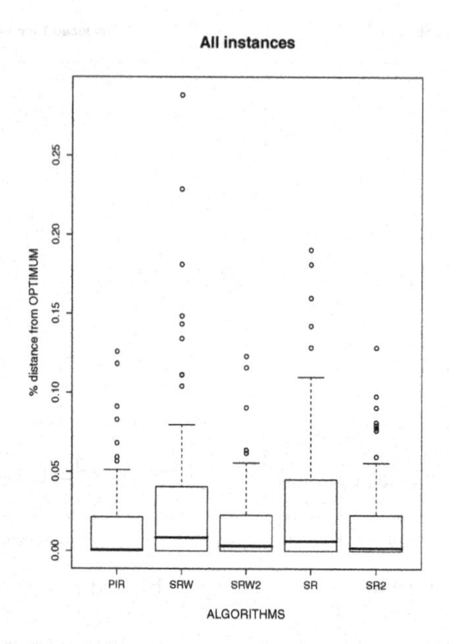

Fig. 4. Aggregate results over all instances. Boxplot of normalized results.

that to achieve better results than PIR we need to develop a more sophisticate communication scheme, that is dependent not only on the instance-size, but also on the run time.

5 Conclusions

The main contribution of this paper is the study of the impact of communication among multiple colonies interconnected with various topologies on the final solution quality reached. We initially restricted the algorithms to the use of a constant communication rate among colonies to exchange the best-so-far solutions. For each topology, with the exception of the parallel independent runs, we have developed two versions: a first one in which the communication is synchronous, and a second one in which the communication is asynchronous. We have shown that all the parallel models perform on average better than the equivalent sequential algorithms (SEQ and SEQ2).

As stated in Section 2, to have a version that was easy to parallelize, we removed from the $\mathcal{MM}$AS implementation the occasional pheromone re-initialization and we used only a best-so-far pheromone update. These modifications result in a stagnation behavior of the sequential algorithm; this stagnation behavior can be avoided to a large extent by parallel independent runs, which also explains its overall good behavior, biasing the performance in favor of PIR over all the other parallel models. We believe that better performance than PIR can be obtained by the parallel models either adding the restarting feature, or implementing communication schemes that avoid early convergence. This second approach could be achieved implementing the acceptance of solutions from other colonies only when they "differ" less than a certain number of components, leading to the creation of groups of colonies that search in different areas of the search space, or by exchanging the solutions with a frequency that depends on both, instance size and run time.

Acknowledgments. This work is supported by COMP^{2}SYS and by the ANTS project. COMP^{2}SYS is a Marie Curie Early Stage Training Site, funded by the European Community's Sixth Framework Programme under contract number MEST-CT-2004-505079. The ANTS project is an Action de Recherche Concertée funded by the Scientific Research Directorate of the French Community of Belgium. M. Dorigo and T. Stützle acknowledge support from the Belgian National Fund for Scientific Research (FNRS), of which they are a Research Director and a Research Associate, respectively. The information provided is the sole responsibility of the authors and does not reflect the opinion of the sponsors. The European Community and the French Community are not responsible for any use that might be made of data appearing in this publication. The authors wish to thank A. L. Christensen for his support during the MPI programming phase.

References

1. Alba, E., ed.: Parallel Metaheuristics: A New Class of Algorithms. Wiley Series on Parallel and Distributed Computing. Wiley-Interscience, Hoboken, NJ (2005)
2. Tanese, R.: Parallel genetic algorithms for a hypercube. In: Proceedings of the second international conference on Genetic Algorithms and their Applications, Hillsdale, NJ, Lawrence Erlbaum Associates, Inc. (1987) 177–183
3. Bullnheimer, B., Kotsis, G., Strauß, C.: Parallelization strategies for the Ant System. In De Leone, R., et al., eds.: High Performance Algorithms and Software in Nonlinear Optimization. Kluwer Academic Publishers, Norwell, MA (1998) 87–100
4. Middendorf, M., Reischle, F., Schmeck, H.: Multi colony ant algorithms. Journal of Heuristics **8**(3) (2002) 305–320
5. Piriyakumar, D.A.L., Levi, P.: A new approach to exploiting parallelism in ant colony optimization. In: International Symposium on Micromechatronics and Human Science (MHS) 2002, Nagoya, Japan. Proceedings, IEEE Standard Office (2002) 237–243
6. Benkner, S., Doerner, K.F., Hartl, R.F., Kiechle, G., Lucka, M.: Communication strategies for parallel cooperative ant colony optimization on clusters and grids. In: Complimentary Proceedings of PARA'04 Workshop on State-of-the-Art in Scientific Computing, June 20-23, 2004, Lyngby, Denmark (2005) 3–12
7. Garey, M.R., Johnson, D.S.: Computers and Intractability / A Guide to the Theory of $\mathcal{NP}$-Completeness. W.H. Freeman & Company, San Francisco, CA (1979)
8. Stützle, T., Hoos, H.H.: $\mathcal{MAX}$–$\mathcal{MIN}$ Ant System. Future Generation Computer System **16**(8) (2000) 889–914
9. Stützle, T.: Parallelization strategies for ant colony optimization. In Eiben, A.E., et al., eds.: Parallel Problem Solving from Nature - PPSN V. Volume 1498 of Lecture Notes in Computer Sciences., Berlin, Germany, Springer-Verlag, (1998) 722–731

10. Dorigo, M., Maniezzo, V., Colorni, A.: Positive feedback as a search strategy. Technical Report 91-016, Dipartimento di Elettronica, Politecnico di Milano, Milan, Italy (1991)
11. Dorigo, M.: Ottimizzazione, apprendimento automatico, ed algoritmi basati su metafora naturale. PhD thesis, Dipartimento di Elettronica, Politecnico di Milano, Milan, Italy (1992)
12. Dorigo, M., Stützle, T.: Ant Colony Optimization. MIT Press, Cambridge, MA (2004)
13. Zlochin, M., Birattari, M., Meuleau, N., Dorigo, M.: Model-based search for combinatorial optimization: A critical survey. Annals of Operations Research **131** (2004) 373–395
14. Grama, A., Gupta, A., Karypis, G., Kumar, V.: Introduction to parallel computing. Second edn. Pearson - Addison Wesley, Harlow, UK (2003)
15. Reinelt, G.: TSPLIB95 `http://www.iwr.uni-heidelberg.de/groups/comopt/-software/tsplib95/index.html` (2004)
16. Conover, W.J.: Practical Nonparametric Statistics. Third edn. John Wiley & Sons, New York, NY (1999)
17. Holm, S.: A simple sequentially rejective multiple test procedure. Scandinavian Journal of Statistics **6** (1979) 65–70
18. Hoos, H., Stützle, T.: Stochastic Local Search: Foundations & Applications. Morgan Kaufmann Publishers Inc., San Francisco, CA (2004)

Placement Constraints and Macrocell Overlap Removal Using Particle Swarm Optimization

Sheng-Ta Hsieh, Tsung-Ying Sun, Cheng-Wei Lin, and Chun-Ling Lin

Intelligent Signal Processing Lab., Department of Electrical Engineering
National Dong Hwa University, Shoufeng, Hualien, Taiwan, R.O.C.
sunty@mail.ndhu.edu.tw

Abstract. This paper presents a macrocell placement constraints and overlap removal methodology using particle swarm optimization (PSO). The authors adopted several techniques along with PSO as to avoid the floorplanning falling into the local minimum and to assist in finding out the global minimum. Our method can deal with various kinds of placement constraints, and consider them simultaneously. Experiments employing MCNC and GSRC benchmarks show the efficiency and robustness of our method for restricted placement and overlap removal obtained by the ability of exploring better solutions. The proposed approach exhibited rapid convergence and led to more optimal solutions than other related approaches, furthermore, it displayed efficient packing with all the constraints satisfied.

1 Introduction

The physical placement of circuits in VLSI chips or *System on Chips* (SoCs) has been given sustained attention in recent years. The major objective of placement is to allocate the modules of a circuit into a chip to optimize some design metric such as chip area and wire length etc. However, in nature, placement is an NP-complete problem. Early research on the placement problem applied force to reduce the overlap betweens cells [1]. For non-slicing structure, many packing representation methods have been proposed in recent years, such as [2,3,4]. On the other hand, [5] and [6] show the generated layouts with cell overlaps. Allowing overlaps during placement process were shown to obtain a better solution, but will increase the cost at the end of for reduce amount of overlap and cannot guarantee entire elimination.

The most recent approach for macrocell overlap removal and placement was conducted by Alupoaei and Katkoori, whose algorithm was based on the ant colony optimization method (ACO) [7]. Each ant generated a placement based on the relative macrocells' positions and information regarding the optimal placement obtained by previous colonies. The disadvantage to this macrocell movement procedure was that the initial relationship between macrocells will influence the final result directly and/or fall into the local minimal.

Although, all the approaches mentioned above have their advantages and disadvantages, they do not cover placement constraint problems. Due to the analog

M. Dorigo et al. (Eds.): ANTS 2006, LNCS 4150, pp. 235–246, 2006.

design, designers will also be interested in a particular kind of placement constraint called symmetry, and some recent literature on this problem can be found in [8] and [9]. The floorplanner in [10] can handle alignment constraint which may arise in bus-based routing. The floorplanners in [11,12,13] can handle pre-place constraint in which some modules are fixed in position. Different approaches are used to handle the different kinds of constraints, but there are no unified methods that can handle all constraints simultaneously.

As opposed to these previously mentioned methods, this paper adopts Particle Swarm Optimization (PSO) with an overlap detection and removal mechanism to search for the optimal placement solution. PSO is a swarm intelligence method that roughly models the social behavior of swarms. The consequence of modeling this social behavior is that the search process allows particles to stochastically return toward previously successful regions in the search space. It has proved to be efficient on a plethora of problems in science and engineering.

We incorporated PSO with a disturbance mechanism [14] to avoid the solution from falling into the local minimum, furthermore, our method can handle different kinds of placement constraints simultaneously, including boundary constrains, pre-place constraint, range constraint, abutment, alignment, and clustering, etc.. Users can dictate a mixed set of constraints and their preferred way of arrangement for the assigned macrocell. Our floorplanner will then be able to place all of them simultaneously.

Section II defines various placement constraint problems definition. Section III describes the original PSO methodology. Section IV presents the proposed methods. Section V presents the experimental results. Aside from the comparisons with ACO, SA and B*-tree in un-constrainted conditions. Section V also exhibits the results of situation involving different kinds of randomly selected multi-constraint in floorplanning. Section VI of the paper contains the conclusion.

2 Placement Constraints

During floorplanning, we gave some information for a set of macrocells, including their width, length and cell numbers. We addressed the floorplanning problem with a number of placement constraints, i.e., besides the module information, we also gave some placement constraints between the modules. Our goal was to plan their positions on a chip such that all the placement constraints can be satisfied and the area and interconnection cost is minimized.

2.1 Cell Definition

We used the notation $c(i, x, y)$ to denote the i_{th}cell's location, where x and y are presented i_{th}cell's position on x-axis and y-axis respectively. Note that the cell positions are defined as the cell's lower left corner. Fig. 1 illustrates these definitions.

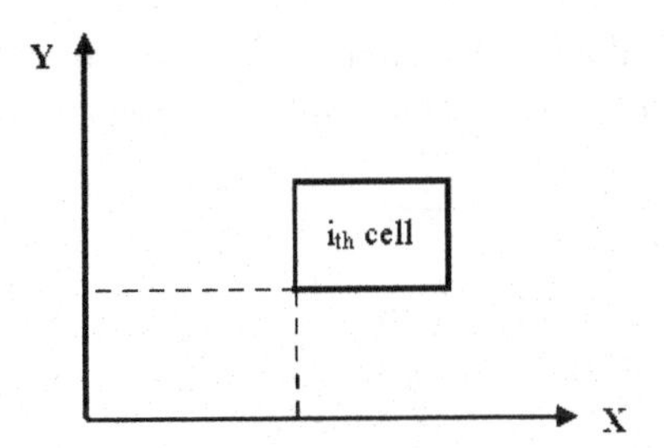

Fig. 1. Cell Definition

2.2 Relative and Absolute Constraints

Placement constraints can be classed as relative and absolute. A relative placement constraint describes the relationship between two modules, and an absolute placement constraint describes the relationship between a module and the chip. i.e., with relative placement constraints, users can restrict the horizontal or vertical distance between two modules to a certain range of values, and with absolute placement constraints is the absolute placement of a module is restricted with respect to the whole chip. Users can restrict the placement of a module such that its distance from the boundary of the chip is within a certain range of values. The modules under both kinds of constraints can be set as restriction conditions for the particle movement in the PSO evolution process. About the detail of the definition and applications will be described in the following section.

2.3 General Use Placement Constraints

Head Using the above specifications for absolute and relative placement constraints, we can describe many different kinds of placement constraints. In this section, we will pick a few commonly used ones and show how each can be specified using a combination of the relative and absolute placement constraints.

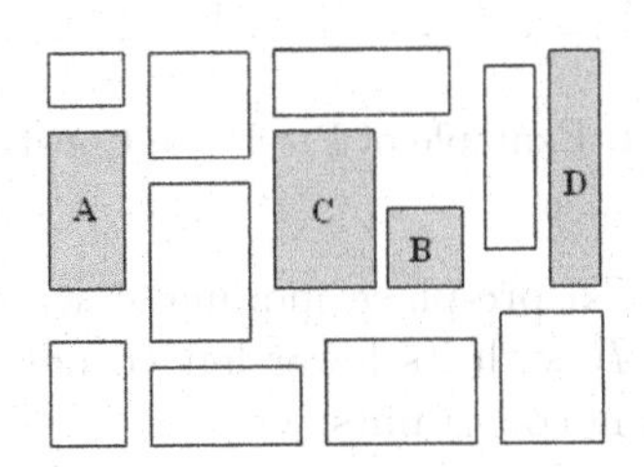

Fig. 2. Example of Alignment Constraint

Alignment: To align module A, B, C and D vertically (Fig. 2), we can impose the following constraints:

$$x_A = x_B = x_C = x_D$$

We restricted the vertical axes of each module to be same; they will thus all align vertically. A similar definition can be applied to their horizontal alignment.

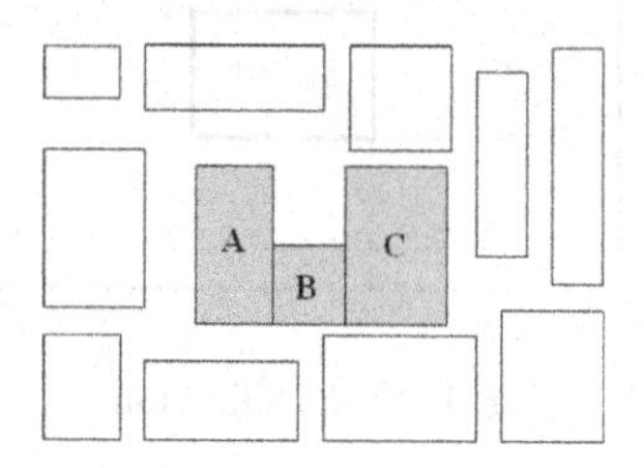

Fig. 3. Example of Abutment Constraint

Abutment: To abut module A, B and$C$ horizontally (Fig. 3), we impose the following constraints:

$$x_B = x_A + w_A$$
$$x_C = x_B + w_B$$
$$y_A = y_B = y_C$$

where w_A and w_B are the widths of module A and B, respectively. In this formulation, the horizontal axes of each module are the same; so they will align horizontally. On the other hand, it restricted the module B to being placed next to module A on the right hand side, and so on. So they will be abutting with each other horizontally.

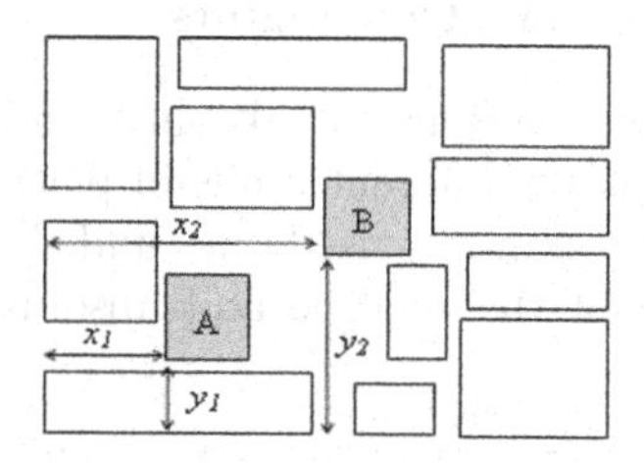

Fig. 4. Example of Pre-Place Constraint

Pre-place Constraint: To pre-place module A with its lower left corner at axis (x_1, y_1) and module B with its lower left corner at axis (x_2, y_2) (Fig. 4), we can impose the following constraints:

$$x_A = x_1, \ y_A = y_1$$
$$x_B = x_2, \ y_B = y_2$$

We restricted module $A$to be x_1 units from the left boundary and y_1 units from the bottom boundary. The similar definition can be applied to module B. Each restricted module will be pre-placed with its lower left corner in the final packing.

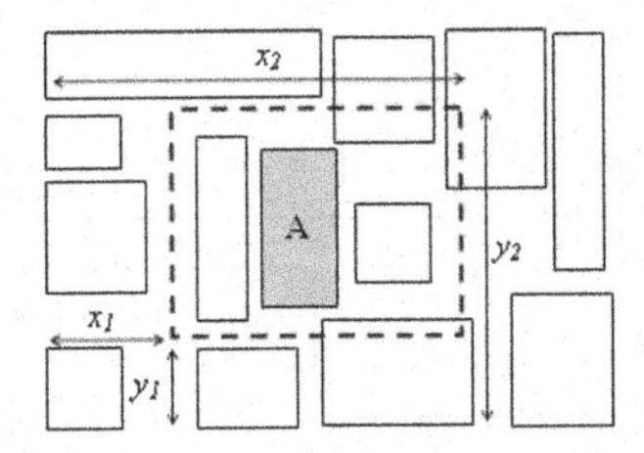

Fig. 5. Example of Range Constraint

Range Constraint: To restrict the position of module A in the range $\{(x_A, y_A)| x_1 \leq x_A \leq x_2, y_1 \leq y_A \leq y_2\}$ (Fig. 5), we can impose the following constraints:

$$x_A = [x_1, x_2],\ y_A = [y_1, y_2]$$

In this formulation, we restrict module A to be x_1 to x_2 units from the left boundary and to be y_1 to y_2 units from the bottom boundary, therefore, module A will be laid in the required rectangular region.

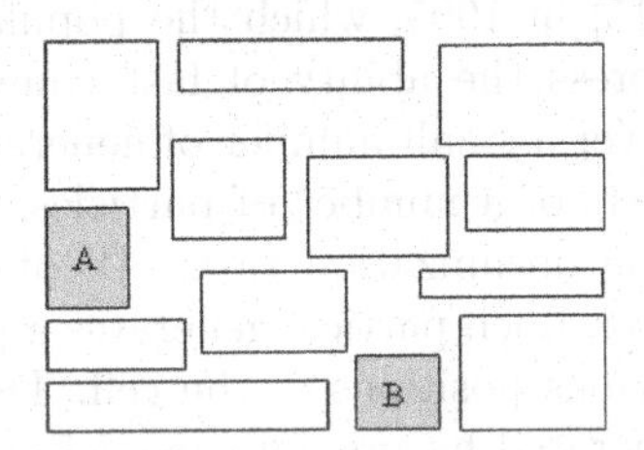

Fig. 6. Example of Boundary Constraint

Boundary Constraint: To place module A along the left boundary and place module B along the bottom boundary of the final packing (Fig. 6), we can impose the following constraints:

$$x_A = x_C + w_C,\ y_A = [y_C, y_C + h_C]$$
$$x_B = x_C + w_C,\ y_B = [y_C, y_C + h_C]$$

In this formulation, we restrict module A to be 0 units from the left boundary, so module A will be abut with the left boundary in the final packing. Module B is restricted to be 0 units from the bottom boundary, so module B will abut with the bottom boundary as required.

Clustering Constraint: Clustering Constraint: To cluster module A and B around C at a distance of most units away vertically. (Fig. 7), we can impose the following constraints:

$$x_A = x_C + w_C,\ y_A = [y_C, y_C + h_C]$$
$$x_B = x_C + w_C,\ y_B = [y_C, y_C + h_C]$$

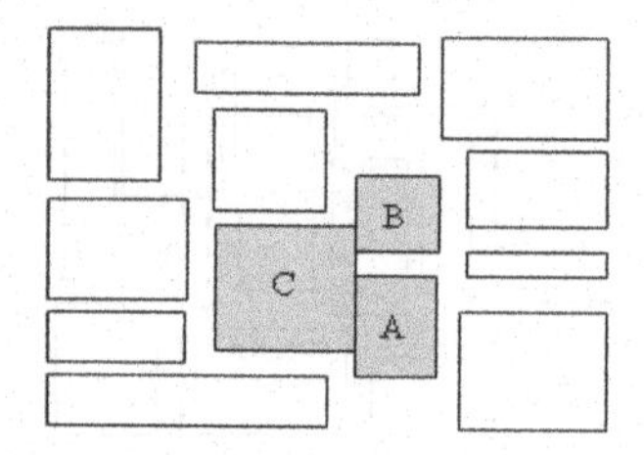

Fig. 7. Example of Cluster Constraint

In this formulation, we restrict the vertical distances of A and B from C to be at most units in vertical directions, so they will cluster around C at a vertical distance of at most h_C units away. We can also restrict the horizontal distances of A and B from Cin a similar way.

3 Particle Swarm Optimization (PSO)

The PSO is a population based optimization technique that was proposed by Kennedy and Eberhart [15] in 1995, which the population is referred to as a *swarm.* The particles express the ability of fast convergence to local and/or global optimal positions over a small number of generations.

A swarm in PSO consists of a number of particles. Each particle represents a potential solution of the optimization task. All of the particles iteratively discover a probable solution. Each particle generates a position according to the new velocity and the previous positions of the cell. This is compared with the best position which is generated by previous particles in the cost function and the best one is kept; i.e., each particle accelerates in the direction of not only the local best solution but also the global best position. If a particle discovers a new probable solution, other particles will move closer to it to explore the region more completely in the process.

Let $s$denote the swarm numbers. In general, there are three attributes, current position x_i, current velocity v_i and past best position Pb_i, for particles in the search space to present their features. Each particle in the swarm is iteratively updated according to the aforementioned attributes. Assuming that the function f is to be minimized so that the dimension consists of $n$particles, the new velocity of every particle is updated by (1).

$$\begin{aligned} v_{i,j}(t+1) = wv_{i,j}(t) + c_1 r_{1,i,j}(t)[Pb_{i,j}(t) - x_{i,j}(t)] \\ + c_2 r_{2,i,j}(t)[Gb_j(t) - x_{i,j}(t)] \end{aligned} \tag{1}$$

where $v_{i,j}$ is the velocity of the ith particle of the jth swarm for all$j \in 1 \dots s$, w is the inertia weight of velocity, c_1 and c_2 denote the *acceleration coefficients*, r_1 and $r_2$are elements from two uniform random sequences in the range (0, 1), and $t$is the number of generations. The new position of the $i$thparticle is calculated as follows:

$$x_i(t+1) = x_i(t) + v_i(t+1) \tag{2}$$

The past best position of each particle is updated by (3).

$$Pb_i(t+1) = \begin{cases} Pb_i(t), & \text{if } f(x_i(t+1)) \geq f(Pb_i(t)) \\ x_i(t+1), & \text{if } f(x_i(t+1)) < f(Pb_i(t)) \end{cases} \tag{3}$$

The global best position Gb found from all particles during previous three steps is defined as:

$$Gb(t+1) = \arg \min_{Pb_i} f(Pb_i(t+1)),\ 1 \leq i \leq n \tag{4}$$

4 PSO Algorithm for Macrocell Overlap Removal and Placement

The first step of using PSO to handle the macrocell overlap removal and placement is defining each module as a swarm that consists of a number of particles. The particle's move will lead the module to find another potential better solution for placement. For the initial state of the placement, each module was randomly generated in the floorplanning and overlap was allowed. After a number of generations, distance between each of the module will shrink, i.e. the chip size will get smaller. The overlap of between two modules will be eliminated by our overlap detection and removal mechanism.

4.1 Handling Placement Constraints by PSO

We consider two general kinds of placement constraints, absolute and relative. For relative placement constraint, users can restrict the horizontal or vertical distance between two modules to a certain value, or to a certain range of values.

We adopted a master-slave concept to define the relationship between cells. One cell is defined as 'master', the other cells as 'slaves'. All the slave cells will be moved only after the master cell has moved. i.e., the master and slave cells will be moved by PSO. Furthermore, the movement strategy of slave cells will also obey the cell's relationship defined by the constraints. For example, if we define cell A and cell B, constrain their vertical alignment, and define cell A as the master cell, then cell B will be moved only after cell A, and the x-axis of cell B is set according to cell A's current position. Absolute placement constraint is similarly specified except it does not involve master-slave concept.

4.2 Extend PSO Searching Space

In the original PSO, the moving vector of velocity would be decided according to the past best solution and the global best solution. This procedure seems make sense in evolutional computing, where all the new generations would inherit the past generations' experience or ability and move all particles around to the global best solution. In dynamic problems, optimal solution may change in each time slot but not fix at a specific position. According to pervious experiment may not be able to find optimal solutions. Thus the algorithm should be modified to meet this requirements. For example, in overlap detection and removal mechanism, all existing macrocells were regarded as forbidden regions. On the other hand, in when searching for better solutions, the modules may not just only move forward but also backward for possible arrangements. Thus, in some cases when using original PSO, the particle may not be able to find a feasible solution in each generation. It could ignore other potential solution in searching space, and spend more time moving towards the global minimum solution.

To enhance the particles' searching ability and save evolution time, we modified the velocity update equation as follow:

$$\begin{aligned} v_{i,j}(t+1) = T\{&wv_{i,j}(t) + c_1 r_{1,i,j}(t)[Pb_{i,j}(t) - x_{i,j}(t)] \\ &+ c_2 r_{2,i,j}(t)[Gb_j(t) - x_{i,j}(t)]\} \end{aligned} \tag{5}$$

where T denotes the *turn-around factor*. Normally, the T would be set at 1 (move forward). The particle's movement will follow the original PSO. If the particle can not find a feasible solution however, the T of the regenerated macrocell will be switched to -1 (move backward) for this generation. When a feasible solution has been found, the T will be restored as 1.

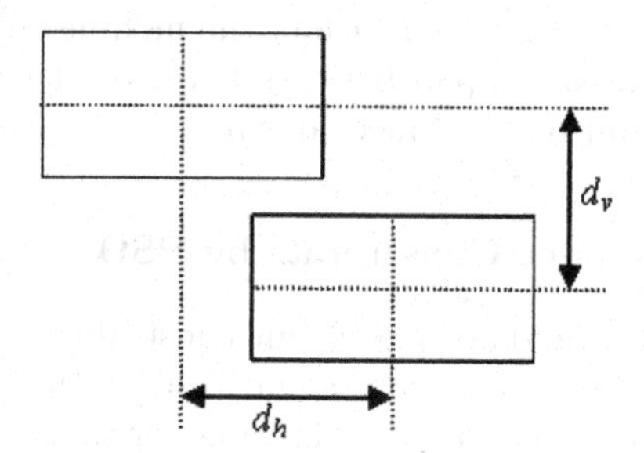

Fig. 8. The distance measure between two macrocells mechanism

4.3 Overlap Detection and Removal Mechanism

After the global best position *Gb* is updated to lead the macrocell to move to a new position, we should detect if the current macrocell overlaps with the existing macrocells. The horizontal distance d_h and vertical distance d_v between two macrocells depicted in Fig. 8 are measured via the macrocells' center. Then, they are compared with the half sum of two macrocell's height s_h and half sum of two macrocell's width s_w to estimate the occurrence of overlap by (6).

$$\text{Overlap} = \begin{cases} 1, \text{ if } d_h < 0.5s_w \text{ and } d_v < 0.5s_h \\ 0, \qquad\qquad \text{otherwise} \end{cases} \tag{6}$$

The positions for the existing macrocell were regarded as forbidden regions. However, if the newly generated macrocell overlapped with forbidden regions, it would be discarded and regenerated until the overlaps were free. While this process requires more time for computation, it guarantees that each macrocell's movement is overlap free.

4.4 PSO for Placement Constraint and Overlap Removal

The minimized chip area and wire length are required and estimated by the objective function (7). The swarm numbers $s$denotes the number of macrocells, and the particle number is defined as $1 \leq i \leq m$. The macrocells were randomly placed on the floorplan and the overlap was allowed initially. The x_i represents the macrocell current position and the initial state of $v_{i,j}$, Pb_i and Gb were set as 0. After that, particles were moved by (5) and (2), and through the constraints handling process, illegal moving vectors were cut off. The overlap detection and removal mechanism will also eliminate the cells that overlap with other macrocell(s). Then, the newly local best position would be updated by (3) and the global best position would be updated by (4). To guarantee that each constrained macrocell would not violate placement constraints, we set all the constrained modules as having higher priority in each movement for earlier generations to ensure that all set constraints were reached, i.e., while other non-constrained modules overlap constrained modules, these modules would be removed and regenerated. Thus, all the particles would keep moving to find a better solution until it reached the goal or met the termination condition.

Table 1. Compared our method with other approaches without placement constraints

Circuits	Cells	Our method (10 Particles)			Ant Colony [7] (100 Colony 200 Ants)		
		Chip Size (mm^2)	Wire Length (mm)	Run Time	Chip Size (mm^2)	Wire Length (mm)	Run Time
ami49	49	51.46	179.28	1m52s	63.32	218.36	7m58s
n_50	50	0.27	12.8	1m40s	0.41	18.01	7m12s
n_100	100	0.24	24.17	3m43s	0.45	40.36	15m01s
n_200	200	0.24	47.74	8m48s	0.46	73.68	45m59s
n_300	300	0.38	89.84	15m43s	0.57	119.43	1h24m12s
		Simulated Annealing			B*-tree [4]		
Circuits	Cells	Chip Size (mm^2)	Wire Length (mm)	Run Time	Chip Size (mm^2)	Wire Length (mm)	Run Time
ami49	49	69.5	237.79	3m4s	50.71	208.38	1m49s
n_50	50	0.5	17.68	1m50s	0.27	14.97	1m40s
n_100	100	0.47	35.10	4m1s	0.27	31.75	11m39s
n_200	200	0.56	72.68	10m55s	0.28	84.01	1h36m12s
n_300	300	0.63	132.91	23m12s	0.49	204.44	5h26m50s

5 Experiments

The experiments in this study employed GSRC and MCNC benchmarks [16] for the proposed placement constraints and overlap removal procedures and compared the results with other related research. All the macrocells were set as hard IP modules and without rotation. The simulation programs were written in MATLAB, and the results were obtained on Pentium 4 1.7 GHz with 512MB RAM. The PSO experiments with w, c_1 and c_2 initializations were 0.12, 0.25 and 0.25, respectively. We ran each floorplanner 10 times and calculated their average outcomes of wire length, chip area and run time.

Table 2. Using our method to deal with placement constraints

Circuits	Cells	8 macrocells under 4 random selected constraints			12 macrocells under 4 random selected constraints		
		Chip Size (mm^2)	Wire Length (mm)	Run Time	Chip Size (mm^2)	Wire Length (mm)	Run Time
ami49	49	51.91	177.67	2m51s	52.21	186.82	3m2s
n_50	50	0.29	13.65	3m21s	0.3	13.86	3m56s
n_100	100	0.27	26.17	5m23s	0.28	31.75	5m50s
n_200	200	0.28	52.47	11m18s	0.28	52.95	12m25s
n_300	300	0.43	97.09	20m6s	0.43	97.79	20m35s

The experiment results of placement without constraint are shown in Table I. Compared with related research, our method is more efficiency in finding better solutions with respect to chip area or wire length and can avoid being trapped in the local minimum. This is due to the ACO method allowing overlap in the initial stage. It not only has to execute the ACO procedure, detect and remove overlaps, but also has to deal with constraint graph for each macrocell. Besides this, in worse case, the time complexity of this method is $O(n^3)$, whose ability shows a feasible result for the floorplan. The convergence of SA method is quite slow. An acceptable solution would be found only after a great deal of computation time. On the other hands, although all the macrocells under B*-tree structure were overlap free, which could seemingly save a lot of time to detect and remove overlap between macrocells, it would spend more time arranging all cells while one cell was removed, exchanged or placed. In a minority of macrocells cases, it could still be more efficiency on placement. Once the cells number was increased, the computation burden of this method would increase nonlinearly, furthermore, it would likely fall into the local minimal. In unconstrained cases, our method expressed more optimal results for wire length and chip area, and shorter run times than ACO, SA and B*-tree.

The placement constraint experimental results are shown in Table II. Even restricted to different numbers of macrocells for placement; our method exhibited reasonable outcomes for chip size, wire length and run times to fit all selected constraints and with solutions that are acceptable. Furthermore, our method has no cases of constraint violations in each experiment. We can also find out that the variation in computation times of our method with different restricted cells is very small. A multi-constraint floorplanning example in GSRC n100 is illustrated in Fig. 9.

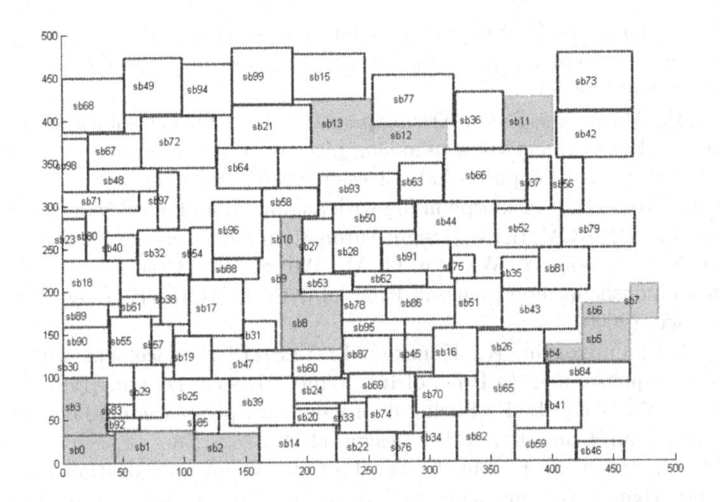

Fig. 9. Cell 0-3 cluster at lower corner of the chip, cell 4, 6 and 7 cluster around cell 5, cell8-10 abut vertically and cell11-12 align horizontally

6 Conclusions

In this paper, we presented a method to handle various placement constraints in floorplanning simultaneously. The experimental results proved that our method can lead to more optimal solutions in reasonable computation times for the hard IP modules in either constrained or unconstrained placement. Several benchmarks were adopted for testing and the results were very reliable. Placements with all the constraints satisfied can be obtained efficiently by our method.

References

1. Quinn, N., Breuer, M.: A forced directed componentplacementprocedure for printed circuit boards. IEEE Trans. on Circuits and Systems, vol. 26. (1979) 377–388.
2. Murata, H., Fujiyoshi, K., Nakatake, S., Kajitani, Y.: VLSI module placement based on rectangle-packing by the sequence-pair. IEEE Trans. on Computer Aided Design, vol. 15. (1996) 1518–1524.
3. Guo, P. N., Takahashi, T., Cheng, C. K., Yoshimura, T.: Floorplanning using a tree representation. IEEE Trans. on Computer-Aided Design, vol. 20. (2001) 281–289.
4. Chang, Y. C., Chang, Y. W., Wu, G. M., Wu, S. W.: B*-Trees: A new representation for nonslicing floorplans. Design Automation Conference. (2000) 458–463.
5. Sigl, G., Doll, K., Johannes, F. M.: Analytical placement: A linear or a quadratic objective function. Design Automation Conference. (1991) 427–432.
6. Mo, F., Tabbara, A., Brayton, R. K.: A force-directed macro-cell place. Computer-Aided Design Conference. (2000) 177–180.
7. Alupoaei, S., Katkoori, S.: Ant Colony System Application to Macrocell Overlap Removal. IEEE Trans. on Very Large Scale Integration (VLSI) Systems, vol. 12. (2004) 1118–1123.

8. Balasa, F., Lampert, K.: Symmetry within the sequence-pair representation in the context of placement for analog design. IEEE Trans. on Computer-Aided Design, vol. 19. (2000) 712–731.
9. Balasa, F., Maruvada, S. C., Krishnamoorthy, K.: Efficient solution space exploration based on segment trees in analog placement with symmetry constraints. in Proc. of Int. Conf. Computer-Aided Design. (2002) 497–502.
10. Tang, X., Wong, D. F.: Floorplanning with alignment and performance constraints. in Proc. of 39th ACM/IEEE Design Automation Conference. (2002) 848–853.
11. Chang, Y. C.,Chang, Y. W., Wu, G. M., Wu, S. W.: B*-trees: A new representation for nonslicing floorplans. in Proc. of 37th ACM/IEEE Design Automation Conference. (2000).
12. Murata, H., Fujiyoushi, K., Kaneko, M.: VLSI/PCB placement with obstacles based on sequence-pair. in Proc. of Int. Symp. Physical Design. (1997) 26–31.
13. Young, F. Y., Wong, D. F.: Slicing floorplans with pre-placedmodules. in Proc. of IEEE Int. Conf. Computer-Aided Design. (1998) 252–258.
14. Sun, T. Y., Hsieh, S. T., Lin, C. W.: Particle swarm optimization Incorporated with disturbance for improving the efficiency of macrocell overlap removal and placement," in Proc. of The 2005 International Conference on Artificial Intelligence (ICAI'05). (2005) 122–125.
15. Eberhart, R. C., Kennedy, J.: A new optimizer using particle swarm theory. in Proc. of 6th Int. Symp. Micro Machine and Human Science. (1995) 39–43.
16. http://www.cse.ucsc.edu/research/surf/GSRC/progress.html

PLANTS: Application of Ant Colony Optimization to Structure-Based Drug Design

Oliver Korb[1], Thomas Stützle[2], and Thomas E. Exner[1]

[1] Theoretische Chemische Dynamik, Universität Konstanz, Konstanz, Germany
{Oliver.Korb, Thomas.Exner}@uni-konstanz.de
[2] IRIDIA, CoDE, Université Libre de Bruxelles, Brussels, Belgium
stuetzle@ulb.ac.be

Abstract. A central part of the rational drug development process is the prediction of the complex structure of a small ligand with a protein, the so-called protein-ligand docking problem, used in *virtual screening* of large databases and lead optimization. In the work presented here, we introduce a new docking algorithm called PLANTS (**P**rotein-**L**igand **ANT S**ystem), which is based on ant colony optimization. An artificial ant colony is employed to find a minimum energy conformation of the ligand in the protein's binding site. We present the effectiveness of PLANTS for several parameter settings as well as a direct comparison to a state-of-the-art program called GOLD, which is based on a genetic algorithm. Last but not least, results for a virtual screening on the protein target factor Xa are presented.

1 Introduction

Finding new drugs is notoriously time-consuming and expensive taking up to 15 years [1] and costing several hundred million dollars. Today's drug discovery process pursued by major pharmaceutical companies begins with the identification of a suitable protein, the target, in whose function a potential drug could interfere to fight a disease. For this target, specific assays are developed, which are then used in *high-throughput screening* experiments to test the biological activity of large databases of possible drug candidates. Molecules with high affinity, so-called lead structures, are then chemically varied (lead optimization cycle) and the most potent ones of the resulting candidates are transferred to the preclinical and finally the clinical development phase. In this way, out of hundreds of thousands to millions of molecules, which have to be synthesized, a drug can be identified. To speed up the process and save money, computer methodologies have become a crucial part of these drug discovery projects, from hit identification to lead optimization. Approaches such as ligand- or structure-based *virtual screening* techniques are widely used in many discovery efforts [2]. One key methodology, the docking of small molecules (ligands) to a protein (receptor), remains a highly active area of research. In this, a complex structure, i.e. the orientation and conformation of the ligand within the active site of the protein, should be predicted. This was first described by Emil Fischer

M. Dorigo et al. (Eds.): ANTS 2006, LNCS 4150, pp. 247–258, 2006.

in terms of the lock-and-key metaphor [3]. To identify the correct pose of a specific ligand and to rank different ligands according to their binding affinity, an estimate of the binding free energy of the complex formation should also be calculated. Altogether, this is called the *protein-ligand docking problem* (PLDP) for which we propose a new algorithm based on ant colony optimization (ACO) [4].

2 Computational Approaches to the Docking Problem

A large variety of different approaches for a solution of the PLDP has been proposed. These can be broadly classified as *fragment-based*, as *stochastic optimization* methods for finding the global minimum, or as *multiconformer docking* approaches. Recent studies [5,6] compared different docking tools on a large test set of experimentally determined complex structures. They reported success rates of 30 to 60 %, where the success rate is defined as the percentage of complexes, for which the predicted structure with the lowest energy is very close (*root mean square deviation* (RMSD) within 2.0 Å) to the experimentally determined structure. This shows that a universal docking tool that has excellent predictive capabilities across many complexes is not available at the moment. Concentrating on stochastic optimization methods, this can be attributed to the *scoring problem* and the *sampling problem.* Given a protein and a ligand structure, a *scoring function* measures the binding strength of the ligand at a specific position of the protein. Currently, there exists no perfect scoring function, which is able to perform correct measurements for all given input structures. But even if there was a perfect scoring function, there would still be the problem, that there is no guarantee that the correct binding mode of the ligand is actually found by the sampling algorithm. Given one of these scoring functions f and the protein's and ligand's degrees of freedom, the PLDP can be formulated as searching for the values to be assigned to the degrees of freedom that globally minimize the scoring function. In the most approaches, the protein is kept rigid, in which case only the ligand's 3 translational, 3 rotational and r torsional degrees of freedom, describing rotations of single bonds that are not part of a ring system, need to be optimized. Thus, the total number of variables, that is the dimension of the optimization problem, equals $n = 6 + r$. In the actual implementations given in the literature (see [7] and references therein), a wide repertoire of optimization strategies is used to find the global minimum corresponding to the complex structure. E.g., genetic algorithms are used in the programs GOLD and AutoDock, Monte Carlo minimization in the programs ICM and QXP, and simulated annealing, evolutionary programming, and tabu search in PRO_LEADS.

3 PLANTS

We present a new algorithm, called Protein–Ligand ANT System (PLANTS) for sampling the search space. PLANTS is based on ACO, a technique that was

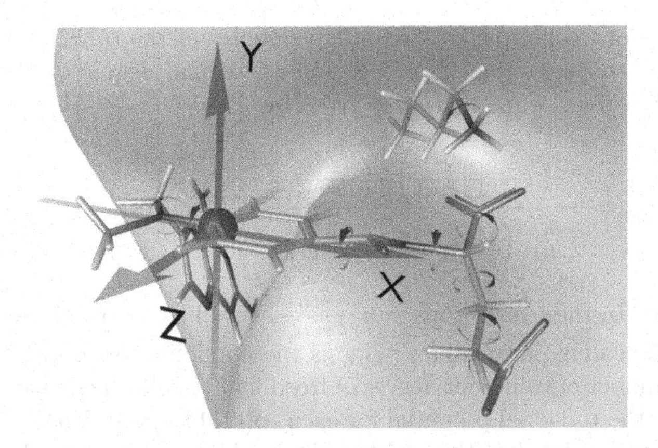

Fig. 1. Degrees of freedom for the docking problem. The origin of the ligand's coordinate system is shown as a sphere. The ligand's translational degrees of freedom are shown as large arrows, which also constitute the axes of rotation. The small arrows mark the ligand's rotatable bonds as well as a rotatable donor group in a single protein side-chain (upper right corner), which originates from the schematic protein surface shown in the background.

not yet tested for tackling the PLDP. PLANTS treats the ligand flexible, which means that there are $6 + r$ degrees of freedom for the ligand as described above. The flexibility of the protein is partially considered by the optimization of the positions of hydrogen atoms that could be involved in hydrogen bonding. Both the ligand's and the protein's degrees of freedom are illustrated in Figure 1. The search space with respect to the ligand's translational degrees of freedom is defined by the size of the binding site given for each protein.

Pheromone model. The displacement of the ligand and the torsion angles are continuous variables. Since ACO was originally designed to tackle combinatorial optimization problems, we decided to discretize the continuous variables such that we can directly apply existing ACO techniques to the problem. To do so, we used for each of the three translational degrees of freedom an interval length of 0.1Å, while for the three rotational degrees of freedom and all torsional degrees of freedom an interval of 1° was taken. Each degree of freedom i has associated a pheromone vector τ_i with as many entries as values result from the discretization. Hence, each pheromone vector associated with rotational or torsional degrees of freedom has 360 entries, while the number of entries of the pheromone vectors corresponding to the three translational degrees of freedom depends on the diameter of the binding site. A pheromone trail τ_{ij} then refers to the desirability of assigning the value j to degree of freedom i.

ACO algorithm. PLANTS is based on $\mathcal{MAX}$–$\mathcal{MIN}$ Ant System ($\mathcal{MM}$AS) [8]. The (artificial) ants construct solutions by choosing, based on the pheromone values and heuristic information, one value for each degree of freedom. The order of the degrees of freedom in the solution construction is arbitrarily fixed, since each degree of freedom is treated independently of the others. The probability that an ant chooses value j for a ligand's torsional degree of freedom i or for each other degrees of freedom k, is given by

$$p_{ij} = \frac{\left(1 - \frac{1}{1+\gamma\cdot\tau_{ij}}\right)\cdot\left(\frac{1}{1+\delta\cdot\eta_{ij}}\right)^{\beta}}{\sum_{l=1}^{n_i}\left(1 - \frac{1}{1+\gamma\cdot\tau_{il}}\right)\cdot\left(\frac{1}{1+\delta\cdot\eta_{il}}\right)^{\beta}} \text{ and } p_{kj} = \frac{\tau_{kj}}{\sum_{l=1}^{n_k}\tau_{kl}}, \tag{1}$$

respectively. In these equations, $\gamma = \frac{200}{\tau_{max_i}}$ as well as $\delta = 0.3$ are experimentally determined scaling parameters, τ_{max_i} is the maximum pheromone value and n_i (n_k) the number of values for degree of freedom i (k). The heuristic information η is given by the torsional potential for each rotatable bond. The rationale behind that information is, that the construction of high energy ligand conformations should be avoided. The nonlinear influence of the pheromone trails on the selection probabilities for the torsional degrees of freedom was chosen to account for the imperfectness of the heuristic information. As usual in $\mathcal{MMAS}$, only one solution is used to deposit pheromone after each iteration; in PLANTS, this is the best solution generated in the current iteration, s^{ib}. The pheromone update is defined as

$$\tau_{ij}(t+1) = (1-\rho)\tau_{ij}(t) + I_{ij}^{ib}(t)\Delta\tau^{ib}(t), \tag{2}$$

where

$$\Delta\tau^{\mathrm{ib}}(t) = \begin{cases} |f(s^{ib})| & \text{if } f(s^{ib}) < 0 \\ 0 & \text{otherwise} \end{cases} \tag{3}$$

and $f(s^{ib})$ is the scoring function value of s^{ib}. For a translational degree of freedom i, $I_{ij}^{ib}(t)$ is one, if s^{ib} assigned a value in $\{j-1, j, j+1\}$ to i; for rotational and torsional degrees of freedom, $I_{ij}^{ib}(t)$ is one if a value in $\{j-2, j-1, j, j+1, j+2\} \bmod n_i$ was taken; otherwise $I_{ij}^{ib}(t)$ is zero. The rationale for the choice of Equation 3 is that our scoring function indicates high affinity by strongly negative energy values, which means that the larger the absolute value the better; positive energies would actually correspond to negative affinity and, hence, do not receive any positive feedback. If $f(s^{ib})$ is positive, no pheromone is deposited in an iteration. The upper pheromone trail limit in PLANTS is set to $\tau_{max} = |f(s^{gb})|/\rho$, where $f(s^{gb})$ is the score of the best solution found since the start of the algorithm. τ_{min} is set using the formulas given in [8] with a setting of $p_{best} = 0.9$.

Local search. As in most applications of $\mathcal{MMAS}$ and, more in general, of ACO algorithms to $\mathcal{NP}$-hard problems, we improve candidate solutions by a local search algorithm. We use the simplex local search algorithm described by Nelder and Mead for continuous function optimization [9]. The simplex algorithm is a geometrically inspired approach, which transforms the points of a given start simplex by using the operations reflection, expansion and contraction until the fractional range from the highest to the lowest point in the simplex with respect to the function value is less than a tolerance value, which we choose as 0.01; for details see [10]. In the simplex algorithm we use $\Delta_{trans} = 2\text{Å}$, $\Delta_{rot} = 90°$ and $\Delta_{tors} = 90°$ as the parameter setting for the construction of the initial simplex. Once all ants have improved their solution, the simplex algorithm is used again to refine the best of these ants. This refinement local search is restarted as long as the improvement in the scoring function through one application of the simplex local search is larger than 0.2.

Algorithmic outline of PLANTS. A high-level outline of the PLANTS algorithm is given in Algorithm 1. Most details of the algorithm follow what is usually done in ACO algorithms; necessary details are explained next. The number of iterations is determined by the formula

$$iterations = \sigma \cdot \frac{10}{m} \cdot (100 + 50 \cdot lrb + 5 \cdot lha), \tag{4}$$

where σ is a parameter used for scaling the number of iterations, m is the colony size, lrb is the number of rotatable bonds and lha the number of heavy atoms in the ligand. Because of the usage of lrb and lha, the number of iterations depends on the properties of the ligand. As can be seen from this formula, very flexible and large ligands get more time for searching than rigid and small ones. The function RefinementLocalSearch applies the refinement local search as described in the previous paragraph and the procedure UpdatePheromones applies the pheromone update as described above and includes also the check regarding the pheromone trail limits. Noteworthy are the diversification features applied by the algorithm. PLANTS memorizes the best solution found, s^{db}, since the last search diversification. If more than 10 iteration-best solutions found in PLANTS since the last search diversification differ from s^{db} by less than $0.02 \cdot |f(s^{db})|$, again a search diversification is invoked. For the search diversification, one of two different possibilities is applied. The first is a pheromone trail smoothing, as proposed in [8], using a smoothing factor of 0.5. If three subsequent smoothings have been applied, the search is restarted by erasing all pheromone trails and resetting them to their initial value. This second type of diversification actually corresponds to a complete restart of the algorithm. Once the algorithm terminates, it returns the best solution found during the whole search process and the set M of all solutions returned by the procedures LocalSearch and RefinementLocalSearch, which are used for further processing by a clustering algorithm.

Clustering algorithm. The clustering algorithm is used as a means of post-processing the output of PLANTS. It first sorts all the solutions in M according to increasing scoring function values. Then it extracts a number of ligand structures given by *rankingStructures*, a parameter which is set typically to 10, that satisfy the condition that the minimal RMSD between any of these extracted solutions is larger than 2 Å. These solutions can then be used for rescoring with other scoring functions in order to increase the chance of finding a ligand conformation that is similar to the experimental binding mode. This feature is especially interesting for virtual screening applications.

Algorithm 1. PLANTS

```
InitializeParametersAndPheromones()
for i = 1 to iterations do
  for j = 1 to ants do
    s_j ← ConstructSolution()
    s*_j ← LocalSearch(s_j)
    M ← M ∪ s*_j
  end for
  s^ib ← GetBestSolution()
  s^ib ← RefinementLocalSearch(s^ib, 0.2)
  M ← M ∪ s^ib
  UpdatePheromones(s^ib)
  if diversificationCriteriaMet then
    ApplySearchDiversification()
  end if
end for
return best solution found, M
```

Empirical scoring function. The empirical scoring function used in PLANTS is a combination of parts of published ones [11,12]. The first part of the intermolecular score is based on a modified version of the *piecewise linear potential* (PLP) scoring function [11]. This part is mainly used to model steric interactions between the protein and the ligand. The second part introduces directed hydrogen bonding interactions between both complex partners as published in GOLD's CHEMSCORE implementation [12]. The intramolecular ligand scoring function consists of a simple clash term and a torsional potential as described in [13]. Additionally, if the ligand's reference point is outside the predefined binding site, a penalty term is added. Throughout this paper, this scoring function will be referred to as CHEMPLP.

4 Parameter Optimization and Validation of PLANTS

The *clean list* of the comprehensive CCDC/ASTEX dataset [14] has been used for the validation of PLANTS. From these 224 complexes, 11 include covalently bound ligands and these had to be removed, because they cannot be handled by PLANTS at the moment. Hence, our test set consists of 213 non-covalently bound complexes, we call *clean list*^{nc}. The number of rotatable bonds of the ligands in *clean list*^{nc} ranges from 0 to 28. For all experiments, the spherical binding site defined for each protein-ligand complex was used to determine the search space for the ligand's translational degrees of freedom. Before docking, the ligand structures were randomized with respect to the translational, rotational and torsional degrees of freedom. The randomized structures were then passed to PLANTS in order to prevent biased parameter settings. Here, we examine the influence of some of PLANTS' parameters. We have chosen a subset of 33 complexes with 0 to 10 rotatable bonds (3 complexes for each number of rotatable bonds) to reduce the high computation times required when testing across the complete test set. We varied the parameters σ, m (the number of ants), ρ and β considering three or four values for each, which resulted in 144 distinct parameter configurations. On each complex, PLANTS was run for 10 independent trials. We measured for each configuration the average success rate, computation time and the average number of function evaluations. The success rate is defined as the percentage of complexes for which the top-ranked docking solution is within 2.0 Å of the experimentally determined binding mode as given in the CCDC/ASTEX dataset. The computation times in this section are given in seconds on a single Pentium 4 Xeon, 2.8 GHz CPU; protein setup time (6 s on average) and ligand setup time (0.01 s on average) are excluded.

The plots in Figure 2 allow for a detailed, graphical analysis of the results (note that the issues discussed below could easily be confirmed when discussed in terms of numerical results and statistical significance – here we are more interested in the general behavior implied by the parameter settings); in the plots, each data point gives the average computation time (x-axis) and the average success rate (y-axis) for one PLANTS parameter configuration.

In a first step, the data were plotted in dependence of the number of ants (parameter m) that was used as a blocking factor; see Figure 2a. As can be clearly seen in this plot, the configurations with only a single ant are clearly dominated by the other configurations using more ants. The high computation times for the configurations with only one ant can be attributed to the larger number of iterations (see Equation 4) and the resulting large number of times the procedure RefinementLocalSearch was executed. Hence, in the further analysis, we exclude the configurations with one ant. As can be seen in Figure 2a, the preferable parameter setting for m appears to be 20 or 50, since mainly these configurations are part of the curve including the non-dominated configurations. Next, the parameter σ was used as a blocking factor and the plot in Figure 2b shows that the points clearly fall into three clusters with respect to the docking time in dependence of the value of σ. This plot (together with the observations made below), also suggests that the parameter σ may be used, as expected, to tune the tradeoff between computation time and solution quality if required. In a next step, we analyzed the data in dependence of the evaporation rate ρ. As shown in Figure 2c, for each value of σ, the four values of ρ define four clearly distinct clusters: With a decrease of the value for ρ, the computation times increase. This effect can be explained, since the higher ρ, the faster will $\mathcal{MM}$AS converge towards the best solutions seen so far; this convergence again typically leads to less iterations of the local search, since it will start from better initial solutions. In general, evaporation factors of $\rho = 0.25$ or $\rho = 0.5$ seem to be favorable when considering both, the success rate and the docking time. In a final step, we examined the influence caused by the value of β, which determines the influence of the heuristic information on the computational results. As can be seen in Figure 2d, the influence of β appears to be minor. This impression can also be confirmed by computing the average success rates and times across all configurations with a same value of β, which are, essentially, all the same. This may be the case

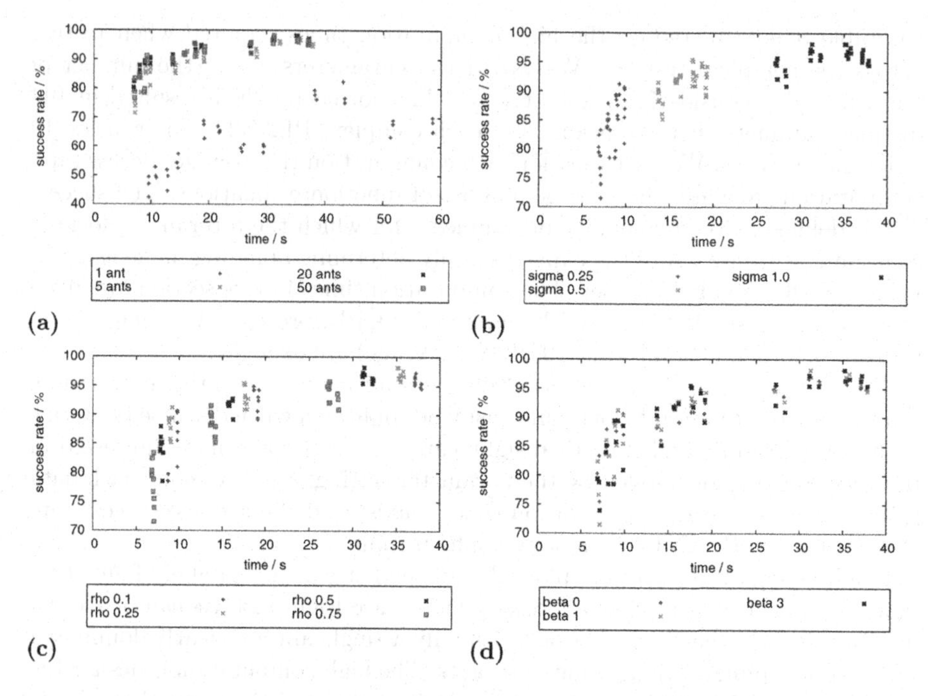

Fig. 2. Influence of different parameter settings on the other parameter configurations with respect to the average success rate and docking time. For further explanations see the text.

because in general a ligand can engage many low energy conformations with respect to the torsional potential, which is used as the heuristic information.

Starting from this analysis, PLANTS was tested with several settings on the whole *clean list*^{nc}. The applied parameter setting as well as the success rate for the (i) top-ranked solution, (ii) up to rank 3 and (iii) up to rank 10 (ranks w.r.t. the solutions in the order as returned by the clustering algorithm—a success is obtained if among these highest ranked ligands we have the desired one) and the average docking time along with the number of scoring function evaluations are presented in Table 1 (see upper part marked with **PLANTS**). As already observed for the subset consisting of 33 complexes, parameter σ controls the tradeoff between success rate and docking time.

The success rates for the top-ranked solutions range from about 63 % at docking times of approximately 25 s ($\sigma = 0.25$) to 75 % at docking times of 290 s ($\sigma = 3$) for each complex, on average. However, because of the high docking time, parameter setting $\sigma = 3$ is not really applicable in virtual screening applications where thousands of ligands may have to be docked. Interestingly, for $\sigma = 0.5$ and $\sigma = 1$ the use of 20 ants seems to be preferable over the 50-ants setting. An explanation for this may be the higher number of iterations carried out by PLANTS with 20 ants (see Equation 4) and the possibly positive benefits of

Table 1. Results on the *clean list*^{nc} for PLANTS and GOLD for selected parameter settings averaged over 25 independent experiments. Standard deviations for the success rates are given in parentheses.

PLANTS								
				success rate (%) up to rank				
σ	*ants*	ρ	β	1	3	10	time (s)	eval. (10^6)
0.25	50	0.25	1	63.86 (1.86)	75.18 (1.73)	80.59 (1.34)	27.01	0.93
0.25	20	0.25	3	63.57 (1.68)	73.71 (2.13)	78.84 (2.16)	25.10	0.86
0.50	50	0.25	3	67.53 (2.22)	78.39 (2.00)	83.31 (1.95)	51.56	1.76
0.50	20	0.25	3	68.90 (1.97)	79.57 (1.76)	84.64 (1.15)	49.27	1.69
1.00	50	0.50	1	71.19 (1.47)	82.40 (1.60)	87.64 (1.40)	88.76	2.99
1.00	20	0.25	3	72.34 (1.27)	83.62 (1.55)	88.62 (1.32)	97.68	3.36
3.00	20	0.25	3	75.19 (1.10)	87.92 (1.11)	92.66 (0.89)	290.13	9.96
GOLD								
				success rate (%) up to rank				
autoscale				1	3	10	time (s)	eval. (10^6)
0.1				67.27 (1.62)	73.75 (1.26)	78.12 (1.47)	42.45	n.a.
0.3				69.43 (1.66)	75.42 (1.65)	81.03 (2.01)	115.21	n.a.
1.0				73.69 (1.44)	78.10 (1.37)	82.35 (1.14)	308.98	n.a.

more pheromone smoothings and restarts. We also compared PLANTS to GOLD (Genetic Optimisation for Ligand Docking) [15], a state-of-the-art docking program that is frequently used in the pharmaceutical industry. Detailed information about GOLD can be found in [15,12]. For the experiments presented in this section, GOLD version 3.0.1 has been employed. The maximum number of GA runs per ligand was set to 10 and *early termination* as well as *cavity detection* was activated. Different time settings for both PLANTS and GOLD were compared with respect to the programs' success rate and docking time. In the case of PLANTS, the CHEMPLP scoring function was used, while the GOLD scoring function was used for GOLD. The results for both programs on the *clean list*^{nc} are presented in Table 1. As can be observed, except for *autoscale* set to 0.1, PLANTS' results dominate those of GOLD; this can be seen by the fact that for PLANTS we always have configurations that achieve higher success rates in shorter time. When considering the success rates up to ranks 3 or 10, we even have that configurations of PLANTS reach higher success rates than the highest achieved by GOLD (of 82.35 %) in about one sixth of the time; this is a very encouraging result for virtual screening applications.

5 Virtual Screening

As mentioned in the introduction, *virtual screening* of large compound libraries is one of the main applications of current docking tools. Therefore, PLANTS was

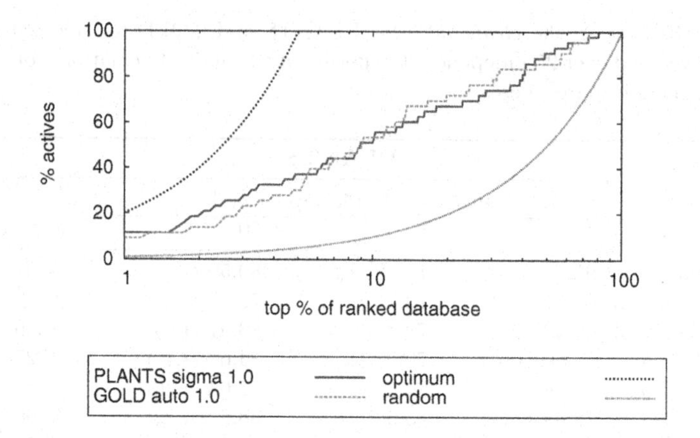

Fig. 3. Enrichments for the *virtual screening* against coagulation factor Xa with PLANTS and GOLD. The plot uses a logarithmic scaling for the x-axis. For further explanations see the text.

also tested with respect to its ability to discriminate between biologically active and inactive ligands. Factor Xa was chosen as the protein, which is a target for antithrombotics, developed to treat imbalances between clotting, clotting inactivation, and thrombolytic processes in the blood coagulation cascade. A database of 43 active and 817 inactive ligands was docked into PDB-entry 1FAX (coagulation factor Xa inhibitor complex) from the CCDC/ASTEX dataset. The 43 active ligands taken from [16] are publicly available. The ZINC database [17] was used to retrieve inactive ligands that approximately match the properties (number of rotatable bonds, hydrogen bond donors, acceptors and heavy atoms) of the active ligands to ensure a screening under realistic circumstances as carried out in the pharmaceutical industry. Prior to docking, all ligands were minimized in vacuo using the MMFF94 force-field [18] to prevent the use of poor ligand geometries during docking. All 860 ligands were then docked with PLANTS using the CHEMPLP scoring function and GOLD using the GOLD scoring function. For both programs the default settings ($\sigma = 1$ for PLANTS and $autoscale = 1$ for GOLD) were used. The computations were carried out on an AMD Opteron processor with 2 GHz. The average docking time per complex was 68.9 s and 297.14 s for PLANTS and GOLD, respectively. However, it may be noted that faster search settings for GOLD may have provided similar results. After the docking, all ligand configurations were ranked according to their scoring function value starting with the best scoring configurations (PLANTS minimizes the scoring function while GOLD maximizes the fitness value). The results of the virtual screening are shown in Figure 3. The x-scale was set logarithmically to emphasize the part of the ranked database that contains the candidate ligands for in vitro tests for biological activity; hence, for an algorithm to be useful in virtual screening it is important that within a small percentage of its top-ranked ligands is an as high as possible percentage of active ligands. The figure shows the percentage of biologically active ligands for each percentage of the ranked database as identified by PLANTS and GOLD as well as the theoretically optimal curve and the curve for a random selection strategy. Both programs perform clearly better than random selection but not as good as the optimal selection. PLANTS performs slightly better up to the 5% top-ranked ligands of the database while GOLD finds more active ligands in relation to PLANTS beyond 10 %; however, this is not relevant for practice.

6 Conclusions

In this study, we presented a new docking algorithm based on the ACO metaheuristic. Several parameter settings were studied to assure high success rates in pose prediction for different timings. Default settings ($\sigma = 1$) are able to reproduce ligand geometries similar to the crystal geometry in about 72 % of the cases at average docking times of 97 seconds. Furthermore, it could be shown that PLANTS is competitive in terms of pose prediction accuracy as well as docking times to the state-of-the-art docking program GOLD, which is based on a genetic algorithm. Last but not least, PLANTS was able to identify biologically active ligands at the top-ranked positions of a ligand database targeting coagulation factor Xa. Besides these promising results, there is still significant space for improvement. Especially the CHEMPLP scoring function used in PLANTS is currently one of the limiting factors. This scoring function could either be improved to model e.g. metal-ligand interactions more appropriately or be replaced by an other scoring function. Additionally, almost the whole receptor except rotatable hydrogen bond donors is currently kept rigid. This is of course a hard approximation which especially influences the results of virtual screenings. In a next step, protein side-chain flexibility will be introduced, which fits well into the proposed ACO algorithm. In this case, simply additional pheromone vectors are introduced for each degree of freedom of the flexible protein side-chains. Because of the high computational demands when considering side-chain flexibility, a port of PLANTS from the CPU to the GPU is planned to exploit the computational power of today's graphics processing units.

Acknowledgments. The authors thank Dr. Peter Monecke and Dr. Gerhard Hessler for helpful discussions and a careful reading of the manuscript. This work was in part supported by a scholarship of the Landesgraduiertenförderung Baden-Württemberg awarded to Oliver Korb. Thomas Stützle acknowledges support of the Belgian FNRS, of which he is a research associate.

References

1. Müller, G.: Medicinal chemistry of target family-directed masterkeys. Drug Discovery Today **8**(15) (2003) 681–691
2. Oprea, T., Matter, H.: Integrating virtual screening in lead discovery. Current Opinion in Chemical Biology **8** (2004) 349–358
3. Fischer, E.: Einfluss der Configuration auf die Wirkung der Enzyme. Chemische Berichte **27** (1894) 2985–2993

4. Dorigo, M., Stützle, T.: Ant Colony Optimization. MIT Press, Cambridge, MA, USA (2004)
5. Kellenberger, E., Rodrigo, J., Muller, P., Rognan, D.: Comparative evaluation of eight docking tools for docking and virtual screening accuracy. Proteins **57**(2) (2004) 225–242
6. Kontoyianni, M., McClellan, L., Sokol, G.: Evaluation of docking performance: Comparative data on docking algorithms. Journal of Medicinal Chemistry **47**(3) (2004) 558–565
7. Taylor, R., Jewsbury, P., Essex, J.: A review of protein-small molecule docking methods. Journal of Computer-Aided Molecular Design **16** (2002) 151–166
8. Stützle, T., Hoos, H.H.: $\mathcal{MAX}$–$\mathcal{MIN}$ Ant System. Future Generation Computer Systems **16**(8) (2000) 889–914
9. Nelder, J.A., Mead, R.: A simplex method for function minimization. Computer-Journal **7** (1965) 308–313
10. Press, W.H., Flannery, B.P., Teukolsky, S.A., Vetterling, W.T.: Numerical Recipes in C: The Art of Scientific Computing. Cambridge University Press (1992)
11. Gehlhaar, D. K.; Verkhivker, G. M.; Rejto, P. A.; Sherman, C. J.; Fogel, D. B.; Fogel, L. J.; Freer, S. T.: Molecular recognition of the inhibitor AG-1243 by HIV-1 protease: conformationally flexible docking by evolutionary programming. Chemistry and Biology **2** (1995) 317–324
12. Verdonk, M.L., Cole, J.C., Hartshorn, M.J., Murray, C.W., Taylor, R.D.: Improved protein-ligand docking using GOLD. Proteins **52** (2003) 609–623
13. Clark, M., III, R.C., van Opdenhosch, N.: Validation of the General Purpose Tripos 5.2 Force Field. Journal of Computational Chemistry **10** (1989) 982–1012
14. Nissink, J., Murray, C., Hartshorn, M., Verdonk, M., Cole, J., Taylor, R.: A new test set for validating predictions of protein-ligand interaction. Proteins **49**(4) (2002) 457–471
15. Jones, G., Willett, P., Glen, R.C., Leach, A.R., Taylor, R.: Development and validation of a genetic algorithm for flexible docking. Journal of Molecular Biology **267** (1997) 727–748
16. Jacobsson, M., Liden, P., Stjernschantz, E., Boström, H., Norinder, U.: Improving struture-based virtual screening by multivariate analysis of scoring data. Journal of Medicinal Chemistry **46**(26) (2003) 5781–5789
17. Irwin, J., Shoichet, B.: ZINC - A Free Database of Commercially Available Compounds for Virtual Screening. Journal of Chemical Information and Modeling **45**(1) (2005) 177–82
18. Halgren, T.: Merck molecular force field. I. Basis, form, scope, parameterization, and performance of MMFF94. Journal of Computational Chemistry **17**(5-6) (1996) 490–519

Rendezvous of Glowworm-Inspired Robot Swarms at Multiple Source Locations: A Sound Source Based Real-Robot Implementation

Krishnanand N. Kaipa, Amruth Puttappa, Guruprasad M. Hegde, Sharschchandra V. Bidargaddi, and Debasish Ghose

Indian Institute of Science, Bangalore, India
krishna@aero.iisc.ernet.in, amruthp@gmail.com, prasad.mh@gmail.com, bidargaddi@ieee.org, dghose@aero.iisc.ernet.in

Abstract. This paper presents a novel glowworm metaphor based distributed algorithm that enables a minimalist mobile robot swarm to effectively split into subgroups, exhibit simultaneous taxis towards, and rendezvous at multiple source locations. The locations of interest could represent radiation sources such as nuclear and hazardous aerosol spills spread within an unknown environment. The glowworm algorithm is based on a glowworm swarm optimization (GSO) technique that finds multiple optima of multimodal functions. The algorithm is in the same spirit as the ant-colony optimization (ACO) and particle swarm optimization (PSO) algorithms, but with several significant differences. A key feature of the GSO algorithm is the use of an adaptive local-decision domain, which is used effectively to detect the multiple optimum locations of the multimodal function. We conduct sound source localization experiments, using a set of four wheeled robots (christened Glowworms), to validate the glowworm approach to the problem of multiple source localization. We also examine the behavior of the glowworm algorithm in the presence of uncertainty due to perceptional noise. A comparison with a gradient based approach reveals the superiority of the glowworm algorithm in coping with uncertainty.

1 Introduction

Localization of multiple radiation sources using mobile robot swarms has received some attention recently [1,2,3,4,5] in the collective robotics community. In particular, the goal of the above problem is to drive groups of mobile agents to multiple sources of a general nutrient profile that is distributed spatially on a two dimensional workspace. This problem is representative of a wide variety of applications that include detection of multiple radiating sources such as nuclear/hazardous aerosol spills and origins of a fire-calamity and localization/decommissioning of hostile transmitters that are scattered over a landscape, by sensing signals radiating from them. For instance, several forest fires at different locations give rise to a temperature profile that peaks at the locations of the fire. Similar phenomenon can be observed in nuclear radiations and electromagnetic radiations

M. Dorigo et al. (Eds.): ANTS 2006, LNCS 4150, pp. 259–269, 2006.

from signal sources. In all the above situations, there is an imperative need to simultaneously identify and neutralize all the multiple sources using a swarm of robots before they cause a great loss to the environment and people in the vicinity. The above problem presents several challenges such as multiplicity of sources, time-varying nature of the sources, dynamically changing environment, and perceptional noise. Thus the objective is to devise local control strategies that allow a swarm of mobile robots – equipped with only rudimentary and noisy sensors – to perform the task of multiple source localization while coping well with the challenges described above.

Multimodal function optimization has been addressed extensively in the recent literature [2,5,6,10]. Most prior work on this topic focussed on developing algorithms to find the global optima of the given multimodal function, while avoiding local optima. However, there is another class of optimization problems which is different from the problem of finding only the global optimum of a multimodal function. The objective of this problem class is to find multiple optima having either equal or unequal function values [2,6,7,8].

In this paper, we describe a novel *glowworm* metaphor based distributed algorithm that enables a *minimalist* mobile robot swarm to split into subgroups, exhibit simultaneous taxis towards, and rendezvous at multiple unknown radiation source locations. The algorithm is based on a glowworm swarm optimization (GSO) technique [2] that finds multiple optima (not necessarily equal) of multimodal functions. The significant difference between our work and most earlier approaches to rendezvous problems is the use of a variable local-decision domain by the agents in order to compute their movements. We show that the glowworm algorithm originally designed for solving optimization problems could be applied directly (with almost no modifications) to the specific collective robotics task of simultaneously localizing multiple sources of interest. The preliminary results of this method were presented in [2]. We have reported the related theoretical foundations, by posing the objective of the glowworm algorithm as a problem of rendezvous of visibility limited mobile-agents at multiple locations, in [3] and the results of a preliminary real-robot-implementation of the glowworm algorithm (using four mobile robots) to achieve a sound source localization task was described in [4]. Comparisons of the GSO algorithm with established nature-inspired algorithms such as ACO and PSO, simulation test results of the GSO algorithm on a series of standard multimodal functions, dependence of GSOs performance on the critical parameters in the algorithm, and simulation results for problems in higher dimensional spaces were reported in [5].

2 The Glowworm Algorithm

2.1 Overview

The GSO algorithm is in the same spirit as the ACO [11] and PSO [13] techniques but is different in many aspects that help in achieving simultaneous detection of multiple local optima of multimodal functions. This is a problem not directly addressed by ACO or PSO techniques. Generally, ACO and PSO techniques are used for locating global optima. However, our objective is to locate as many of the peaks as possible. This requirement is the main motivation for formulating the GSO technique. In the GSO algorithm, the agents are initially deployed randomly in the objective function space. Similar to how agents in the ACO technique use pheromonal deposits to effect stigmergic communication among the members of the ant colony, the agents in our algorithm carry a luminescence quantity called luciferin along with them for a similar purpose. Agents are thought of as glowworms that emit a light whose intensity of luminescence is proportional to the associated luciferin. Each glowworm uses the luciferin to (indirectly) communicate the function-profile information at its current location to the neighbors. The glowworms depend on a adaptive local-decision domain, which is bounded above by a circular sensor range, to compute their movements. Each glowworm selects a neighbor that has a luciferin value more than its own, using a probabilistic mechanism, and moves towards it. That is, they are attracted to neighbors that glow brighter. These movements, that are based only on local information, enable the glowworms to split into subgroups and exhibit a simultaneous taxis-behavior towards the optimum locations leading to the detection of multiple optima of the given objective function. Comparisons of the GSO algorithm with ACO and PSO techniques are summarized in Tables 1 and 2 (Refer to [5] for a more detailed description).

Table 1. ACO versus GSO

	Standard ACO	**GSO**
1	Effective in *discrete* setting [11]	Applied to *continuous* domain
2	Global optimum or multiple global optima of equal value [9,10]	Multiple optima of *equal* or *unequal* values
	Special variant of ACO [12]	
1	Cannot be applied when ants (agents) have limited sensing range	Useful for applications where robots have limited sensor range
2	*Global* information used	*Local* information used
3	Pheromones associated with *paths* from nest to regions	Luciferin carried by and associated with *glowworms*
4	Pheromone information used to select *regions*	Luciferin information used to select *neighbors*
5	Shifting of selected region's center in a *random* direction	*Deterministic* movements towards selected neighbor

2.2 Algorithm Description

The GSO algorithm starts by placing the glowworms randomly in the workspace so that they are well dispersed. Initially, all the glowworms contain an equal quantity of luciferin ($\ell_j(0) = \ell_0, \forall j$). Each iteration consists of a luciferin-update phase followed by a movement-phase based on a transition rule.

Luciferin-update phase: The luciferin update depends on the function value at the glowworm position and so, even though all glowworms start with the same

Table 2. PSO versus GSO

	PSO	GSO
1	Direction of movement based on previous best positions	Agent movement along line-of-sight with a neighbor
2	Dynamic neighborhood based on k nearest neighbors	Local decision domain based on varying range
3	Neighborhood range covers the entire search space	Maximum range hard limited by finite sensor range
4	Limited to numerical optimization models	Effective detection of multiple peaks/sources in addition to numerical optimization tasks

luciferin value during the initial iteration, these values change according to the function values at their current positions. The luciferin update rule is given by:

$$\ell_j(t+1) = (1-\rho)\ell_j(t) + \gamma J_j(t+1) \tag{1}$$

where, ρ is the luciferin decay constant ($0 < \rho < 1$) and γ is the luciferin enhancement constant and $J_j(t)$ represents the value of the objective function at agent j's location at time t.

While the first term in (1) represents the decaying nature of the luciferin that indirectly allows the agents to escape inferior regions and move towards promising regions of the objective functions space, the second term represents the fitness of the agent j's location at time $t+1$.

Movement-phase: During the movement-phase, each glowworm decides, using a probabilistic mechanism, to move towards a neighbor that has a luciferin value more than its own. For each glowworm i, the probability of moving towards a neighbor j is given by:

$$p_j(t) = \frac{(\ell_j(t) - \ell_i(t))}{\sum_{k \in N_i(t)} (\ell_k(t) - \ell_i(t))} \tag{2}$$

where, $j \in N_i(t), N_i(t) = \{j : d_{i,j}(t) < r_d^i(t); \ell_i(t) < \ell_j(t)\}$, t is the time (or step) index, $d_{i,j}(t)$ represents the euclidian distance between glowworms i and j at time t, $\ell_j(t)$ represents the luciferin level associated with glowworm j at time t, $r_d^i(t)$ represents the variable local-decision range associated with glowworm i at time t, and r_s represents the radial range of the luciferin sensor. Let the glowworm i select a glowworm $j \in N_i(t)$ with $p_j(t)$ given by (2). Then the discrete-time model of the glowworm movements can be stated as:

$$x_i(t+1) = x_i(t) + s\left(\frac{x_j(t) - x_i(t)}{\|x_j(t) - x_i(t)\|}\right) \tag{3}$$

where s is the step size.

Local-decision range update rule: Since we assume that *a priori* information about the objective function is not available, in order to detect multiple peaks, the sensor range must be made a varying parameter. For this purpose, we associate each agent i with a local-decision domain whose radial range r_d^i is dynamic in nature ($0 < r_d^i \leq r_s^i$). A suitable function is chosen to adaptively update the local-decision domain range of each glowworm. This is given by:

$$r_d^i(t+1) = \min\{r_s, \max\{0, r_d^i(t) + \beta(n_t - |N_i(t)|)\}\} \tag{4}$$

where, β is a constant parameter and n_t is used as a threshold parameter to control the number of neighbors.

3 Modelling of Perceptional Noise

We examine the behavior of the algorithm in the presence of uncertainty due to perceptional noise. We use a gaussian distribution $N(\mu, \sigma_f^2)$ with mean μ and variance σ_f^2 to model the sensor noise $\omega(t)$. Since the noise introduced affects the quality of measurements of the function values, (1) is modified as follows:

$$\ell_j(t+1) = (1-\rho)\ell_j(t) + \gamma(1+\omega(t))J_j(t+1) \tag{5}$$

Note from (5) that the noise is made a function of the sensor measurements at each agent's location. We compare the glowworm algorithm with the following gradient-based algorithm and noise introduced into the gradient measurements:

Let $\Delta_{(x,y)}(t) = \begin{pmatrix}\Delta_x & \Delta_y\end{pmatrix}^T$ represent the gradient vector measured at the location (x, y) by a glowworm i for some $i = 1 \cdots n$ at time t, with $\omega_x(t)\Delta_x$ and $\omega_y(t)\Delta_y$ being the noise levels in the measurements of Δ_x and Δ_y, respectively. Let $\omega_x(t)$ and $\omega_y(t)$ follow a gaussian distribution $N(\mu, \sigma_g^2)$ with mean μ and variance σ_g^2. Therefore, the noisy gradient measurements are given by:

$$\begin{aligned}\tilde{\Delta}_x &= (1+\omega_x(t))\Delta_x \\ \tilde{\Delta}_y &= (1+\omega_y(t))\Delta_y\end{aligned} \tag{6}$$

From (6), the gradient based movement model of each glowworm i is given by:

$$X_i(t+1) = X_i(t) + \frac{\delta}{|\tilde{\Delta}_{(x,y)}|}\begin{pmatrix}\tilde{\Delta}_x & \tilde{\Delta}_y\end{pmatrix}^T \tag{7}$$

where $X_i = \begin{pmatrix}x_i & y_i\end{pmatrix}$ and $|\tilde{\Delta}_{(x,y)}| = \sqrt{\tilde{\Delta}_x^{\,2} + \tilde{\Delta}_y^{\,2}}$ From (5) and (6), it is clear that while noise is present in sensing of values of the function profile in the case of glowworm algorithm, noise occurs in gradient measurements in the latter case. Therefore, the comparison of performance between the two algorithms is not straightforward. For this purpose, we derive a simple relation between the variances involved in both the cases. Figure 1 (a) shows the variation of J with respect to x. The quantities $\Delta J_{\min}$ and $\Delta J_{\max}$ represent the minimum and maximum values of $\tilde{\Delta} J$ for a corresponding value of $\Delta x = x_2 - x_1$, where $\tilde{\Delta} J$

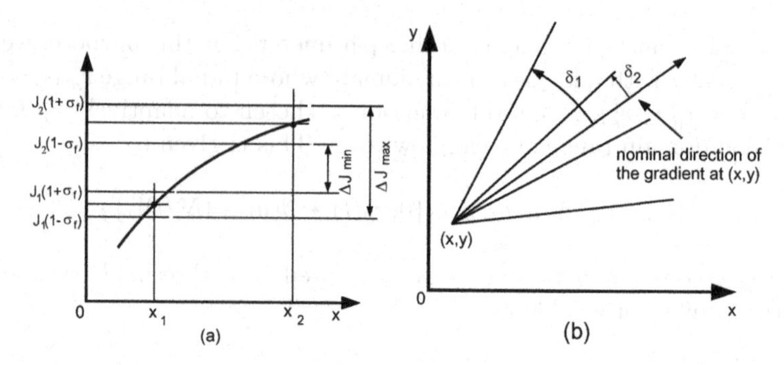

Fig. 1. a) Graph of J versus x serving to find the relationship between variance in the function values and the variance in the gradients computed by using a finite-difference method. b) $\pm\delta_1$ and $\pm\delta_2$ represent the angular dispersions in the directions of the nominal gradient when the same amount of noise is introduced into the function values (glowworm algorithm) and gradient values (gradient algorithm), respectively.

represents the difference between noisy function values measured at x_1 and x_2. From the figure we have,

$$\Delta J_{\min} = J_2 - J_1 - \sigma_f(J_1 + J_2)$$
$$\Delta J_{\max} = J_2 - J_1 + \sigma_f(J_1 + J_2) \tag{8}$$
$$\Rightarrow = \frac{\Delta \tilde{J}}{\Delta x} \in \left(\frac{\Delta J}{\Delta x} - \sigma_f \frac{J_1 + J_2}{\Delta x}, \frac{\Delta J}{\Delta x} + \sigma_f \frac{J_1 + J_2}{\Delta x} \right) \tag{9}$$

Assuming $J_1 \approx J_2$ for small variation in the function values with respect to small deviation in x, we get

$$\sigma_g \approx \left(\frac{2J_1}{\Delta x} \right) \sigma_f \tag{10}$$

From (10), it is clear that a small variance in the function value leads to a very large variance in gradient value obtained by a finite-difference method, especially as $\Delta x \to 0$. In our simulations, we use variance values in the gradient algorithm that are of the same order as the variance values in the glowworm algorithm. This means that we are actually dealing with far less noise in the gradient algorithm than in the glowworm algorithm. The gradient-cone interpretation given in Figure 1 (b) explains the above fact in the following manner. Here, $\pm\delta_1$ and $\pm\delta_2$ represent the angular dispersions in the directions of the nominal gradient when the same amount of noise is introduced into the function values (glowworm algorithm) and gradient values (gradient algorithm), respectively. Clearly, $\delta_1 > \delta_2$. Note that for small values of Δx, the angle of the outer cone can actually become much larger, leading to the condition $\delta_1 \gg \delta_2$.

4 Simulation Experiments

Results demonstrating the capability of glowworm algorithm to capture multiple peaks of a number of complex multimodal functions have been reported in [3,5] for constant and variable local-decision range cases with no noise. We have shown that when constant decision range is used, the number of peaks captured decreases with increase in the value of decision range. Interestingly, when the decision-range is made adaptive, even though $r_d^i(0)$ is chosen to be greater than the maximum distance between the peaks, all the peaks are captured.

In this work, we initially demonstrate the capability of the glowworm algorithm to capture multiple peaks of a multimodal function, under perfect sensing conditions. Later, we report the algorithm's performance in the presence of noise.

4.1 Performance of the Algorithm in the Absence of Noise

We consider the following function to model the multimodal nature of the sources in the environment:

$$\begin{aligned} J(x,y) &= 3(1-x)^2 \exp(-(x^2)-(y+1)^2) \\ &\quad - 10(x/5 - x^3 - y^5)\exp(-x^2-y^2) \\ &\quad - (1/3)\exp(-(x+1)^2-y^2) \end{aligned} \tag{11}$$

The function $J(x,y)$ consists of a set of three peaks (Figure 2 (a)) at locations $(-0.0093, 1.5814)$, $(1.2857, -0.0048)$, and $(-0.46, -0.6292)$. The nominal values of various parameters used in the simulations are shown in Table 3 where n is the number of glowworms. A set of 50 glowworms are randomly deployed in a two-dimensional workspace of size $6X6$ square units.

Table 3. List of parameters used in the simulations

Function	n	ρ	γ	β	r_s^i	s
J	50	0.4	0.6	0.01	3	0.01

Figure 2 (b) shows the emergence of the solution when the local-decision domain range is made to vary according to (4). A value of $r_d^i(0) = 3$ is chosen during the simulation. Figure 2 (c) shows the co-location (at final time) of all the glowworms on the iso-value contours of the multimodal function.

4.2 Performance of the Algorithm in the Presence of Noise

We consider the same multimodal function given in (11) and the noise models described in the previous section to test the algorithmic response to uncertainty conditions. The noise $\omega(t)$ follows a gaussian distribution $N(0, \sigma_f^2)$ with zero mean and variance σ_f^2. We use the mean minimum distance to the sources $dmin_{av}$ as a performance metric in our simulations. This quantity gives a very useful insight in the present context, considering the multiplicity of the sources and

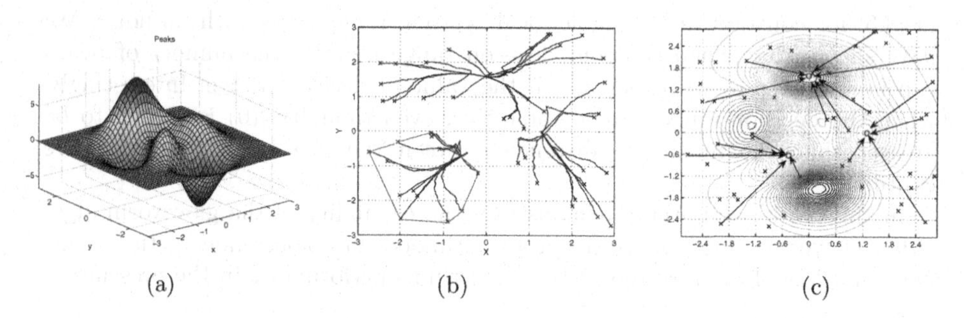

(a) (b) (c)

Fig. 2. a) Function $J(x, y)$ with local maxima at $(-0.0093, 1.5814)$, $(1.2857, -0.0048)$, and $(-0.46, -0.6292)$. b) Emergence of solution. c) Initial placement and final co-location of glowworms on the iso-value contour of the objective function.

the need to evaluate the deviation of the agent's final locations from the source locations. In particular, $dmin_{av}$ is given by:

$$dmin_{av} = \frac{1}{n}\sum_{i=1}^{n} \min\{d_{i1}, \cdots, d_{im}\} \tag{12}$$

where $d_{ij} = \|X_i - S_j\|$, $i = 1, \cdots, n$, $j = 1, \cdots, m$, X_i is glowworm i's location and S_j is source j's, and m is the number of source locations.

Figure 3 (a) shows the plots of $dmin_{av}(t)$, for several values of standard deviation σ_f, averaged over a set of 20 simulation trials for each σ_f, in the case of the glowworm algorithm. The standard deviation of $dmin_{av}(t)$ in each case is relatively very less when compared to that of the introduced noise ($\approx 0.02 \times \sigma_f$). The glowworm algorithm shows good performance with fairly high noise levels. There is graceful degradation of performance only with significant increase in levels of measurement noise. Whereas, gradient based algorithm degrades rather fast in the presence of noise (Figure 3 (b)). Note that the behavior of $dmin_{av}$ in the case of glowworm algorithm, that occurs when $\sigma_f = 10$, is relatively better than the behavior of $dmin_{av}$ in the case of gradient based algorithm, that occurs when $\sigma_g = 0.1$. Multimodal optimization could also be achieved by distributing the agents uniformly in the solution space and allowing them to perform gradient descent to the local peaks, using no communication at all. However, we notice that a random initial placement results in most of agents moving away from the peaks and settling at the edges of the solution space. This is evident in Figure 3 (b) where the mean minimum distance to the peaks $dmin_{av}(t)$ doesn't reduce to zero even in the absence of noise ($\sigma_g = 0$).

5 Glowworms

Four small-wheeled robots christened Glowworms (named after the glowworm algorithm) were built to conduct our experiments [4]. Each Glowworm has been designed to provide features of basic mobility on plain/smooth surfaces, obstacle sensing, relative localization/identification of neighbors, and infrared-based

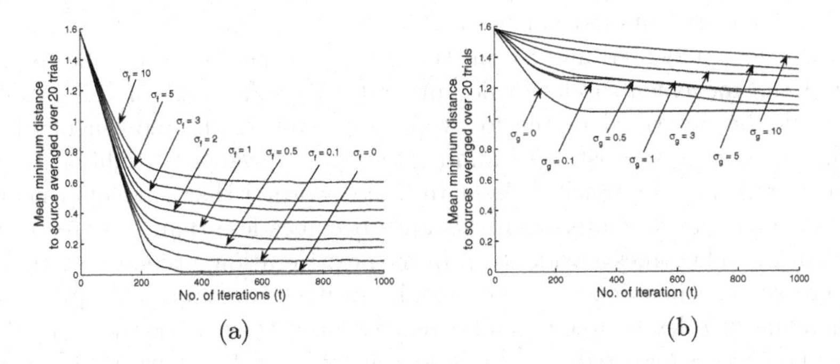

(a) (b)

Fig. 3. a) Plots of $dmin_{av}(t)$ (averaged over 20 trials), for various values of σ_f, in the case of glowworm algorithm. b) Plots of $dmin_{av}(t)$ (averaged over 20 trials), for various values of σ_g, in the case of gradient based algorithm.

luciferin glow/reception. A circular array of sixteen infrared transmitters placed radially outward is used as the glowworm's beacon to obtain a near circular emission pattern around the robot. The glow consists of an infrared light modulated by an 8-bit serial binary signal that is proportional to the Glowworm's luciferin value at the current sensing-decision-action cycle (Refer to [4] for a description of the Glowworm hardware modules).

6 Sound Source Localization

In this experiment, the glowworms localize a sound source which is a loud speaker activated by a square wave signal of frequency 28 Hz. A microphone based sound sensor enables each Glowworm to measure the intensity of sound at its current location. We place a Glowworm (A) near the source and a dummy Glowworm (B) away from the source which is kept stationary but made to emit luciferin proportional to the intensity measurement at its location. Since A is already located at the source, it doesn't get a direction to move. Therefore, it remains stationary at the source location. Initially, since B is in the vicinity of C (while A is not), it moves towards B. However, as it reaches closer to B it senses A and hence, changes direction in order to move towards A. Since D is closer to A, it makes deterministic movements towards A at every step. In this manner, the glowworms localize the sound source eventually. Snapshots from a video of the above experiment are shown in Figure 4.

A dummy glowworm was placed at the source location only for the purpose of demonstration and *is* not a necessary requirement for the algorithm working. According to the glowworm algorithm, a glowworm with the maximum luciferin at a particular iteration remains stationary during that iteration. Ideally, the above property leads to a dead-lock situation when all the glowworms are located such that the peak-location lies outside the convex-hull formed by the glowworm positions. Since the agent movements are restricted to the interior region of the convex-hull, all the glowworms converge to a glowworm that attains maximum luciferin value during its movements within the convex-hull. As a result, all the glowworms get co-located away from the peak-location. However, the discrete nature of the movement-update rule automatically takes care of this problem which could be described in the following way. During the movement phase, each glowworm moves a distance of finite step-size s towards a neighbor. Hence, when a glowworm i approaches closer to a glowworm j that is located nearest to a peak such that the inter-agent distance becomes less than s, i crosses the position of j and becomes a leader to j. In the next iteration, i remains stationary and j crosses the position of i thus regaining its leadership. This process of interchanging of roles between i and j repeats until they reach the peak. This phenomenon was supported by simulations in [2] and [5]. However, since real robots cannot move over each other, the robot's behavior should be modified such that it emulates the above phenomenon when it encounters another robot.

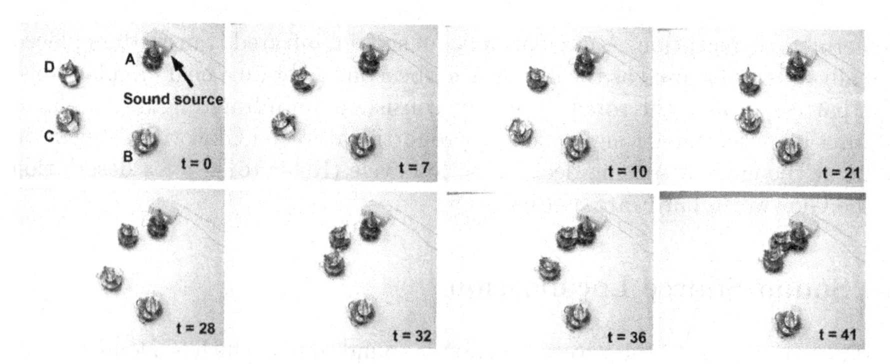

Fig. 4. Demonstration of a sound-source localization task

7 Conclusions

We describe a novel glowworm algorithm that allows a group of minimalist mobile robots to simultaneously localize multiple source locations of interest. The low-cost constraint on various hardware modules of the robots used for our experiments gives rise to considerable amount of noise in their perceptional capabilities, offering a good platform to test the performance of algorithm in the presence of sensor-noise. Simulations show that the algorithm exhibits good performance in the presence of fairly high noise levels. We observe graceful degradation only with significant increase in levels of perceptional noise. A comparison between the glowworm algorithm and the gradient based algorithm reveals the superiority of the glowworm algorithm in coping with uncertainty. Results of the sound-localization experiment conducted with a set of four glowworms support the claim that the algorithm performs fairly well under uncertainty. The response of the algorithm in the presence of a forbidden region was described in [2] using simulations. It was shown that the agents not only escape entry into, but also take a de tour about the forbidden region and eventually reach the source location. However, we need to conduct the same experiment using real robots in order to determine additional behaviors, like obstacle avoidance, that are needed to achieve a similar response. Our future work involves a thorough qualitative and quantitative comparison of GSO with other competing nature-inspired algorithms.

Acknowledgements. The authors wish to acknowledge partial financial support from MHRD and from DRDO-IISc Mathematical Engineering program.

References

1. Cui, X., Hardin, C.T., Ragade, R.K., Elmaghraby, A.S.: A swarm approach for emission sources localization. In: Proceedings of the 16th International Conference on Tools with Artificial Intelligence. (2004) 424–430
2. Krishnanand, K.N., Ghose, D.: Detection of multiple source locations using a glowworm metaphor with applications to collective robotics. In: Proceedings of IEEE Swarm Intelligence Symposium. (2005) 84–91
3. Krishnanand, K.N., Ghose, D.: Theoretical foundations for multiple rendezvous of glowworm-inspired mobile agents with variable local-decision domains. In: Proceedings of American Control Conference. (2006) (to appear)
4. Krishnanand, K.N., Amruth, P., Guruprasad, M.H., Sharschchandra, V.B., Ghose, D.: Glowworm-inspired robot swarm for simultaneous taxis towards multiple radiation sources. In: Proceedings of IEEE International Conference on Robotics and Automation. (2006) (to appear)
5. Krishnanand, K.N., Ghose, D.: Glowworm-inspired swarms with adaptive local-decision domains for multimodal function optimization. In: Proceedings of IEEE Swarm Intelligence Symposium. (2006) (to appear)
6. Goldberg, D.E., Richardson, J.: Genetic algorithms with sharing for multimodal function optimization. In: Genetic Algorithms and their Applications ICCGA87. (1987) 41–49
7. Leung, K.S., Liang, Y.: Adaptive elitist-population based genetic algorithm for multimodal function optimization. In: GECCO 2003, LNCS 2723 (E. Cantu-Paz et al., eds.). (2003) 1160–1171
8. Deb, K.: Multi-objective optimization using evolutionary algorithms. John Wiley & Sons (Asia) Pte. Ltd., (2003)
9. Parsopoulos, K.E., Vrahatis, M.N.: On the computation of all global minimizers through particle swarm optimization. IEEE Transactions on Evolutionary Computation (June 2004)
10. Parsopoulos, K.E., Vrahatis, M.N.: Modification of the particle swarm optimizer for locating all the global minima. Artificial Neural Networks and Genetic Algorithms (2001)
11. Dorigo, M., Stützle, T.: Ant Colony Optimization. MIT Press (2004)
12. Bilchev, G., Parmee, I.C.: The ant colony metaphor for searching continuous design spaces. In: Proceedings of AISB Workshop on Evolutionary Computing Lecture Notes in Computer Science. (1995) 25–39
13. Kennedy, J., Eberhart, R.: Particle swarm optimization. In: Proceedings of IEEE International Conference on Neural Networks. (1995) 1942–1948

Replicating Multi-quality Web Applications Using ACO and Bipartite Graphs*

Christopher B. Mayer[1], Judson Dressler[1], Felicia Harlow[1], Gregory Brault[1], and K. Selçuk Candan[2]

[1] Department of Electrical and Computer Engineering
Air Force Institute of Technology,** Wright-Patterson AFB, OH, USA
{chris.mayer, judson.dressler, felicia.harlow, gregory.brault}@afit.edu

[2] Computer Science and Engineering Department,
Arizona State University, Tempe, AZ, USA
candan@asu.edu

Abstract. This paper presents the application of the Ant Colony Optimization (ACO) meta-heuristic to a new NP-hard problem involving the replication of multi-quality database-driven web applications (DAs) by a large application service provider (ASP). The ASP must assign DA replicas to its network of heterogeneous servers so that user demand is satisfied at the desired quality level and replica update loads are minimized. Our ACO algorithm, AntDA, for solving the ASP's replication problem has several novel or infrequently seen features: ants traverse a bipartite graph in both directions as they construct solutions, pheromone is used for traversing from one side of the bipartite graph to the other and back again, heuristic edge values change as ants construct solutions, and ants may sometimes produce infeasible solutions. Testing shows that the best results are achieved by using pheromone and heuristics to traverse the bipartite graph in both directions. Additionally, experiments show that AntDA outperforms several other solution methods.

1 Motivation and Related Work

An Application Service Provider (ASP) (e.g., Akamai or ASP-One) is a company that specializes in hosting web applications on behalf of clients. Database-driven web applications (DAs) are a particular kind of web application hosted by an ASP in which responses to user requests are built, in part at least, by querying a database. The database allows the DA to customize responses based on user input and enables the DA's content to change dynamically. Some DAs can provide multiple freshness/quality levels of service in order to meet the needs of different types of users [1,2]. For example, consider an on-line brokerage where users have been grouped into two categories: high-quality and low-quality. High-quality users expect very fresh (timely) quotes. Low-quality users, on the other

* Research funded by NSF grant 998404-0010819000.

** The views expressed herein are those of the authors and do not reflect the official policy or position of the U.S. Air Force, Dept. of Defense, or the U.S. Government.

M. Dorigo et al. (Eds.): ANTS 2006, LNCS 4150, pp. 270–281, 2006.

hand, are pleased with the default or moderately-fresh content. Note that low-quality users can be satisfied by high-quality results, but the reverse is not true.

In order to furnish each DA with enough processing capacity to meet demand, an ASP replicates the DAs on its network of servers. For a DA replica to function, the replica's local database needs to be updated with fresh content and/or kept synchronized. The load on a server hosting a DA replica has two components: (1) the **request load** incurred by responding to user requests and (2) the **update load** stemming from database changes. To simplify real DA behavior, request complexity is assumed to be independent of quality level. Thus the request load for each quality offered by a DA is solely dependent on the number of requests received. However, since database freshness determines a replica's quality level, higher quality levels require more frequent database updates. Thus, for a given DA, update loads increase with quality level.

The ASP must establish replicas so that demand for each quality of each DA is met and update burdens (costs) are minimized. Obviously, the ASP cannot allow any server to become overloaded. That is, the sum of the request loads and update loads of the replicas hosted by a server cannot exceed the server's capacity. We have chosen the sum of all replica update loads, **system update burden**, as the cost to be minimized as it also indirectly minimizes the number of replicas and network bandwidth. The ASP's replication problem, the **DA Replication Problem** (DA*rep*), is formalized as an mixed integer linear program (MILP) in Fig. 1. This problem is shown to be NP-hard in [3]. DA*rep* is similar to several multi-commodity facility location problems [4,5] and multidimensional knapsack problems [6,7]. Though algorithms for solving these similar problems have been proposed, the differences are significant enough that they cannot be easily adapted for DA*rep*.

In this paper, inspired by the success of multidimensional knapsack ACO algorithms [8,9] and other ACO algorithms for solving problems similar to DA*rep* [10,11,12,13], we have implemented an ACO-style algorithm called AntDA that has several *novel or rarely seen features.*

1.1 Ant Colony Optimization

Ant Colony Optimization (ACO) is a meta-heuristic based on the natural behavior of real ants and their *seeming* ability to efficiently solve (what we can call) minimization problems obeying just a few simple rules. In ACO, the minimization problem structure is represented as a (directed or undirected) graph and agents acting as ants traverse this graph's edges as they link the graph's vertices together to construct a solution. After finding a *solution*, pheromone is deposited on the edges used in its solution in proportion to the solution's quality. That is, edges critical to the best solutions receive more pheromone than other edges. Pheromone evaporation makes edges less attractive over time and, hence, acts to weed out solutions that are not preferred.

To move from one vertex to another, each ant computes a value for outgoing edges by combining the edge's pheromone concentration and heuristic desirability. The ant randomly selects an outgoing edge to traverse with higher-valued

The Database-driven Application Replication Problem

Definitions: The ASP has m servers, $S = \{1, \ldots, m\}$. Each server $s \in S$ has a processing capacity denoted by C_s. The ASP has n customer-provided DAs to be hosted, $D = \{1, \ldots, n\}$, on its m servers. Each $d \in D$ operates at one or more service quality levels, $Q_d = \{1, \ldots, q, \ldots, qmax(d)\}$, where $qmax(d)$ is the highest level offered by DA d. There is a request load for each quality of each DA: $rl_d = \{rl_{d,1}, \ldots, rl_{d,q}, \ldots, rl_{d,qmax(d)}\}$. For each service quality of a DA d there is a certain update load required to maintain that service quality: $ul_d = \{ul_{d,1}, \ldots, ul_{d,q}, \ldots, ul_{d,qmax(d)}\}$. Let $x_{s,d,q} \in \{0,1\}$ be a binary variable that indicates that server s is hosting a replica of a certain $\langle d, q \rangle$ pair where "$\langle d, q \rangle$ pair" is shorthand for "quality q of DA d." Let $\lambda_{s,d,q} \in [0,1]$ denote the fraction of the request load of a $\langle d, q \rangle$ pair, $rl_{d,q}$, assigned to server s. The update load experienced by server s depends on the quality level of the DA replicas it hosts:

$$ul_s = \sum_{d \in D} \sum_{q \in Q_d} x_{s,d,q} \cdot ul_{d,q}. \tag{1}$$

Server s's request load is the fraction of each $\langle d, q \rangle$ pair's request load sent to it:

$$rl_s = \sum_{d \in D} \sum_{q \in Q_d} \lambda_{s,d,q} \cdot rl_{d,q}. \tag{2}$$

Objective and Constraints: The ASP seeks an assignment of DA replicas to servers that minimizes the system-wide update burden, UB, subject to the four constraints enumerated below.

$$\min UB = \min \sum_{s \in S} \sum_{d \in D} \sum_{q \in Q_d} ul_{d,q} \cdot x_{s,d,q}, \tag{3}$$

1. Request load for each quality, q, of each DA, d is satisfied: $\sum_{s \in S} \lambda_{s,d,q} = 1$.
2. For each server s, a replica of DA d operates at no more than one quality: $\sum_{q \in Q_d} x_{s,d,q} \leq 1$.
3. A server's processing capacity cannot be exceeded: $ul_s + rl_s \leq C_s$.
4. Requests processed by a replica must meet or exceed the request's quality expectation, q_r: $\sum_{q_p | q_p, q_r \in Q_d \wedge q_p \geq q_r} x_{s,d,q_p} \geq \lambda_{s,d,q_r}$.

Fig. 1. DA*rep* formalized as an mixed integer linear program

edges selected more often. Making random decisions using both pheromone (representing past good solutions) and heuristics (which guides ants in the absence of pheromone) encourages exploration near known good solutions and allows for enough variability that distant solutions are unlikely to go unnoticed. The books [14] and [15] contain excellent explanations of the ACO meta-heuristic.

Structurally, ACO algorithms usually consist of a doubly-nested loop. The outer loop controls the number of trials executed. A trial represents a separate solving of the problem. The inner loop governs the time steps in each trial. During a time step each ant traverses the graph and constructs a solution, ants deposit pheromone on the graph's edges, and pheromone evaporates. The updated graph is then used as the basis for the next time step. Being a stochastic process, ACO requires multiple trials each consisting of many time steps.

ACO algorithms have been implemented for many NP-complete combinatorial problems such as traveling salesman [16,17], job-shop scheduling [18], graph coloring [19,20], sequential ordering [21], and multidimensional knapsack [8,9]. Many of the ACO algorithms cited above produce solutions that compare favorably with the best-known solutions for the aforementioned problems.

1.2 Contributions of the Paper

As stated before, in this paper, ACO is adapted to solve the a new NP-hard **DA Replication Problem** (DA*rep*). The resulting algorithm, AntDA, has several novel or rarely seen features that set it apart from other ACO implementations:

- Ants traverse a bipartite graph consisting of a set vertices representing the DAs (things to be assigned) and a set of vertices representing the servers (where things can be assigned). Ants move back-and-forth between the two sets of vertices along directed edges as they construct solutions. Other ACO algorithms for solving problems of a bipartite nature use either a bipartite graph in which ants travel in only one direction (e.g., [9,11,12]) or transform the problem so that vertices represent potential assignments (fully enumerate the solution space) and are all of the same type (e.g., [13,20]).
- Pheromone on DA-to-server edges and server-to-DA edges is used for traversing the graph in both directions. Although not the first to use pheromone for dual purposes [14,15,20], AntDA is one of the few that do. Moreover, experiments show that pheromone in both directions produces the best solutions in terms of both update load (cost) and convergence rates.
- Heuristic values of the edge selection equations change as the ants construct their solutions. Other ACO algorithms with this feature include [8,9].
- In DA*rep* ants may produce infeasible solutions. Since an infeasible solution may be a stepping-stone to feasible solutions, ants with infeasible solutions can also deposit pheromone. Unlike [12] where infeasible solutions pay a constant penalty in terms of pheromone deposited, AntDA's penalty is proportional to the infeasibility.

The rest of this paper is organized as follows. Section 2 presents the AntDA algorithm and the Server-Filling local optimization heuristic. Experimental results of AntDA's performance and the impact of bi-directional bipartite pheromone and the Server-Filling optimization heuristic (a local optimizer introduced just for AntDA) are given in Section 3.

2 AntDA: An ACO Algorithm for DA*rep*

In AntDA, ants operate on a bipartite graph representing an instance of DA*rep* (Fig. 2). The graph, $G = (V, E)$, consists of a set of vertices, V, and edges connecting vertices, E. The vertices are divided into two groups, DQ and S, such

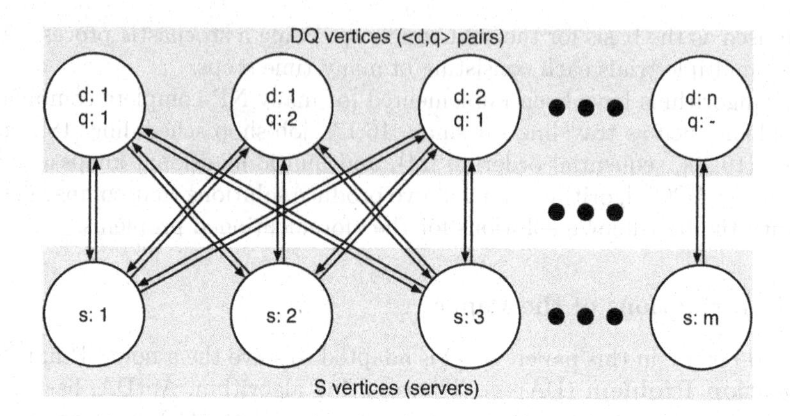

Fig. 2. The bipartite problem graph used by AntDA. Although all $\langle d, q\rangle$ pairs are connected to servers via directional edges and vice-versa, single non-directional edges are shown here for simplicity.

that $V = DQ \cup S$ and $DQ \cap S = \emptyset$. Each vertex in DQ represents a $\langle d, q\rangle$ pair. The vertices in S represent the servers. Directed edges connect each vertex in DQ with each vertex in S and vice versa: $E = \{(dq, s), (s, dq) \mid dq \in DQ \wedge s \in S\}$. Each (dq, s) edge has a pheromone level, $\tau_{dq,s}$, that represents the learned desirability of having server s handle request load for $\langle d, q\rangle$ pair dq. Similarly each (s, dq) edge has an associated pheromone level, $\tau_{s,dq}$, that is used by ants to select the next $\langle d, q\rangle$ pair to be assigned to a server. Each edge $e \in E$ begins a trial with a default level of pheromone, τ_0.

Within a time step, ants construct solutions by moving back and forth between vertices in DQ ($\langle d, q\rangle$ pairs) and vertices in S (servers). An ant works on a solution until either server capacity is exhausted or all request load has been assigned to the servers. Ants work independently (maintain their own solution spaces) but start each time step with identical graphs. When all ants have constructed a solution, pheromone is deposited, pheromone evaporates, each ant's graph is updated with the new pheromone values, and then the next time step begins. Ants are randomly placed at server vertices at the start of a time step.

The following subsections provide the details of AntDA. Note that the description uses terms defined in the DA*rep* formulation of Fig. 1.

2.1 Transitioning from $\langle d, q\rangle$ Pairs to Servers

When at vertex dq, ant k must find a server on which it will create a replica and/or assign request load for the $\langle d, q\rangle$ pair represented by dq. Let S^k_{dq} be the set of servers (vertices) on which request load of the $\langle d, q\rangle$ pair represented by vertex dq can be assigned. Let $rem(C_s) = C_s - rl_s - ul_s$ represent the remaining (unused) capacity of server s. Server s can be assigned by ant k if the net change in update load on s caused by its hosting DA d at quality q is less than $rem(C_s)$ (i.e., s will have capacity to handle request load for the $\langle d, q\rangle$ pair). This is the **server hosting condition**.

The probability that ant k selects edge (dq, s) for traversal is

$$p^k_{dq,s}(t) = \begin{cases} \frac{[\tau_{dq,s}(t)]^\alpha \cdot [\eta_{dq,s}]^\beta}{\sum\limits_{s \in S^k_{dq}} [\tau_{dq,s}(t)]^\alpha \cdot [\eta_{dq,s}]^\beta}, & \text{when } s \in S^k_{dq} \\ 0, & \text{when } s \notin S^k_{dq}. \end{cases} \quad (4)$$

$\tau_{dq,s}(t)$ is the pheromone concentration on edge (dq, s) at time step t. α and β are constants governing the relative importance of pheromone to the heuristic desirability, $\eta_{dq,s}$, of traveling along edge (dq, s)

$$\eta_{dq,s} = \frac{rem(C_s)}{\sum\limits_{s \in S^k_{dq}} rem(C_s)}. \quad (5)$$

The heuristic is based on the idea that greedily selecting the server with the most remaining capacity should reduce the number of replicas created and, thus, update burden produced. Note that $\eta_{dq,s}$ values are not constant. Instead, they change as server capacity is consumed.

After selecting edge (dq, s) the ant moves from vertex dq to vertex s. Once at s, the ant checks to see if a replica of DA d exists on server s. If not, one is created at service level q by setting $x_{s,d,q} = 1$. If a replica of d already exists on s at a lower quality $r < q$, the replica's quality level (update load) is increased by setting $x_{s,d,r} = 0$ and $x_{s,d,q} = 1$. After establishing a replica at the correct quality level, the ant assigns as much remaining (unassigned) request load of the $\langle d, q \rangle$ pair, $rem(rl_{d,q})$, to s as possible, stopping when $rem(rl_{d,q}) = 0$ or $rem(C_s) = 0$. The request load assignment is accomplished by setting $\lambda_{s,d,q}$ to the fraction of the $\langle d, q \rangle$ pair's request load, $rl_{d,q}$ that s will handle. Any time an assignment is made, the server's remaining capacity, $rem(C_s)$, is decreased based on the amount of update load of the replica and request load assigned.

After creating a replica of DA d at quality level q on server s, the ant invokes following two-step **Server-Filling (SF)** heuristic.

1. SF first tries to avoid the creation of extra replicas of d by finding other qualities of d that will completely fit on s. More specifically, SF looks for another quality $r \in Q_d$ such that all of $rem(rl_{d,r})$ can be assigned to s. SF assigns the highest r found by setting $x_{s,d,*}$ and $\lambda_{s,d,*}$ variables as needed, repeating with lower qualities of d if possible. Note that replica quality may have to be increased since it may be that $r > q$ and hence $ul_{d,r} > ul_{d,q}$. This is done by setting $x_{s,d,q} = 0$ and $x_{s,d,r} = 1$. Let y be the highest quality of d assigned to s at the end of this step.
2. If s still has spare capacity after step 1, SF looks for the highest quality $u \in Q_d$ of DA d such that $u < y$ and assigns as much request load of quality u as possible to the replica (by setting $\lambda_{s,d,u}$).

The SF heuristic is an optional, but beneficial, part of AntDA; SF reduced update burdens by over 4% versus not using it (Section 3).

2.2 Moving from Servers to $\langle d, q \rangle$ Pairs

An ant at vertex s must decide which $\langle d, q \rangle$ pair should be assigned next. Let DQ_s^k be the set of dq vertices which are still capable of being assigned to servers. A $\langle d, q \rangle$ pair can be assigned if

1. it has some amount of unassigned request load ($rem(rl_{d,q}) > 0$), and
2. there is at least one server $s \in S$ for which the server hosting condition (first paragraph of Section 2.1) is satisfied with respect to the $\langle d, q \rangle$ pair.

If $DQ_s^k = \emptyset$, the ant's walk terminates. Otherwise, the probability that ant k selects edge (s, dq) is given by:

$$p_{s,dq}^k(t) = \begin{cases} \frac{[\tau_{s,dq}(t)]^\alpha \cdot [\eta_{s,dq}]^\beta}{\sum\limits_{s \in DQ_s^k} [\tau_{s,dq}(t)]^\alpha \cdot [\eta_{s,dq}]^\beta}, & \text{when } dq \in DQ_s^k \\ 0, & \text{when } dq \notin DQ_s^k. \end{cases} \tag{6}$$

$\tau_{s,dq}(t)$ is the pheromone concentration on (s, dq). α and β again control the relative importance of pheromone and heuristic desirability, in this case $\eta_{s,dq}$:

$$\eta_{s,dq} = \frac{ul_{d,q} \cdot rem(rl_{d,q})}{\sum\limits_{dq \in DQ_s^k} ul_{dq} \cdot rem(rl_{d,q})}. \tag{7}$$

Dividing $ul_{d,q} \cdot rem(rl_{d,q})$ by a server's remaining capacity, $rem(C_s)$, estimates the update burden incurred by creating replicas on servers of size $rem(C_s)$. Eq. (7) is an appropriate heuristic since it prefers $\langle d, q \rangle$ pairs most likely to produce high update burdens (no matter which servers are used). Note that $\eta_{s,dq}$ values change as the ant constructs its solution.

After making its selection, the ant traverses the edge to the selected $\langle d, q \rangle$ pair and then transitions back to a server node (Section 2.1).

2.3 Pheromone Update Rule

When each ant has constructed a solution, it is time to deposit pheromone.

By finding a solution, an ant has assigned values for the $x_{s,d,q}$ and $\lambda_{s,d,q}$ variables from the formal version of DA*rep* given in Fig. 1. Since better solutions have lower update burdens, the amount of pheromone deposited should be inversely proportional to a solution's update burden. However, low update burdens are not always better – some ants' solutions may be infeasible (i.e., they do not assign all request load). Therefore, we differentiate between feasible and infeasible solutions when deciding how much pheromone to deposit on the edges used in a solution: Let $UB_k(t)$ be the update burden of ant k's solution after time step t as computed by (3). We adjust $UB_k(t)$ to account for infeasible assignments as follows:

$$UB'_k(t) = \frac{UB_k(t)}{\left(\frac{RL_k(t)}{RL}\right)^\omega} \tag{8}$$

where $RL_k(t) = \sum_{d \in D} \sum_{q \in Q_d} \lambda_{s,d,q} \cdot rl_{d,q}$ is the amount of request load assigned by ant k in time step t, $RL = \sum_{d \in D} \sum_{q \in Q_d} rl_{d,q}$ is the total amount of system-wide request load, and ω is a constant scaling factor.

Once $UB'_k(t)$ has been determined, it is used to calculate the amount of new pheromone ant k will deposit. We allow the ants with the m best solutions to deposit pheromone after each time step. More specifically, if edge $e \in E$ was traversed in the ith best solution and $i \leq m$, then the amount of pheromone deposited on e by the ant that produced the ith best solution is

$$\Delta^i_e(t) = \frac{\gamma}{UL'_i(t)} \tag{9}$$

where γ is a constant. We set $\gamma = 1$ during our experiments as we found it had little, if any, impact on performance. If an edge e was not used by ant i, then $\Delta^i_e(t) = 0$.

Let $\Delta\tau_e(t) = \sum_{i=1}^{m} \Delta^i_e(t)$ be the amount of new pheromone to be deposited on edge e because of the m solutions chosen. The amount of pheromone on the edges in graph G is then updated as is typically done in ACO [14,15]:

$$\tau_e(t+1) \leftarrow (1-\rho) \cdot \tau_e(t) + \rho \cdot \Delta\tau_e(t) \tag{10}$$

Note that pheromone updates apply to both kinds of edges: those leading from DQ to S and from S to DQ.

3 Experimental Validation of AntDA

This section presents the results of experiments that compare AntDA with other solution methods, reveals the importance of the Server-Filling (SF) optimization heuristic, and demonstrates the importance of pheromone and heuristics on ants traversing the bipartite graph.

Experimental Configuration. Unless otherwise noted, AntDA was run with the parameter values and conditions shown in Table 1. Not reported in this paper, our efforts to tune the parameters revealed that, for the most part, changing any particular value had only minor effects. Only the number of ants allowed to deposit pheromone, m, was observed to have a significant impact on solution values and convergence rates. Setting m to be 10% of the number of ants increased the convergence rate six-fold and improved solutions by 5% versus the worst case in which pheromone was deposited on all tours. Test cases were subjected to fifty trials of 400 time steps each.

Results for performance comparisons were obtained by using a random assignment algorithm, `Random`, a greedy algorithm, `Greedy`, and the `LINGO` solver [22]. `Random` randomly picks a $\langle d, q \rangle$ pair with non-zero remaining request load and assigns it to a random server capable of hosting it. `Random` reports the best solution found out of 1000 trials. The `Greedy` algorithm makes assignments by choosing the $\langle d, q \rangle$ pair with the highest predicted update burden. This is the

Table 1. Parameter and condition values for AntDA experiments

$\alpha = 1$	$\beta = 8$	$\rho = 0.8$	$\#\ \text{Ants} = \|DQ\| + \|S\|$
$\gamma = 1$	$\omega = 4$	$\tau_0 = 0.1$	$m = \lfloor \#\ \text{Ants} \cdot 0.1 \rfloor$

Table 2. AntDA solutions compared to other methods. Smaller numbers are better.

Case #	Random	LINGO	Greedy	AntDA Min	AntDA Max	AntDA Avg	AntDA Std Dev
1	1057	1025	842	784	800	793.86	5.77
2	1135	1168	884	811	824	817.94	2.94
3	1048	1066	788	764	771	766.06	2.78
4	1099	1199	849	813	822	818.98	2.02
5	1137	913	867	811	823	814.64	2.60

same as computing the numerator of (7) for each $\langle d, q \rangle$ pair still capable of being assigned to some server. The $\langle d, q \rangle$ pair selected is then assigned to the server with the most remaining capacity. The `Greedy` algorithm also executes the Server-Filling optimization heuristic. Both `Random` and `Greedy` algorithms stop once all request load has been assigned or when no more request load can be assigned to any server. `LINGO` solves DA*rep* using the MILP formulation of Fig. 1. Although MILP solvers such as `LINGO` are the only known method besides complete enumeration that can find guaranteed optimal solutions, execution times can be prohibitive. Therefore, we allotted four hours for `LINGO` to work on problems. This was sufficient time for `LINGO` to produce feasible, but not necessarily optimal, solutions and provides a notion of DA*rep*'s complexity.

Due to space limitations, experimental results are shown for just five test cases (hypothetical DA*rep* instances) involving five DAs of three quality levels each.[1] The update load for quality levels 1, 2, and 3 were randomly generated within the bounds of $[5, 14]$, $[16, 25]$, and $[27, 36]$, respectively. The update loads for quality three make it impossible to assign request load to the smallest of the servers. Request load increased in conjunction with quality level and was randomly generated in the ranges of $[34, 100]$, $[133, 200]$, $[233, 300]$, respectively. Once the DAs were generated, servers were added to complete each test case. Servers were added in sets of five servers having sizes $\{25, 50, 75, 100, 125\}$ until the `Greedy` algorithm produced feasible solutions. The number of vertices in the test case graphs were either 70 or 75.

Results and Analyses. Table 2 shows the performance of AntDA and the other solution methods for the five test cases. Clearly AntDA produces better solutions than the other three methods.

[1] Although we have a suite of over 70 test cases with three or fewer quality levels per DA and varying update loads and request loads, we display only the most difficult cases. Results for the other test cases are similar to those shown here.

Table 3. The effect of the Server-Filling heuristic on AntDA update burdens. Smaller numbers are better.

	Server-Filling Off				Server-Filling On				difference of
Case #	Min	Max	Avg	Std Dev	Min	Max	Avg	Std Dev	min solutions
1	823	840	830.76	4.58	784	800	793.86	5.77	4.74%
2	841	878	863.56	6.94	811	824	817.94	2.94	3.57%
3	793	820	805.66	7.63	764	771	766.06	2.78	3.66%
4	852	875	862.54	7.91	813	822	818.98	2.02	4.58%
5	852	891	870.64	9.81	811	823	814.64	2.60	4.81%

Note that, while it produces the lowest update burdens, AntDA has higher solution times than `Random` and `Greedy`; `Greedy` can produce a solution in milliseconds, `Random` needed about 1.5 minutes, while AntDA took just under four hours to complete a test case.

We, however, also note that AntDA found its best solution by the 18th time step on average with a standard deviation of 10.17. Using the rule-of-thumb that 95% of values fall within two standard deviations of their mean, this means that AntDA will find its best answer within 50 time steps $(18 + (2 \cdot 10.17) \leq 50)$ 95% of the time for the five test cases. Constraining the number of time steps as such brings AntDA down to a half hour per test case (about 30 seconds per trial), which is a realistic time bound in DA replication scenarios.

The next set of results show the impact of the Server-Filling (SF) optimization heuristic on AntDA. Table 3 clearly shows that using SF results in lower minimum, maximum, and average update burdens versus not using it. Comparing just minimum solution values, SF caused solution values to improve by 3.57% to 4.81% (rightmost column of Table Table 3). In all but test case #1, SF's solutions also experienced much lower standard deviations.

The use of a bipartite graph in which the ants follow pheromone and heuristics in order to traverse from one set of vertices to the other and back raises the question of how important pheromone and heuristics are to AntDA performance. Table 4 shows the impact of all possible cases of turning pheromone and heuristics off for the fifth of the five test cases. Similar results were obtained from the other four test cases. For example, line 1 of Table 4 shows performance results when pheromone is turned off when moving from an application vertex to a server (the pheromone components (4) are removed). Similarly, line 4 shows the impact of ignoring heuristics when moving from a server to a $\langle d, q \rangle$ pair caused by removing the heuristic components of (6).

Line 9 of Table 4 contains the statistics for normal AntDA operation. Note that lines 2, 4, and 5 come close to meeting the minimum update burden (UB) of line 9 yet have higher maximum and average update burdens. Also, lines 4 and 5 converge on their best solution (rightmost column) after 221.22 and 148.30 time steps on average, respectively, which is much worse than the normal mode of operation (line 9). Line 2 fares much better with respect to convergence rate and minimum update burden, but still has a higher maximum and average update burden than the normal mode. Altogether, Table 4 indicates that using

Table 4. The importance of pheromone and heuristics on AntDA update burdens. Smaller numbers are better.

Line	Pheromone Off	Heuristic Off	Min	Max	Avg	Std Dev	Conv. Avg.
1	*DQ*-to-*S*	-	838	866	864.78	9.13	198.55
2	*S*-to-*DQ*	-	812	836	824.48	4.83	4.84
3	-	*DQ*-to-*S*	830	887	854.16	14.36	44.22
4	-	*S*-to-*DQ*	815	864	841.68	9.55	221.22
5	both	-	815	826	822.24	2.60	148.30
6	-	both	835	896	863.40	11.23	121.56
7	*DQ*-to-*S*	*S*-to-*DQ*	874	874	858.20	9.25	204.04
8	*S*-to-*DQ*	*DQ*-to-*S*	833	973	911.24	23.08	17.30
9	-	-	811	822	816.26	2.90	25.28

pheromone and heuristics to make edge selections in both directions (normal operating mode) gives the best overall performance.

4 Conclusion

In this paper the Ant Colony Optimization (ACO) meta-heuristic was successfully applied to a new NP-hard problem, DA Replication Problem(DA*rep*), in which an application service provider must replicate multi-quality database-driven web applications on its network of servers at minimal cost. The ACO formulation for DA*rep*, AntDA, is the first to use a fully bipartite graph. Ants deposit and follow pheromone on directed edges connecting a set of application vertices with a set of server vertices and then back from servers to applications. Other interesting aspects of AntDA include dynamically changing heuristic values and the possibility of infeasible solutions. Experiments showed that AntDA outperforms several other solution methods. Moreover, the use of the Server-Filling optimization heuristic by AntDA decreased costs by over 4%. Tests in which pheromone and heuristic values were removed from the ants' decision-making process as they traveled the bipartite graph showed that pheromone and heuristics in both traversal directions delivers the best performance.

References

1. Bright, L., Raschid, L.: Using latency-recency profiles for data delivery on the web. In: VLDB. (2002) 550–561
2. Cherniack, M., Galvez, E.F., Franklin, M.J., Zdonik, S.: Profile-driven cache management. In: Proceedings of the International Conference on Data Engineering (ICDE), IEEE Computer Society (2003) 645–656
3. Mayer, C.B.: Quality-based Replication of Freshness-Differentiated Web Applications and Their Back-end Databases. PhD thesis, Arizona State University (2005)
4. Mazzola, J.B., Neebe, A.W.: Lagrangian-relaxation-based solution procedures for a multiproduct capacitated facility location problem with choice of facility type. European Journal of Operational Research **115** (1999) 285–299

5. Pirkul, H., Jayaraman, V.: A multi-commodity, multi-plant, capacitated facility location problem: Formulation and efficient heuristic solution. Computers and Operations Research **25**(10) (1998) 869–878
6. Dawande, M., Kalagnanam, J., Keskinocak, P., Ravi, R., Salman, F.S.: Approximation algorithms for the multiple knapsack problem with assignment restrictions. Combinatorial Optimization **4**(2) (2000) 171–186
7. Shachnai, H., Tamir, T.: Noah's bagels - some combinatorial aspects. In: International Conference on Fun with Algorithms. (1998)
8. Alaya, I., Solnon, C., Ghédira, K.: Ant algorithm for the multidimensional knapsack problem. In: International Conference on Bioinspired Optimization Methods and their Applications. (2004) 63–72
9. Leguizamon, G., Michalewicz, Z.: A new version of ant system for subset problems. In: Proceeding of the 1999 Congress on Evolutionary Computation, IEEE Press (1999) 1459–1464
10. Cordón, O., Fernández de Viana, I., Herrera, F.: Analysis of the best-worst ant system and its variants on the QAP. In: Third International Workshop on Ant Algorithms (ANTS 2002). Volume 2463 of Lecture Notes in Computer Science., Springer (2002) 228–234
11. Foong, W.K., Maier, H.R., Simpson, A.R.: Ant colony optimization for power plant maintenance scheduling optimization. In: Genetic and Evolutionary Computation Conference (GECCO), ACM (2005) 249–256
12. Lourenço, H.R., Serra, D.: Adaptive search heuristics for the generalized assignment problem. Mathware and Soft Computing **9**(2) (2002) 209–234 On-line journal. Articles available at: http://docto-si.ugr.es/Mathware/ENG/mathware.html.
13. Stützle, T., Dorigo, M.: ACO algorithms for the quadratic assignment problem. In: New Ideas in Optimization. McGraw-Hill (1999) 33–50
14. Bonabeau, E., Dorigo, M., Theraulaz, T.: Swarm Intelligence: From Natural to Artificial Systems. Oxford University Press, New York (1999)
15. Dorigo, M., Stützle, T.: Ant Colony Optimization. The MIT Press (2004)
16. Dorigo, M., Maniezzo, V., Colorni, A.: Ant System: Optimization by a colony of cooperating agents. IEEE Transactions on Systems, Man, and Cybernetics–Part B **26**(1) (1996) 29–41
17. Eyckelhof, C.J., Snoek, M.: Ant systems for a dynamic TSP. In: Third International Workshop on Ant Algorithms (ANTS 2002). Volume 2463 of Lecture Notes In Computer Science., Springer (2002) 88–99
18. Blum, C., Sampels, M.: An ant colony optimization algorithm for shop scheduling problems. J. of Mathematical Modelling and Algorithms **34**(3) (2004) 285–308
19. Comellas, F., Ozón, J.: An ant algorithm for the graph colouring problem. In: First International Workshop on Ant Colony Optimization (ANTS '98). (1998)
20. Costa, D., Hertz, A.: Ants can colour graphs. J. of the Operational Research Society **48** (1997) 295–305
21. Gambardella, L.M., Dorigo, M.: An ant colony system hybridized with a new local search for the sequential ordering problem. INFORMS Journal on Computing **12**(3) (2000) 237–255
22. LINDO Systems, Inc.: Lingo (2006) http://www.lindo.com.

Restoration Performance vs. Overhead in a Swarm Intelligence Path Management System

Poul E. Heegaard[1] and Otto J. Wittner[2]

[1] Telenor R&D* and Department of Telematics
Norwegian University of Science and Technology, Norway
poulh@item.ntnu.no
[2] Centre for Quantifiable Quality of Service in Communication Systems**
Norwegian University of Science and Technology, Trondheim, Norway
wittner@q2s.ntnu.no

Abstract. CE-ants is a distributed, robust and adaptive swarm intelligence strategy for dealing with path management in communication networks. This paper focuses on various strategies for adjusting the overhead generated by the CE-ants as the state of the network changes. The overhead is in terms of number of management packets (ants) generated, and the adjustments are done by controlling the ant generation rate that controls the number ants traversing the network. The link state events considered are failure and restoration events. A simulation scenario compares restoration performance of rate adaptation in the source node with rate adaptation in the intermediate nodes close to the link state events. Implicit detection of failure events through monitoring ant parameters are considered. Results indicate that an implicit adjustment in the source node is a promising approach with respect to restoration time and the number of ants required.

1 Introduction

Paths between all source destination pairs in a communication network should be chosen such that an overall good utilisation of network resources is ensured, and hence high throughput, low loss and low latency achieved. The available spare capacity in the network must be utilised in such a manner that a failure results in a minimum disturbance on traffic flows. The combinatorial optimisation aspects of this task are typically NP-hard, see for instance [1]. Nevertheless, considerable knowledge has been acquired for planning paths in networks and insight and practical methods for obtaining such paths by mathematical programming are available. For an overview, see the recently published book by Piro and Medhi [2] and references therein. Several stochastic optimisation techniques have been

* This work was partially supported by the Future & Emerging Technologies unit of the European Commission through Project BISON (IST-2001-38923).
** Centre for Quantifiable Quality of Service in Communication Systems, Centre of Excellence appointed by The Research Council of Norway, funded by the Research Council, NTNU and UNINETT. http://www.q2s.ntnu.no

M. Dorigo et al. (Eds.): ANTS 2006, LNCS 4150, pp. 282–293, 2006.

proposed [3,4,5,6]. Common to these are that they deal with path finding as an optimisation problem where the "solution engine" has a global overview of the problem and that the problem is unchanged until a solution is found. To ensure system robustness and be in-line with the "Internet philosophy" path management is required to be truly *distributed* and *adaptive.* However, one should be aware that applying truly distributed decision-making typically yields solution which are less fine tuned with respect to optimal resource utilisation.

In addition to finding good paths, proper path management requires that: a) the set of operational paths should be continuously updated as the traffic load changes, b) new paths should become almost immediately available between communication nodes when established paths are affected by failures, and c) new or repaired network elements should be put into operation without unnecessary delays. Near immediate and robust fault handling advocates distributed local decision-making on how to deal with failures.

Schoonderwoerd & al. introduced a system applying multiple agents with a behaviour inspired by ants to solve problems in telecommunication networks [7]. Their system belongs to a group of systems today known as *swarm intelligence* [8] systems, and has been studied further by others, see for instance [9,10,11] and references therein. Self-management by swarm intelligence is a candidate to meet the aforementioned requirements and to overcome some of the drawbacks of the current path and fault management strategies. Heegaard, Wittner, and Helvik [12] describes CE-ants (cross-entropy ants), a swarm intelligent based method for dealing with path management in communication networks. This paper focuses on reducing the overhead in terms of the number of management packets (ants) generated in a CE-ants based system. The system applies adaptive multi-path load sharing and stochastic routing, denoted CE-ants with *adaptive path* in [12], which results in robust and adaptive forwarding. The number of ants are reduced by controlling the ant generation rate according to the link state of the network.

A description of the general system model applied in the paper is given in Section 2 followed by a brief presentation of the foundations of CE-ants with adaptive path in Section 3. A set of different rate adaptation strategies are presented in Section 4, and results from a comparative study of their performance are given in Section 5. The study involves simulations of a scenario where a nationwide communication infrastructure is applied. Future work and concluding remarks are found in Section 6.

2 System Model

The system is a bidirectional graph $G(v,l)$ where v is the set of nodes and l is the set of links. The objective is to find a virtual connection $VC_{[s,d]}$ between source node s and destination node d with a minimum *cost*, e.g. minimum delay between s and d. $VC_{[s,d]}$ is realised by one or more paths, $\omega_{[s,d]} = \{s, \cdots, d\} \in \Omega_{[s,d]}$ where $\Omega_{[s,d]}$ is the set of all feasible paths between s and d. A link $l_{[i,j]}$ is specified by its end nodes i and j. The end nodes of link closest to the source and closest to the destination of path $\omega_{[s,d]}$ are denoted the *lower* node and the *upper* node respectively. See Figure 1 for an illustration.

Discovery and maintenance of paths for virtual connections are handled by the CE-ants method and is governed by determining the *cost* of paths. In this paper link cost is delay that includes queuing and processing delay in a node, and transmission and propagation delay of a link. The cost of a link is denoted $c(l)$ and will vary with traffic load and over time. The cost of a path is $c(\omega) = \sum_{\forall l \in \omega} c(l)$. The cost of a virtual connection is either $c(VC) = \min_{\omega \in \Omega} c(\omega)$ when the minimum cost path is used, or $c(VC) = \sum_{\omega \in \Omega} p(\omega) \cdot c(\omega)$ when stochastic routing is applied over the feasible paths $\omega \in \Omega$ with path probability $p(\omega)$. The near optimal or best paths are denoted the *preferred paths* of a VC.

The CE-ants system described in this paper is generating signalling packets, denoted *ants,* at VC source s with rate λ_t at time t. The suffix t is ignored in the following for notation simplicity. The initial rate λ_0 in the exploration phase is $\lambda_0 > \lambda > \lambda_s$ where λ_s is the ant generating rate in steady state. The ants are identified by their destination d and a species identity id, i.e. a tuple $< d, id >$. The ant species identity refers to an ant species with a designated task, in this paper typically finding the minimum cost path from s to d.

3 Cross Entropy Ants (CE-ants)

A CE-ants system is a swarm intelligent system originally inspired by the foraging behaviour of ants. The idea is to have a number of simple ant-like mobile agents, denoted just *ants* in this paper, iteratively searching for paths in a network. Having searched for and found a path, an ant backtracks and revisits all nodes in the path. The ant leaves markings at each node. The markings, denoted *pheromones,* resembling the chemicals left by real ants during ant trail development. The strength of the pheromones left depends on the quality of the path found. Hence, nodes hold distributions of pheromones pointing toward their neighbour nodes. A new ant in its searching phase selects the next node to visit stochastically based on the pheromone distribution seen in the visited node. Using such trail marking ants, together with evaporation of pheromones, the overall process converges quickly towards having the majority of the ants following the trails that tend to be a near optimal path. Due to limited space the foundations for CE-ants are only outlined in the following paragraphs. See [11] for more details.

In [6] Rubinstein presents an algorithm that finds near optimal solutions to hard combinatorial problems by an iterative method. The algorithm is founded on the recognition of that finding the optimal solution by random selection is an extremely rare event. Rubinstein represents the total allocation of pheromones in a network by a probability matrix P_t where an element $P_{t,ij}$ reflects the normalised intensity of pheromones pointing from node i toward node j. An ant's stochastic search for a sample path resembles a Markov Chain selection process based on P_t. By importance sampling in multiple iterations Rubinstein alters the transition matrix ($P_t \rightarrow P_{t+1}$) and increases, as mentioned, certain probabilities such that ants eventually find near optimal paths with high probabilities. Cross entropy is applied to ensure efficient alteration of the matrix. To speed up the process further, a performance function weights the path qualities such that high quality paths have greater influence on the alteration of the matrix. A "temperature" parameter γ_t in the performance function controls the focus of the overall search towards high quality paths. Focus is tightened gradually, i.e. $\gamma_{t-1} > \gamma_t > \gamma_{t+1}$, to avoid local optima (cf. simulated annealing.)

A distributed and asynchronous version of Rubinstein's CE algorithm, referred to as *CE-ants*, is developed and described in details in [13]. On the contrary to Rubinstein's CE algorithm where all ants must produce sample paths before P_t can be updated, in CE-ants an optimised update of P_t can be performed immediately after every new path ω_t (where t is the t'he ant) is found, and a new probability matrix P_{t+1} can be generated. Hence CE-ants may be viewed as an algorithm where search ants evaluate a path found (and re-calculate γ_t) right after they reach their destination node, and then immediately return to their source node backtracking the reverse path. During backtracking, pheromones are placed by updating the relevant probabilities in the transition matrix. Due to the compact autoregressive schemas applied in a CE-ants system, the system becomes both computationally efficient, requires limited amounts of memory and is simple to implement.

In [14] elitism is introduced in the CE-ants system to reduce the overhead by reducing the number of ant updates. The basic idea is that only the ants that find paths with cost values amongst the best (so far) should lead to an update and send backtracking ants. This has been shown in [14] to reduce the overhead and the convergence time. The elitism criterion is a function of the "temperature" and does not introduce any additional parameters and it is self-tuning and handles dynamic network conditions.

In [12] different path management strategies based on CE-ants were discussed. This paper adopts the *adaptive path strategy* that is designed for fast restoration and adaptation to both link failures and changes in traffic load.

4 Rate Adaptation Strategies

On most link state events the multi-path CE-ants method with load sharing, denoted *adaptive path* in [12], will almost immediately provide alternative paths to the virtual connections that are effected. The performance of the VC management depends on the number of ants sent. It is a trade-off between reactive path management and overhead in terms of ant messages. This paper studies a set of strategies where the refresh rate λ in stable state is rather low, but is increased as link state events are detected. This rate adjustment can be initiated either in the source of the virtual connections, or locally in the end-points where the link state event takes place, see Figure 1 for an illustration. (Figure 1 is further described in the following sections.)

The remaining of this section presents the metrics applied for performance evaluation, and introduces a reference strategy denoted *fixed rate* and two alternative rate adaptation strategies denoted *implicit adaptation* and *local adaptation*.

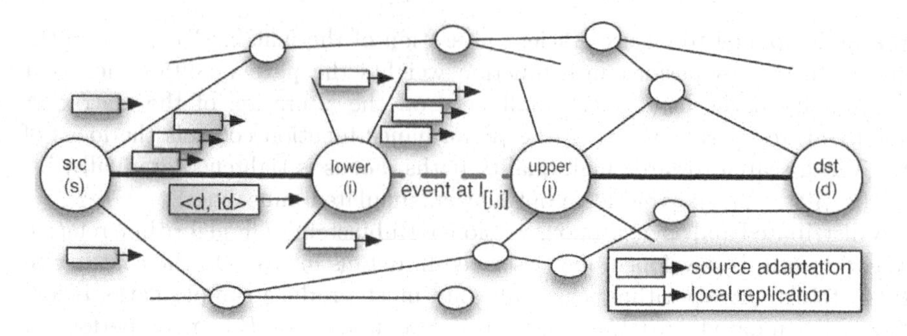

Fig. 1. Illustration of the two main rate adaptation schemes: implicit adaptation and local adaptation

4.1 Performance Metrics

The rate adaptation strategies evaluated in Section 5 are compared by applying the following performance metrics. Note that chances in link states are termed *link state events,* and that shutting down and restoring links generate such events during the simulation experiments.

1. t_r - the *path detection time*, is the time to detect a working path for the virtual connection. This is the time it takes from a link state event t_e occur to a new path from the source to the destination of a virtual connection is found, i.e. to the first packet from source s arrives at destination d.
2. t_o - the *convergence time*, is the time it takes to converge to a new near optimal, or at least good, solution after a link state event. A VC is considered converged when 80% of all packets follow the same path.
3. n - the number of ants generated per VC per time unit.
4. A - steady state service availability. The *service* is a virtual connection with guaranteed packet delay less than $t_{\max}$ [sec.]. If the guarantee is violated the service is not available.

4.2 Fixed Rate

The path finding ability of our CE ants based system is robust and adaptive when the ant rate is high. When no rate change actions are taken on a link state event, the rate will stay high also in steady state when the paths have converged and the VC is established. For such a system configuration the management overhead is independent of the dynamics in the system. However, the obvious downsides are an unnecessary high overhead in stable situations if the rate is high, or too long convergence times (t_o) if the rate is low.

In Section 5 three levels of fixed rates are considered, *high rate* with 500 ants/s, *medium rate* with 200 ants/s, and *low rate* with 50 ants/s. These rates may in general seem high, especially since they concern signalling traffic for managing a single VC. However, in systems realising CE-ants path management, a single

VC would typically carry a number of traffic flows. The overall data rate of a VC would far exceed the ant-signalling rate. Due to the nature of the CE-ants system piggy-backing ant information (typically less than 50 bytes) onto data packets is also an option, and by that a limited number of separate (non-piggy-backed) signalling packets would be required.

4.3 Implicit Adaptation

The convergence time t_o depends on ant generation rate λ from the VC source, and the corresponding update rate λ_e from the VC destination. To reduce t_o, λ must be increased. However, to minimise the overhead, λ should only be increased after a link state event and decreased again on convergence to steady state.

A link state event can either be signalled explicitly towards the source node or handled by implicit notification. We consider implicit notification only since explicit notification would require a separate notification protocol, carefully designed to avoid route flapping (i.e. avoid sending notifications on transient link failures), to avoid unnecessary overhead (i.e. avoid sending notifications on link state events not affecting the performance of the VC in question), and to ensure that notification messages are not lost (i.e. ensure that rate change take place when it should).

The link state events along a preferred path can be implicitly detected in the source node s by monitoring changes in the difference between the outgoing ant rate λ (forward ants) and the incoming ant rate λ_e (backtracking elite ants). Based on the fact that in a steady state $\lambda \approx \lambda_e$ while in a failure state $\lambda_e \approx 0$, we design the following autoregressive expression for λ

$$\lambda \leftarrow \lambda_0 \cdot \min(\varepsilon, \frac{\lambda - \hat{\lambda}_e}{\lambda}) \tag{1}$$

where λ_0 is the initial intensity, $\varepsilon \in (0, 1]$ is a parameter that ensures a minimum ant generator rate. $\hat{\lambda}_e$ is an estimator of λ_e produced by recording the time epochs, t_k, for the k'th incoming ant and averaging over a window of the last α epochs,

$$\hat{\lambda}_{e,t_k} = \alpha/(t_k - t_{k-\alpha}) \tag{2}$$

where the memory term, α, should be between 10 and 30 to enable quick adaptation to changes in incoming ant rates. In a scenario with no link state events, the system will converge towards a near optimal solution, and ε may be set to 0 to force the generation of ants to stop on convergence. However, in a system with link state events changing the operational conditions, a minimum of ants must be generated to detect changes, hence $\varepsilon > 0$. The optimal value of ε depends on the frequency of the system dynamics, the memory factor (β) configured in the CE-ants system, the topology, the expected number of hops in a path, the number of (equal) solutions and the number of parallel VC setups. In Section 5 λ_0 is set to 500 ants/s, i.e. equal to *high rate*, ε is set to 0.1, which results in a minimum rate of 50 ants/s, i.e. equal to *low rate*.

4.4 Local Adaptation

An alternative to notifying the source node of a link state event, is to increase the rate of outgoing ants at the *lower* node of the link effected by the state change. At the *lower* node, ants $< d, id >$ that are coming from node s and heading towards node d are replicated in n_r copies and sent towards d on alternative links (including the incoming link). See Figure 1 for an illustration. The replication of $< d, id >$ ants can be initiated immediately after the detection of a link state event, or when the next forward ant has arrived after the event was detected. The advantage with the former is fast reaction time, while the latter will avoid unnecessary replication of ants in low impact nodes as well as damp the effect of transient link state events. The latter approach is applied in Section 5.

The number of replica, n_r, and the replication rate, λ_r, must be set to reflect the impact a detected link state event has on the performance of the VC maintained.

Replication Rate. The rate of forwarding ants λ_f in the lower node can be estimated as in (2) by recording the time epochs, t_k, for k'th incoming ant and averaging over a window of the last α epochs

$$\hat{\lambda}_{f,t_k} = \alpha/(t_k - t_{k-\alpha}) \tag{3}$$

where the memory term, α, should be between 10 and 30 to enable quick adaptation to changes in node visit rates. The replication rate λ_r could be close to λ_f when the lower node is part of a single preferred path, and close to 0 otherwise. However, to enable fast recovery in the simulations in Section 5, ants are transmitted back-to-back as soon as they are replicated, i.e. $\lambda_r = \mathit{link\,bandwidth}/\mathit{ant\,packet\,size}$.

Replication Number. When the replication is done on arrival of a forwarding ant, a fixed number of replica, n_r, are sent. The number of replica reflects the importance of the lower node by

$$n_{r,t} = n_0(1 + (\nu - 1)\frac{\lambda_{f,t}}{\lambda}) \tag{4}$$

where n_0 is the minimum number of replica (design parameter), and ν is the number of edges in the lower node. If the lower node is along the preferred path, then $\lambda_{f,t} \approx \lambda$ and $n_{r,t} \approx \nu n_0$, otherwise $n_{r,t} \approx n_0$. In Section 5 the generation rate λ at source node s is set to 50 ants/s, i.e. equal to *low rate*, and n_0 is set to 10 ants which results in replication burst sizes for 10 to approximate 200, i.e. maximum burst rates of at least the size of *high rate*.

5 Case Studies of a National-Wide Internet Topology

To compare the different rate adaptation strategies proposed in Section 4, a series of simulation experiments have been conducted applying the backbone topology of a Norwegian Internet provider. The topology, illustrated in Figure 2, consists of a core network with 10 core routers in a sparsely meshed topology, ring based edge networks with a total of 46 edge routers, and a dual homing access network with 160 access routers. The relative transmission capacities are 1, 1/4 and 1/16 for core, edge and access links, respectively.

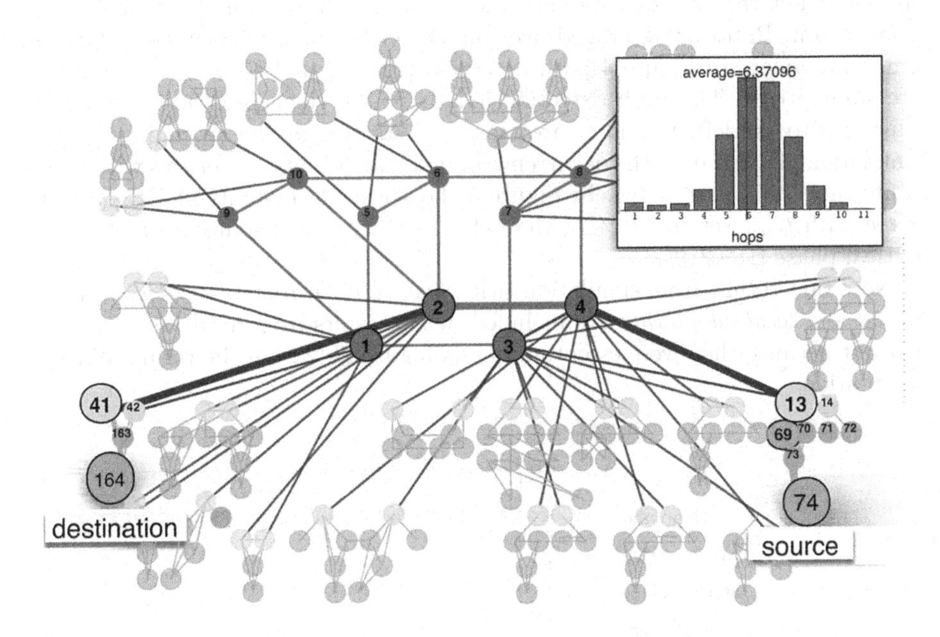

Fig. 2. The simulated backbone network. A virtual connection $VC_{[74,164]}$ is established and monitored. The preferred path in stable phase is $\omega^{(1)}_{[74,164]} = \{74, 69, 13, 4, 2, 41, 164\}$. To the right, the distribution of the number of hops of all paths between access nodes is given. The average number of hops is 6.37. The preferred path $\omega^{(1)}_{[74,164]}$ of $VC_{[74,164]}$ consists of 6 hops.

The management of a single virtual connection, $VC_{[74,164]}$, and its corresponding paths are studied in details. As indicated in Figure 2 $VC_{[74,164]}$ is a typical connection in the topology applied. The paths of $VC_{[74,164]}$ are exposed to link state events like failures and restoration. Data traffic is routed according to the (multi)paths provided by the management algorithm. The performance of the data traffic stream between node 74 and 164 is monitored while different rate adaptation strategies are applied. The simulation series runs over a period [0,300] simulated real-time seconds divided in a number of phases where different link state events occur affecting links along the preferred paths of $VC_{[74,164]}$. In phase [0,10] the network is initialised followed by a stable phase [10,90] where all links are operational. In phase [90,120] $l_{[2,4]}$has failed and in phase [120,150] $l_{[1,42]}$ has also failed. $l_{[2,4]}$ is restored in phase [150,180]. The most critical phase is [180,210] where $l_{[2,4]}$, $l_{[1,42]}$ and $l_{[1,3]}$ have failed simultaneously. Finally, in phase [210,240] $l_{[2,4]}$ is restored again.

All results presented are based on 20 simulation replications. Figure 3 shows the rate of ants in ants per second for the different rate adaptation strategies. To the upper right the rate of ants sent from the source are shown, and to the lower left the rate of ants backtracking and leading to updates of routing information. Rates are averaged over one-second windows and 95%-confidence limits are given. In Figure 4 path detection times t_r and convergence times t_o are summarized. The path detection axis is to the right and the convergence time axis to the left. t_r and t_o are measured starting at the beginning of each simulation phase (i.e. at the occurrence of link state events). Averages with 95%-confidence limits are shown. In Figure 5 the service availability of $VC_{[74,164]}$ is given with t_{max} set to 110% of the delay of the optimal solution in the most critical phase [180,210].

Note that results from simulations with *high rate*, *medium rate*, *implicit adaptation* and *local adaptation* are included, while results from fixed *low rate* were left out because they were essentially indistinguishable from the results of local adaptation.

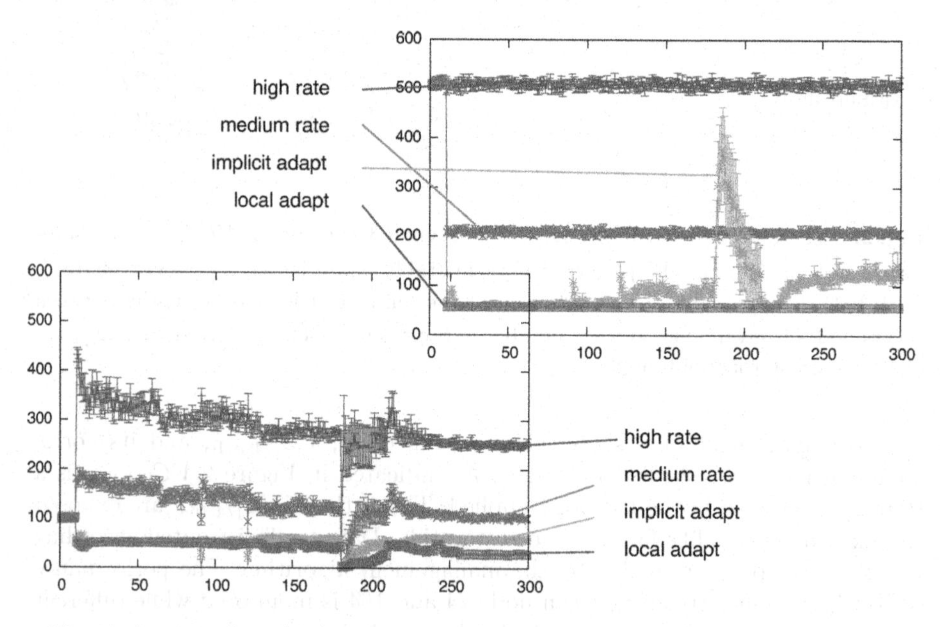

Fig. 3. Ant rate (ants/sec) sent forward (upper right) and backwards (lower left)

From the results in Figure 4 it can be observed that the path detection times are significantly less than the convergence times, averaged over all less than 0.1%. This implies that service interruptions are very short. It is hardly noticeable on the interval service (un)availability estimates in Figure 5.

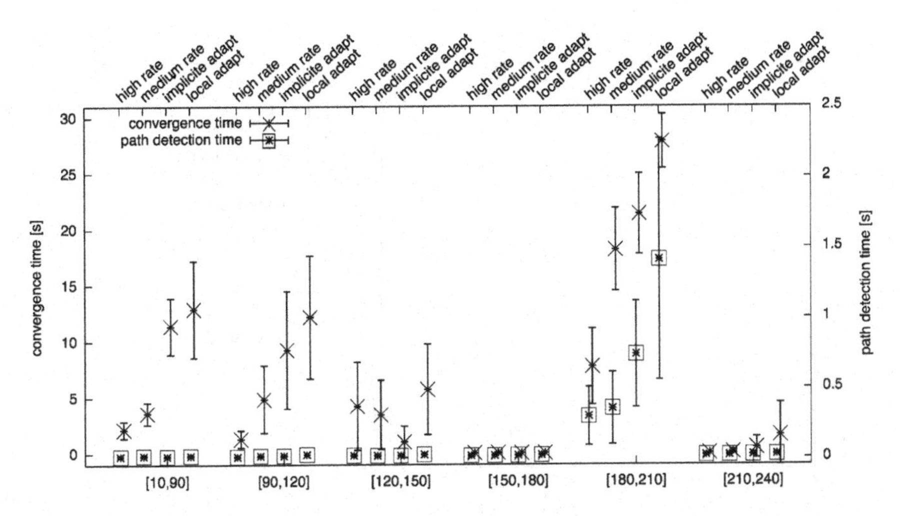

Fig. 4. Path detection and convergence times for all schemes

The implicit adaptation and local adaptation strategies generate significantly fewer ants than the high and medium fixed rate strategies (Figure 3). However, in the critical phase [180,210] a significant number of ants are required to find a set of new good paths. In this phase the source rate is adjusted to exceed the medium fixed rate.

It was expected that the higher (overall) rate the better the performance, and this hold for most of the phases. The only exception is in phase [120,150] where $l_{[1,42]}$ has failed. In Figure 4, it can be seen that the implicit adaptation strategy is very good and outperforms the fixed rate strategies. A possible explanation is that the fixed rate strategies in the phase preceding [120,150] have strong convergence to solutions that include $l_{[1,42]}$. Strong convergence in one phase would increase the convergence times in the following phase if the preferred solution is affected by the link state event between the phases. A strategy that has a high ant generation rate will have strong convergence in terms of strong pheromone values. This further motivates the use of rate adaptive strategies, and further testing of these are required.

Compared to the fixed medium rate strategy, the implicit adaptation strategy generates on average of 50% less ants, and shows very good performance compared to the medium rate. As can be seen in Figure 5, the overall service availability and restoration times are almost identical, while the convergence times of the implicit adaptation strategy is 18% better. However, in the critical phase [180,210], the medium fixed rate has 5% better service availability and 17% shorter convergence times.

The current implementation of the local adaptation strategy generates too small and short-lived bursts to make an impact on the path detection and convergence times. The local adaptation strategy has overhead and performance equal to the fixed low rate strategy (the results are not included in this paper).

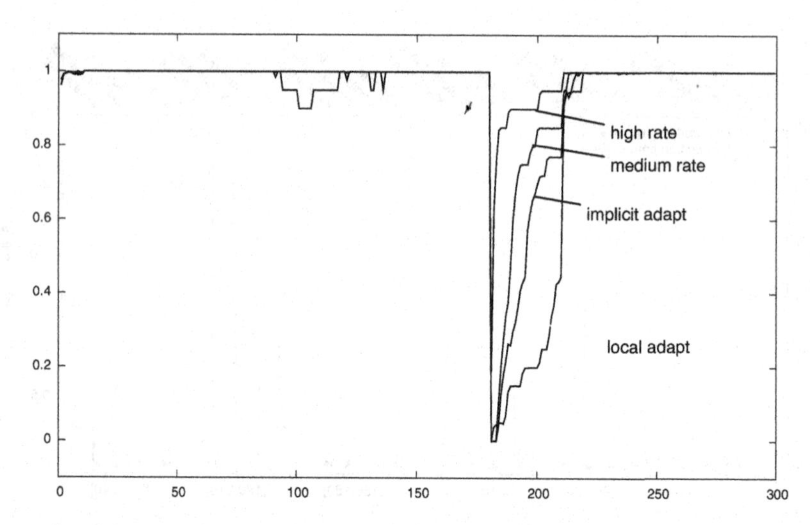

Fig. 5. Availability of $VC_{[74,164]}$ where the service level requirement $t_{\max}$ is set to 10% above the delay of the optimal path in period [180,210] with three link failures

6 Concluding Remarks

CE-ants is a distributed, robust and adaptive swarm intelligence system for dealing with path management in communication networks, and is based on Cross Entropy for stochastic optimisation. In this paper different strategies are studied to control and reduce the overhead in terms of number of management packets (denoted ants) generated in this CE-ants system. Overhead reduction is achieved by controlling the ant generation rate and the number of *elite* ants that update routing information in the network. A series of link state events are simulated and the performance of four strategies compared. Results indicate that an implicit adjustment in the source is a promising approach with respect to path detection and convergence times and the number of ants required. A 50% reduction of overhead may be achieved without loss of performance.

More case studies should be conducted with different transient periods and combinations of link state events, with multiple VCs, background traffic, changing topology, and comparisons with non-swarm based systems. It is also of interest to study how the local adaptations strategy can be made more responsive in order to reduce the path detection and convergence times even further. A combination of local adaptation of the rate and implicit adaptation of the source rate is expected to be a robust strategy. It would also be interesting to study scenarios where differentiated service and dependability requirements lead to differentiated restoration strategies.

Flooding a network under global overload should be avoided, hence strategies where ant information is piggy-backed on data-packets will be considered as well as strategies where a maximum capacity for ant packets is set. Since the CE-ants system tends to be robust to loss of management information, dropping ant packets on congestion may also be an option.

References

1. Ball, M.O.: Handbooks in Operation Research and Management Science, Network Models. Volume 7. North Holland (1995)
2. Pioro, M., Medhi, D.: Routing, Flow and Capacity Design in Communication and Computer Networks. Morgan Kaufmann Publishers (2004)
3. Kirkpatrick, S., Gelatt, C.D., Vecchi, M.P.: Optimization by Simulated Annealing. Science 220 (1983) 671–680
4. Glover, F., Laguna, M.: Tabu Search. Kluwer Academic (1997)
5. Goldberg, D.: Genetic Algorithms in Search, Optimization and Machine Learning. Addison Wesley (1998)
6. Rubinstein, R.Y.: The Cross-Entropy Method for Combinatorial and Continuous Optimization. Methodology and Computing in Applied Probability (1999) 127–190
7. Schoonderwoerd, R., Holland, O., Bruten, J., Rothkrantz, L.: Ant-based Load Balancing in Telecommunications Networks. Adaptive Behavior **5**(2) (1997) 169–207
8. Bonabeau, E., Dorigo, M., Theraulaz, G.: Swarm Intelligence: From Natural to Artifical Systems. Oxford University Press (1999)
9. Di Caro, G., Dorigo, M.: AntNet: Distributed Stigmergetic Control for Communications Networks. Journal of Artificial Intelligence Research **9** (1998) 317–365
10. Wittner, O., Helvik, B.E.: Distributed soft policy enforcement by swarm intelligence; application to load sharing and protection. Annals of Telecommunications **59**(1-2) (2004) 10–24
11. Wittner, O.: Emergent Behavior Based Implements for Distributed Network Management. PhD thesis, Norwegian University of Science and Technology, NTNU, Department of Telematics (2003)
12. Heegaard, P.E., Wittner, O., Helvik, B.E.: Self-managed virtual path management in dynamic networks. In Babaoglu, O., Jelasity, M., Montresor, A., van Moorsel, A., van Steen, M., eds.: Self-* Properties in Complex Information Systems. Lecture Notes in Computer Science, LNCS 3460, Springer-Verlag (2005) 417–432
13. Helvik, B.E., Wittner, O.: Using the Cross Entropy Method to Guide/Govern Mobile Agent's Path Finding in Networks. In: Proceedings of 3rd International Workshop on Mobile Agents for Telecommunication Applications, Springer Verlag (2001)
14. Heegaard, P.E., Wittner, O., Nicola, V.F., Helvik, B.E.: Distributed asynchronous algorithm for cross-entropy-based combinatorial optimization. In: Rare Event Simulation and Combinatorial Optimization (RESIM/COP 2004), Budapest, Hungary (2004)

Solving a Bi-objective Flowshop Scheduling Problem by Pareto-Ant Colony Optimization

Joseph M. Pasia[1,2], Richard F. Hartl[2], and Karl F. Doerner[2]

[1] Department of Mathematics, University of the Philippines-Diliman
Quezon City, Philippines
jmpasia@up.edu.ph

[2] Department of Management Science, University of Vienna, Vienna, Austria
{richard.hartl, karl.doerner}@univie.ac.at

Abstract. In this paper we investigate the performance of pareto ant colony optimization (PACO) in solving a bi-objective permutation flowshop problem. We hybridize this technique by incorporating path relinking (PR) in four different ways. Several test instances are used to test the effectiveness of the different approaches. Computational results show that hybridizing PACO with PR improves the performance of PACO. The hybrid algorithms also show competitive results compared to other state of the art metaheuristics.

1 Introduction

In the permutation flowshop scheduling n jobs have to be sequentially processed on m machines. Each job i has release date r_i, a due date d_i, and a processing time t_{ij} on machine j. The performance of each possible schedule is measured according to some functions like makespan which is defined as $f_1 = \max_i\{s_{im}+t_{im}\}$ and total tardiness $f_2 = \sum_i^n \max(s_{im}+t_{im}-d_i, 0)$ where s_{im} is the schedule time of job i in the last machine m. In this paper, we try to find job sequences such that two objectives, (i) to minimize makespan and (ii) to minimize total tardiness, are accomplished. We call this problem as bi-objective flowshop scheduling problem (BOFSP). In general, there is no single solution that simultaneously accomplish the objectives of a bi-objective optimization problem. Hence, the Pareto optimal set or sometimes called the set of nondominated (efficient) solutions is considered. We say that a solution x is Pareto optimal or nondominated solution if there exists no other feasible solution y such that $f_k(y) \leq f_k(x)$, for $k = 1, 2$ and $f_k(y) < f_k(x)$ for some k. Otherwise, we say that x is dominated by y and we denote this by $y \succ x$.

BOFSP is an NP-hard problem since makespan minimization has been proven NP-hard for more than two machines [1]. Furthermore, the minimization of total tardiness for one machine has been proven NP-hard as well [2]. Therefore, the use of metaheuristics is appropriate.

The Ant Colony Optimization (ACO) [3,4] metaheuristic is inspired by the ability of real ants to determine the shortest path that leads to their food source.

M. Dorigo et al. (Eds.): ANTS 2006, LNCS 4150, pp. 294–305, 2006.

When applied to a scheduling problem, this constructive metaheuristic selects the next unscheduled job to be appended in the partial schedule based on the cost of adding the job (heuristic information) and the desirability of the job (pheromone). After all the ants have constructed their solutions, all or some of them update the pheromone according to their fitness values. This will allow the succeeding ants to select a better path.

ACO has been used successfully in solving several single objective combinatorial problems like in [5,6,7] and it also has become an alternative approximation method for solving a wide range of multiobjective combinatorial optimization problems. It has been applied in scheduling [8,9], portfolio selection [10,11], activity crashing [12], engineering [13,14,15], travelling salesman [16] and in vehicle routing and transportation [17,18] problems.

To solve the BOFSP, two implementations of ant colony optimization are proposed in this paper. The first method is a version of Pareto Ant Colony Optimization (PACO) proposed by Doerner et al. in [10]. The second implementation is also PACO but a Path Relinking (PR) approach is incorporated. Furthermore, four ways of implementing PR are examined.

This paper is organized as follows. Section 2 describes the implementation of PACO in BOFSP. Section 3 presents the numerical results of the study and Sect. 4 provides a short conclusion and future direction of the study.

2 Solution Procedures

2.1 Pareto Ant Colony Optimization

Solution Construction. In PACO [10], each objective function f_k has its own pheromone matrix $[\tau_{ij}]^k$ which describes the desirability of scheduling job i at position j [7,19]. Hence we store the pheromone values in a job×position pheromone matrix. Given the set $\mathcal{U}$ of unscheduled jobs, $i \in \mathcal{U}$ is selected for the jth position according to the *pseudo-random-proportional rule* given by

$$i = \begin{cases} \arg\max_{i \in \mathcal{U}} \sum_{k=1}^{2} w_i^k \left[\sum_{l=1}^{j} \tau_{il}^k\right]^\alpha \cdot \left[\eta_{ij}^k\right]^\beta & \text{if } q \leq q_0 \\ \hat{i} & \text{otherwise} \end{cases}, \tag{1}$$

where the random variable $\hat{i}$ has a probability distribution given by

$$\Pr(i) = \begin{cases} \dfrac{\sum_{k=1}^{2} w_i^k \left[\sum_{l=1}^{j} \tau_{il}^k\right]^\alpha \left[\eta_{ij}^k\right]^\beta}{\sum_{u \in \mathcal{U}} \left(\sum_{k=1}^{2} w_u^k \left[\sum_{l=1}^{j} \tau_{ul}^k\right]^\alpha \left[\eta_{uj}^k\right]^\beta\right)} & i \in \mathcal{U} \\ 0 & \text{otherwise} \end{cases}. \tag{2}$$

Note that this rule is a straightforward extension for multiple pheromone matrices of the rule used in Ant Colony System [20] . The parameters α and β determine the relative influence of the pheromone and heuristic information, respectively, η_{ij}^k is the heuristic information of objective k, $w_i^k \in [0, 1]$ are weights which are uniformly distributed such that $\sum_{k=1}^{2} w_i^k = 1$. In each iteration, every ant is assigned its own weight vector. The parameter q is a random number uniformly distributed in the interval [0, 1], and $q_0 \in [0, 1]$ is a parameter that defines the intensification and diversification properties of the algorithm. Higher value of q_0 prefers intensification over diversification while lower q_0 prefers otherwise. Observe that the summation rule is applied in evaluating the pheromone values, i.e., a job is selected based on the pheromone values up to j^{th} position. This rule was introduced in [21] and was implemented in [9,7,22].

Heuristic Informations. Each objective function f_k has its own heuristic information η_{ij}^k that measures the "attractiveness" of assigning job i to position j. For the objective of minimizing the total tardiness, the heuristic information is expressed by

$$\eta_{ij} = \frac{1}{\max_{i \in \mathcal{U}} (d_i,\ \text{makespan}_j) - \text{makespan}_{j-1}} \tag{3}$$

where makespan_j is the makespan of the partial schedule with j jobs. This heuristic information is an extension of the same heuristic information used in [9].

The heuristic information of makespan minimization makes use of the Liu and Reeve's (LR) heuristic [23]. LR heuristic uses an index function to construct a schedule which minimizes the total flowtime. It consists of two parts, one for measuring the influence of the machine idle time and another for measuring the effects on the completion times of later jobs.

Let S be the sequence of a partial schedule having s jobs and $C(i, h)$ be the completion time on machine h of a job $i \in \mathcal{U}$ if it is scheduled to be the $(s + 1)^{th}$ job. Then the expression for the weighted machine idle time between the processing of job i and the sth job is given by:

$$IT_{is} = \sum_{l=2}^{m} w_{ls} \max\left(C(i, l-1) - C([s], l),\ 0\right) \ , \tag{4}$$

where the weights $w_{ls} = \frac{m}{l+s(m-l)/(n-2)}$ and $[s]$ is the index of the job at the sth position.

For the second component of LR, calculate the average of the processing times of jobs in $\mathcal{U}$ on each machine j and consider these as the processing times t_{pj} of an artificial job p. The values of the processing times are given as follows:

$$t_{pj} = \sum_{q \in \mathcal{U},\, q \neq i} t_{qj}/(n - s - 1) \ . \tag{5}$$

Then the artificial flow time AT_{is} is given by the sum of the completion time of job i and the completion time of the artificial job p which is assumed to be processed after job i:

$$AT_{is} = C(i, m) + C(p, m) \ . \tag{6}$$

Finally, the index function f_{is} for choosing job $i \in \mathcal{U}$ as the $(s + 1)^{th}$ job to be scheduled is the aggregate of the weighted machine idle time and total artificial flow time:

$$f_{is} = (n - s - 2)IT_{is} + AT_{is} \ . \tag{7}$$

The job with the smallest index function value is preferred. Observe that the machine idle time has more weight when selecting the early jobs of the schedule. This is necessary since the accuracy of average processing times in measuring the effects unscheduled jobs is poor due to the fact that there are many unscheduled jobs [23].

Local Pheromone Update. Two types of pheromone updates are applied in PACO. The first update is a local pheromone update which occurs after an ant completes its solution. This update evaporates the pheromone values along the edges visited by the ant in order to allow the succeeding ants to explore other edges. It updates the pheromone matrices according to (8)

$$\tau_{ij}^k = (1-\rho)\cdot\tau_{ij}^k + \rho\cdot\tau_{\min} \qquad \forall k \tag{8}$$

where ρ is the evaporation rate (with $0 \leq \rho \leq 1$) and $\tau_{\min}$ is a small number that serves as the minimum value of the pheromone values.

Local Search. After locally updating the pheromone, each ant undergoes local search. A move operator is applied, i.e., a job is selected and inserted in another position. If the solution created after the move is not dominated by the set of nondominated solutions generated by this local search, this solution becomes the new solution of the ant. The process of move is repeated until a maximum number of iterations has been reached.

The set $\mathcal{L}$ of all nondominated solutions found by the local search updates the external set $\mathcal{G}$ which contains all efficient solutions found so far by the algorithm. The solutions in $\mathcal{L}$ that are not dominated by the solutions in $\mathcal{G}$ are stored in $\mathcal{G}$ and the solutions in $\mathcal{G}$ that are dominated by $\mathcal{L}$ are removed.

Global Pheromone Update. The second update is the global pheromone update. The pheromone values of an objective are updated by the iteration's best and second-best schedules (after local search) with respect to that objective. This global pheromone update rule is defined as follows:

$$\tau_{ij}^k = (1-\rho)\cdot\tau_{ij}^k + \rho\cdot\triangle\tau_{ij}^k \ , \tag{9}$$

where the quantity $\triangle\tau_{ij}^k$ is given by

$$\triangle\tau_{ij}^k = \begin{cases} 15, & (i,j) \text{ is an edge in the best and } 2^{\text{nd}} \text{ best schedules of objective } k \\ 10, & (i,j) \text{ is an edge only in the best schedule of objective } k \\ 5, & (i,j) \text{ is an edge only in the } 2^{\text{nd}} \text{ best schedule of objective } k \ . \end{cases} \tag{10}$$

2.2 PACO with Path Relinking

Path Relinking (PR) is a population based method which is a generalization of an evolutionary approach called scatter search. This method links good solutions and the trajectories that result from the connections comprise the new solutions. Linking the solutions may be performed either in decision or objective space.

In this study, the path relinking approach is incoporated in PACO before the global pheromone update of each iteration. The iteration-best solutions of each objective are linked to some of the solutions in $\mathcal{G}$. Two ways of doing this are examined. The first method links the iteration-best of objective f_1 (f_2) in the direction of the best solution in $\mathcal{G}$ with respect to objective f_2 (f_1). The other method is to connect each of the iteration-best towards a randomly selected solution in $\mathcal{G}$. Figure 1(a) illustrates the link in the first method while Fig. 1(b) represents the link in the second method. Note that in Fig. 1(b) the target solutions can be any of the solutions in $\mathcal{G}$.

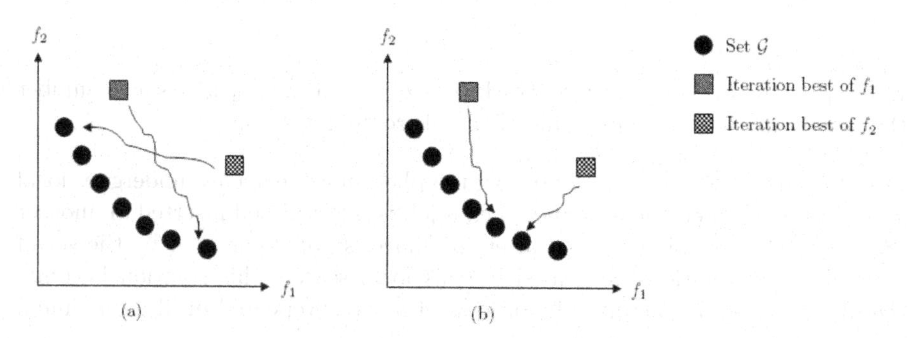

Fig. 1. Direction of the links in PR

There are two strategies of linking two solutions. The first strategy applies the concept of *longest common substring* (LCS). This strategy is implemented according to the implementation of such strategy in [24]. From the starting solution x, a move neighborhood operator generates paths that connect this solution to the target solution y. Only the neighbor solutions that increase LCS with the target solution y are considered.

The other strategy uses *Hamming distance* (HD) to guide the construction of the paths that connect solution x to solution y. Hence the distance between solution x and solution y is the number of positions in which the two solutions have different assigned jobs. The solutions along the path from solution x to solution y have decreasing values of HD to the target solution y.

In both strategies, there are several possible paths that connect one solution to another. Instead of exploring all these paths, only the paths that generate solutions which are nondominated by all the neighborhood are exploited. However, there are still plenty of these paths. Thus a random aggregation of the objectives is applied to select a single path[1].

The set $\mathcal{P}$ of all nondominated solutions generated in PR undergo local search described in Sect. 2.1. Furthermore, the set $\mathcal{P}$ and the nondominated solutions evaluated in the local search update the set $\mathcal{G}$.

[1] Basseur et al. applied the same mechanism in [24].

In summary, we tested four different ways of incorporating PR in PACO. The first implementation combines the concept of LCS and the random selection of the target solutions (PLCSR) and the second implementation combines LCS and the non-random selection of the target solutions (PLCSN). The third method uses HD and the random selection (PHDR) and the last method applies HD and the non-random selection (PHDN).

3 Numerical Results

The proposed algorithms are tested using eight instances which we divided into two groups, small and medium-size test instances. The small-size instances have 20 jobs while the medium-size instances have 50 jobs. The number of machines are either 5, 10 or 20. These instances are taken from some Taillard benchmarks [25] which were extended into bi-objective case in [26]. The performances of algorithms are compared to three other metaheuristics which have been applied to the same instances. These algorithms are the hybridization of a Genetic Algorithm by a Memetic Algorithm (GA+MA) introduced in [27], an Adaptive Genetic/Memetic Algorithm (AGMA) proposed in [28], and the Pareto Iterated Local Search (PILS) described in [29,30]. One common characteristic of these three algorithms is the presence of several neighborhood structures. GA+MA applied two mutation operators, AGMA had four mutation operators while PILS used three neighborhood structures.

We performed all our methods on a personal computer with a Pentium 3.2 Ghz processor, 1 Gb RAM, and the operation system Windows XP; the algorithms were coded in C++ and compiled using Borland 5.5.1 compiler.

3.1 Evaluation Metrics

The traditional approach of assessing the performances of different algorithms for solving bi-objective problems is by graphical visualization of their Pareto fronts. However this approach is not effective when the graphs overlap each other. Hence the use of quantitative measures has become the standard approach. Following [31], this study considered two quantitative measures namely, the *distance from reference set* and the *coverage metric.*

Metric 1. Czyzak and Jaszkiewicz [32] proposed the distance from reference set (DRS). Suppose $\mathcal{R}$ is the reference set and $\mathcal{A}$ is an approximation set. Then DRS takes the average distance from each reference point $x \in \mathcal{R}$ to its closest neighbor in $\mathcal{A}$. Hence its formula is given by:

$$DRS = \frac{1}{|R|} \sum_{y \in \mathcal{R}} \min_{x \in \mathcal{A}} \{\mathbf{c}(x, y)\} \tag{11}$$

where $\mathbf{c}(.,.)$ is any distance in the objective space. In this study, we considered the Tschebycheff metric given by:

$$\mathbf{c}(x, y) = \max_{k} \{0, \frac{1}{\triangle_k} \cdot |f_k(x) - f_k(y)|\} \ . \tag{12}$$

where Δ_k is the range of objective f_k in the reference set. In the evaluation, the reference set is defined as the set of all nondominated solutions extracted from all the Pareto solutions found by all the methods.

Metric 2. The coverage metric $\mathcal{C}$ proposed by Zitzler and Thiele in [33] assigns numbers to pairs of approximation sets, say A and B, by giving the fraction of B that are covered by A. The formula for this metric is given by:

$$\mathcal{C}(A,B) = \frac{|\{b \in B : \exists a \in A \text{ such that } a \succeq b\}|}{|B|} \quad . \tag{13}$$

where $a \succeq b$ if $a \succ b$ or $a = b$. If $\mathcal{C}(A,B) = 1$ and $\mathcal{C}(B,A) < 1$ then A is said to be better than B and if $\mathcal{C}(A,B) = 1$ and $\mathcal{C}(B,A) = 0$ then A dominates B.

3.2 Analysis

Five runs with different random seeds were performed for each of the different PACO variants and each test instance. For small instances with 5 machines, each run of PACO is terminated after 300 generations have been performed and 400 generations for instances with 10 and 20 machines. For the medium instances, PACO makes 600 generations. In order to have comparable run times, the four PACO variants only perform 150 generations. The settings for the other parameters are: $\alpha = 1,\ \beta = 1,\ \rho = 0.05,\ \tau_0 = 1, \tau_{\text{min}} = 0.00001,$ number of ants $=$ 20, $q_0 = 0.75$.

Tables 1, 3 and 4 show the results of the DRS and $\mathcal{C}$ metric of the algorithms. The test instance $T\_N\_M\_n$ corresponds to Taillard's benchmarks having N jobs and M machines. The number n indicates the test instance number.

In Table 1, one may observe the following: First, incorporating path relinking improves the performance of PACO; on average, the distances of the hybrid algorithms from the reference sets are smaller compared to the distances of PACO. Significant improvements are realized when one uses the concept of HD in path relinking. On the other hand, although PACO may have better DRS values in some test instances than that of PLCSN and PLCSR, the latter methods perform much better in large instances (50 jobs with 10 and 20 machines).

Second, it may seem that PLCSN and PLCSR do not offer much big improvement for PACO but looking at the average computational speed of the runs given in Table 2, it is clear that these algorithms converge more quickly. It should be noted that for larger instances, PLCSN and PLCSR only need about 50% of the average runtime of PACO to generate the much improved results. Perhaps increasing the number of generations of PLCSN and PLCSR for the other instances will greatly outperform PACO in terms of DRS. In addition, it is also worth noting that the results of PHDN, which are superior than that of PACO, were generated at smaller running time than PACO.

Third, one may observe that AGMA is the best algorithm having zero DRS in all instances and this may indicate that the solutions of AGMA coincide with the reference set. Fourth, the simple implementation PACO is superior than GA+MA and PILS in almost all small instances. PACO's DRSs for these

Table 1. Evaluation on metric *DRS*

Instance	PACO	PHDN	PHDR	PLCSN	PLCSR	GA+MA	AGMA	PILS
T_20_5_1	**0.0000**	**0.0000**	**0.0000**	**0.0000**	**0.0000**	0.0956	**0.0000**	**0.0000**
T_20_5_2	**0.0000**	**0.0000**	**0.0000**	**0.0000**	**0.0000**	0.0468	**0.0000**	0.0370
T_20_10_1	0.0061	**0.0017**	0.0065	0.0058	0.0077	0.0166	**0.0000**	0.0053
T_20_10_2	0.0148	**0.0115**	0.0129	0.0158	0.0158	0.0448	**0.0000**	0.0308
T_20_20_1	0.0044	**0.0003**	0.0028	0.0042	0.0032	0.0249	**0.0000**	0.0169
T_50_5_1	0.0861	0.0388	**0.0254**	0.1492	0.1556	0.0987	**0.0000**	0.0290
T_50_10_1	0.1088	0.0437	**0.0386**	0.0623	0.0584	0.0818	**0.0000**	0.0712
T_50_20_1	0.1098	0.0787	**0.0504**	0.0718	0.0732	0.7120	**0.0000**	0.0733
Average	0.0413	0.0218	**0.0171**	0.0387	0.0393	0.0601	**0.0000**	0.0330

instances are also not that far from the reference set as shown by its small DRS values. But for a much better performances and for medium-size instances, one may consider incorporating PR. For instance, PHDR outperforms both GA+MA and PILS in all but one test instances (it has the same DRS value as PILS in T_20_5_1).

Finally, the use of random target solution in path relinking seems more beneficial in terms of DRS value and computational speed. This implementation converges faster than when we select the target solution by following some rigid rule like using the best solution of each objective found so far. The advantage of setting random target solution is more prominent in path relinking that uses HD.

Table 2. Average running time in seconds

Instance	PACO	PHDN	PHDR	PLCSN	PLCSR
T_20_5_1	119.7	107.5	92.4	69.2	66.6
T_20_5_2	114.5	88.0	98.1	66.3	66.2
T_20_10_1	333.3	331.5	275.8	222.7	206.5
T_20_10_2	319.2	331.5	265.9	182.8	183.2
T_20_20_1	671.6	709.2	599.5	425.6	468.4
T_50_5_1	12241.7	7755.2	6248.4	3814.9	3662.0
T_50_10_1	23792.5	31454.2	19566.2	10299.6	9772.7
T_50_20_1	46470.6	60204.9	40910.8	26091.8	21598.3

Tables 3 and 4 summarize the coverage metric between different algorithms and they validate most of the findings of DRS. In Table 3, all of the nondominated points generated by PACO in all medium size instances are dominated by PHDR. PACO is also dominated by PHDN and PLCSN in the T_50_10_1 instance and by PLCSN and PLCSR in the T_50_20_1 instance. The small row-values and high column-values of PACO, especially in the medium size instances, also indicate that using the PACO variants generate better results. Thus, this confirms the advantage of incorporating PR in PACO when solving BOFSP for medium size instances.

In Table 4, only PHDR is compared against the other metaheuristics to simplify the comparison. PHDR is chosen to represent the PACO variants since it performs relatively better than the others. This table shows that PHDR outperforms GA+MA and PILS in most cases. For example in the small instances, PHDR covers at least 92% of PILS but PILS only covers at least 82% of PHDR. For the medium instances, PILS performs better than PHDR in T_50_5_1 but PHDR is much better in T_50_10_1 and T_50_5_1 covering at least 80% of PILS. Unlike GA+MA and PILS, PHDR is not dominated by AGMA in all instances. GA+MA and PILS are dominated by AGMA in instances T_50_10_1 and T_50_5_1 respectively.

Table 3. Evaluation on Metric $\mathcal{C}$

T_20_5_1	PACO	PHDN	PHDR	PLCSN	PLCSR
PACO	-	1.000	1.0000	1.0000	1.0000
PHDN	1.0000	-	1.0000	1.0000	1.0000
PHDR	1.0000	1.0000	-	1.0000	1.0000
PLSCN	1.0000	1.0000	1.0000	-	1.0000
PLSCR	1.0000	1.0000	1.0000	1.0000	-

T_20_5_2	PACO	PHDN	PHDR	PLCSN	PLCSR
PACO	-	1.000	1.0000	1.0000	1.0000
PLCSR	1.0000	-	1.0000	1.0000	1.0000
PLCSN	1.0000	1.0000	-	1.0000	1.0000
PHDR	1.0000	1.0000	1.0000	-	1.0000
PHDN	1.0000	1.0000	1.0000	1.0000	-

T_20_10_1	PACO	PHDN	PHDR	PLCSN	PLCSR
PACO	-	0.9286	0.9231	0.9024	0.9211
PHDN	0.9750	-	0.8974	0.9024	0.9211
PHDR	0.9750	0.9286	-	0.9512	0.9737
PLCSN	0.9750	0.9524	0.9744	-	1.0000
PLCSR	0.9250	0.9048	0.9231	0.9268	-

T_20_10_2	PACO	PHDN	PHDR	PLCSN	PLCSR
PACO	-	0.8966	0.9310	0.9643	0.9643
PHDN	0.9655	-	0.9655	1.0000	1.0000
PHDR	1.0000	0.9655	-	1.0000	1.0000
PLCSN	0.9655	0.9310	0.9310	-	1.0000
PLCSR	0.9655	0.9310	0.9310	1.0000	-

T_20_20_1	PACO	PHDN	PHDR	PLCSN	PLCSR
PACO	-	0.8974	0.9500	0.9737	0.9500
PHDN	1.0000	-	1.0000	1.0000	1.0000
PHDR	1.0000	0.9487	-	1.0000	1.0000
PLCSN	0.9474	0.8718	0.9000	-	0.9000
PLCSR	0.9474	0.8974	0.9500	0.9474	-

T_50_5_1	PACO	PHDN	PHDR	PLCSN	PLCSR
PACO	-	0.1000	0.0000	0.2857	0.1250
PHDN	0.9167	-	0.2000	0.2857	0.5000
PHDR	1.0000	0.9000	-	0.8571	1.0000
PLCSN	0.8333	0.5000	0.3000	-	0.3750
PLCSR	0.9167	0.6000	0.3000	0.5714	-

T_50_10_1	PACO	PHDN	PHDR	PLCSN	PLCSR
PACO	-	0.0000	0.0000	0.0000	0.1282
PHDN	1.0000	-	0.4186	0.8276	0.6154
PHDR	1.0000	0.5625	-	0.8621	0.7179
PLCSN	1.0000	0.2188	0.1395	-	0.2308
PLCSR	0.9189	0.375	0.3023	0.7931	-

T_50_20_1	PACO	PHDN	PHDR	PLCSN	PLCSR
PACO	-	0.0172	0.0000	0.0000	0.0000
PHDN	0.9800	-	0.0612	0.5000	0.2093
PHDR	1.0000	0.9483	-	0.8750	0.9070
PLCSN	1.0000	0.6724	0.1633	-	0.5349
PLCSR	1.0000	0.5517	0.1837	0.6667	-

4 Conclusion and Future Direction

In this study we have shown that PACO is a competitive method for solving bi-objective flowshop scheduling problems; it is easy to implement and yielded good results. In addition, one may greatly improve its performance by incorporating path relinking mechanism in it. The different hybrid approaches generated better results at a shorter computational time and yielded more significant improvements in medium-size instances. Their performances are even better relative to the performances of some recent state of the art metaheuristics. These hybrid approaches are indeed very promising and further studies must be conducted in order to exploit the potential of PACO and PR in solving the BOFSP. Hence in the future, we will investigate other ways of incorporating the path relinking mechanism. We believe that there is a big area for improvement in the hybridization of PACO and PR and the methods we presented here are a good start in developing a more efficient and robust algorithm.

Table 4. Evaluation on Metric $\mathcal{C}$

T_20_5_1	PHDR	GA+MA	AGMA	PILS
PHDR	-	1.0000	1.0000	1.0000
GA+MA	0.7500	-	0.7500	0.7500
AGMA	1.0000	1.0000	-	1.0000
PILS	1.0000	1.0000	1.0000	-

T_20_5_2	PHDR	GA+MA	AGMA	PILS
PHDR	-	0.6667	1.0000	1.0000
GA+MA	1.0000	-	0.6667	0.6000
AGMA	1.0000	1.0000	-	1.0000
PILS	0.8333	0.8571	0.8333	-

T_20_10_1	PHDR	GA+MA	AGMA	PILS
PHDR	-	1.0000	0.9286	0.9268
GA+MA	0.3846	-	0.3571	0.3659
AGMA	1.0000	1.0000	-	1.0000
PILS	0.9048	1.0000	0.9286	-

T_20_10_2	PHDR	GA+MA	AGMA	PILS
PHDR	-	1.0000	0.9355	1.0000
GA+MA	0.3448	-	0.3226	1.0000
AGMA	1.0000	1.0000	-	1.0000
PILS	0.8966	0.8387	0.8387	-

T_20_20_1	PHDR	GA+MA	AGMA	PILS
PHDR	-	1.0000	0.9250	0.9714
GA+MA	0.5000	-	0.5000	0.6000
AGMA	1.0000	1.0000	-	1.0000
PILS	0.8250	0.9000	0.7750	-

T_50_5_1	PHDR	GA+MA	AGMA	PILS
PHDR	-	1.0000	0.2000	0.3333
GA+MA	0.0000	-	0.0000	0.0000
AGMA	1.0000	1.0000	-	1.0000
PILS	0.8000	1.0000	0.5000	-

T_50_10_1	PHDR	GA+MA	AGMA	PILS
PHDR	-	0.3636	0.0333	0.9091
GA+MA	0.2790	-	0.0833	0.6364
AGMA	1.0000	1.0000	-	1.0000
PILS	0.0000	0.0000	0.0000	-

T_50_20_1	PHDR	GA+MA	AGMA	PILS
PHDR	-	0.8182	0.0111	0.8095
GA+MA	0.1429	-	0.0000	0.3968
AGMA	1.0000	1.0000	-	1.0000
PILS	0.2245	0.5682	0.0000	-

Acknowledgements. This research was supported in part by the Austrian Exchange Service.

References

1. Lenstra, J., Kan, A., Brucker, P.: Complexity of machine scheduling problems. Annals of Discrete Mathematics **1** (1977) 343–362
2. Du, J., Leung, J.: Minimizing total tardiness on one machine is np-hard. Mathematics of operations research **15** (1990) 483–495
3. Dorigo, M., Maniezzo, V., Colorni, A.: Ant System: Optimization by a colony of cooperating agents. IEEE Transactions on Systems, Man, and Cybernetics - Part B **26**(1) (1996) 29–41
4. Dorigo, M., Di Caro, G.: The ant colony optimization meta-heuristic. In Corne, D., Dorigo, M., Glover, F., eds.: New Ideas in Optimization. McGraw-Hill, New York (1999) 11–32
5. Blum, C.: Beam-ACO - hybridizing ant colony optimization with beam search: an application to open shop scheduling. Computers & OR **32** (2005) 1565–1591
6. Doerner, K., Gronalt, M., Hartl, R.F., Reimann, M., Strauss, C., Stummer, M.: Savingsants for the vehicle routing problem. In Cagnoni, S., et al., eds.: Proceedings of Applications of Evolutionary Computing : EvoWorkshops 2002. Volume 2279 of Lecture Notes in Computer Science. (2002) 11–20
7. Merkle, D., Middendorf, M., Schmeck, H.: Ant colony optimization for resource-constrained project scheduling. IEEE Transactions on Evolutionary Computation **6** (2002) 333–346
8. Gravel, M., Price, W., Gagné, C.: Scheduling continuous casting of aluminum using a multiple objective ant colony optimization metaheuristic. European Journal of Operational Research **143** (2002) 218–229

9. Guntsch, M., Middendorf, M.: Solving multi-criteria optimization problems with population-based aco. In Fonseca, C., Fleming, P., Zitzler., E., Deb, K., Thiele., L., eds.: Proceedings of the 2nd International Conference on Evolutionary Multi-Criterion Optimization. Volume 2632 of Lecture Notes in Computer Science. (2003) 464–478
10. Doerner, K., Gutjahr, W., Hartl, R., Strauss, C., Stummer, C.: Pareto ant colony optimization: A metaheuristic approach to multiobjective portfolio selection. Annals of Operations Research **131** (2004) 79–99
11. Doerner, K., Gutjahr, W.J., Hartl, R.F., Strauss, C., Stummer, C.: Pareto ant colony optimization in multiobjective project portfolio selection with ILP preprocessing. European Journal of Operational Research **171** (2006) 830–841
12. Doerner, K., Gutjahr, W.J., Hartl, R.F., Strauss, C., Stummer, C.: (Nature-inspired metaheuristics in multiobjective activity crashing) To appear: Omega.
13. Mariano, C., Morales, E.: A multiple objective ant-q algorithm for the design of water distribution irrigation networks. Technical report, Instituto Mexicano de Tecnología del Agua (1999)
14. Shelokar, P., Jayaraman, V., Kulkarni, B.: Ant algorithm for single and multi-objective reliability optimization problems. Quality and Reliability Engineering International **18** (2002) 497–514
15. Shelokar, P., Jayaraman, V., Kulkarni, B.: Multiobjective optimization of reactor-regenerator system using ant algorithm. Petroleum Science and Technology **21** (2003) 1167–1184
16. García-Martínez, C., Cordón, O., Herrera, F.: An empirical analysis of multiple objective ant colony optimization algorithms for the bi-criteria tsp. In Dorigo, M., et al., eds.: Proceedings of ANTS 2004. Volume 3172 of Lecture Notes in Computer Science. (2004)
17. Gambardella, L., Taillard, E., Agazzi, G.: MACS-VRPTW: A multiple ant colony system for vehicle routing problems with time windows. In Corne, D., Dorigo, M., Glover, F., eds.: New Ideas in Optimization. McGraw-Hill, New York (1999) 63–76
18. Iredi, S., Merkle., D., Middendorf, M.: Bi-criterion optimization with multicolony ant algorithms. In Zitzler, E., et al., eds.: Proceedings of First International Conference on Evolutionary Multi-criterion Optimization. Volume 1281 of Lecture Notes in Computer Science. (2001) 359–372
19. Stützle, T.: An ant approach to the flow shop problem. In: Proceedings of EU-FIT98, Aachen (1998) 1560–1564
20. Dorigo, M., Gambardella, L.: Ant colonies for the traveling salesman problem. Biosystems **43**(2) (1997) 73–81
21. Merkle, D., Middendorf, M.: An ant algorithm with a new pheromone evaluation rule for total tardiness problems. In: Proceedings of the EvoWorkshops 2000. Volume 1803 of Lecture Notes in Computer Science. (2000) 287–296
22. Rajendran, C., Ziegler, H.: Ant-colony algorithms for permutation flowshop scheduling to minimize makespan/total flowtime of jobs. European Journal of Operational Research **155** (2004) 426–438
23. Liu, J., Reeves, C.: Constructive and composite heuristic solutions to the p$\|\sum C_i$ scheduling problem. European Journal of Operational Research **132** (2001) 439–452
24. Basseur, M., Seynhaeve, F., Talbi, E.: Path relinking in pareto multi-objective genetic algorithms. In Coello, C., Aguirre, A., Zitzler, E., eds.: Evolutionary Multi-criterion Optimization, EMO'2005. Volume 3410 of Lecture Notes in Computer Science. (2005) 120–134

25. Taillard, E.: Benchmarks for basic scheduling problems. European Journal of Operational Research **64** (1993) 278–285
26. Talbi, E., Rahoual, M., Mabed, M., Dhaenens, C.: A hybrid evolutionary approach for multicriteria optimization problems: Application to the flowshop. In Zitzler, E., et al., eds.: Evolutionary Multi-criterion Optimization. Volume 1993 of Lecture Notes in Computer Science. (2001) 416–428
27. Basseur, M., Seynhaeve, F., Talbi, E.: Design on multi-objective evolutionary algorithms to flow-shop scheduling problem. In: Congress on Evolutionary Computation. Volume 2., Piscataway, IEEE Service Center (2002)
28. Basseur, M., Seynhaeve, F., Talbi, E.: Adaptive mechanisms for multi-objective evolutionary algorithms. In: IMACS multiconference, Computational Engineering in Systems Applications (CESA'03), Piscataway, IEEE Service Center (2003) S3-R-00-222.
29. Geiger, M.: MOOPPS - An optimization system for multiobjective production scheduling. In: The Sixth Metaheuristic International Conference (MIC'05), Vienna, Austria (2005)
30. Geiger, M.J.: "Personal Communication".
31. Zitzler, E., Thiele, L., Laumanns, M., Fonseca, C., Fonseca, V.: Performance assessment of multiobjective optimizers: An analysis and review. IEEE Trans. Evolutionary Computation **7** (2003) 117–132
32. Czyzack, P., Jaszkiewicz, A.: Pareto simulated annealing - a metaheuristic technique for multiple-objective combinatorial optimization. Journal of Multi-Criteria Decision Analysis **7** (1998) 34–47
33. Zitzler, E., Thiele, L.: Multiobjective optimization using evolutionary algorithms - a comparative case study. In Eiben, et al., eds.: Fifth International Conference on Parallel Problem Solving from Nature. Volume PPSN-V. (1998) 292–301

Traffic Patterns and Flow Characteristics in an Ant Trail Model

Alexander John[1], Andreas Schadschneider[1], Debashish Chowdhury[2], and Katsuhiro Nishinari[3]

[1] Institut für Theoretische Physik, Universität zu Köln, Germany
aj@thp.uni-koeln.de, as@thp.uni-koeln.de
[2] Department of Physics, Indian Institute of Technology, Kanpur, India
debch@iitk.ac.in
[3] Department of Aeronautics and Astronautics, Faculty of Engineering
University of Tokyo, Japan
tknishi@mail.ecc.u-tokyo.ac.jp

Abstract. We have constructed a minimal cellular automaton model for simulating traffic on preexisting ant trails. Uni- as well as bidirectional trails are investigated and characteristic patterns in the distribution of workers on the trail are identified. Some of these patterns have already been observed empirically. They give rise to unusual flow characteristics which are also discussed. The question of possible functions of the observed traffic patterns and the resulting flow characteristics will be treated for simplified biological scenarios.

1 Introduction

The use of cellular automaton models is well established in various fields. Applications reach from purely physical systems [1], like models of surface growth to multi-agent systems like vehicular traffic [2,3,4]. The Nagel-Schreckenberg model [2] and its extensions [3] have proved to be very useful tools for improving flow e.g. by reducing the frequency of jams in vehicular traffic. Similar results also exist in pedestrians dynamics [5,6].

In biology various traffic-like problems are known. On a microscopic scale, namely in a living cell [7,8], the movement of motor proteins can be found. On a larger scale the behaviour of social insects has attracted a lot of interest [9,10,11,12]. Beside pure behavioural biology, colonies of social insects can be seen as a multi-agent system, facing and solving various kinds of problems [11,12,13]. For that reason the social insect metaphor has already been employed with great success in computer sciences [14,15]. Recently concepts like chemotaxis [16], known from ants have been adopted for incorporating the mutual interaction in models of pedestrians dynamics [6]. Implementing these interactions by a 'virtual chemotaxis' mechanism allows to reproduce the collective phenomena observed empirically in the dynamics of large crowds. But also the investigation of ants themselves employing computer simulations is of interest. This is done

M. Dorigo et al. (Eds.): ANTS 2006, LNCS 4150, pp. 306–315, 2006.

for various reasons. One is the rich physics displayed even by simple cellular automaton models [17,18]. From a biologists' point of view, some mechanisms like the very complex chemical communication system are difficult to access experimentally [16,19]. Although the formation of trails can be reproduced quite well, the organisation of traffic flow itself seems to be understood [13,20] less clearly.

Taking account of these aspects we have developed a minimal cellular automaton model for traffic flow on preexisting uni- and bidirectional ant trails [21,22]. We will focus here on aspects like possible advantages of certain spatio-temporal organisations of workers on the trail and the resulting flow characteristics. Complementary aspects like non-equilibrium physics or application to pedestrian dynamics are treated in [23,24,25].

2 The Model

We employ a one-dimensional cellular automaton model for bidirectional trails which also includes the unidirectional case. In absence of counterflowing ants in one direction the full bidirectional model reduces to the one for unidirectional trails. So we first start constructing the simpler unidirectional model which will finally be extended to the bidirectional case.

One important way of characterising a traffic-like system is the fundamental diagram. As important properties like mutual blocking depend on the number of agents in the system, average velocity V and flow F are plotted versus density ϱ. Although both are equivalent due to the hydrodynamic relation $F = \varrho V$ the uni- as well as the bidirectional model will exhibit features, which can be discused most easily using both descriptions. In addition we will make use of space-time plots, showing the spatio-temporal distribution of ants on the trail.

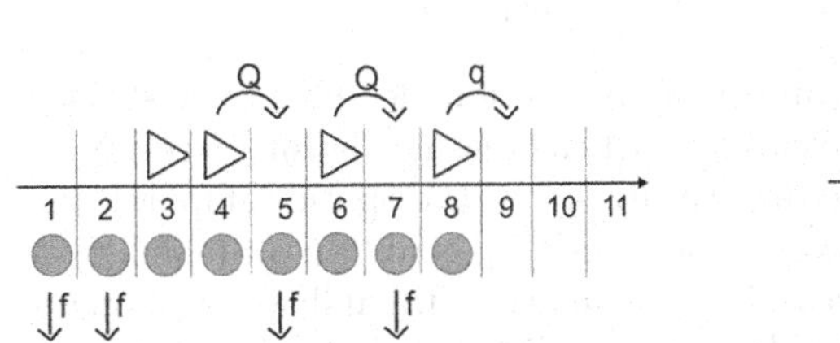

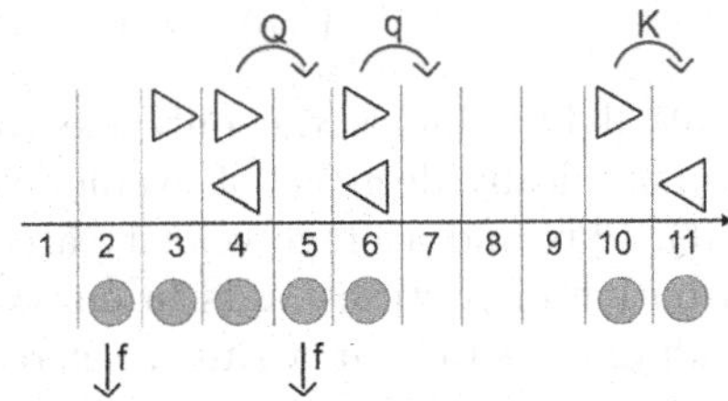

Fig. 1. Illustration of the definitions for the uni- and bidirectional model: ants moving to the right (▷), ants moving to the left (◁), pheromone marks (•). Although a natural trail is two dimensional, movement itself takes places in one dimension. With respect to reality the trail is mapped onto a one-dimensional lattice with L sites. The movement of ants is incorporated by the stochastic hopping of N particles along the lattice. Java applets can be found at *www.thp.uni-koeln.de/ant-traffic.*

2.1 Unidirectional Case

The model for the unidirectional case [21] can be seen as an extension of the so-called totally asymmetric simple exclusion process (TASEP) [26,3] a well studied stochastic model for non-equilibrium physics. So we start constructing the unidirectional model by defining the TASEP. Adding certain extensions will finally lead to the unidirectional ant trail model. But the TASEP will still be recovered for an appropriate choice of parameters.

TASEP Regime. The TASEP is a model well-known in computational and non-equilibrium physics [26,3]. Particles hop in one fixed direction from one lattice-site i to the next one $i+1$ (see Fig. 1 left). Each site can be occupied only by one particle (exclusion principle). Also time evolution is discrete ($\Delta t \to 0$ for $L \to \infty$) . In each update step, one site is chosen at random.

If site i is occupied and $i+1$ is empty, hopping takes places with probability q. If site $i+1$ is blocked, i.e. occupied by another particle, nothing happens (exclusion principle) and another site is chosen randomly. A particle leaving site L will hop to site 1. So effectively hopping takes places on a ring. More complicated treatments of the lattice boundaries are possible and can have a crucial impact on the physics of the system. But the main effects in our models are quite stable against a change of boundary conditions [18]. Also the definition of the update procedure itself can have a strong influence. For modelling vehicular traffic a so called time-parallel update is widely used which performs a synchronous update of all lattice sites [3]. As only the occupation of the lattice at the moment of updating is used, a synchronous update incorporates time latencies of the drivers. But in case of ants the perception range is limited to their immediate environment. So we make use of the random-sequential update, described above. But unlike for some models of vehicular traffic [2,3] the generic effects of the model do not depend on the choice of a particular update procedure.

In case of the TASEP with random sequential dynamics, flow F and average velocity V as a function of density ϱ are given by:

$$V(\varrho) = q(1-\varrho), \qquad F(\varrho) = q\varrho(1-\varrho). \tag{1}$$

Mutual blocking is the only mechanism of interaction, leading to a strictly monotonically decrease of average velocity with increasing density (see Fig. 2 left). Flux and average velocity are directly linked to the spatio-temporal distribution of particles, ants in our case. The density ϱ gives the probability of finding a particle at a site i, whereas $1-\varrho$ gives the probability of finding a particular site being unoccupied. Neglecting correlations, the average velocity is just the product of the hopping probability and the probability of finding site $i+1$ being unoccupied. The corresponding distribution of particles is obviously homogenous on time-average (see Fig. 3 left) as particles are distributed randomly at any instance of time.

Cluster Regime. The model for unidirectional ant-trails [21] is constructed as an extension of the TASEP. This is done in order to incorporate the attractive impact of pheromone marks which are one of the most important forms of communication in ant colonies [16,19]. Now each lattice site can also be occupied by a pheromone-mark and/or one ant (see Fig. 1 left). If only a pheromone mark but no ant is present at site $i+1$, an ant at site i will hop to site $i+1$ with

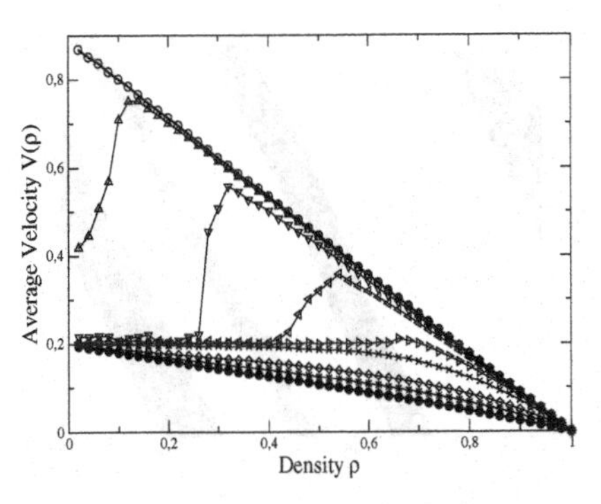

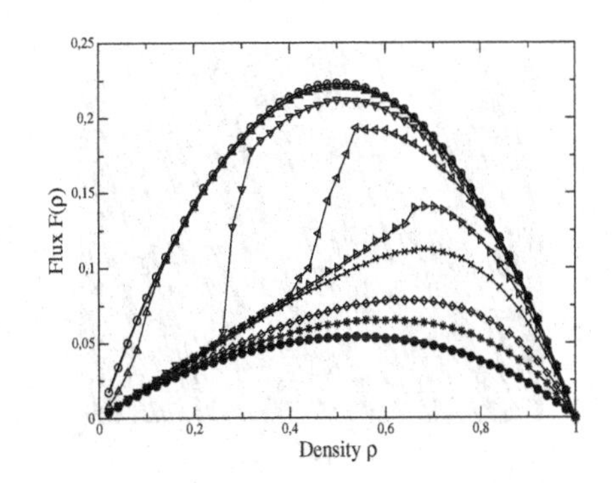

Fig. 2. Fundamental diagrams for the unidirectional model: $Q = 0.9$, $q = 0.2$ and $f = 0(\circ)$, $0.0002(\triangle)$, $0.0008(\triangledown)$, $0.002(\triangleleft)$, $0.008(\triangleright)$, $0.02\ (\times)$, $0.08(\diamond)$, $0.2(*)$, $1(\bullet)$. In case of $f = 1$ and $f = 0$ the average velocity $V(\varrho)$ and flux $F(\varrho)$ are exactly known from the TASEP. At low to intermediate densities, average velocity stays constant for a suitable choice of evaporation probability f.

probability Q. If neither an ant nor a mark are present, hopping will take place with probability $q < Q$. Also the evaporation of pheromones is incorporated. If one site is marked but no ant is present, this mark will be removed (evaporated) with probability f. If an ant is present and the site is chosen for updating, the mark will not evaporate, reflecting the reinforcement of the trail by the ants.

In the case $f = 0$ ($f = 1$) pheromones will never (immediately) evaporate. The TASEP is recovered with hopping probability Q (q) (see Fig. 2). But more interesting properties arise in case of $f \neq 0, 1$. The most surprising feature is the non-monotonic dependence of the average velocity on density for small evaporation rates f (see Fig. 2 left).

At low to intermediate densities, the average velocity stays constant. Beyond a certain threshold value a sharp increase of velocity can be seen. In the regime of high densities, the monotonic decrease known from the TASEP is found. Both regimes are a generic effect of the incorporation of pheromone marks. Each particle is followed by a trace of marks. A succeeding ant perceiving this trace will hop with probability $Q > q$. If the preceding ant sees no pheromone mark, it will hop with probability $q < Q$. Finally the faster ants will catch up with the slower one, forming a moving cluster (see Fig. 3 right). Similar results are known from systems with particle-wise [27] defects where each particle j has its own individual hopping probability q_j. Corresponding to the existence of clusters, the average velocity stays constant $v = q$. At very low evaporation rates or at high densities (small average distance), all ants perceive pheromone marks. So on average their hopping probabilities reaches the same value Q, leading to TASEP-like features. Also the corresponding distribution of ants becomes homogeneous (see Fig. 3 left).

These rules reflect not only similarities, but also the differences between ant trails and vehicular traffic. One is the lack of some kind of velocity memory. In our model, ants reach their walking speed within one single update step. This is similar to the motion of pedestrians where the walking speed is also reached

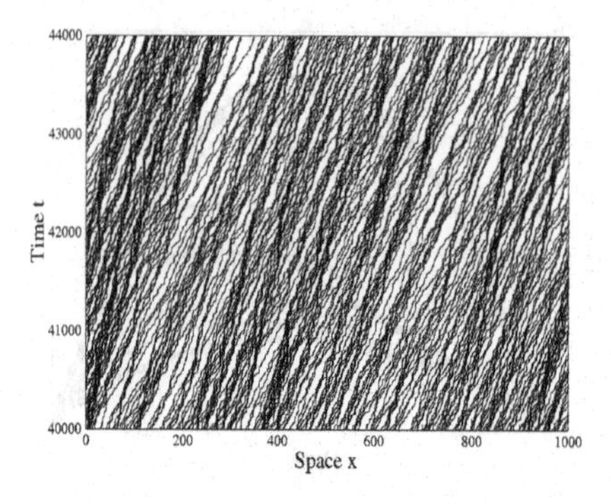

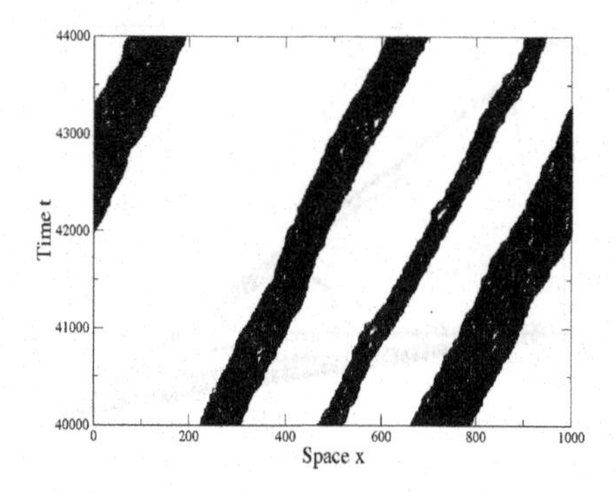

Fig. 3. Space-time plots for the unidirectional model: $Q = 0.9$, $q = 0.2$, $f = 0.002$, $\varrho = 0.2$. On the left the TASEP case ($p = 0.9$) is shown. Beside fluctuations, particles are distributed randomly. The plot on the right shows the stationary state of the cluster case in the unidirectional ant trail model.

within a time less than 1 sec. In vehicular traffic cars cannot accelerate instantaneously to the maximum velocity and so velocity is only increased gradually. Furthermore the behaviour of ants is much more homogeneous [22] than that of drivers on a highway, where also a mixture of different vehicles with different maximum speeds, acceleration capabilities etc. influences the behaviour. As a consequence of the speed homogeneity, overtaking is rarely observed on ant trails and has therefore not been incorporated into our model.

The stochasticity in the model has mainly two reasons. First many influencing factors are not known or difficult to include directly since the model would become too complicated. Thus they are included in a statistical sense through probabilities for a certain behaviour. The second reason lies in the behaviour of the ants themselves. It seems as if they possess some intrinsic stochasticity depending on the evolution of the particular species [16]. This is also widely used in applications [14,15,20]. In vehicular traffic stochasticity is used to incorporate fluctuations in the driver's behaviour which can lead to spontaneously formed phantom jams [2].

2.2 Bidirectional Case

For extending the unidirectional model to the bidirectional case several models have been proposed [25,28,29]. As one common requirement they should reduce to the unidirectional model in case of vanishing counterflow. The extension discussed here is achieved just by adding one lattice for ants moving in the opposite direction (see Fig. 1 right). Both lattices for ants are treated with the same set of rules. For simplicity also the number of ants is the same in both directions. Already in this special case, the generic properties of this model can be observed. Ants in both directions share the same lattice for pheromones. Although this already induces one mechanism of coupling (interaction) a more direct one is also used. Ants facing each other in opposite directions hop with probability K (see Fig. 1 right). Overall one deals with three hopping probabilities depending on

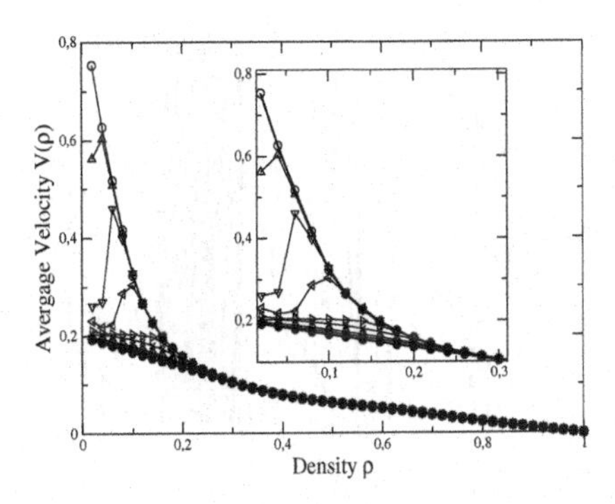

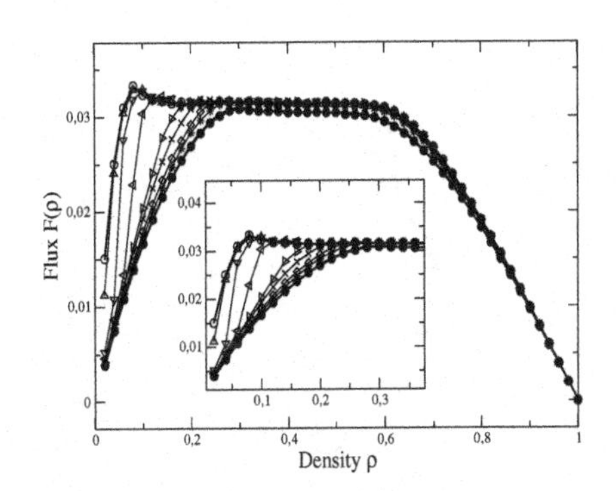

Fig. 4. Fundamental diagrams for the unidirectional model: $Q = 0.9$, $q = 0.2$ and $f = 0(\circ)$, $0.0002(\triangle)$, $0.0008(\triangledown)$, $0.002(\triangleleft)$, $0.008(\triangleright)$, $0.02\ (\times)$, $0.08(\diamond)$, $0.2(*)$, $1(\bullet)$. The average velocity shows the same non-monotonicity already known from the unidirectional model. The main property of the bidirectional model can only be observed in flux. For intermediate densities the flux is nearly independent of the density ϱ and the evaporation rate f.

the occupation of the nearest neighboring site in hopping direction. With respect to reality workers facing each other in opposite directions have to slow down due to information exchange or just in order to avoid collisions [10,13]. So we choose hopping probabilities satisfying $K < q < Q$. Overall this is a crucial difference to most models of vehicular traffic which usually neglect the coupling of lanes in different directions [3]. One related example is the organisation of traffic flow at bottlenecks e.g. inside a nest [12]. In that case traffic flow is organised such that the number of encounters inside the bottleneck, namely a narrow corridor, is reduced. In our model the main effect originates from head-on encounters. Also some similarities to models of pedestrians dynamics can be drawn [3,5]. Under crowded conditions separated lanes for each direction are formed dynamically. These lanes are stabilised by incorporating the desire of pedestrians to reduce the number of encounters with others in the counterflow.

Cluster Regime. At low densities the average velocity roughly shows the same behaviour already known from the unidirectional case (see Fig. 5 left), including the anomalous density dependence (see Fig. 4 left). As the average lifetime of the pheromone-marks is determined by the mean distance of ants in both directions, this regime only exists at very low densities. With increasing density the lifetime of the marks increases and they obviously loose their function as they are present at any site.

Plateau Regime. The generic property of this model is found in flux and cannot be observed directly in the average velocity. For all values of f, the flux roughly stays constant over certain density regimes depending on f (see Fig. 4 right). Due to the behaviour of the average velocity one observes a shift of the beginning of the plateau in flux to lower densities with decreasing f. But at intermediate to high densities, there is obviously no dependence on f. Also the

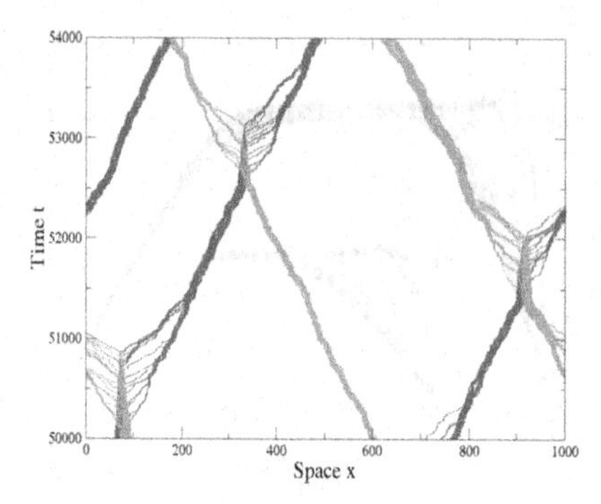

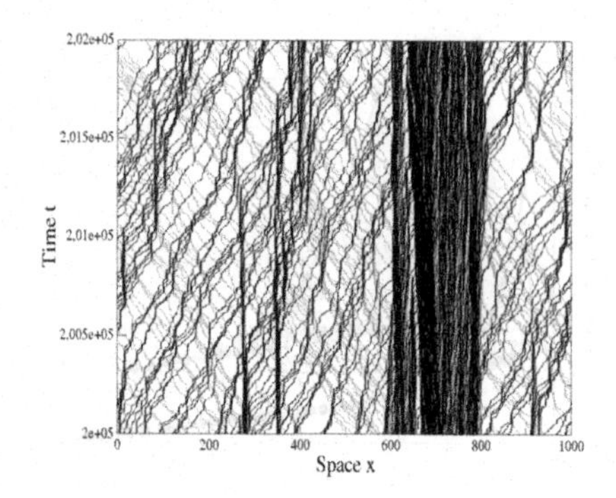

Fig. 5. Space-time plots for the bidirectional model: $Q = 0.9$, $q = 0.2$, $K = 0.1$, $f = 0.002$. The left plot shows the formation of small moving clusters for $\varrho = 0.03$. With increasing density this regime vanishes and a large localised cluster appeares (plotted for $\varrho = 0.2$).

value of flux seems to be nearly independent of ϱ in that regime. It obviously stays constant. Unlike in the unidirectional model this effect is not caused by the pheromone marks. It has its origin in the mutual hindrance by counterflowing ants. Similar effects are known from systems with lattice-wise disorder [30] where the hopping probabilities depend on the position. In our case we find several localised clusters of ants on the lattices in both directions (see Fig. 5 right). They form some kind of dynamically induced defect, as ants facing this clusters move with reduced hopping probability K.

3 Summary and Discussions

We have introduced minimal cellular automaton models for traffic on uni- and bidirectional ant trails. The stationary state in both models was characterised by the spatio-temporal distribution of ants on the trail. We have shown that this distribution is linked to the fundamental diagram, describing the flow properties of the system. The predictions of both descriptions can be tested experimentally [22]. First observations for one particular species of ants seem to confirm at least the spatio-temporal patterns (see Fig. 6), produced by the models qualitatively. On a quantitative level, measuring time headways, velocity distributions and also fundamental diagrams will be the next step [10,22]. In general, clustering phenomena seem to be a common feature in ant colonies [16].

Overall the question regarding an possible advantage of organising traffic in the discussed way arises. Both models produce moving clusters for an appropriate choice of parameters. In the unidirectional model this happens for a relatively wide density regime. The existence of these clusters corresponds to the regime of constant average velocity. In the TASEP the average velocity decreases strictly monotonous with increasing density as ants are distributed homogeneously along the trail. So moving in clusters enables the ants the keep on moving at a higher velocity in comparison to a homogenous distribution in TASEP. This is obviously

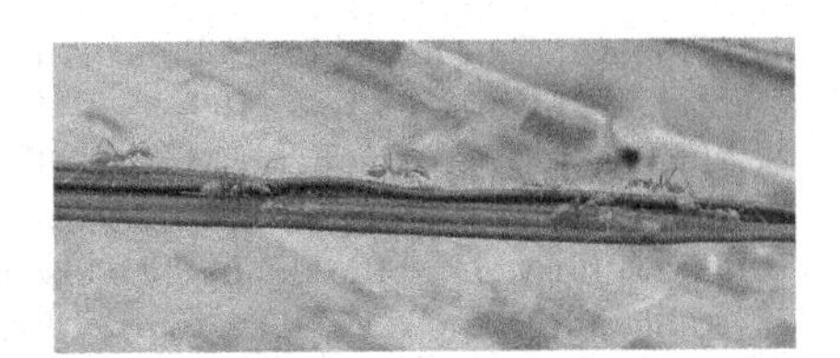

Fig. 6. These photographs show a bidirectional trail of weaver ants (Oecophylla smaragdina) on a cable at the Centre for Ecological Sciences of the Indian Institute of Science in Bangalore. In absence of counterflow, workers are moving in clusters along the cable. In presence of counterflow, head-on encounters are used to exchange information.

achieved by reducing mutual blocking. One might argue that the fundamental diagram shows a maximum value of average velocity. This maximum is attained at the point of sharp increase from the cluster- to the homogenous distribution. Nevertheless a maximum of the average velocity at this point would be quite unstable. Even small fluctuations in density would lead to large fluctuations in the average velocity. In comparison, the cluster-regime is quite stable against fluctuations in density. With respect to natural systems, clustering leads to an decrease of travel time. Unidirectional trails occur for example at the beginning or the end of a swarm raid [16,9,10]. At the beginning a food source has to be occupied as fast as possible. Like at the end of the raid, the density of workers on the trail is quite low. Thus an minimisation of travel time also minimises the time of being extremely vulnerable.

In the bidirectional case, moving clusters only occur at very low densities. But the underlaying mechanism is still the same as in the unidirectional model. At intermediate to high densities the main feature namely the localised cluster, leading to a constant value of flow emerges. Although we have restricted our investigations to the case of equal numbers of ants in both directions, the main effect is not changed. In a natural system one might encounter bidirectional trails during the exploitation of a food source [16,9,10]. In this situation flow is the crucial quantity. The outbound flow determines the number of ants, travelling to the food source, whereas inbound flow is related to the amount of food carried back to the nest. Strongly different values of inbound and outbound flow would lead to a too lage or too small number of ants at the source. So a constant flow which is roughly independent of density fluctuations also ensures a constant number of workers at a particular destination.

Acknowledgments. The work of one of the authors (AJ) has been supported by the German Academic Exchange Service (DAAD) through a joint Indo-German research project. We would also like to thank M. Burd, P. Chakroborty, R. Gadagkar, B. Hölldobler, A. Kunwar and T. Varghese for informative and stimulating discussions.

References

1. Chopard, B., Droz, M.: Cellular Automata Modeling of Physical Systems. Cambridge University Press (1998)
2. Nagel, K., Schreckenberg, M.: A cellular automaton model for freeway traffic. J. Phys. I France **2** (1992) 2221
3. Chowdhury, D., Santen, L., Schadschneider, A.: Statistical physics of vehicular traffic and some related systems. Phys. Rep. **329** (2000) 199–329
4. O'Loan, O.J., Evans, M.R., Cates, M.E.: Jamming transition in a homogeneous one-dimensional system: The bus route model. Phys. Rev. E **58** (1998) 1404
5. Kirchner, A., Schadschneider, A.: Simulation of evacuation processes using a bionics-inspired cellular automaton model for pedestrian dynamics. Physica A **312** (2002) 260
6. Schadschneider, A., Kirchner, A., Nishinari, K.: From ant trails to pedestrian dynamics. Applied Bionics and Biomechanics **1** (2003) 11
7. Howard, J.: Mechanics of motor proteins and the cytoskeleton. Sinauer Associates (2001)
8. Chowdhury, D., Schadschneider, A., Nishinari, K.: Physics of transport and traffic phenomena in biology: from molecular motors and cells to organisms. Phys. Life Rev. **2** (2005) 318–352
9. Burd, M., Aranwela, N.: Head-on encounter rates and walking speed of foragers in leaf-cutting ant traffic. Insectes Soc. **50** (2003) 3–8
10. Burd, M., Archer, D., Aranwela, N., Stradling, D.J.: Traffic dynamics of the leaf-cutting ant, atta cephalotes. American Natur. **159** (2002) 283–293
11. Dussutour, A., Fourcassié, V., Helbing, D., Deneubourg, J.L.: Optimal traffic organization under crowded condition. Nature **428** (2004) 70
12. Dussutour, A., Deneubourg, J.L., Fourcassié, V.: Temporal organization of bi-directional traffic in the ant lasius niger (l.). Jrl. Exp. Biol. (2005)
13. Couzin, I.D., Franks, N.R.: Self-organized lane formation and optimized traffic flow in army ants. Proc. Roy. Soc. London B **270** (2003) 139–146
14. Bonabeau, E., Dorigo, M., Theraulaz, G.: Swarm Intelligence: From Natural to Artificial Systems. Oxford University Press (1999)
15. Bonabeau, E., Dorigo, M., Theraulaz, G.: Inspiration for optimization from social insect behaviour. Nature **406** (2000) 39–42
16. Hölldobler, B., Wilson, E.O.: The Ants. Cambridge, Belknap (1990)
17. Chowdhury, D., Nishinari, K., Schadschneider, A.: Self-organized patterns and traffic flow in colonies of organisms: from bacteria and social insects to vertebrates. Phase Trans. **77** (2004) 601
18. Kunwar, A., John, A., Nishinari, K., Schadschneider, A., Chowdhury, D.: Collective traffic-like movement of ants on a trail: dynamical phases and phase transitions. J. Phys. Soc. Jpn. **73** (2004) 2979
19. Moffet, M.W.: Ants that go with the flow: A new method of orientation by mass communication. Naturwissenschaften **74** (1987) 551–553
20. Camazine, S., Deneubourg, J.L., Franks, N.R., Sneyd, J., Theraulaz, G., Bonabeau, E.: Self-organization in Biological Systems. Princeton University Press (2001)
21. Chowdhury, D., Guttal, V., Nishinari, K., Schadschneider, A.: A cellular-automata model of flow in ant-trails: Non-monotonic variation of speed with density. J. Phys. A: Math. Gen. **35** (2002) L573–L577
22. John, A., et al.: (in preparation)

23. Schadschneider, A., Chowdhury, D., John, A., Nishinari, K.: Anomalous fundamental diagrams in traffic on ant trails. In Hoogendoorn, S., Luding, S., Bovy, P., Schreckenberg, M., Wolf, D., eds.: Proceedings of Traffic and Granular Flow'03, Springer (2003)
24. John, A., Kunwar, A., Namazi, A., Chowdhury, D., Nishinari, K., Schadschneider, A.: Traffic on bi-directional ant-trails: Fundamental diagrams and coarsening behaviour. In Kühne, R., Pöschel, T., Schadschneider, A., Schreckenberg, M., Wolf, D., eds.: Proceedings of Traffic and Granular Flow'05, Springer (2005)
25. John, A., Kunwar, A., Namazi, A., Chowdhury, D., Nishinari, K., Schadschneider, A.: Traffic on bi-directional ant-trails. In Gattermann, P., Waldau, N., Schreckenberg, M., eds.: Proceedings of Pedestrian and Evacuation Dynamics'05, Springer (2005)
26. Schütz, G.M.: Exactly solvable models for many-body systems far from equilibrium. In Domb, C., Lebowitz, J., eds.: Phase Transitions and Critical Phenomena, Vol. 19, Academic Press (2000)
27. Krug, J., Ferrari, P.: Phase transitions in driven diffusive systems with random rates. J. Phys. A **29** (1996) L465
28. John, A., Schadschneider, A., Chowdhury, D., Nishinari, K.: Collective effects in traffic on bi-directional ant-trails. J. Theor. Biol. **231** (2004) 279
29. Kunwar, A., Chowdhury, D., Schadschneider, A., Nishinari, K.: Competition of coarsening and shredding of clusters in a driven diffusive lattice gas. to appear in J. Stat. Mech. (2006)
30. Tripathy, G., Barma, M.: Steady state and dynamics of driven diffusive systems with quenched disorder. Phys. Rev. Lett. **78** (1997) 3039

A Continuous Particle Swarm Optimization Algorithm for Uncapacitated Facility Location Problem

Mehmet Sevkli[1] and Ali R. Guner[2]

[1] Department of Industrial Engineering, Fatih University, Istanbul, Turkey
msevkli@fatih.edu.tr
[2] Department of Industrial and Manufacturing Engineering
Wayne State University, Detroit, MI, USA
arguner@wayne.edu

Abstract. In this paper, a continuous Particle Swarm Optimization (*PSO*) algorithm is presented for the Uncapacitated Facility Location (*UFL*) problem. In order to improve the solution quality a local search is embedded to the *PSO* algorithm. It is applied to several benchmark suites collected from OR-library. The results are presented and compared to the results of two recent metaheuristic approaches, namely Genetic Algorithm(*GA*) and Evolutionary Simulated Annealing (*ESA*). It is concluded that the *PSO* algorithm is better than the compared methods and generates more robust results.

1 Introduction

The Particle Swarm Optimization (*PSO*) is one of the recent metaheuristics invented by Eberhart and Kennedy [1] based on the metaphor of social interaction and communication such as bird flocking and fish schooling. In *PSO*, the potential solutions, so-called particles, move around in a multi-dimensional search space with a velocity, which is constantly updated by the particle's own experience and the experience of the particle's neighbors or the experience of the whole swarm. *PSO* has been successfully applied to a wide range of applications such as function optimization, neural network training, task assignment, and scheduling problems.

Location problems are one of the most widely studied problems in NP-hard [2] combinatorial optimization problems thus there is a very rich literature in operations research (OR)[3]. In addition, the bank account location problem, network design, vehicle routing, distributed data and communication networks, computer network design, cluster analysis, machine scheduling, economic lot sizing, portfolio management are some instances without facilities to locate problems that is modelled as an *UFL* problem in the literature.

UFL problems have been studied and examined extensively by various attempts and approaches. All important approaches relevant to *UFL* problems can be classified into two main categories: exact and metaheuristics based algorithms. There is a variety of exact algorithms to solve the *UFL* problem, such

M. Dorigo et al. (Eds.): ANTS 2006, LNCS 4150, pp. 316–323, 2006.

as branch and bound [4], linear programming, Lagrangean relaxation [5], dual approach (DUALLOC) of Erlenkotter [6] and the primal-dual approaches of Körkel [7]. Although DUALLOC is an exact algorithm, it can also be used as a heuristic to find good solutions. It is obvious that since the UFL problem is NP-hard [2] exact algorithms may not be able to solve large practical problems efficiently. There are several studies to solve UFL problem with metaheuristics. Some of recent studies are tabu search [8,9], genetic algorithms [10], neighborhood search [11], and simulated annealing [12].

In an UFL problem there are a number of sites, n and a number of customers, m. Each site has a fixed cost fc_i. There is a transport cost from each site to each customer c_{ij}. There is no limit of capacity for any candidate site and the whole demand of each customer has to be assigned to one site. We are asked to find the number of sites to be established and specify those sites such that the total cost will be minimized(1). The mathematical formulation of UFL can be stated as follows [2]:

$$Z = min \left(\sum_{j=1}^{m} \sum_{i=1}^{n} c_{ij}.x_{ij} + \sum_{i=1}^{n} fc_i.y_i \right) . \tag{1}$$

subject to

$$\sum_{i=1}^{n} x_{ij} = 1 . \tag{2}$$

$$0 \leq x_{ij} \leq y_i \in \{0;1\} . \tag{3}$$

where

$$i = 1, ..., n; j = 1, ..., m;$$

$$x_{ij} : \text{the quantity supplied from facility } i \text{ to customer } j;$$

$$y_i : \text{whether facility } i \text{ is established } (y_i = 1) \text{ or not } (y_i = 0).$$

Constraint (2) makes sure that all demands have been met by the open sites, and (3) is to keep integrity. Since it is assumed that there is no capacity limit for any facility, the demand size of each customer is ignored, and therefore (2) established without considering demand variable.

The organization of this paper is as follows. Section 2 introduces the proposed PSO algorithm for UFL together with the details about local search procedure. In Section 3 the experimental results are provided and Section 4 presents the conclusions driven.

2 PSO Algorithm for UFL

2.1 A Pure PSO Algorithm

The PSO algorithm proposed here for the UFL problems considers each particle based on three key vectors; position (X_i), velocity (V_i), and open facility (Y_i).

$X_i = [x_{i1}, x_{i2}, x_{i3}, , x_{in}]$ denotes the ith position vector in the swarm, where x_{ik} is the position value of the ith particle with respect to the kth dimension $(k = 1, 2, 3, , n)$. $V_i = [v_{i1}, v_{i2}, v_{i3}, ..., v_{in}]$ denotes the ith velocity vector in the swarm, where v_{ik} is the velocity value of the ith particle with respect to the kth dimension. $Y_i = [y_{i1}, y_{i2}, y_{i3} ... y_{in}]$ represents the opening or closing facilities identified based on the position vector(X_i), where y_{ik} represents opening or closing the kth facility of the ith particle. For an *n-facility* problem, each particle contains n number of dimensions.

Initially, the position(x_{ij}) and velocity(v_{ij}) vectors are generated randomly and uniformly as continuous sets of values between (-10.0,+10.0) and (-4.0,+4.0) respectively that is consistent with the literature[13]. The position vector $X_i = [x_{i1}, x_{i2}, x_{i3}, ..., x_{in}]$ corresponds to the continuous position values for n facilities, but it does not represent a candidate solution to calculate a total cost. In order to create a candidate solution, a particle, the position vector is converted to a binary variables, $Y_i \leftarrow X_i$, which is also a key element of a particle. In other words, a continuous set is converted to a discrete set for the purpose of creating a candidate solution, particle. The fitness of the ith particle is calculated by using open facility vector (Y_i). For simplicity, $f_i\,(Y_i \leftarrow X_i)$ is from now on be denoted with f_i.

In order to ascertain how to derive an open facility vector from position vector, an instance of 5-facility problem is illustrated in Table 1. Position values are converted to binary variables using following formula:

$$y_i = \lfloor |x_i mod2| \rfloor \ . \tag{4}$$

In equation (4) a position value is first divided by 2 and then the absolute value of the remainder is floored; and the obtained integer number is taken as an element of the open facility vector. For example, fifth element of the open facility vector, y_5, can be found as follows: $\lfloor |-5.45 mod2| \rfloor = \lfloor |-1.45| \rfloor = \lfloor 1.45 \rfloor = 1$.

Table 1. An illustration of deriving open facility vector from position vector for a 5-facility to 6-customer problem

	Particle Dimension(k)				
i_{th} Particle Vectors	1	2	3	4	5
Position Vector(X_i)	1.8	3.01	-0.99	0.72	-5.45
Velocity Vector(V_i)	-0.52	2.06	3.56	2.45	-1.44
Open Facility Vector (Y_i)	1	1	0	0	1

For each particle in the swarm, a personal best, $P_i = [p_{i1}, p_{i2}, p_{i3}, ..., p_{in}]$, is defined, whereby p_{ik} denotes the position value of the ith personal best with respect to the kth dimension. The personal bests are determined just after generating Y_i vectors corresponding to their fitness values. In every generation,t, the personal best of each particle is updated if a better fitness value is obtained. Regarding the objective function, $f_i\,(Y_i \leftarrow X_i)$, the fitness value for the personal best of the ith particle, P_i , is denoted by f_i^{pb}. The personal best vector is initialized with position vector ($P_i = X_i$) at the beginning. Where $P_i = [p_{i1}, p_{i2}, p_{i3}, ..., p_{in}]$ is the position vector and the fitness values of the personal bests are equal to the fitness of positions,$f_i^{pb} = f_i$.

Then, the best particle with respect to fitness value in the whole swarm is selected with the name global best and denoted as $G_i = [g_1, g_2, g_3, ..., g_n]$. The global best, $f_g = f(Y \leftarrow G)$, can be obtained as the best of personal bests over the whole swarm, $f_g = min\{f_i^{pb}\}$, with its corresponding position vector, X_g, which is to be used for $G = X_g$, where G $= [g_1 = x_{g1}, g_2 = x_{g2}, g_3 = x_{g3}, , g_n = x_{gn}]$ and $Y_g = [y_{g1}, y_{g2}, y_{g3}, ..., y_{gn}]$ denotes the Y_i vector of the global best found.

Afterwards, the velocity of each particle is updated based on its personal best and the global best in the following way(5):

$$v_{ik}^{(t+1)} = \left(w.v_{ik}^{(t)} + c_1 r_1 \left(p_{ik}^{(t)} - x_{ik}^{(t)}\right) + c_2 r_2 \left(g_k^{(t)} - x_{ik}^{(t)}\right)\right) . \tag{5}$$

where, w is the inertia weight to control the impact of the previous velocity on the current one. In addition, r_1 and r_2 are random numbers between [0,1] and c_1 and c_2 are the learning factors, which are also called social and cognitive parameters respectively. The next step is to update the positions with (6).

$$x_{ik}^{(t+1)} = x_{ik}^{(t)} + v_{ik}^{(t+1)} . \tag{6}$$

After getting position values updated for all particles, the corresponding open facility vectors are determined with their fitness values in order to start a new iteration if the predetermined stopping criterion is not satisfied. In this study, we apply the *gbest* model of Kennedy and Eberhart [14], which is elaborated in the pseudo code given below.

```
Begin
  Initialize particles positions
  For each particle
    Find open facility vector (4)
    Evaluate(1)
    Do{
      Find the personal best and the global best
      Update velocity(5) and position(6) vectors
      Update open facility vector (4)
      Evaluate(1)
      Apply local search(in (PSO_LS))
    }While (Termination)
End
```

Fig. 1. Pseudo code of *PSO* algorithm for UFL problem

2.2 PSO with Local Search

Apparently, *PSO* conducts such a rough search that it produces premature results, which do not offer satisfactory solutions. For this purpose, it is inevitable to embed a local search algorithm into PSO so as to produce more satisfactory solutions. In this study, we have employed local search to neighbors of the global best position vector. For the UFL problem, flip operator is employed as a neighborhood structure. Flip operator can be defined as picking one position value of the global best randomly, and then changing its value with using following:

$$g_i = \begin{cases} 0 \leq \rho \leq 1 \ , g_i + 1 \\ 0 \leq \rho \leq 1 \ , g_i - 1 \end{cases}$$

Where ρ is a uniformly generated number between 0 and 1. This function is used for opening a new facility or closing an open one. The local search algorithm applied in this study is sketched in Figure 2. The global best found at the end of each iteration of PSO is adopted as the initial solution by local search algorithm. In order not to loss the best found and to diversify the solution, the global best is randomly modified in which two facilities are flipped based on both random parameters generated, η and κ. Then, flip operator is applied as long as it gets better solution. The final produced one is evaluated and replaced with the old global best if it is better than the initial one.

```
Begin
  Set global best position vector (Y_g) to s_0
  Modify s_0 based on η, κ and set to s
  Set 0 to loop
  Apply Flip to s and get s_1
    if f(s_1) ≤ f(s)
      Replace s with s_1
    else loop=loop+1
  Until (loop < n )
  if f(s) ≤ f(s_0)
    Replace Y_g with s
End.
```

Fig. 2. Pseudo code for local search

3 Experimental Results

This experimental study has been completed in two stages; PSO and PSO_{LS}. Experimental results provided in this section is carried out with two algorithms over 15 benchmark problems that are taken from the OR Library [15], a collection of benchmarks for OR studies. The benchmarks are introduced in Table 2 with their sizes and the optimum values. Although the optimum values are known, it is really hard to hit the optima in every attempt of optimization. Since the main idea is to test the performance of PSO algorithm with UFL benchmark, the results are provided in Table 2 with regard to solution quality indexes: Average Relative Percent Error ($ARPE$) which is as defined in (7), Hit to Optimum Rate (HR) and Computational Processing Time(CPU).

$$ARPE = \sum_{i=1}^{R} \left(\frac{(H_i - U) \times 100}{U_i} \right) / R \tag{7}$$

where H_i denotes the ith replication solution value whereas U is the optimal value provided in literature and R is the number of replications. HR provides

Table 2. Experimental results gained with PSO and PSO_{LS}

			PSO			PSO_{LS}		
Problem	Size	Optimum	$ARPE$	HR	CPU	$ARPE$	HR	CPU
*Cap*71	16 × 50	932615.75	0.05	0.87	0.12	0.00	1.00	0.01
*Cap*72	16 × 50	977799.40	0.07	0.80	0.16	0.00	1.00	0.01
*Cap*73	16 × 50	1010641.45	0.06	0.63	0.27	0.00	1.00	0.01
*Cap*74	16 × 50	1034976.98	0.07	0.73	0.21	0.00	1.00	0.01
*Cap*101	25 × 50	796648.44	0.14	0.53	0.67	0.00	1.00	0.08
*Cap*102	25 × 50	854704.20	0.15	0.40	0.85	0.00	1.00	0.02
*Cap*103	25 × 50	893782.11	0.16	0.20	1.08	0.00	1.00	0.07
*Cap*104	25 × 50	928941.75	0.18	0.70	0.47	0.00	1.00	0.02
*Cap*131	50 × 50	793439.56	0.75	0.07	4.26	0.00	1.00	0.57
*Cap*132	50 × 50	851495.33	0.78	0.00	4.56	0.00	1.00	0.18
*Cap*133	50 × 50	893076.71	0.73	0.00	4.58	0.00	1.00	0.42
*Cap*134	50 × 50	928941.75	0.89	0.10	4.15	0.00	1.00	0.09
CapA	100 × 1000	17156454.48	22.01	0.00	13.64	0.00	1.00	3.03
CapB	100 × 1000	12979071.58	10.75	0.00	16.58	0.00	1.00	5.18
CapC	100 × 1000	11505594.33	9.72	0.00	24.18	0.02	0.50	8.43

the ratio between the number of runs yielded the optimum and the total numbers of experimental trials. The higher HR the better quality of solution, while the lower $ARPE$ the better quality. Obviously, spent CPU for both algorithms are obtained when the best value is got over 1000 iterations for PSO and 250 iterations for PSO_{LS}. All algorithms and other related software were coded in *Borland C++ Builder6* and run on an *Intel Pentium IV 2.6 GHz* PC with *256MB* memory. The parameters used for the PSO algorithm are as follows: The size of the population set to the number of facilities, the social and cognitive parameters were taken as $c1 = c2 = 2$ consistent with the literature[13]. Inertia weight, w, is taken as a random number between 0.5 and 1. For every benchmark suite both algorithms are conducted for 30 replications.

The performance of PSO does not look that impressive as the results produced within the range of time over 1000 iterations. The PSO without local search found 10 optimal solution whereas the PSO with local search algorithm found 15 among 15 benchmark problems. The $ARPE$ index for PSO is very high for *CapA*, *CapB* and *CapC* benchmarks and none of the attempts hit the optimum value. It may be possible to improve the quality of solutions by carrying on with PSO for further number of iterations, but, then the main idea and useful motivation of employing the heuristics, getting better quality within shorter time, will be lost. This fact imposed that it is essential to empower PSO_{LS} algorithm to mature the proposed PSO.

The performance of PSO_{LS} algorithm looks very impressive compared to PSO algorithm with respect to all three indexes of solution quality. HR is 1.00 which means 100% of the runs yield with optimum for all benchmark except *CapC*. The experimental study is carried out as a comparative work with a GA

Table 3. Summary of results gained from different algorithms for comparison

	Deviation from Optimum			Average CPU		
Problem	GA	ESA	PSO_{LS}	GA	ESA	PSO_{LS}
*Cap*71	0.00	0.00	0.00	0.287	0.041	0.010
*Cap*72	0.00	0.00	0.00	0.322	0.028	0.010
*Cap*73	0.00033	0.00	0.00	0.773	0.031	0.010
*Cap*74	0.00	0.00	0.00	0.200	0.018	0.010
*Cap*101	0.00020	0.00	0.00	0.801	0.256	0.080
*Cap*102	0.00	0.00	0.00	0.896	0.098	0.020
*Cap*103	0.00015	0.00	0.00	1.371	0.119	0.070
*Cap*104	0.00	0.00	0.00	0.514	0.026	0.020
*Cap*131	0.00065	0.00008	0.00	6.663	2.506	0.570
*Cap*132	0.00	0.00	0.00	5.274	0.446	0.180
*Cap*133	0.00037	0.00002	0.00	7.189	0.443	0.420
*Cap*134	0.00	0.00	0.00	2.573	0.079	0.090
CapA	0.00	0.00	0.00	184.422	17.930	3.030
CapB	0.00172	0.00070	0.00	510.445	91.937	5.180
CapC	0.00131	0.00119	0.00	591.516	131.345	8.430

introduced by Jaramillo et al. [10] and a ESA proposed by Aydin and Fogarty [12].The results of the first two approaches and PSO_{LS} are summarized in Table 3. The performance of PSO_{LS} algorithm looks more impressive compared to the both GA and ESA in both two indexes. Especially, in respect of CPU time PSO_{LS} much more robust than GA. Comparing with ESA, especially for *CapA*, *CapB* and *CapC* ESA consumes more CPU time than PSO_{LS} algorithm. In conclusion, we can say that PSO_{LS} algorithm is more robust than not only pure PSO algorithm but also both GA and ESA.

4 Conclusion

In this paper, a PSO and a PSO_{LS} algorithm applied to solve UFL problems. The algorithm has been tested on several benchmark problem instances and optimal result are obtained in a reasonable computing time. The results of PSO_{LS} are compared with the results of two recent metaheuristic approaches, namely Genetic Algorithm and Evolutionary Simulated Annealing. It is concluded that the PSO_{LS} algorithm is better than the compared methods and generating more robust results. In addition, to the best of our knowledge, this is the first application of PSO algorithm reported for the UFL in the literature.

References

1. Eberhart, R.C., Kennedy, J.: A New Optimizer Using Particle Swarm Theory. In Proc. of the 6th Int. Symposium on Micro Machine and Human Science, Nagoya Japan (1995) 39–43
2. Cornuéjols, G., Nemhauser, G.L.,Wolsey, L.A.: The Uncapacitated Facility Location Problem. Discrete Location Theory Wiley-Interscience, New York (1990) 119-171
3. Mirchandani, P.B., Francis, R.L. (eds.): Discrete Location Theory. Wiley-Interscience, New York (1990)
4. Klose, A.: A Branch and Bound Algorithm for an UFLP with a Side Constraint. Int. Trans. Opl. Res. **5** (1998) 155-168
5. Barcelo, J., Hallefjord, A., Fernandez, E. and Jrnsten, K.: Lagrengean Relaxation and Constraint Generation Procedures for Capacitated Plant Location Problems with Single Sourcing. O R Spektrum **12** (1990) 79-78
6. Erlenkotter, D.: A dual-based procedure for uncapacitated facility location. Operations Research **26** (1978) 992-1009
7. Körkel, M.: On the Exact Solution of Large-Scale Simple Plant Location Problems. European J. of Operational Research,**39(1)**(1989) 157-173
8. K.S. Al-Sultan, Al-Fawzan, M.A.: A tabu search approach to the uncapacitated facility location problem. Annals of Operations Research **86** (1999) 91-103
9. Laurent, M. and Hentenryck, P.V., A Simple Tabu Search for Warehouse Location. European J. of Operational Research **157** (2004) 576-591
10. Jaramillo, J.H., Bhadury, J., Batta, R.: On the Use of Genetic Algorithms to Solve Location Problems. Computers & Operations Research **29** (2002) 761-779.
11. Ghosh, D.: Neighborhood Search Heuristics for the Uncapacitated Facility Location Problem. European J. of Operational Research **150** (2003) 150-162
12. Aydin, M.E., Fogarty, T.C.: A Distributed Evolutionary Simulated Annealing Algorithm for Combinatorial Optimization Problems. J. of Heuristics **10** (2004) 269-292
13. Shi, Y. Eberhart, R.: Parameter selection in particle swarm optimization. In Evolutionary Programming VIZ: Proc. EP98. Springer-Verlag, New York (1998) 591-600.
14. Kennedy, J., Eberhart, R.C., Shi, Y.: Swarm intelligence. Morgan-Kaufmann, San Francisco (2001).
15. Beasley J.E.: OR-Library: http://people.brunel.ac.uk/~mastjjb/jeb/info.html (2005)

A Direct Application of Ant Colony Optimization to Function Optimization Problem in Continuous Domain

Min Kong and Peng Tian

Shanghai Jiaotong University, Shanghai, China
kongmin@sjtu.edu.cn, ptian@sjtu.edu.cn

Abstract. This paper proposes a direct application of Ant Colony Optimization to the function optimization problem in continuous domain. In the proposed algorithm, artificial ants construct solutions by selecting values for each variable randomly biased by a specific variable-related normal distribution, of which the mean and deviation values are represented by pheromone modified by ants according to the previous search experience. Some methods to avoid premature convergence, such as local search in different neighborhood structure, pheromone re-initialization and different solutions for pheromone intensification are incorporated into the proposed algorithm. Experimental setting of the parameters are presented, and the experimental results show the potential of the proposed algorithm in dealing with the function optimization problem of different characteristics.

1 Introduction

Ant Colony Optimization (ACO) is a stochastic meta-heuristic for solutions to combinatorial optimization problems. Although ACO has been proved to be one of the best meta-heuristics in some combinatorial optimization problems [1,2], the application to the function optimization problem appears a little difficult, since the pheromone laying method is not straightforward, specially in the continuous domain. There have been several ant-based algorithms proposed for the function optimization problem, they are CACO [3,4,5,6], API [7], CIAC [8,9], the extended ACO [10], and BAS [11].

This paper proposes DACO, a Direct application of ACO for tackling the function optimization problem in continuous domain. DACO is motivated and developed from Socha [10], specially by the idea of using the normal distribution. Different from other ACO applications to the function optimization problems, pheromone in DACO are directly associated to the mean and deviation value of a specific normal distribution for every variable. Artificial ants construct solutions by generating stochastically the value of every variable according to the normal distribution. Experimental results over various benchmark problems show the potential of DACO for solving function optimization problems of different characteristics.

M. Dorigo et al. (Eds.): ANTS 2006, LNCS 4150, pp. 324–331, 2006.

2 ACO in Continuous Domain

2.1 Solution Construction

In DACO, a number of artificial ants are managed in searching the continuous domain of the problem instance at each iteration. Each artificial ant constructs a solution $x = [x_1, \cdots, x_n]$ by generating a real number for each variable x_i stochastically decided by a special normal distribution $N(\mu_i, \sigma_i^2)$, of which the characteristics are modified by the ants during iterations in a form of pheromone maintenance procedure.

DACO incorporates a number of n normal distributions, with each normal distribution $N(\mu_i, \sigma_i^2), i \in \{1, \cdots, n\}$ associated to each variable x_i, where n is the number of variables in the function optimization problem. And two kinds of pheromone are incorporated: one is associated to $\mu = [\mu_1, \cdots, \mu_n]$, the other is associated to $\sigma = [\sigma_1, \cdots, \sigma_n]$. The amount of pheromone associated with μ and σ directly represent the value of μ and σ respectively. In the following, we will use the term μ and σ as the pheromone representation as well.

The solutions generated according to the normal distribution are checked and modified to fit into the constraint range:

$$x_i = \begin{cases} a_i & x_i < a_i \\ x_i & a_i \leq x_i \leq b_i \\ b_i & x_i > b_i \end{cases} \tag{1}$$

where b_i and a_i are the top and lower limit of variable x_i.

2.2 Pheromone Update

Initially, pheromone are set as $\mu_i(0) = a_i + rand(i)(b_i - a_i)$, and $\sigma_i(0) = (b_i - a_i)/2$ for $i = 1, \cdots, n$, where $rand(i)$ is a randomly generated number with an even distribution in the range of $[0, 1]$.

After all the ants have finished their solution construction, the pheromone evaporation phase is performed first, in which all the pheromone evaporate:

$$\begin{aligned} \mu(t) &\leftarrow (1 - \rho)\mu(t - 1) \\ \sigma(t) &\leftarrow (1 - \rho)\sigma(t - 1) \end{aligned} \tag{2}$$

where $\rho \in [0, 1]$ is the pheromone evaporation parameter and t is the iteration number.

Then, the pheromone intensification procedure is performed as:

$$\begin{aligned} \mu(t) &\leftarrow \mu(t) + \rho x \\ \sigma(t) &\leftarrow \sigma(t) + \rho|x - \mu(t - 1)| \end{aligned} \tag{3}$$

where $x = [x_1, \cdots, x_n]$ is the solution used for pheromone intensification, which is normally the global best solution S^{gb} found by the previous search.

3 Methods to Avoid Premature Convergence

Local Search. Local search has been testified to be of great improvement in ACO metaheuristic [1,2], the result of a local search is often a better solution found in the neighborhood of the current solution, where the structure of the neighborhood is specially designed. In DACO to the continuous function optimization problem, a simple local search is designed, in which a random bidirectional movement is performed one by one for every variable. The local search for every variable x_i contains two phases: the first phase increase x_i with a random distance d_i, and continues until a worse solution is found or the constraint limitation are violated. The second phase reverse the process by decrease x_i with a random distance d_i. The random distance d_i moved at every step is biased by σ_i:

$$d_i = rand() * \sigma_i \tag{4}$$

The local search process is performed to the iteration best solution S^{ib} and the global best solution S^{gb} during every iteration cycle.

Pheromone Re-initialization. The main idea is to re-initialize all the pheromone once the algorithm gets near to premature convergence. In DACO, a convergence factor, cf, is defined to monitor the status of premature convergence:

$$cf = \frac{\sum_{i=1}^{n} \frac{2\sigma_i}{b_i - a_i}}{n} \tag{5}$$

Initially, since σ_i is set as $\sigma_i = (b_i - a_i)/2$, thus that $cf = 1$; when the algorithm gets into some local optima, σ will be decreased to be close to 0, such that $cf \to 0$. According to the definition of cf, we can see that as the algorithm gets near to a local optima, cf changes from 1 to 0. We can set a threshold cf_r as a trigger for the re-initialization: once $cf \leq cf_r$, the algorithm is considered to be in some state of premature convergence, then a re-initialization procedure will be performed, in which the pheromone μ is set equal to S^{gb}, σ_i is set to the initial value $(b_i - a_i)/2$, followed directly by a pheromone intensification phase using all the previous re-initialized best solutions S^{rb}s as a weighted combination for pheromone intensification:

$$\begin{aligned} \mu &\leftarrow \mu + \rho \sum_{j=1}^{n_r} w_j S^{rb}(j) \\ \sigma &\leftarrow \sigma + \rho \sum_{j=1}^{n_r} w_j |S^{rb}(j) - S^{gb}| \end{aligned} \tag{6}$$

where n_r is the number of previous S^{rb}, w_j is the weight for the jth S^{rb}, which is calculated according to the fitness value of every S^{rb}:

$$w_j = \frac{f(S^{rb}(j))}{\sum_{l=1}^{n_r} f(S^{rb}(l))} \tag{7}$$

Using Different Solutions for Pheromone Intensification. Using different solution for pheromone intensification is another method for diversification. We use the method introduced by Blum and Dorigo [12], where different combination of the iteration best solution S^{ib}, re-initialized best solution S^{rb} and global best solution S^{gb} are selected for the pheromone intensification according to the current status of convergence, that is, a weighted combination of solutions is selected depending on the value of cf. The details of the selecting strategy is described in Tab.1, where w_{ib}, w_{rb} and w_{gb} are the weights for the solution S^{ib}, S^{rb} and S^{gb} respectively. cf_is are the threshold parameters.

Table 1. Pheromone Intensification Strategy for DACO

	$cf > cf_1$	$cf \in [cf_2, cf_1)$	$cf \in [cf_3, cf_2)$	$cf \in [cf_4, cf_3)$	$cf \in [cf_r, cf_4)$
w_{ib}	1	2/3	1/3	0	0
w_{rb}	0	1/3	2/3	1	0
w_{gb}	0	0	0	0	1

4 Experimental Results

In all the experiments performed in this paper, a test is considered to be successful and stopped if it observes the following condition:

$$|f^* - f_{known\_best}| < \epsilon_1 \cdot |f_{known\_best}| + \epsilon_2 \tag{8}$$

where f^* is the optimum found by DACO, f_{known_best} is the global optimum of the test function. ϵ_1 and ϵ_2 are two accuracy parameters, which are set to be: $\epsilon_1 = \epsilon_2 = 10^{-4}$.

4.1 Primary Experiments for Parameter Setting

We set the parameters one by one with other parameters being fixed with the default values: $m = 10$, $\rho = 0.1$. The test functions used are R2, SM, GP, MG and ST. Due to space limitation, all the test functions used in this paper are not listed, for more detailed description, please refer to [8] and [10].

Parameter Setting for the cf_i Threshold. To get a good cf_i setting, a set of 25 combinations of cf_i are designed for test considering a wide range of different combinations. Figure.1 shows the test results, which includes the accumulated best and average evaluation numbers for all the test functions over 100 runs.

From the test results we can see that several combinations of the cf_i settings have better performance than others. Among them, the combination 9 is the best, which is $cf_i = [1.0, 0.5, 0.1, 0.05, 0.01]$.

Parameter Setting for m. Parameter m is tested with a range from 1 to 20 for all the 5 test functions. Figure.2 shows the test results of average evaluation number over 100 runs for each test function. Overall, more m needs more evaluation number, but a number of 1 or 2 do not definitely get the best performance. For the following experimental tests, we will use $m = 3$ as the parameter setting.

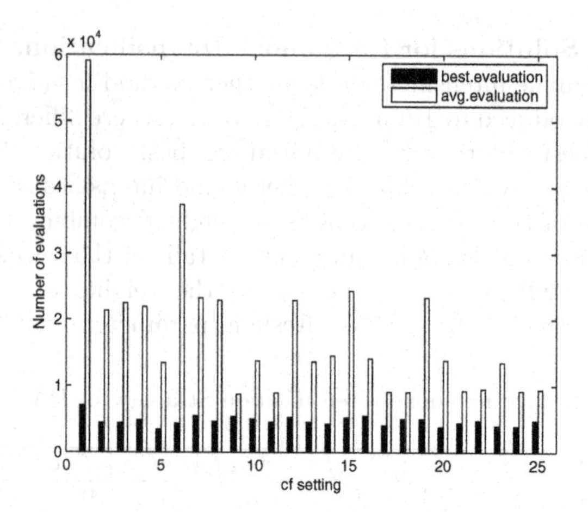

Fig. 1. Test Results of cf setting

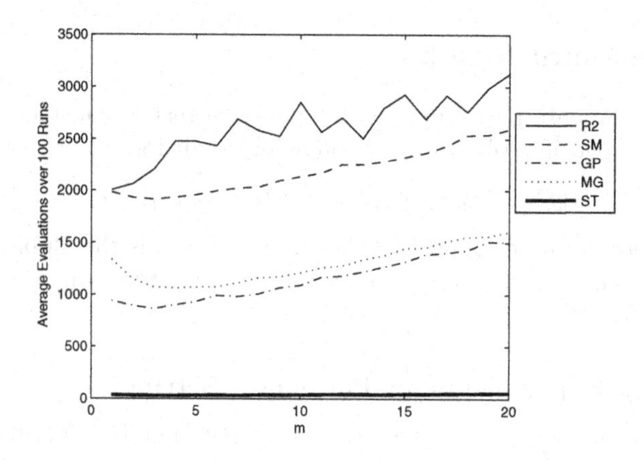

Fig. 2. Test Results of Parameter m

Parameter Setting for ρ. Parameter ρ is tested with a range from 0.01 to 1 with a step of 0.01. The accumulated average evaluation of the five test function is presented in Figure.3. From Figure.3 we can see that the proper setting of ρ would be 0.85.

4.2 Comparison with Other Algorithms

Table 2 displays the test results of DACO comparing to other Ant Systems in continuous domain. Some information of the test results of other Ant Systems are unavailable, so the relative places are blank. The brackets indicate the tests are performed under a fixed evaluation number. For all the test functions, DACO outperforms other Ant Systems both in success rate and the average

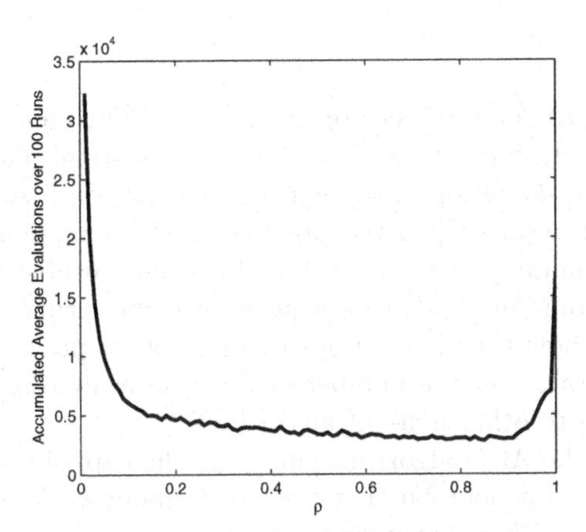

Fig. 3. Test Results of Parameter ρ

Table 2. Comparison of DACO with Other Ant Systems

f	CACO		API		CIAC		ExtendedACO		DACO	
	% ok	evals	% ok	evals	% ok	evals	% ok	evals	% ok	evals
$R2$	100	6842		[10000]	100	11797		2905	100	1946.77
SM	100	22050		[10000]	100	50000		695	100	264.87
GP	100	5330			56	23391		364	100	229.53
MG	100	1688			20	11751			100	169.49
ST		[6000]			94	28201			100	33.33

Table 3. Comparison of DACO with Non-Ant Systems

f	DACO	Non-Ant Systems		
		CGA [13]	ECTS [14]	ESA [15]
SM	264.87	750	338	
GP	229.53	410	231	783
$R2$	1946.77	960	480	796
$Z2$	92.28	620	195	15820
$H_{3,4}$	164.45	582	548	698

evaluation number. For all the functions tested, DACO finds all the known-optimal solutions.

Table 3 displays the comparison results of DACO with some other non-ant algorithms. Except the test function R2, DACO outperforms all the other non-ant algorithms in all the rest of test functions.

Overall, DACO greatly improves the ability of ACO in quickly finding the optimal solutions of function optimization problems with different characteristics, it can be considered as a optional method in dealing with function optimization problems.

5 Conclusions

This paper presented DACO, a direct application of ACO to handle continuous optimization problems. In the proposed version of the system, pheromone trails represent the mean and deviation of the normal distribution for every variable, and they are modified according to the previous search experience. Ants construct solutions by randomly selecting values for each variable biased by the normal distribution. Some methods to escape from local optima are presented. Experimental results show that DACO outperforms other previous ant related algorithms in success rate and the number of function evaluation, and is quite competitive comparing to other non-ant algorithms.

The new version of the ACO algorithm improves the capabilities of the ACO to the function optimization field. Further research is under work for the principle parameter setting and quicker convergence speed.

References

1. Bonabeau, E., Dorigo, M., Theraulaz, G.: Swarm Intelligence: From Natural to Artificial Systems. Oxford University Press (1999)
2. Dorigo, M., Stützle, T.: Ant Colony Optimization. MIT Press (2004)
3. Bilchev, G., Parmee, I.: The ant colony metaphor for searching continuous design spaces. In Fogarty, T., ed.: AISB Workshop on Evolutionary Computing, Springer-Verlag (1995) 25–39
4. Bilchev, G., Parmee, I.: Constrained optimization with an ant colony search model. In: 2nd International Conference on Adaptive Computing in Engineering Design and Control. (1996) 26–28
5. Mathur, M., Karale, S., Priye, S., Jyaraman, V., Kulkarni, B.: Ant colony approach to continuous function optimization. Ind. Eng. Chem. Res. **39**(10) (2000) 3814–3822
6. Wodrich, M., Bilchev, G.: Cooperative distributed search: the ant's way. Control & Cybernetics **3** (1997) 413–446
7. Monmarche, N., Venturini, G., Slimane, M.: On how pachycondyla apicalis ants suggest a new search algorithm. Future Generation Computer Systems **16**(8) (2000) 937–946
8. Dreo, J., Siarry, P.: A new ant colony algorithm using the heterarchical concept aimed at optimization of multiminima continuous functions. In Dorigo, M., Di Caro, G., Samples, M., eds.: Third International Workshop, ANTS 2002. LNCS 2463, Springer (2002) 216–221
9. Dreo, J., Siarry, P.: Continuous interacting ant colony algorithm based on dense heterarchy. Future Generation Computer Systems **20**(5) (2004) 841–856
10. Socha, K.: Aco for continuous and mixed variable optimization. In Dorigo, M., Birattari, M., Blum, C., eds.: ANTS 2004. LNCS 3172, Springer (2004) 25–36
11. Kong, M., Tian, P.: A binary ant colony optimization for the unconstrained function optimization problem. In: 2005 International Conference on Computational Intelligence and Security (CIS'05). LNAI 3801, Springer (2005) 682–687
12. Blum, C., Dorigo, M.: The hyper-cube framework for ant colony optimization. IEEE Transactions on Systems, Man and Cybernetics, Part B **34**(2) (2004) 1161–1172

13. Chelouah, R., Siarry, P.: A continuous genetic algorithm designed for the global optimization of multimodal functions. Journal of Heuristics **6** (2000) 191–213
14. Chelouah, R., Siarry, P.: Enhanced continuous tabu search. In Voss, S., Martello, S., Osman, I., Roucairol, C., eds.: Meta-heuristics: advances and trends in local search paradigms for optimization. Chapter 4. Kluwer Academic Publishers (1999) 49–61
15. Siarry, P., Berthiau, G., Durbin, F., Haussy, J.: Enhanced simulated annealing for globally minimizing functions of many continuous variables. ACM Transactions on Mathematical Software **23**(2) (1997) 209–228

A Parallel ACO Approach Based on One Pheromone Matrix*

Qiang Lv[1], Xiaoyan Xia[2], and Peide Qian[1]

[1] School of Computer Science and Technology, Soochow Univeristy, P.R. China
[2] Jiangsu Provincial Key Lab of Computer Information Processing Technology
Suzhou, P.R. China
{qiang, xyxia, pdqian}@suda.edu.cn

Abstract. This paper presents and implements an approach to parallel ACO algorithms. The principal idea is to make multiple ant colonies share and utilize only one pheromone matrix. We call it SHOP (SHaring One Pheromone matrix) approach. We apply this idea to the two currently best instances of ACO sequential algorithms (MMAS and ACS), and try to hybridize these two different ACO instances. We mainly describe how to design parallel ACS and MMAS based on SHOP. We present our computing results of applying our approach to solving 10 symmetric traveling salesman problems, and give comparisons with the relevant sequential versions under the fair computing environment. The experimental results indicate that SHOP-ACO algorithms perform overall better than the sequential ACO algorithms in both the computation time and solution quality.

1 Introduction

Being a successful metaheuristic, Ant Colony Optimization (ACO) performs excellently in solving most combinatorial optimization problems (COPs). Because the ants in one colony behave in fact in parallel and the parallel computing platform is gradually popular, the research of parallel ACO has its inherent advantages. It is natural that studying parallel ACO has been active in recent years.

There are a lot of categories of strategies of parallel metaheuristics. Parallel ACO strategies also fall into some of them, e.g., fine-grained and coarse-grained parallel. It seems that the fine-grained parallel strategies have not been proved suitable for parallel ACO. On the contrary most literature of parallel ACO show that the coarse-grained strategies are much better than the existing fine-grained strategies.

In order to reduce the communications between the processors, Bullnheirmer et al. [1] presented a new PAPI (Partially Asynchronous Parallel Implementation). In PAPI, after a fixed number of iterations the multi-colony exchanges

* Supported by 211 Project of Soochow University, PRC.

M. Dorigo et al. (Eds.): ANTS 2006, LNCS 4150, pp. 332–339, 2006.

pheromone. Middendorf et al. [2] categorized parallel ant colonies into heterogeneous colony and homogeneous colony. They studied and compared four different communication models between ant colonies. Stützle [3] presented a simple strategy for parallel ACO, that is to make the ant colonies run independently on processors. Talbi et al. [4,5] applied the master/slave approach to parallel AS to solve QAP. The master thread collects the solutions from the parallel ant colonies, and broadcasts the update information to every colony for their pheromone matrices. Randall et al. [6] proposed five strategies of parallel ACO. They implemented parallel ACS by the parallel ants strategy and tested on some small scale TSP instances. Chu et al. [7] also made p ACS colonies run on p processors, with three communication models between these colonies. Delisle et al. [8] put forward the only work on parallel ACO on a shared memory architecture using OpenMP. Their target problem was a large scale industrial scheduling problem. They concluded that a shared memory architecture also provided a good platform for parallel ACO.

We draw some outlines, which are important for this paper, from the literature as follows:

- Coarse-grained parallel ACO strategy performs better than fine-grained strategy.
- All current parallel ant colonies keep their own data structures, especially pheromone matrices.
- Being guided by the traditional strategies of parallel metaheuristics, the literature of parallel ACO focused on the communication models between the parallel search threads.
- The experimental results analysis focused more on time speedup, less on solution quality.
- Most of the work is done on connected computers, only the work of [8] is on SMP via openMP.

The biological attributes and the design of ACO suggest that parallel ACO is a natural research trend. The original idea of ant algorithms employs cooperations between ants in one colony to fulfill the optimization tasks. This paper introduces a new approach of employing cooperations between the colonies to finish the tasks. It is obvious that ant colonies can work in parallel. We call our approach SHOP (SHaring One Pheromone matrix), which makes p ant colonies run on p processors with the same sharing pheromone matrix.

This paper is organized as follows: Section 1 is the brief introduction and related work; Section 2 is the description of the design of SHOP strategy; Section 3 is the experimental results and comparison analysis; and Section 4 is the conclusion and our future agenda.

2 Design of SHOP

2.1 General Ideas

We do think the communication model of parallel metaheuristics is the common problem for all parallelization issues, and therefore it is not the feature problem of parallel ACO. Being a metaheuristic, parallel ACO also may pay much attention to the parallel strategy of the cooperations between the parallel ant colonies. Pheromone matrix is the critical data structure of ACO, which accumulates the experiences of ants tours and guides ants to search towards solutions with good qualities. The reasons for varieties of ACO algorithms are due to the different design of pheromone and update policies of the pheromone matrix, problem-dependent of course. As far as the current research of parallel ACO, all designs assign an independent pheromone matrix to each colony. So all the ants within one colony or across the parallel colonies are consistent from the view of pheromone consistency.

Our approach also exploits the pheromone matrix for all parallel colonies. Inspired by the work [3,8], we make independent ant colonies share the same pheromone matrix. We describe the design of how to apply SHOP to the two best ACO algorithms, MMAS and ACS, and try to hybridize the two. As a result we naturally adopt SMP as our parallel computing platform. The programming environment is Pthread, and the target problem is TSP. After initializing data, each ant colony runs on an independent processor with sharing the same pheromone matrix. We tackle TSP by both the cooperations between the ants within a colony (ACO mechanism) and the cooperations between the colonies (our SHOP mechanism).

2.2 Parallel MMAS (PMMAS)

PMMAS distributes ant colonies to parallel processors, running each colony as one thread with the same pheromone matrix. After getting a feasible solution, the colony updates the common pheromone matrix as MMAS usually behaves, except for the synchronization with other colonies.

The master thread is responsible for PMMAS initialization. Although not necessarily, all colonies have the same parameter settings as in MMAS.

Each ant within its host colony constructs its solution in parallel based on the sequential MMAS algorithm, that is, each processor runs the same MMAS algorithm in parallel, and optimizes the solution found by 3-opt local search.

In PMMAS design, each search thread updates the pheromone matrix exactly once after every iteration, which is the original update scheme of MMAS. For p colonies, each colony updates the pheromone synchronously and uses its own iteration-best and best-so-far solution. p best-so-far ants in every generation update the pheromone matrix: on their tours the pheromone are increased, on other tours the pheromone are decreased p times. In the ideal situation, the update speed of pheromone on every edge is p times that of its sequential version, though we have the same total number of ants in sequential version.

An alternative pheromone update method is to create a pheromone update thread. Additionally we can build two structures, colony-best-so-far ant and colony-iteration-best ant, to record proper ants' solutions. Thus after every generation the colony-iteration-best ant is responsible for the pheromone update. According to the original pheromone update rule of MMAS, after every fixed number of generations the pheromone update thread chooses the colony-best-so-far ant to release a certain amount of pheromone on its tour, i.e., $\Delta\tau_{ij} = 1/L_{cb}$ where L_{cb} stands for the length of the tour which was found by the colony-best-so-far ant. Meanwhile, according to evaporation factor ρ, there is a decrease of pheromone on all nearer adjacent edges which connect to each node of the colony-best-so-far tour. We do not evaporate pheromone of all edges in order to avoid the unnecessary computation and therefore the algorithm efficiency is improved.

Like MMAS, PMMAS also limits the pheromone values into an interval. But we make a little modification on how to determine the pheromone interval $[\tau_{min}, \tau_{max}]$, with $\tau_{min} = \tau_{max}/(2n)$ unchanged but $\tau_{max} = 1/(\rho L_{cb})$ where L_{cb} is the length of the tour which was found by the colony-best-so-far ant.

Pheromone re-initialization is a creative idea of MMAS. Such mechanism works also well for PMMAS. When one of the parallel colonies meets the re-initialization conditions, the best-so-far solution will be recorded (if the solution is worthy to do so) as the colony-best-so-far and the pheromone matrix will be initialized to τ_{max}. Then this colony restarts to construct solutions as usual. If the other colonies also meet the re-initialization conditions but get less better solution than the colony-best-so-far solution, they restart by updating the new initial pheromone value produced by the other colonies. In this way, PMMAS takes much more advantage of re-initialization efficiency than MMAS.

2.3 Parallel ACS (PACS)

In addition to releasing pheromone on the tour of the iteration-best ant, ACS is designed to increase pheromone on the visited edges while an ant is constructing its tour. Such local updating of pheromone will directly influence the subsequent tours of other ants within the same iteration.

PACS has almost the same parallel idea as PMMAS. Multi-colony runs in parallel the sequential ACS algorithm with the same settings (although same settings are not necessary.). The results for PACS algorithm are obviously different from those for sequential ACS. Because of the shared pheromone matrix, after the ants have finished the local pheromone update, the subsequent ants in the same colony, as well as the ants in other parallel colonies, will be affected in making their selection based on the pheromone matrix.

Compared with the parallel strategy in [7], PACS makes the information feedback more effective between colonies. After an ant in one colony has updated the pheromone on one edge, ants in other colonies will immediately sense the change, and make some adjustments for their future tours or even for their next cities.

Whether such perturbation is for or against the direction towards the optimal solution, there is no theoretical answer now. But one point seems to be confirmed that the whole colony moves towards a good direction because PACS matches much more the nature behavior of real ants than ACS does. We therefore declare that the performance of PACS will not be worse than ACS, because real ants release pheromone in a real time way. We need more empirical study to justify it. The negative side of PACS is that it will need too much synchronization caused by very frequent access to pheromone matrix for the demand of pheromone update.

2.4 Parallel Hybridizing MMAS and ACS (MMACS)

MMAS and ACS have their own unique features respectively, and have been successfully applied to COPs. MMACS tries to parallel hybridize the two algorithms and take the advantages of the two. Of course such hybridization finally will be focused on the design of pheromone and its update mechanism.

To make the description easily understood, we only discuss the situation of two threads running two colonies in parallel, with MMAS and ACS respectively.

First, we have to make some adjustments for combining the pheromone settings of MMAS and ACS . For MMAS, we have $\tau_{ij} \in P_1 = [\tau_{min}, \tau_{max}]$, $\tau_{max} = 1/(\rho L_{cb})$, $\tau_{min} = \tau_{max}/(2n)$, $\tau_0 = 1/(\rho L_{best})$. For ACS, $\tau_{ij} \in P_2 = [\tau_0, 1/L_{cb}]$, $\tau_0 = 1/(\rho L_{best})$, where L_{cb} varies dynamically with the running of algorithm. When the solution evolves towards good quality, L_{cb} is decreasing but with a lower bound of L_{opt}. Thus the upper limit of P_2 is up to $1/L_{opt}$ and that of P_1 up to $1/(\rho L_{opt})$. Because ρ is a small positive real number, we have $\tau_{min} > \tau_0, \tau_{max} > 1/L_{opt}$. So for MMACS we take $\tau_{ij} \in P_1 \cap P_2 = [\tau_{min}, 1/L_{cb}]$, and the initial values $\tau_0 = \tau_{max} = 1/(\rho L_{best})$. Evaporation factor ρ takes the same settings as it has in MMAS and ACS respectively.

In MMACS, the parallel ACS colony is also applied the same re-initializing mechanism as MMAS. Just record the current result and expect the algorithm to find a better one to replace it. ACS benefits from this and has more opportunities to find a better solution or even the best one, and its performance and efficiency of the intensification and diversification are therefore improved. The condition of re-initialization here is as the same as in PMMAS, i.e., one of the two colonies reaches the presetting iteration number or convergence estimation.

Besides increasing the pheromone on the shortest tour after every iteration, the ACS colony of MMACS releases some pheromone on the corresponding edges at each move during the construction procedure, and this will immediately affect selection behavior of the MMAS colony of MMACS and direct the MMAS towards better solutions. Vice versa, the MMAS colony affects the ACS colony positively by re-initializing pheromone matrix.

3 Experimental Results

3.1 General Description

In this section when we compare the solution quality of the SHOP-ACO algorithms with that of the sequential ACO algorithms, we take the average solution of each algorithm running k (=10 in all our experiments) times under the fair running-time. By saying fair running-time, we mean that when we assign t_{max} to the compared sequential algorithms, respectively we restrict the running-time of our SHOP-ACO algorithms to t_{max}/p, where p is the number of colonies or parallel processors. Thus, we unify the time-cost of parallel algorithms and sequential algorithms into a fair criterion. Accordingly we assign $1/p$ the number of the ants of the sequential colony to every parallel colony.

From the view of taking computing resources, SHOP-ACO algorithms use less memory resource than the sequential algorithms. So theoretically speaking, SHOP wins if it achieves almost the same solution quality as the sequential algorithms.

The computing environment for conducting the experiments here is IBM p-server with 2 Power5-CPU (1.5GHz) and 6.0GB memory. The operating system is AIX 5.3 and the programming tool is PThread+gcc. Because of this hardware environment, we just implement two parallel ant colonies in our experiments. 10 typical problem instances randomly chosen from the TSPLIB [9] are divided into small scale (less than 1000 cities) group and large scale (less than 6000 cities) group.

The target sequential ACO algorithms are SMMAS (Sequential MMAS) and SACS (Sequential ACS). The common parameter settings, except for indicating explicitly, are as follows: $\alpha = 1$, $\beta = 2$, $\rho = 0.2$, $\lambda_{branch} = 1$, and $q_0 = 0.98$ (if ACS-concerned).

SHOP does not focus on parallelizing the local search and ACO algorithms. SHOP adopts the same strategy about local search as that of ACS and MMAS, that is, PMMAS, PACS and MMACS all adopt 3-opt local search.

3.2 Comparison with the Sequential Algorithms

Table 1 lists some of the results collected from our experiment.

From Table 1 we can see, all the optimal of the small scale instances can be caught by all the 5 algorithms. So let us focus on the average quality of solutions each algorithm achieved. Figure 1 indicates e_{avg} plots of the 5 algorithms testing on small scale problems over 10 runs. From Fig. 1 we have clearly found that PACS performs better than SACS, but SMMAS is the best. It seems to us that transfusing the mechanism of local pheromone update into SMMAS does not work very well because SMMAS is designed to have only global pheromone update.

Table 1 tells us that for the large scale TSP instances the 5 algorithms can hardly find the optimal. The only little exception is the result on *pr2392* - only PMMAS found the optimal. So let us focus on the excess of the best solution found by these algorithms. Fig. 2 indicates e_{bst} plots of the 5 algorithms testing on large scale problems over 10 runs. This time we have noticed that PMMAS performs better than others. It seems to us that integrating some sort of perturbations into SMMAS will make PMMAS perform better on large scale problems.

So based on the above illustration we conclude that SHOP algorithms overall achieve comparatively better solution quality than sequential ones. It is not reasonable to interpret such performance as being caused by the memory-sense tasks [10], because the space complexity of SHOP-ACO algorithms is almost p times the size of the sequential ones. On the contrary the synchronization of SHOP-ACO takes additional time in SHOP implementations. By now we temporarily attribute the performance to SHOP strategy. Of course we need more numerical analysis to justify this statement. We hope that SHOP strategy affects positively on the improvement of solution quality.

Table 1. SHOP testing on 10 TSP instances

instance	algorithm	$Iter_{avg}$	Sol_{std}	t_{avg}	e_b	e_{avg}	instance	$Iter_{avg}$	Sol_{std}	t_{avg}	e_b	e_{avg}
lin318	pmmas	210	37.82	4.54	0	0.000393	rl1304	594.4	420.74	101.54	0	0.001782
(20)	smmas	289.7	36.05	5.33	0	0.000271	(30)	1022.3	171.07	149.53	0	0.00042
(20)	pacs	88.5	52.18	2.85	0	0.001651	(300)	539.5	339.84	93.69	0	0.001415
	sacs	106.9	61.3	1.72	0	0.002099		296.7	492.93	42.08	0	0.00177
	mmacs	158.8	75.93	2.94	0	0.001228		441.3	293.34	69.19	0	0.001469
pcb442	pmmas	504.2	42.31	10.51	0	0.002371	fl1577	980.8	28.76	130.5	0.002427	0.004571
(20)	smmas	906	66.53	17.01	0	0.00139	(30)	1859.9	13.62	202.51	0.002742	0.003564
(40)	pacs	516	68.63	9.58	0	0.001069	(300)	717.6	82.34	99.21	0.000315	0.002683
	sacs	487	93.98	7.6	0	0.001719		1661.6	67.54	174.19	0.000225	0.002068
	mmacs	825.8	66.18	13.25	0	0.001617		733.9	449.39	205.88	0.003029	0.004685
d657	pmmas	362.3	34.77	35.87	0.00002	0.00094	pr2392	448.9	1724.63	223.86	0	0.008381
(20)	smmas	1394.4	23.87	64.92	0.00002	0.000599	(50)	722.6	599.85	271.68	0.000251	0.003131
(100)	pacs	617.5	31.09	30.71	0.00002	0.000961	(500)	500.6	727.36	230.21	0.001754	0.00427
	sacs	684.7	28.23	24.64	0.000368	0.001288		1143.4	518.74	391.93	0.000635	0.002961
	mmacs	515.2	43.34	20.77	0.00002	0.001157		553.9	117.25	404.54	0.006604	0.012029
gr666	pmmas	437.4	171.18	41.65	0.000156	0.00078	fl3795	704.1	1087.25	467.18	0.007264	0.023735
(20)	smmas	1101	88.57	44.41	0	0.000229	(50)	1085.5	72.5	681.78	0.007924	0.011577
(100)	pacs	320.4	215.81	29.76	0.000166	0.000843	(1000)	366.2	193.95	415.14	0.003441	0.009568
	sacs	1664.3	299.04	79.1	0.00232	0.003382		1591	54.34	885.81	0.000382	0.002468
	mmacs	833.2	188.91	34.81	0	0.000983		553.9	117.25	404.54	0.006604	0.012029
rat783	pmmas	826.9	8.08	45.6	0	0.00117	rl5915	87.7	7699.91	515.12	0.029273	0.034468
(20)	smmas	1848.3	3.87	76.89	0	0.000284	(100)	493.5	1638.72	1956.81	0.006905	0.011835
(140)	pacs	625.1	5.22	35.11	0	0.000784	(2000)	75	2288.76	793.3	0.015997	0.022486
	sacs	1715	11.7	62.43	0	0.002044		348.8	1054.5	1441.59	0.01914	0.022142
	mmacs	1213.5	12.91	51.71	0.001022	0.002714		256.5	7345.55	864.57	0.006811	0.015136

The first number in bracket indicates the number of ants of the sequential colony. The second is the maximum computing time allowed (in seconds).
$Iter_{avg}$: Average iterations for getting all the best solutions. Sol_{std} : Standard deviation of all the best solutions. t_{avg} : Average time for getting all the best solutions. e_b : The excess from the optimal of the Best solution. e_{avg} : The excess from the optimal of Sol_{avg}.

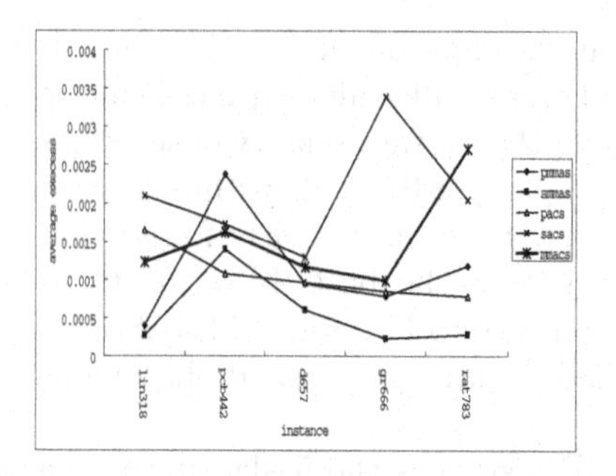

Fig. 1. The excess of the average quality comparison between 5 ACO algorithms on small scale TSP instances

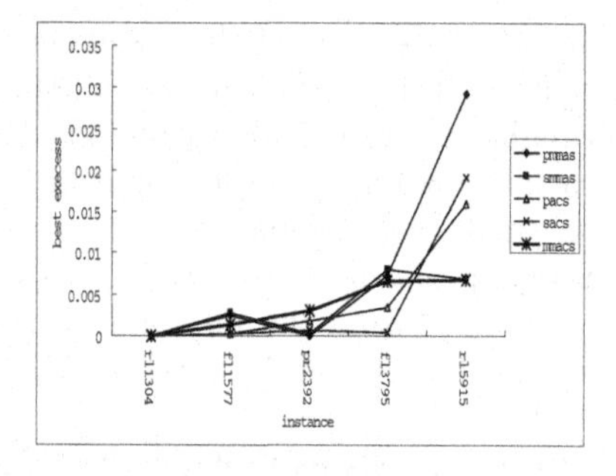

Fig. 2. The excess of the best quality comparison between 5 ACO algorithms on large scale TSP instances

4 Conclusion and Future Work

SHOP presents a very simple but efficient parallel approach to ACO. Future work will be carried on, such as,

- More fine synchronization. The current implementation of SHOP just locks the pheromone matrix as a whole when one ant needs to access even one element of the matrix. This will be a heavy negative factor for the performance. Obviously, it is not necessary for ants to lock the matrix as a whole. But more control logic will have to be introduced if more fine synchronization is implemented. So the next question will be:

- Investigating how synchronization affects the performance. More detailed analysis in terms of quantity relationship between synchronization and performance will be required to guide a fine synchronization.
- With the number of processors increasing, does our results show the same dominance? How about groups sharing pheromone matrix?
- Are the results reported here fortuitous or certain? We are applying SHOP implementation to more applications, such as learning Bayes Network, MCP, QAP, MKP, Chinese Word Segmentation and so on.

Our work illustrates that the biological features of ACO provide us much more research ideas on parallel hybrid metaheuristics. The behavior of parallel ACO is naturally close to the original behavior of natural ants.

References

1. Bullnheimer, B., Kotsis, G., Strauß, C.: Parallelization strategies for the ant system. Applied Optimization **24** (1998) 87–100
2. Middendorf, M., Reischle, F., Schmeck, H.: Multi colony ant algorithms. Journal of Heuristics **8**(3) (2002) 305–320
3. Stützle, T.: Parallelization strategies for ant colony optimization. 5th International Conference on Parallel Problem Soving for Nature, Lecture Notes in Computer Science **1498** (1998) 722–731
4. Talbi, E.G., Roux, O., Fonlupt, C., Robillard, D.: Parallel ant colonies for combinatorial optimization problems. Lecture Notes in Computer Science **1586** (1999) 239–247
5. Talbi, E.G., Roux, O., Fonlupt, C., Robillard, D.: Parallel ant colonies for the quadratic assignment problem. Future Generation Computer Systems **17** (2001) 441–449
6. Randall, M., Lewis, A.: A parallel implementation of ant colony optimization. Journal of Parallel and Distributed Computing **62** (2002) 1421–1432
7. Chu, S.C., Roddick, J.F., Pan, J.S., Su, C.J.: Parallel ant colony systems. International Syposium on Methodologies for Intelligent Systems, Lecture Notes in Computer Science **2871** (2003) 279–284
8. Delisle, P., Krajecki, M., Gravel, M., Gagné, C.: Parallel implementation of an ant colony optimization metaheuristic with openmp. International Conference on Parallel Architectures and Compilation Techniques (2001)
9. http://www.iwr.uni-heidelberg.de/groups/comopt/software/TSPLIB95/.
10. Fischer, D.: On super linear speedups. Parallel Computing **17** (1991) 695–697

An ACO-Based Clustering Algorithm

Yucheng Kao and Kevin Cheng

Tatung University, Taipei, Taiwan
ykao@ttu.edu.tw

Abstract. Data clustering is one of important research topics of data mining. In this paper, we propose a new clustering algorithm based on ant colony optimization, called Ant Colony Optimization for Clustering (ACOC). At the core of the algorithm we use both the accumulated pheromone and the heuristic information, the distances between data objects and cluster centers of ants, to guide artificial ants to group data objects into proper clusters. This allows the algorithm to perform the clustering process more effectively and efficiently. Due to the nature of stochastic and population-based search, the ACOC can overcome the drawbacks of traditional clustering methods that easily converge to local optima. Experimental results show that the ACOC can find relatively good solutions.

1 Introduction

The data clustering problem considers grouping a data set into several non-empty subsets of some similarity. Many clustering methods have already been devised, for examples, partitioning methods, hierarchical methods, density-based methods, grid-based methods, and model-based methods [1]. In this paper, we will focus on the partitioning methods, especially on the k-means algorithm. In this algorithm, k data objects are randomly selected to represent initial cluster centers, and then each data object is assigned into its closest clusters, based on the distance between the data and the cluster center. After assignments, cluster centers are determined again by computing the mean value of the objects for each cluster. This process iterates until the cluster centers do not change anymore. Although a good clustering quality can be obtained, the k-means algorithm has some drawbacks [2]. For example, the selection of initial cluster centers may affect the goodness of clustering results. Besides, the k-means algorithm sometimes falls into the trap of local minima.

To overcome the weaknesses of traditional partitioning methods, this paper proposes a new data clustering algorithm which is based on the concept of ant colony optimization. The rest of this paper is organized as follows. Section 2 presents a brief introduction to ant algorithms and their applications to data clustering problems. Section 3 provides the details of the proposed clustering algorithm, ACOC. Section 4 reports experimental results and analysis. A conclusion is drawn in Section 5.

M. Dorigo et al. (Eds.): ANTS 2006, LNCS 4150, pp. 340–347, 2006.

2 Ant Algorithms

Ant algorithms are a class of the algorithms based on artificial swarm intelligence, which is inspired by the collective behavior of social insects. For real ants, they have two important types of behavior: foraging and clustering. The algorithms of ant colony optimization (ACO) have their origins in the ant foraging behavior. They were proposed by Dorigo *et al.* [3,4] and are useful in solving discrete optimization problems. The ant clustering algorithms originated from the studies of ant clustering of dead bodies. They were introduced by Deneubourg *et al.* [5] and improved by Handl *et al.* [6] and are mainly applied to solve data clustering problems.

In the ACO algorithm, an artificial ant colony simulates the pheromone trail following behavior of real ants. Artificial ants move on a synthetic map (construction graph [9]) representing a specific problem to construct solutions successively. The artificial pheromone that corresponds to the record of routes taken by the ant colony is accumulated at run-time through a learning mechanism. Individual ants concurrently collect necessary information, stochastically make their own decisions, and independently construct solutions in a stepwise procedure. The information required for making a decision at each step include pheromone concentration, problem data and heuristic function values. The pheromone laid on the path belonging to the iteration-best solution will be positively increased to become more attractive in the subsequent iterations. Because of autocatalytic and collective behavior, ACO can effectively and efficiently solve a wide class of combinatorial optimization problems [9]. In this paper, we will apply the ACO to the data clustering problem.

The data clustering problem is an NP-hard problem when the number of clusters is larger than three [8]. In the past, different meta-heuristic approaches have been applied to solve the data clustering problem. For example, Al-Sultang introduced a tabu search-based algorithm [7] and Maulik *et al.* proposed a genetic algorithm-based approach [10]. They all showed that their own algorithms were superior to the traditional k-means method. An ACO-based algorithm for data clustering problems was first introduced by Shelokar *et al.* [11]. They did not give a name to their algorithm, so we call it the Shelokar algorithm hereafter. The algorithm mainly relies on pheromone trails to guide ants to select a cluster for each data object, and a local search is required to randomly improve the iteration-best solution before updating pheromone trails. They defined that the trail value, τ_{ij}, at node (i, j) represents the pheromone concentration of object i associated to the cluster j. In the algorithm artificial ants visit data objects one by one in sequence and select clusters for data objects by considering pheromone information only. Their experimental results showed that the algorithm can effectively solve a variety of clustering problems. However, based on our experimental observation, premature convergence may occur, infeasible solutions that contain empty clusters may be produced, and longer computation times are often required. These problems may result from that the algorithm relies on pheromone trails only.

In this paper, attempts were made to improve the Shelokar algorithm by introducing the concept of dynamic cluster centers in the ant clustering process and by considering pheromone trails and heuristic information together at each solution construction step.

3 ACOC Algorithm

In this section we mathematically formulate the data clustering problem, conceptually introduce our ACO-based clustering algorithm and briefly describe the computing procedure of the proposed algorithm.

3.1 Mathematical Formulation

We model the data clustering problem as a clustering optimization problem. Given a data set containing m data objects with n attributes and a predetermined number of clusters (g), the proposed algorithm has to find out an optimal cluster configuration such that the total sum of clustering errors for all data objects can be minimized. Equation (1) is the objective function. Constraint (3) states that each data object belongs to only one cluster, and constraint (4) states that no cluster is empty.

$$\text{Minimize } J(W,C) = \sum_{i=1}^{m}\sum_{j=1}^{g} w_{ij} \left\| X_i - C_j \right\| \tag{1}$$

$$\text{Where } \left\| \mathrm{X}_i - \mathrm{C}_j \right\| = \sqrt{\sum_{v=1}^{n} (x_{iv} - c_{jv})^2} \tag{2}$$

Subject to

$$\sum_{j=1}^{g} w_{ij} = 1, \qquad i = 1, ..., m \tag{3}$$

$$\sum_{j=1}^{g} w_{ij} \geq 1, \qquad j = 1, ..., g \tag{4}$$

$$w_{ij} = \begin{cases} 1, & \text{if data } i \text{ is clustered into cluster } j \\ 0, & \text{otherwise} \end{cases} \qquad i = 1, ..., m \text{ and } j = 1, ..., g \tag{5}$$

$$C_j = \frac{\sum_{i=1}^{m} w_{ij} X_i}{\sum_{i=1}^{m} w_{ij}}, \qquad j = 1, ..., g \tag{6}$$

Where, $\mathbf{x}_i$ is the vector of ith data object and $\mathbf{x}_i \in \mathrm{R}^n$; x_{iv} is the value of vth attribute of ith data object, $\mathbf{c}_j$ is the vector of jth cluster center and $\mathbf{c}_j \in \mathrm{R}^n$; c_{jv} is the value of vth attribute of jth cluster center, w_{ij} is the associated weight value of $\mathbf{x}_i$ with $\mathbf{c}_j$, $\mathbf{X}$ is the data matrix of size $m \times n$, $\mathbf{C}$ is the cluster-center matrix of size $g \times n$, and $\mathbf{W}$ is the weight matrix of size $m \times g$.

Basic Idea. In the ACOC algorithm, the solution space is modeled as a graph of object-cluster node matrix. The number of rows equals m, and the number of columns equals g. Each node denoted by $\mathrm{N}(i,\ j)$ means that data object i would be assigned to cluster j. Artificial ants can stay only one of g nodes for each object. Fig. 1 illustrates an example of construction graphs for clustering problems, where hollow circles denote unvisited nodes and solid circles represent visited nodes. A string is used to represent solutions built by ants. Considering the clustering result of Fig. 1, the corresponding solution string is (2, 1, 2, 1, 3, 3).

On the graph, each ant moves from one node to other, deposits pheromone on nodes, and constructs a solution in a stepwise way. At each step, an ant randomly selects an ungrouped object and adds a new node to its partial solution by considering both pheromone intensity and heuristic information. The memory list ($\mathbf{tb}^k$) can prevent a data object from being clustered more than once by an ant. When the memory list is full, it means that the ant completes solution construction. The moving sequence of the example in Fig. 1 is marked by the numbers next to the dotted arcs.

In the ACOC, ants deposit pheromone on nodes. The nodes with stronger pheromone would be more attractive to ants. The ACOC uses a pheromone matrix (**PM**) to store pheromone values. The heuristic information indicates the desirability of assigning a data object to a particular cluster. It is obtained by calculating the reciprocal of the Euclidean distance between the data object to be grouped and each cluster center of some ant. The nodes with higher heuristic values would be more likely to be selected by ants. Each ant carries a cluster-center matrix ($\mathbf{C}^k$) to store its own cluster centers and updates them right after each clustering step.

3.2 Computing Procedure

The complete procedure of ACOC is described as follows.

Step 1 Initialize the pheromone matrix: The elements of the pheromone matrix (**PM**) are set to arbitrarily chosen small values (τ_0).

Step 2 Initialize all ants: Start a new iteration. Reset the memory list ($\mathbf{tb}^k$), cluster center matrix ($\mathbf{C}^k$) and weight matrix ($\mathbf{W}^k$) for each ant, where $k = 1\sim R$. R is the total number of ants, $R \leq m$.

Step 3 Select data object i: Each ant randomly selects a data object, i, that is not in its memory list.

Step 4 Select cluster j: To determine j for a selected i, two strategies, exploitation and exploration, can be applied. The former is to allow ants to move in a greedy manner to a node whose product of pheromone level and heuristic value is the highest (see equation (7)). The latter is to allot probabilities to candidate nodes, and then let an ant choose one of them in a stochastic manner according to equation (8). The more promising a node, the higher probability it has. Ants choose one of these strategies by using equation (7) with a priori defined probability q_0 and a randomly generated probability q. Based on equations (7) and (8), ants can determine the value of j. Note that $\eta_{ij}^k = 1/d^k(i,\ j)$ and is the

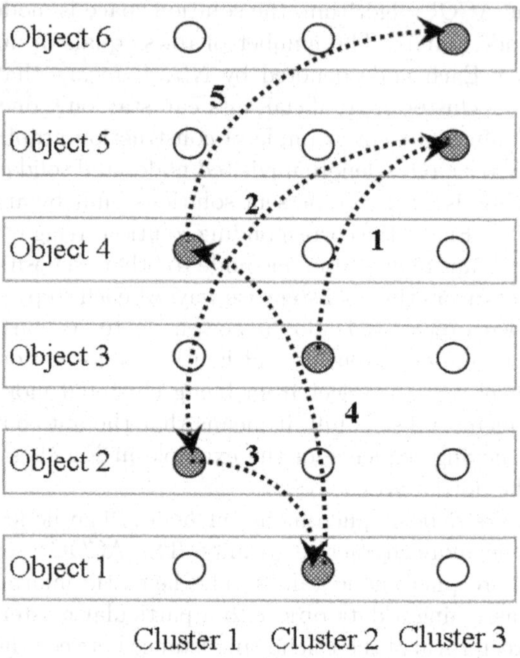

Fig. 1. Construction graph for ACOC

heuristic value of N(i, j) for ant k. The distance between object i and center j of ant k, $d^k(i, j)$, is defined in equation (9).

$$j = \begin{cases} \arg\max_{u \in Ni}\{[\tau(i,u)][\eta^k(i,u)]^\beta\} & \text{if } q \le q_0 \\ S & \text{otherwise} \end{cases} \tag{7}$$

where Ni is the set of g nodes belonging to data object i, and the value of S is chosen according to equation (8).

$$P^k(i,j) = \frac{[\tau(i,u)]\,[\eta^k(i,u)]^\beta}{\sum_{j=1}^{g} [\tau(i,u)]\,[\eta^k(i,u)]^\beta} \tag{8}$$

where β is the parameter specifying the relative weight of η^k_{ij}, $\beta > 0$.

$$d^k(i,j) = \sqrt{\sum_{v=1}^{n} \left(x_{iv} - c^k_{jv}\right)^2} \tag{9}$$

where c^k_{jv} refers to the value of attribute v of cluster center$j$ of ant k.

Step 5 Update ants' information: Update the memory list ($\mathbf{tb}^k$), weight matrix ($\mathbf{W}^k$, use equation (5)) and cluster center matrix ($\mathbf{C}^k$, use equation (6)) of each ant.

Step 6 Check memory list of each ant: Check if the memory list of each ant is full. If it is not, then go back to step 3; otherwise, go to step 7.

Step 7 Calculate objective function values: Calculate the objective function value of each ant, J^k, by using equation (1). After that, rank R ants (solutions) in the ascending order of J^k values. The best solution is called iteration-best solution. It is compared with the best-so-far solution, and the better one will be the new best-so-far solution.

Step 8 Update pheromone trails: Update the pheromone matrix, **PM**. The global updating rule is applied, and only the elitist ants are allowed to add pheromone at the end of each iteration. The trail updating equation is defined below.

Where ρ is the pheromone evaporation rate, $(0.0 < \rho < 1.0)$, t is the iteration number, r is the number of elitist ants, and $\Delta\tau_{ij}^h = 1/J^h$.

Step 9 Check termination condition: If the number of iterations exceeds the maximum iteration number, then stop and output the best-so-far solution, otherwise go to step 2.

4 Numerical Experiments

The proposed algorithm has been coded in Visual Basic and executed on a Pentium M 1.6 GHz notebook computer. The performance of the proposed ant algorithm was evaluated by testing on four datasets. For the artificial datasets, the two problems reported in Shelokar [11] were selected in this paper. The first dataset is composed of three clusters and each cluster has 50 data objects, while the second dataset has six clusters with 25 data objects in each cluster. For the real life datasets, iris and wine were selected from the website of UCI repository of machine learning databases [12].

The study compared the performance of the k-means, Shelokar and ACOC. For each test problem, these three algorithms were performed 10 times individually. The parameter values used in Shelokar and ACOC were: $R = 10, r = 1, q_0 = 0.0001$, τ_0 = uniform distribution [0.7, 0.8], $\beta = 2.0$, $\rho = 0.1$, local search rate = 0.01 and the maximum number of iterations = 1000. Besides, we modified the way of pheromone updating used in the Shelokar algorithm to avoid solution premature.

Table 1 summarizes the quality and efficiency of solutions for all selected problems obtained by three algorithms, respectively. For each test problem, the best, average and worst objective function values found in 10 distinct runs are sorted in the table. The corresponding computation times to attain the best solution in 10 runs are also listed.

For the three algorithms, the best experimental results have been put as bold in Table 1. The results show that the ACOC algorithm is the best in terms of the quality of solutions for all selected problems. The standard deviations of ACOC are also very small. On the other hand, the ACOC algorithm is better than the Shelokar algorithm but worse than the k-means algorithm in terms of the processing time required attaining the best solutions. Fig. 2 illustrates that the ACOC finds very good solutions very quickly, compared to the Shelokar algorithm,. The data values are obtained from one of 10 runs of each method applied over the first dataset. However, the ACOC has larger standard deviations of processing time (see Table 1), and it is worth solving this problem in the future.

Table 1. Comparison of three algorithms applied to four test problems

		Objective function value				CPU time(seconds)			
K-means	g	Best	Avg	Worst	Stdev	Best	Avg	Worst	Stdev
Prob. 1	3	144.69	148.47	156.28	5.42	0.01	0.012	0.03	0.006
Prob. 2	6	109.50	113.59	127.10	5.14	0.01	0.015	0.03	0.007
Iris	3	97.33	99.84	122.28	7.89	0.001	0.01	0.02	0.007
Wine	3	16555.68	16900.54	18436.96	730.79	0.01	0.03	0.06	0.014
Shelokar	g	Best	Avg	Worst	Stdev	Best	Avg	Worst	Stdev
Prob. 1	3	144.46	146.20	150.73	2.25	112.34	13.05	13.84	0.44
Prob. 2	6	199.68	207.02	211.95	3.48	9.21	12.09	13.52	1.77
Iris	3	97.22	97.78	98.74	0.56	14.58	15.53	16.10	0.54
Wine	3	16860.17	18022.16	19818.06	817.40	31.50	32.70	33.57	0.77
ACOC	g	Best	Avg	Worst	Stdev	Best	Avg	Worst	Stdev
Prob. 1	3	144.46	144.46	144.46	0.00	1.51	4.12	10.55	3.05
Prob. 2	6	108.75	109.34	114.44	1.79	2.08	11.47	16.75	4.41
Iris	3	97.22	97.22	97.23	0.00	3.58	8.52	19.22	4.59
Wine	3	16530.54	16530.54	16530.54	0.00	4.88	19.04	43.37	11.69

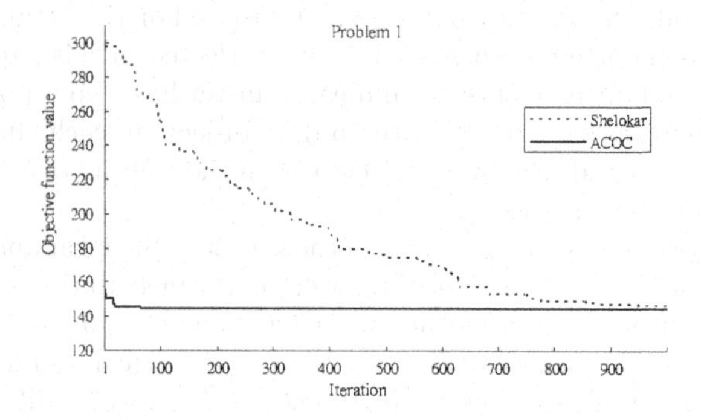

Fig. 2. Comparison of ACOC with Shelokar

5 Conclusion

This paper proposes a new algorithm based on ant colony optimization to solve the data clustering problem. In the ACOC algorithm, the solution space is modeled as an array-based construction graph on which artificial ants randomly move from node to node to construct solutions successively. Through many ants concurrently and stochastically searching for the best solution, the possibility of falling into local optima is decreased. At an iteration level, each ant will gradually form its own cluster centers for calculating heuristic values. When selecting the next node, ants consider not only the pheromone levels but also heuristic values of candidate nodes. This allows ants to find better solution in a shorter time. Numerical experiments showed that the ACOC algorithm can find better solutions than its counterparts and is faster than the Shelokar algorithm in terms of the processing time required.

References

1. Han, J. and Kamber, M.: Data mining: Concepts and Techniques. Morgan Kaufmann Publisher, San Francisco (2001)
2. Peña, J. M., Lozano, J. A., Larrañaga, P.: An empirical comparison of four initialization methods for the K-Means algorithm. Pattern Recognition Letters 20 (1999) 1027-1040
3. Dorigo, M., Maniezzo V., Colorni, A.: Ant System: Optimization by a Colony of Cooperating Agents. IEEE Trans. Sys. Man Cyb. B 26 (1996) 29-41
4. Dorigo, M., Gambardella, L.: Ant Colony System: A Cooperative Learning Approach to the Traveling Salesman Problem. IEEE Trans. Evol. Comp. 1 (1997) 53-66
5. Deneubourg, J. L., Goss, S., N., Sendova-Franks, A., Detrain, C, Chretien, L.: The dynamics of collective sorting robot-like ants and ant-like robots. Proc. Of the 1st Conf. on Sim. of Adaptive Behavior (1991) 356-363
6. Handl, J., Knowles, J., Dorigo, M.: Ant-based clustering and topographic mapping. Artificial Life 12 (2006) 35-61
7. Al-Sultan, K. S.: A tabu search approach to the clustering problem. Pattern Recognition 28 (1995) 1443-1451
8. Welch, J. W.: J. Stat. Comput. Simulat. 15 (1983) 17-25
9. Dorigo, M., Stützle, T.: Ant Colony Optimization. MIT Press, Cambridge, MA (2004)
10. Maulik, U., Bandyopadhyay, S.: Genetic algorithm-based clustering technique. Pattern Recognition 33 (2000) 1455-1465
11. Shelokar, P. S., Jayaraman, V. K., Kulkarni, B. D.: An ant colony approach for clustering. Analytica Chimica Acta 509 (2004) 187-195
12. UCI Repository of Machine Learning Databases. (1998) `http://www.ics.uci.edu/~mlearn/MLRepository.html`

An Adaptive Search Heuristic for the Capacitated Fixed Charge Location Problem

Harry Venables[1] and Alfredo Moscardini[2]

[1] Sunderland Business School, University of Sunderland, United Kingdom
harry.venables@sunderland.ac.uk
[2] School of Computing & Technology, University of Sunderland, United Kingdom
alfredo.moscardini@sunderland.ac.uk

Abstract. The Capacitated Fixed Charge Location Problem (CFCLP) consists of selecting a set of facilities that must completely supply a set of customers at a minimum cost. The CFCLP is *NP-hard* thus solution methods are often obtained by using sophisticated techniques. However, if a set of facilities is known a priori then the CFCLP reduces to a transportation problem (TP). Although this can be used to derive solutions by randomly selecting sufficient facilities to be fixed open and noting any cost improvements, it is perceived as a poor technique that does not guarantee solutions near the optimal. This paper presents an adaptive sampling algorithm using Ant Colony Optimization (ACO). We hypothesize that random selection of facilities using ACO will generate at least near-optimal solutions for the CFCLP. Computational results for a series of standard benchmark problems are presented which appear to confirm this hypothesis.

1 Introduction

The Capacitated Fixed Charge Location Problem (CFCLP) considers the problem of obtaining a subset from a set of potential facilities that have to supply a set of customers at a minimum cost, where each customer has an associated demand to be met and each facility has a finite amount of supply available. Facility location has applications in various domains: public sector, private sector and in environmentally sensitive areas [1,2,3].

There are various solution techniques available for the CFCLP which fall into two classes: exact or approximate methods. A review of most of the common techniques is given in [4]. A great deal of success in solving location problems has been achieved using Lagrangean relaxation combined with various local search strategies [5,6] and more recently [7,8]. These methods are usually designed for specific types of CFCLPs and are dependent on problem size and definition of the allocation variables. However, these techniques often involve high levels of sophisticated skills to develop and implement successfully.

The aim of this paper is to develop a method in which Ant Colony Optimization (ACO) plays a principal role in solving the CFCLP. The methodology is iterative and consists of three phases. The first is an adaptive construction

M. Dorigo et al. (Eds.): ANTS 2006, LNCS 4150, pp. 348–355, 2006.

phase concerned with obtaining facilities required to be opened that are selected according to their likelihood of being in a solution. The second phase attempts to make improvements to the current solution using a "drop" local search procedure. The third phase is to update certain variables associated with ACO. The whole process is repeated until some stopping criteria are met. The ACO algorithm used in the construction and update phases is adapted from the *MAX-MIN* Ant System [9,10].

2 Formulation of the CFCLP

A fixed charge location problem incurs some one off charges associated with opening a facilities and has a similar structure to the Transportation Problem (TP). If the set of facilities is known a priori then the problem often gives an unbalanced TP. Let us define:

$m =$ the number of potential facilities,
$n =$ the number of customers,
$q_j =$ the demand of customer j,
$Q_i =$ the capacity at facility i,
$c_{ij} =$ the cost of supplying all of the demand from customer j to facility i,
$f_i =$ the fixed charge associated with opening facility i,
$x_{ij} =$ the fraction of the demand of customer j supplied from facility i,

and the decision variable associated with opening a facility i

$$y_i = \begin{cases} 1 \text{ if facility } i \text{ is opened,} \\ 0 \text{ otherwise.} \end{cases}$$

The capacitated fixed charge location problem is defined as

$$\min \quad z = \sum_{i=1}^{m} \sum_{j=1}^{n} c_{ij} x_{ij} + \sum_{i=1}^{m} f_i y_i \tag{1}$$

such that

$$\sum_{i=1}^{m} x_{ij} = 1 \qquad j = 1, \ldots, n\,. \tag{2}$$

$$\sum_{j=1}^{n} q_j x_{ij} \le Q_i \qquad i = 1, \ldots, m\,. \tag{3}$$

$$x_{ij} \le y_i \qquad i = 1, \ldots, m, \quad j = 1, \ldots, n\,. \tag{4}$$

$$y_i \in \{0, 1\} \qquad i = 1, \ldots, m\,. \tag{5}$$

$$0 \le x_{ij} \le 1 \qquad i = 1, \ldots, m, \quad j = 1, \ldots, n\,. \tag{6}$$

where (1) is the objective function used to minimize the total fixed and transportation costs associated with facility and allocation variables y_i and x_{ij}. Constraint (2) ensures that the demand q_j of each customer j is satisfied, (3) ensures that an open facility i does not supply more than its capacity Q_i, (4) further strengthens (3) by only allowing the assignment of customer j to a facility i that is open, (5) is the integral condition, concerned with a facility i being selected as opened or closed and finally (6) refers to the fractional assignment condition that allows the demand of customer j to be allocated to more than one facility.

3 Ant Colony Optimization for the CFCLP

This study attempts to use the ACO paradigm, based on the behaviour of foraging ants [9,11,12], to solve the CFCLP by exploiting pheromone intensities corresponding to opening facilities using a *MAX-MIN* Ant System (MMAS) algorithm [10]. Although little literature is available on the use of ACO to solve capacitated location problems, MMAS was employed successfully to the Generalized Assignment Problem (GAP) by integrating an iterative local search procedure using a single ant approach [13]. GAP is a derivative of the $0-1$ CFCLP when facilities are fixed open. This study considers the effects of a single ant in obtaining a solution to the CFCLP. We represent the CFCLP as a fully connected graph whose nodes are facilities. Initially all facilities are closed and a single ant tour consists of walking on the graph so that each facility is visited only once; to determine if facilities should remain closed or be opened. Ants are guided to facilities by some combined pheromone intensity, τ_i, and a priori information about the problem instance that is known as *visibility*, η_i, [9].

If the potential of a facility being in the optimal is obtainable directly from the problem instance then this would provide a useful measure for an ant's visibility. One way of dealing with this issue is to consider relaxing the integral constraint (9) as illustrated in [14]. To achieve this the CFCLP is transformed into the Source Induced Fixed Charge Transportation Problem (SIFCTP), then a "total opportunity-cost" concept based on [15] is applied to the relaxed version, which allows visibility values associated with facilities to be calculated prior to use of the MMAS algorithm.

Derivation of the Ant's Visibility

It is possible to rewrite the CFCLP as a SIFCTP by redefining c_{ij} to be the cost of assigning a unit of demand from customer j to facility i, and x_{ij} becomes the amount of demand assigned from j to i. The objective function in (1) is unchanged whereas, the demand constraint (2) becomes

$$\sum_{i=1}^{m} x_{ij} = q_j \qquad j = 1, \ldots, n\,. \tag{7}$$

The capacity constraint (3) is now

$$\sum_{j=1}^{n} x_{ij} \leq Q_i \qquad i = 1, \ldots, m\,. \tag{8}$$

The strengthening constraint (4) associated with only assigning demand to open facilities is no longer required as a fixed cost is only incurred if some demand is assigned to a facility.

$$y_i = \begin{cases} 1 \text{ if } \sum_{j=1}^{n} x_{ij} > 0, \\ 0 \text{ otherwise.} \end{cases} \quad (9)$$

Finally, the fractional assignment condition given in (6) is replaced with

$$x_{ij} \geq 0 \quad i = 1, \ldots, m, \quad j = 1, \ldots, n. \quad (10)$$

The relaxed integral condition (9) is $y_i = \sum_{j=1}^{n} x_{ij}/m_{ij}$ where $m_{ij} = \min(q_j, Q_i)$ and is written into the objective function (1) to give an unbalanced TP as used by [14]:

$$\min z = \sum_{i=0}^{m} \sum_{j=0}^{n} C_{ij} x_{ij} \,, \quad (11)$$

subject to (7), (8) and (10) with unit transportation costs

$$C_{ij} = c_{ij} + f_i/m_{ij} \,. \quad (12)$$

Although this relaxed version could be easily solved, integral relaxation is seen as unreliable [16]. However, we use these relaxed costs combined with a method originally designed to give an approximate solution to the TP [15] to derive an ant's visibility. The approximate TP method initially calculates a "total opportunity-cost" matrix T_{ij}, where i and j are supply and demand points. Allocations are then made by considering the smallest T_{ij} values and assigning as much demand as allowed to the supply point without exceeding its capacity until all demand is assigned. The matrix is calculated:

$$T_{ij} = E_{ij} + F_{ij} \qquad i = 1, \ldots, m, \quad j = 1, \ldots, n. \quad (13)$$

Where E_{ij} and F_{ij} are the unit cost matrix row and column opportunity costs;

$$E_{ij} = c_{ij} - c_{ij^*} \qquad i = 1, \ldots, m, \quad j^* = \operatorname{argmin}_{j \in J} \{c_{ij}\} \,, \quad (14)$$

and

$$F_{ij} = c_{ij} - c_{i^*j} \qquad j = 1, \ldots, n, \quad i^* = \operatorname{argmin}_{i \in I} \{c_{ij}\} \,. \quad (15)$$

Rather than obtain the approximate solution to the TP described in (11) by using the method of [15], we consider the notion that demand assignments are assigned by using the lowest T_{ij} values. Thus an ant's visibility is related to those facilities having low T_{ij} values as they are more likely to be in the optimal solution. Define

$$\eta_i = \frac{1}{T_i} \qquad i = 1, \ldots, m. \quad (16)$$

Where

$$T_i = \sum_{j=1}^{n} T_{ij} \qquad i = 1, \ldots, m. \quad (17)$$

Equation (16) refers to the ant's visibility η_i. Effectively, those facilities with a greater visibility value are more likely to be found in the optimal solution.

MAX-MIN Ant System

Each facility i has a pheromone τ_i and a visibility η_i value made available to an ant. Moves on the graph are made from facility to facility according to a probabilistic decision rule based upon the pheromone intensity and visibility of those moves permitted. Each move is examined to see if an improvement in the objective function is observed; if so, then the corresponding facility is set to be open. This approach is essentially an ACO based ADD heuristic, where facilities are added to the current solution if an overall cost improvement occurs.

The selection process for making a move on the graph is based upon a pseudo-random proportional rule as used by [9]. A move is made with some probability q_0 and the facility with the greatest combined pheromone and visibility is chosen, otherwise a facility is chosen by using a probability function. The parameter q_0 predetermines the level of exploitation of strongest combined values whereas its complement allows for exploration of the solution space. Each move to a new potential facility i is determined using the decision rule:

$$i = \begin{cases} \text{argmax}_{l \in L} \left\{ [\tau_l]^\alpha [\eta_l]^\beta \right\}, & \text{if } q \leq q_0, \\ \text{I}, & \text{otherwise.} \end{cases} \tag{18}$$

Where q is a random variable uniformly distributed in $[0, 1]$, $0 \leq q_0 \leq 1$, L is the set of unopened facilities, and I is a random variable that is selected according to the following probability distribution:

$$p_i = \frac{[\tau_i]^\alpha [\eta_i]^\beta}{\sum_{l \in L} [\tau_l]^\alpha [\eta_l]^\beta} . \tag{19}$$

Where α and β are parameters corresponding to the influential roles of pheromone intensity τ_i and visibility information η_i. Once a move is made on the graph, based on this and previous moves, we approximate the corresponding TP using the method of [15] to determine if a facility is to be opened. A facility is opened if there is reduction in the objective function. A complete tour or iteration is when all facilities have been visited and their status determined.

Once a tour is complete then pheromone update takes place, which includes some evaporation and deposit of pheromone at each facility. Pheromone persistence at a facility is denoted by a parameter, $0 < \rho \leq 1$, that represents the rate of evaporation at each iteration which allows the algorithm to "forget" poor tours [9]:

$$\tau_i \leftarrow (1 - \rho)\tau_i \qquad i = 1, \ldots, m . \tag{20}$$

Pheromones are deposited on those facilities belonging to the best tour to-date:

$$\tau_i \leftarrow \tau_i + \Delta\tau_i^{best} \qquad i = 1, \ldots, m , \tag{21}$$

where $\Delta\tau_i^{best} = 1/z^{best}$ and z^{best} is the overall cost of the best tour. Upper and lower limits τ_{max} and τ_{min} are placed on the pheromones in an attempt to avoid convergence to a local optimum. These are set as $\tau_{max} = 1/\rho z^{best}$

and $\tau_{min} = \tau_{max}/a$ where a is a parameter. Also, τ_{max} is updated whenever an improvement is made in the best overall cost z^{best}. If the procedure begins to converge to a local optimum, or there is no improvement in the overall cost after a chosen number of iterations, then the pheromones are reset to the current value of τ_{max}. This is an attempt to encourage a new exploratory search away from the region of stagnation [10]. The method we use to measure stagnation is based on [9] which observes:

$$\frac{\sum_{\tau_i \in T} min\{\tau_{max} - \tau_i, \tau_i - \tau_{min}\}}{m} \to 0\,, \tag{22}$$

as the algorithm approaches stagnation, where T is the set of pheromones for the current tour and m is the number of facilities.

Local Improvement Phase

During the construction phase facilities are either fixed open or remain closed. This may result in some facilities being fixed open early on in the process that may later only play a minor role in accommodating customer demand. Improvements may be made locally by closing one or more facilities in the current solution by using a DROP heuristic. The facility that gives the best improvement is set to be closed. Thus, if the current solution has facilities $Y = \{y_i \mid y_i \in \{0,1\}\}$ with a total cost $z\,(Y)$ we select facility $i^* \mid y_{i^*} = 1$ that gives the least total cost and reset $y_{i^*} = 0$ and Y accordingly by approximating as series of TPs using the method of [15]. This process is repeated until no further improvements can be made.

4 Computational Experience

Experiments for a series of benchmark capacitated location problems whose optimal solution are known were used to test our hypothesis. They were coded in C++ and experiments were executed on a Dell Inspiron 8600 with a 1.60 GHz Pentium M processor and 786Mb RAM. The problems used are available from the OR-Library (http://people.brunel.ac.uk/ mastjjb/jeb/info.html). The following parameters were found to be robust to small changes: $\alpha = 2.5$, $\beta = 0.8$, $\rho = 0.06$, $q_0 = 0.5$ and $a = 2n$, where n is the number of customers. The number of iterations in each experiment was limited to two hundred as testing displayed little significant change in the best solution beyond this value. Pheromones are initialized to $1/\rho z_0$, where z_0 is an initial feasible solution obtained using a technique based on [15], they are reset to the current value of τ_{max} should there be no overall improvement after fifty iterations.

The results obtained support the use of the local search "drop" heuristic as out of 31 problems considered solutions were generated for 23 in a shorter time, 26 required fewer iterations and 19 benefited from the local search giving improved objective values, which is reinforced with lower errors from known optimal values. Without the "drop", the average error was 5.83% with a standard deviation

Table 1. Non-local search (Z_{NLS}) and facility "drop" local search (Z_{drop}) results for a selection of problem instances

			Z_{NLS}			Z_{drop}	
Problem	m×n	secs	iters	% err	secs	iters	% err
cap81	25×50	4.777	30	15.34%	4.267	5	11.48%
cap82		4.907	31	15.04%	3.605	4	9.75%
cap83		4.947	32	14.33%	5.88	9	9.09%
cap131	50×50	61.368	89	0.61%	129.216	115	0.33%
cap132		61.218	95	0.41%	11.877	10	0.41%
cap133		57.763	94	0.70%	2.593	2	0.26%
cap134		89.74	160	0.10%	55.149	82	0.10%

of 11.33%, whereas corresponding values including the "drop" were 3.16% and 3.94% respectively, which indicates significant improvements and greater reliability. Best and worst case performances were 0.10% and 12.45%. Solutions to 74% of the problems achieved an error of less than 5%, whilst 52% achieved an error of less than 1% indicating some worthy merit in the ACO application with local search. Table 1 displays the results obtained for two classes of problems as examples of good and poor performance.

5 Conclusions

The following conclusions are based on the results presented in the previous section. The results indicate convergent behaviour of the ACO algorithm and near optimal solutions within 1% error for most of the problems tested, which suggests the algorithm does have the potential to solve CFCLPs. The algorithm would benefit from employing an interchange heuristic in the improvement phase similar to those used by [5,6].

We adopted a visibility strategy based on the relaxation of facility integral constraints combined with a "total opportunity-cost" method which was aided by reformulating the CFCLP to a SIFCTP. Interestingly, linear relaxation of the integral condition for capacitated location problems was previously found it to be unreliable [16]. However, on the contrary, we found this to be useful in determining solutions because during parameter setting tests, better solutions were obtained with the inclusion of visibility. The strategy we adopted approximates a TP for every step in the construction and local search phases which is likely to be computationally inefficient. This is reflected in the CPU times.

Our results are certainly encouraging and thus we postulate that our hypothesis "random selection of facilities using ACO will generate at least near-optimal solutions for the CFCLP" is achievable in most of the cases. However, we do note that a better local search algorithm may reinforce this hypothesis. Future research should concentrate on addressing these issues raised concerning solution improvement and efficiency.

References

1. Daskin, M.: Network and Discrete Location: Models, Algorithms and Applications. John Wiley and Sons, Inc., New York (1995)
2. Drezner, Z., ed.: Facility Location. A Survey of Applications and Methods. Springer, New York (1995)
3. Love, R., Morris, J., Wesolowsky, G.: Facilities Location: Models and Methods. North Holland, New York (1998)
4. Sridharan, R.: Invited review: The capaciated plant location problem. European Journal of Operational Research **87** (1995) 203–213
5. Beasley, J.: Lagrangean heuristcs for location problems. European Journal of Operational Research **65** (1993) 383–399
6. Agar, M., Sahli, S.: Lagrangean heuristics applied to variety of large capacitated plant location problems. Journal of the Operational Research Society **49**(10) (1998) 1072–1084
7. Ahuja, R., Orlin, J., S.Pallottino, Scaparra, M., Scutellà, M.: A multi-exchange heuristic for the single-source capacitated facility location problem. Management Science **50**(6) (2004) 749–760
8. Holmberg, K., Ronnqvist, D., Yuan, D.: An exact algorithm for the capacitated facility location problem with single sourcing. European Journal of Operational Research **113** (1999) 544–559
9. Dorigo, M., Stützle, T.: Ant Colony Optimization. The MIT Press, Cambridge, MA (2004)
10. Stützle, T., Hoos, H.: The $\mathcal{MAX}-\mathcal{MIN}$ Ant System. Future Generation Computer Systems **16**(8) (2000) 889–914
11. Bonabeau, E., Dorigo, M., Theraulaz, G.: Swarm Intelligence: From Natural to Artificial Systems. Oxford University Press, New York (1999)
12. Theraulaz, G., Bonabeau, E.: A brief history of stigmergy. Artificial Life **5** (1999) 97–116
13. Lourenço, H., Serra, D.: Adaptive search heuristics for the generalized assignment problem. Mathware & Soft Computing **9** (2002) 209–234
14. Adlakha, V., Kowlaski, K.: A simple algorithm for the source-induced fixed-charge transportation problem. Journal of the Operational Research Society **55**(12) (2004) 1275–1280
15. Kirca, O., Satir, A.: A heuristic for obtaining an initial solution for the transportation problem. Journal of the Operational Research Society **41**(9) (1990) 865–867
16. Baker, B.: Linear relaxation of the capacitated warehouse location problem. Journal of the Operational Research Society **33** (1982) 475–479

An Ant Colony System for the Open Vehicle Routing Problem

Xiangyong Li and Peng Tian

Antai College of Economics & Management
Shanghai Jiaotong University, Shanghai, P.R. China
lixiangyong@163.com, ptian@sjtu.edu.cn

Abstract. This paper studies the open vehicle routing problem (OVRP), in which the vehicles do not return to the starting depot after serving the last customers or, if they do, they must make the same trip in the reverse order. We present an ant colony system hybridized with local search for solving the OVRP (ACS-OVRP). Additionally, a Post-Optimization procedure is incorporated in the proposed algorithm to further improve the best-found solutions. The computational results of ACS-OVRP compared to those of other algorithms are reported, which indicate that the ACS-OVRP is another efficient algorithm for solving the OVRP.

1 Introduction

The open vehicle routing problem (OVRP) is a special variant of the standard vehicle routing problem (VRP). The most important feature consists in that the route of VRP is *hamiltonian cycle*, whereas that of OVRP is *hamiltonian path*. Such a different can be attributed to that the vehicles in the OVRP are not required to return to the depot, or if they are required to do so, they must return exactly along the same trip in the reverse order. The OVRP is a basic distribution management problem that can be used to model many real-life problems. It can be encountered in many real-word problems, for example, a delivery company without its vehicle fleet contracts its delivery to the hired vehicles. In such instance, the delivery company is not concerned with whether the vehicles return the depot and does not pay any traveling cost between the last delivery customer and the depot. It can be modeled as a OVRP. Other applications fitting the OVRP framework include the newspaper home delivery problem[1] etc..

The earliest publication about the OVRP can trace back to the article by Schrage in 1981[2]. But in the following 20 years since 1981, OVRP received little study and there is no related publication. Since 2000 some researchers began to study the solutions and several methods have been developed. Sariklis and Powell[3] proposed a two-stage heuristic, i.e. clustering and routing phases. In the clustering phase, the customers were first assigned to the clusters by taking into account vehicle capacity constraints and then the clusters were improved by reassignments of customers among these clusters. In the second phase, the clusters are transformed into an open vehicle route by solving a minimum spanning tree

M. Dorigo et al. (Eds.): ANTS 2006, LNCS 4150, pp. 356–363, 2006.

problem (MSTP). A tabu search algorithm was presented in[4]. Brandão generated initial solutions by using the nearest neighbor heuristic and the heuristic based on a pseudo lower bound. The initial solution was then submitted to either the nearest neighbor method or the unstringing and stringing procedure to improve the solution cost. The neighborhood structure was defined based on insert and swap operators. Fu et al.[5] built another tabu search. The initial solution was generated by a farthest first heuristic and the neighborhood structure was defined on four different neighborhood moves, i.e., vertex reassignment, vertex swap, 2-Opt, and trail swap. Tarantilis et al.[6] developed an adaptive memory programming algorithm for the OVRP. The set of the OVRP solutions was stored in an adaptive memory that was dynamically updated during the search process. The sequences of vertices of these solutions were periodically extracted from the adaptive memory, giving a larger weight to the routes belonging to the best solution. The algorithm had two phases. In pool generation phase, the initial pool of routes was generated using the weighted savings. The solutions were then improved using a standard tabu search. In pool exploitation phase, promising sequences of vertices of the solution are extracted, a solution was generated and improved using tabu search, and the set of solutions was updated. In[7], a list based threshold accepting (LBTA) was presented for solving the OVRP. It was an annealing-based method using a list of threshold values to guide intelligently local search. Local search are performing by using 2-opt, 1-1 exchanges (swap two customers from either same or different routes), and 1-0 exchanges (move a customer from its position on one route to another position on the same route or a different route). Recently Li et al.[1] develops a variant of the record-to-record travel algorithm developed to handle the very large standard VRP[8] to solve the OVRP. The record-to-record travel was a deterministic variant of simulated annealing. For the detailed description, we refer the reader to Li et al.[8,1].

Because of the intrinsic complexity, it is impractical to find an optimum for many combinatorial optimization problems, e.g. OVRP in a moderate computation cost. A reasonable choice is to apply metaheuristic to quickly produce good solutions. One of the most successful metaheuristics is Ant Colony Optimization (ACO), which was a common framework for the existing applications and algorithmic variants of ant algorithms. Ant Colony System (ACS), a particular instance of ACO, has proved to be competitive compared with other metaheuristics. In this paper, we apply ACS to OVRP and propose an ACS hybridized with local search and a post-optimization procedure for solving the OVRP (ACS-OVRP). The rest of this paper is organized as follows. Section 2 describes ACS-OVRP in detail. The experimental results are reported in section 3. Section 4 concludes this paper.

2 Ant Colony System for the OVRP

2.1 Problem Definition

The OVRP is a relaxation of the classical VRP. From a point of view of graph, the OVRP can be defined by a complete weighted graph G=(V, E), where

$V = \{0, 1, \cdots, n\}$ is the node set, and $E = \{(i,j)|i,j \in V, i \neq j\}$ is the edge set. Node 0 is the depot, and $C = \{1, 2, \cdots, n\}$ denotes customer set. n is the number of customers. Each arc (i, j) is associated with a traveling distances d_{ij}. Each customer i, has a fixed demand q_i, to be delivered and a service time δ_i. Each vehicle serves a subset of customers on its route, which begins at the depot and ends at the last customer. Each vehicle has the same capacity Q and the traveling cost constraint L. The objective of the OVRP is to determine a set of minimum routes with minimum total traveling cost, which satisfy the constraints: (i) Each vehicle starts at the depot. It doesn't return to the depot, or if it do so, it must go back to the depot along the same trip in the opposite order; (ii) The service of each customer can only be fulfilled by one vehicle; (iii) The total demand of each route can not exceed the vehicle capacity; (iv) The total traveling cost of each route can not exceed the restriction L. The OVRP has multiple objectives, i.e. minimizing not only the number of vehicles required, but also the corresponding total traveled distanced. In general, a hierarchical objective is considered in the literatures, the number of routes is the primary objective and then the total travel distance is minimized for the obtained number of vehicle routes.

2.2 Ant Colony System for Solving the OVRP

Ant Colony System (ACS) is first proposed by Dorigo and Gambardella[9] and has proved to be one of the most promising ACO metaheuristics. In this section, we propose an algorithm based on ACS for OVRP. It works as follows. At each iteration, first a set of artificial ants probabilistically build the solutions, exploiting a given pheromone model. Then the generated solution can be improved by applying local search. After each ant builds its solution, the best-so-far solution is used to update the pheromone trail. Unlike AS, a local pheromone update procedure is included in ACS. The main procedures are iterated until termination condition is met. Additionally, a post optimization procedure is implemented to further improve the obtained optima. The main procedures are as follows.

Solution Construction. In ACS, m ants concurrently build the solutions of the OVRP by exploiting a probability model indicated by pheromone trail and heuristic information. Each ant starts from the node 0 (the depot) and probabilistically choose the next node until all customer nodes have been visited. When ant k is located at customer node i, it probabilistically chooses next city j to visit in the set of feasible nodes by using a pseudorandom proportional rule[9]:

$$j = \begin{cases} \arg\max_{l \in N_i^k} \{\tau_{il}^{\alpha} \eta_{il}^{\beta}\}, & \text{if } q \leq q_0 \\ J, & \text{if } q > q_0 \end{cases} \tag{1}$$

J is the customer node determined over the following probability distribution:

$$p_{ij}^k = \frac{\tau_{ij}^{\alpha} \eta_{ij}^{\beta}}{\Sigma_{l \in N_i^k} \tau_{il}^{\alpha} \eta_{il}^{\beta}}, \qquad \text{if } j \in N_i^k \tag{2}$$

where N_i^k is the set of all feasible nodes j still to be visited. τ_{ij} is the pheromone trail indicating the desirability of visiting customer j directly after customer i. η_{ij} is the heuristic information and it shows the heuristic desirability of choosing customer j as the next city. In the context of OVRP, we consider $\eta_{ij} = 1/d_{ij}$ as in TSP, where d_{ij} is the travel distance between node i and j. α and β are two parameters indicating the relative importance of the pheromone trail and the heuristic information. q is a random number uniformly distributed in the interval $[0,1]$; q_0 is a control parameter and $0 \leq q_0 \leq 1$. From equation (1), the node j maximizing the product $\tau_{il}^{\alpha}\eta_{il}^{\beta}$ will be chosen as the next city with probability q_0. While with probability $1-q_0$, the city j is chosen with the probability defined in equation (2). q_0 is an important parameter, which controls the tradeoff between the exploration and exploitation[9].

Local Search. The vast literatures indicate that ACO hybridized with a local search can obtain better solution. In this section, we consider some local search algorithm embedded in the ACS-OVRP. Our implementation of local search oscillates between the inter-route as well as intra-route improvements. For the intra-improvement, we use a method similar to 2-Opt for each vehicle route of the OVRP. When implementing 2-Opt to each vehicle route, the termination condition used is the best-accept (BA) strategy, i.e. all the neighbors are examined until no improvement can be obtained. In order to get an inter-route improvement, an exchange similar to 2-Opt*[10] is used. But it is modified in consideration of the property of OVRP, namely the route is not a tour. Another local search is swap operator, which performs by exchanging the customer in one route with other customer in another route. In the implementation for the OVRP, one customer or two adjacent customers are considered for swapping. We also use another local search, relocate operator. The basic idea is to eject a small sequence of customers (here we consider one customer) at the current location and try to improve the solution by reinserting the sequence at another location. Relocate operator is suitable for inter-route and intra-route improvements.

Except for the 2-Opt, other operators are implemented with FA criteria, i.e. they will stop if the first improvement is obtained. Local search is implemented in a random order. After the ants finish solution construction, we first produce a random permutation of integers 1-4 and then the local search is implemented in turn according to the permutation. Here we gradually increase the number of ants applying local search, to obtain a tradeoff between solution quality and computation time. ACS-OVRP first allows only the iteration-best ant to apply local search. Then the number *Elitist_num* of ants applying local search is increased by one every *num_iter* iterations and a upper bound is set to 10. When triggering local search, the ants are first ranked in terms of travel cost, and then only the *Elitist_num* best-ranked ants are chosen to apply local search.

Pheromone Trail Update. In the ACS-OVRP, the pheromone trail is updated by two updating procedures, local updating and global updating.

The local pheromone trail update rule is triggered immediately after the ant chooses a next city. The updating equation is as follows:

$$\tau_{ij} = (1-a) \cdot \tau_{ij} + a \cdot \tau_0 \quad (3)$$

where a $(0 < a \leq 1)$ is the evaporation rate. τ_0 is the initial pheromone trail and a good choice is to set $\tau_0 = 1/(n \cdot L_f)$. L_f is the length of initial solution generated by the nearest neighbor heuristic, and n is the number of customers. The effect of local updating is to reduce the pheromone trail on the visited arcs at current iteration, making these arcs less attractive for the following ants.

After all ants finish constructing their solutions the best-so-far ant is then allowed to deposit additional pheromone trail. The basic idea is that the information of the best-so-far solution is indicated by the pheromone trail and the arcs included in the best-so-far solution will be biased by other ants in the following iterations. The updating equation is as follows:

$$\tau_{ij} = (1-\rho) \cdot \tau_{ij} + \rho / L_{bs} \quad (4)$$

where L_{bs} is the length of the best found solution. $0 < \rho \leq 1$ is pheromone evaporation rate. In this updating equation, both pheromone evaporation and new pheromone deposit are only applied to the arc (i, j) included in the best solution. In consideration of property of OVRP, pheromone trail on the arc connecting the last customer and the depot, i.e. the arc $(i, 0), i \in C$, will not be updated.

Post-Optimization. In our ACS-OVRP, we consider a post-optimization procedure to further improve the best found solutions. When the termination condition of the main loop is met, the post-optimization procedure is performed to further improve the solution. Our Post-Optimization procedure consists of the above four operators mentioned in the subsection *"Local search"*. The operators are implemented in different orders to obtain the best possible improvement.

3 Experiment and Results

3.1 Experiment Setting

In this section, ACS-OVRP was tested on a set of problems. The problems C1-C14 were taken from Christofides et al.[11] and the data can be downloaded from the web (http://people.brunel.ac.uk/~mastjjb/jeb/info.html). The properties of the problems are summarized in Table 1. L is the maximum route length restriction which is the original route length restriction multiplied by 0.9 because the original problems were designed for VRP, not for OVRP[4,5]. For the vehicle without driving time limit, we set $L = \infty$. n is the number of customers. Q is the vehicle capacity. K_{min} is the estimated minimum number of route by $\lceil \Sigma_{i=1}^{n} d_i / Q \rceil$. δ is service time. All the experiments were implemented on a Pentium IV 1.8GHz with Matlab 7.0 simulation platform.

In the preliminary experiment, we checked the performance of ACS-OVRP under different parameters and found that ACS-OVRP performed better under the following parameters: $\rho = a = 0.1$, number of ants, 20; the size of candidate list, $\lceil n/5 \rceil$; the maximum iterations, 300. $num\_iter = 50$, $\alpha = 1$, and $\beta = 2$.

Table 1. Characteristics of the testing problems

No	n	Q	L	δ	K_{min}	No	n	Q	L	δ	K_{min}
C1	50	160	∞	0	5	C8	100	200	207	10	8
C2	75	140	∞	0	10	C9	150	200	180	10	12
C3	100	200	∞	0	8	C10	199	200	180	10	16
C4	150	200	∞	0	12	C11	120	200	∞	0	7
C5	199	200	∞	0	16	C12	100	200	∞	0	10
C6	50	160	180	10	5	C13	120	200	648	50	7
C7	75	140	144	10	10	C14	100	200	936	90	10

3.2 Computational Results and Comparison with Other Algorithms

We tested the performance of ACS-OVRP and compared it with other methods in the literatures. The algorithms compared with ACS-OVRP are as follows. SP: the method of cluster first, route second in[3]. TSB1 and TSB2: tabu search algorithm in[4]. TSB1 and TSB2 denote the tabu search algorithms with different initial solutions, namely the initial solutions produced by K-tree with unstringing and stringing, and by nearest neighbor heuristic. AMP: adaptive memory programming algorithm (AMP)[6]. BATA: the backtracking adaptive threshold accepting algorithm[12]. LBTA: the list based threshold accepting (LBTA) algorithm[7]. TSF and TSR: tabu search algorithm with different initial solutions[5]. In TSF, the initial solutions are produced by a farthest first heuristic. The ones of TSR are randomly generated. We note that some results are not correct in[5] because of some errors in program. We got the corrected results and reported them here. ORTR: the record-to-record travel algorithm[1]. All methods consider the number of routes as the primary objective. As in Brandão[4], the reported solutions contain only the traveling cost without service time.

We first studied the effect of post-optimization procedure. The results are reported in Table 2. It can be seen that the effect of the post-optimization is slightly significant and the obtained improvement is quite obvious, especially for large-scale testing problem. In particular, the solution quality of problem C1 and C2 can not be further improved. It is probably due to the fact that ACS-OVRP has found the best solutions for these problems. But it is an interesting observation that the solution quality of big problem, e.g. C4 and C5, has been improved by using post-optimization procedure. The results show that it is valuable to incorporate post-optimization procedure to further improve the solution quality.

Table 2. Summary of the effects of Post-Optimization

	C1	C2	C3	C4	C5	C6	C7
Post-Opt	5/416.1	10/571.7	8/649.0	12/748.4	16/1017.3	6/413.0	11/568.5
No Post-Opt	5/416.1	10/571.7	8/654.9	12/917.3	16/1083.2	6/413.1	11/568.5
	C8	C9	C10	C11	C12	C13	C14
Post-Opt	9/647.9	14/764.2	17/903.1	7/685.3	10/536.3	11/903.8	11/593.1
No Post-Opt	9/653.7	14/770.5	17/936.4	7/692.0	10/538.3	12/924.4	11/597.6

Table 3. Comparison of the optimums produced by ACS-OVRP and other algorithms

Problem	SP	TSB1	TSB2	AMP	BATA
C1	5/488.20	5/416.1	5/438.2	6/412.96	6/412.96
C2	10/795.33	10/574.5	10/584.7	11/564.06	11/564.06
C3	8/815.04	8/641.6	8/643.4	9/641.77	8/642.42
C4	12/1034.14	12/740.8	12/767.4	12/735.47	12/736.89
C5	16/1349.71	16/953.4	16/1010.9	17/877.13	**16/879.37**
C6		**6/412.96**	6/416.00		
C7		**10/634.54**	11/580.97		
C8		**9/644.63**	9/652.09		
C9		**13/785.2**	14/827.6		
C10		17/884.63	17/946.8		
C11	7/828.25	7/683.4	7/713.3	10/679.38	9/679.60
C12	10/882.27	10/535.1	10/543.2	**10/534.24**	**10/534.24**
C13		**11/943.66**	11/994.26		
C14		11/597.3	12/651.92		
Problem	LBTA	TSF	TSR	ORTR	ACS-OVRP
C1	6/412.96	5/416.1	**5/416.06**	**5/416.06**	**5/416.06**
C2	11/564.06	10/569.8	**10/567.14**	**10/567.14**	10/571.70
C3	9/639.57	8/641.9	8/643.05	**8/639.74**	8/649.02
C4	12/733.68	12/742.4	12/738.94	**12/733.13**	12/748.40
C5	17/870.26	17/879.9	17/878.95	16/924.96	16/1017.28
C6		6/413.0	**6/412.96**	**6/412.96**	**6/412.96**
C7		11/568.5	11/568.49	11/568.49	11/568.49
C8		9/648.0	9/647.26	**9/644.63**	9/647.94
C9		14/767.1	14/761.28	**14/756.38**	14/764.15
C10		17/904.1	17/903.10	**17/876.02**	17/903.10
C11	10/678.54	7/717.2	7/724.46	**7/682.54**	7/685.32
C12	**10/534.24**	10/537.8	10/534.71	**10/534.24**	10/536.33
C13		12/917.9	12/922.28	12/896.50	12/903.82
C14		11/660.2	11/600.66	**11/591.87**	11/593.08

In Table 3, we compared the best found solutions of ACS-OVRP with those of other algorithms. The best solution corresponding to each problem among all the algorithms is marked in bold face. The results show that:

(a) No algorithm can dominate other algorithms in terms of the number of the vehicles and the corresponding traveling cost over all the problems. But ORTR performs well over most of the problems.

(b) ACS-OVRP outperforms some algorithms in some problems. For example, for C2 ACS-OVRP found better solutions with less vehicle than AMP, BATA and LBTA. Similar results can be observed on C3 and C5. Especially for problem C11, the best solutions found by AMP and LBTA respectively have 3 and 2 vehicles more than the one by ACS-OVRP. It can also be seen that our proposed algorithm outperforms TSR and TSF on some problems e.g. C5, C11, C13 and C14.

(c) ACS-OVRP appears to be very competitive with other algorithms on many problems. But for some large problem, for example, C5 and C10, ORTR outperforms our approach. The best found solution of ACS-OVRP has the same fleet size as that of ORTR, but higher traveling cost. It may be due to that ORTR is specially designed for large-scale VRPs and OVRPs.

4 Conclusions

In this paper, we propose a metaheuristic coupled ACS with local search for the OVRP. Moreover, a Post-Optimization procedure is introduced to further improve the solutions. The effects of Post-Optimization procedure were proved by the computation results. Finally, we compare the performance of ACS-OVRP with those of other algorithms. The comparison results show that ACS-OVRP is an efficient algorithm for solving the OVRP. During the implementation, we found that the computation cost of local search accounted for a large proportion of total time consumption because of their intrinsic computation complexity. How to speed up the implementation of local search in ACS-OVRP and other efficient local search need to be studied in the future work.

References

1. Li, F., Golden, B., Wasil, E.: The open vehicle routing problem: Algorithms, large-scale test problems, and computational results. Computers and Operations Research (In press)
2. Schrage, L.: Formulation and structure of more complex realistic routing and scheduling problem. Networks **11** (1981) 229–232
3. Sariklis, D., Powell, S.: A heuristic method for the open vehicle routing problem. The Journal of the Operational Research Society **51**(5) (2000) 564–573
4. Brãndao, J.: A tabu search algorithm for the open vehicle routing problem. European Journal of Operational Research **157**(3) (2004) 552–564
5. Fu, Z., Eglese, R., Li, L.: A new tabu search heuristic for the open vehicle routing problem. Journal of the Opeartional Research Society **56**(2) (2005) 267–274
6. Tarantilis, C., Diakoulaki, D., Kiranoudis, C.: Combination of geographical infromation system and efficient routing algorithms for real life distribution operations. European Journal of Opeartions Research **152**(2) (2004) 437–453
7. Tarantilis, C., Ioannou, G., Kiranoudis, C., Prastacos, G.: Solving the open vehicle routing problem via a single parameter metaheuristic algorithm. Journal of the Operational Research Society **56** (2005) 588–596
8. Li, F., Golden, B., Wasil, E.: Very large-scale vehicle routing: new test problems, algorithms, and results. Computers and Operations Research **32**(5) (2005) 1165–1179
9. Dorigo, M., Stützle, T.: Ant Colony Optimization. The MIT Press, Massachusetts (2004)
10. Potvin, P., Rousseau, J.: An exchange heuristic for routing problem with time windows. Journal of the Operational Research Society **46**(1433-1446) (1995)
11. Christofides, N., Mingozzi, A., Toth, P.: The vehicle routing problem. In Christofides, N., Mingozzi, A., Toth, P., Sandi, C., eds.: Combinatorial optimization. Chichester, UK: Wiley (1979) 315–338
12. Tarantilis, C., Ioannou, G., Kiranoudis, C., Prastacos, G.: A threshold accepting approach for the open vehicle routing. RAIRO Opeartions Research **38** (2004) 345–360

An Ant-Based Approach to Color Reduction

Avazeh Tashakkori Ghanbarian and Ehasanollah Kabir

Department of Electrical Engineering, Tarbiat Modarres University, Tehran, Iran
a_t_ghanbarian@hotmail.com

Abstract. In this article a method for color reduction based on ant colony algorithm is presented. Generally color reduction involves two steps: choosing a proper palette and mapping colors to this palette. This article is about the first step. Using ant colony algorithm, pixel clusters are formed based on their colors and neighborhood information to make final palette. A comparison between the results of the proposed method and some other methods is presented. There are some parameters in the proposed method which can be set based on user needs and priorities. This increases the flexibility of the method.

1 Introduction

Researchers in the field of artificial intelligence are inspired by the nature for many years. We can point to neural networks, evolutionary algorithms and artificial immune system as well-known examples. Also some algorithms have been introduced based on swarm behavior studies [1,2].The first clustering method based on ant colony was introduced by Deneubourg et al. in 1990 [2]. They modeled ants as simple agents which move randomly in a square grid. Data items are scattered randomly on this grid and can be picked up, moved and dropped by ants. In this model ants are likely to pick up the items which are surrounded by dissimilar items and have tendency toward dropping them near similar items. By iterating these actions, the distribution of items on the grid will change. This distribution is used as a feedback and by repeating these operations, the items are clustered on the grid.

In 1994 Lumer and Faieta expanded Deneunbourg's model for data analysis [3]. They presented a modified method on numeric data and improved convergence time. They described data items through numeric vectors and used Euclidean distance to calculate the distance between them. Generally this algorithm can be used on any data set to which a function can be declared as a measurement of dissimilarity. Also they deployed a short-term memory to memorize last transmitted items and the position they were dropped. Indeed they introduced the inhomogeneous populations of agents for the first time which are the agents with different individual initialized parameters.

After Lumer and Faieta, Kuntz et al. proposed a method using ant-based clustering for graph partitioning [4]. Also other studies proceeded ant-based clustering applicability in document retrieval and visualization. In [5] Deneunbourg's basic algorithm was deployed on web page categorization and visualization. Handl and

M. Dorigo et al. (Eds.): ANTS 2006, LNCS 4150, pp. 364–371, 2006.

Meyer used ant-based clustering to create topic maps dynamically [5]. Other approaches combine ant colony with other methods like c-means. Ant-based clustering makes initial clusters for c-means instead of having no primary information. On the other hand c-means easily handles misplaced or free items and improves quality [6]. In another approach real ants' behavior is simulated by a hybridization of ant-based clustering and fuzzy if-then rules [7]. [8] proposes a new model based on ants' chemical recognition system to cope with unsupervised clustering problems.

Image color reduction or quantization is one of the basic image processing techniques. Color reduction has two main steps: 1- K colors are chosen from the original ones to form the color palette. 2- The output image is reconstructed using this palette. Obviously quality of the output image depends on the first step, palette design. In this article a method for palette design with pre-determined number of colors using ant-based clustering is introduced. The rest of the paper is organized as follows. Section 2 presents the principal characteristics of the proposed method and also introduces the color reduction algorithm. Parameters are described in Section 3. Section 4 compares the results with 5 other methods and finally Section 5 draws the conclusion.

2 The Proposed Method

In this section general arrangement, block diagram and the algorithm of the proposed method is introduced. Figure 1 shows the block diagram of the proposed method. In the first phase, a modified ant-based clustering algorithm is applied to image pixels. The second phase is dedicated to choosing a representative for each cluster. In the third phase, pixels are mapped to the formed palette and color reduction is completed. The simplest method is used for mapping pixels to the palette. In this method the closest item to each color in the palette replaces it in the final reconstructed image. Modified ant-based clustering algorithm is discussed in the following.

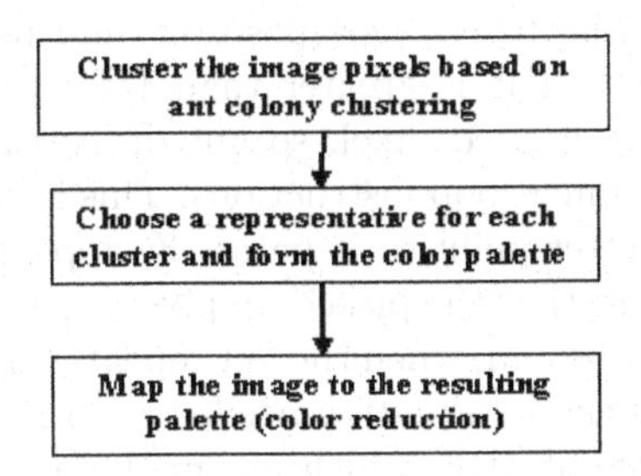

Fig. 1. Color reduction method

2.1 Modified Ant-Based Clustering

This method is based on the algorithm presented by Handl [5]. This algorithm is modified to apply to color clustering and color reduction problem. The most important modification is that in the database related applications, first the data items are randomly spread on the grid. However in the proposed method the algorithm is applied on the image itself. One advantage is that color and adjacency information are considered simultaneously in the clustering process. Functions used in this method are introduced as follows.

$$d(i,j) = |R_i - R_j| + |G_i - G_j| + |B_i - B_j| \tag{1}$$

$$f(i) = \max(0, \frac{1}{\delta^2} \sum_j (1 - \frac{d(i,j)}{\alpha})) \tag{2}$$

d (i,j) defines dissimilarity between items i and j in RGB space. f(i) is a neighbourhood function for pixel i. Parameter α determines the influence of dissimilarity function on f(i). δ is the radius of perception. In this method, δ is set to 1 so that 8 pixels around each pixel are considered in calculating f(i). Some neighbours to i may be empty since their pixels have been picked up previously. Consequently less items are involved in calculating f(i), thus f(i) is smaller in sparse areas than it is in dense areas. Likewise similarity in dense areas is more than the similarity in sparse areas; therefore ants prefer to drop their items in dense areas rather than in sparse areas. This leads to condensing areas.

The main algorithm is presented in Alg.1. The first step includes initialization of the parameters. These parameters are discussed in details later. Then the algorithm's main loop starts. In the beginning of each iteration, ant picks up an item randomly, then it searches its memory to find a similar item. The best match for pixel i is the one with the least dissimilarity to it. If dissimilarity is less than the memory threshold, MTh, ant will directly drop the pixel in place of the best match and update its memory. Otherwise it will start the inner loop. This loop starts with taking a step towards the best neighbour, which is a pixel with the lowest d(i,j). This method has chosen due to meaningfulness of pixel neighbourhood in the image. Then neighbourhood function,f(i), is calculated for the carried pixel, according to the new position, and its value is compared to the drop threshold, DTh. If it is greater than DTh , the current position is suitable for dropping the carried pixel, so ant drops the pixel. Otherwise ant keeps on performing the inner loop instructions. This is repeated until ant drops the carried pixel on a better point or it passes K steps, in this case the carried pixel is deleted, in the case that the picked up pixel is a noisy pixel or the ant has chosen a wrong path. At the end, when the ant repeats the outer loop sufficiently, its memory is filled with the information needed to form the palette. Simply we will use the heaps' representatives which are in the ant's memory to form the palette.

In our method, unlike Handl's method, which allows one item per cell, there can be more than one item in one cell. In other words in this method each cluster is created by gathering data items in a cell, making a heap. The idea of heaps have also been used in some ant-based clustering approaches [6, 7].Using heaps will lead to simplicity in dropping items, spatial discrimination of clusters and

```
1- Initialize ant parameters
2- For #iteration do
3- Pick up a random pixel
4- Check memory for similar item
5- If similar item exists then update the memory and drop the item
6-else repeat
        (a) Move to the best match neighbor
        (b) If f(i) > DTh then drop pixel
   until item has been dropped or K moves has done
7-End for
```

Alg. 1. The proposed algorithm

simpler color assignment at the end of the algorithm (Alg. 1). As a consequence random picking will be used not only to increase the variety of the pixels seen by the ant but also to decrease the execution time.

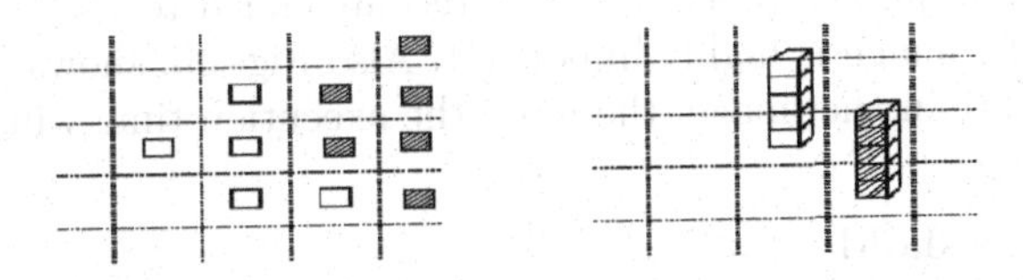

Fig. 2. Two clusters made in Handl method are shown on the left. The items of two clusters locating beside each other can be hardly recognized. Two clusters created by the proposed method are shown in the right.

3 Parameters Analysis

As described before, the proposed algorithm starts with random picking of the image pixels which is simply a sampling phase. The result is directly affected by this phase. We used large number of pixels in all samples to avoid oscillating result due to different samples. Furthermore in order to increase the precision and reliability, the average of some consecutive results is used in analyses. 4 images which have been used for analysis are shown in Fig. 5, 3 of these images are from USC[1] database and the last one, Flower image, is an arbitrary image chosen as a representative of images with large number of colors. In diagrams demonstrated in this section, reduced size images have been used because average result of at least 5 runs should be calculated for each point shown in the diagrams. ACDSee 3.1 with lanczose filter and highest quality was deployed for size reduction of images to 256 × 256. However reported AQE[2] is very close or equal to AQE of the original images in every case. The test platform was a Pentium III 800 MHz with 256 MB RAM. The algorithm was implemented by Visual Basic 6.0. For statistical comparison we used AQE as a measure for the amount of distortion

[1] http://sipi.usc.edu/services/database/Database.html

[2] Average Quantization Error.

in the quantized image. If A(i,j,k) is k'th color element of (i,j) pixel of original image and B(i,j,k) is the corresponding color element in the quantized image, then AQE is defined as follows:

$$AQE = \frac{1}{MN} \sum_{i=1}^{M} \sum_{j=1}^{N} \sqrt{\sum_{k=1}^{3} (A(i,j,k) - B(i,j,k))^2} \tag{3}$$

3.1 Parameter α

Parameter α in Eq. 2, determines the influence of dissimilarity function on f(i). Mean results of 5 runs are shown in Figs. 3a. and 3b. Other parameters are MTh=35 and DTh=0.6 and ant memory size is 64. When α increases, the number of steps ant passes before dropping the item is reduced. This is because the dissimilarity influence on f(i) is reduced and the ant is more likely to find a similar pixel sooner. On the other hand Fig. 3a. shows that increasing α has little effect on the quality. Therefore it is recommended to use bigger α unless the quality of the image is highly important. Also Fig. 3b. shows that the more different colors are in the image, the more the execution time will be.

3.2 Drop Threshold

Drop threshold, DTh, determines whether a point is a proper point for dropping or not (Alg. 1). As it was demonstrated for α, different values between 0.1 and 1.1 was tested for DTh which is a meaningful range. Results showed that ant must pass more steps due to increment of DTh, because it should find more similar pixel each time which gets eventually harder. This leads to longer execution. On the other hand, increasing the DTh does not have much effect on result's quality and AQE remains almost constant.

3.3 Memory Threshold

Memory threshold, MTh, determines whether the picked up pixel is similar enough to its closest item in the memory or not (Sec. 2). Figs. 3c. and 3d. show MTh effect on the mean of AQE and execution time for 5 runs. Other parameters are set to α=1 and DTh=0.6. In most images AQE is reduced until MTh=35 and image quality improves. Just in Lenna image, the quality improves until MTh=20 then it deteriorate gradually. That's because the color spectrum is limited in Lenna image and lower thresholds result in least distortion. In other cases, due to increasing MTh, more pixels can be dropped by using items in ant's memory which leads to more memory updates. The more the items in memory be updated, the better the cluster representative will be and finally the AQE will be reduced. Also dropping pixels by remembering items in the memory reduces the execution time. On the other hand by increasing MTh more than 35, the pixels which should form separate memory items are mixed with the previous items in memory so AQE increases. Thus MTh=35 results in better quality of images.

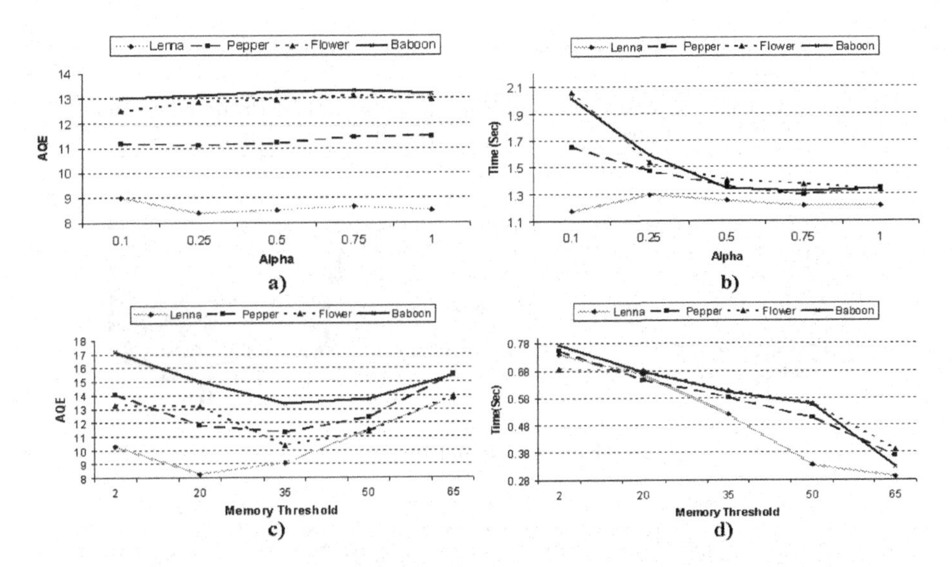

Fig. 3. a) Effect of α on AQE (MTh=35, DTh=0.6) b) Effect of α on execution time (MTh=35, DTh=0.6) c) MTh effect on AQE (α=1, DTh=0.6) d) MTh effect on execution time in seconds (α=1, DTh=0.6)

4 Experimental Results

In this section the results of the proposed algorithm are compared with 4 other color reduction methods. Fig. 4 shows the results of Octree, Center-cut, C-Means, Bing's method and our method. Also Table 1 lists AQE of these results for more precise comparison. All results reported here for other methods are from reference [10]. It is worth noting that all parameters of the algorithm, as shown in Fig. 6, are set based on the experiments on 4 images of Fig. 5. As it can be seen in Fig. 4 and Table 1 resulting images from Octree, center-cut and C-means methods show obvious degradations. This is mainly because these methods do not use pixel neighborhood information in clustering. Also Bing's method does not use neighborhood information efficiently enough to produce high quality results with the least amount of distortion (Table 1).

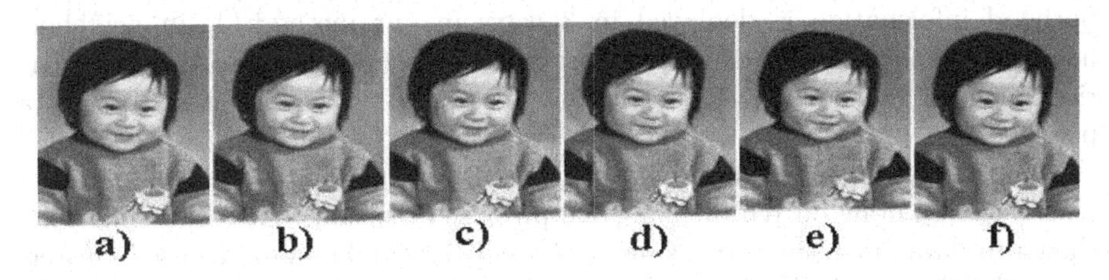

Fig. 4. Comparison of color quantization methods a) Original Image b) Octree c) center-cut d) C-means e) Bing's method f) proposed method (as α=1, DTh=0.6, MTh=35)

Table 1. AQE comparison of 5 methods

Method	Octree	Center-cut	C-Means	Bing's Method	Proposed Method
AQE	14.26	12.08	13.78	13.36	10.52

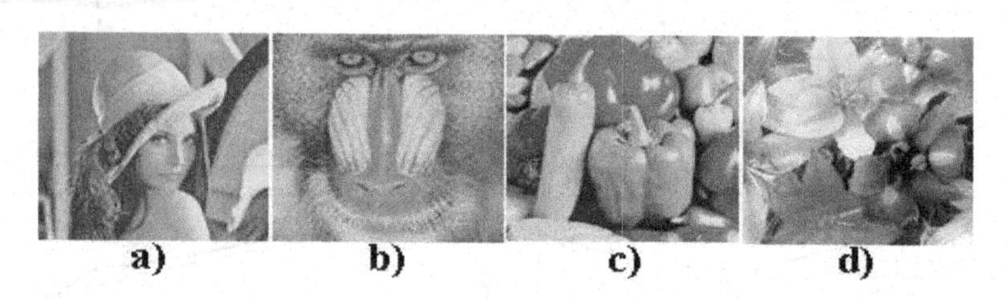

Fig. 5. Original Images

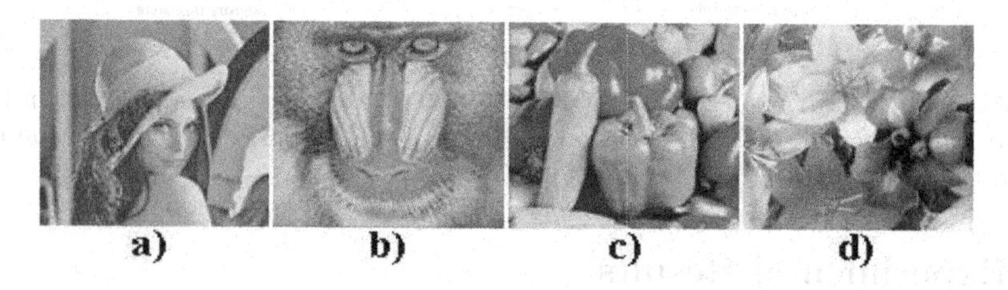

Fig. 6. Result images reduced to 64 colors by the proposed algorithm (α=1, DTh=0.6, MTh=35, X=20, Y=10

5 Conclusion

In this paper ant-based clustering ability for color reduction was demonstrated. Parameters in this method can be simply set to consider user priorities. This increases flexibility of the algorithm and also makes it easy for nonprofessional users. Also with some changes, like just replacing the dissimilarity function, this solution can be extended to reduce colors in gray scale images.

Basically using color information together with pixels' neighborhood information in the proposed method is the main cause of improving results. Neighborhood information is deployed in 3 ways in our method: 1. in neighborhood function. 2. in ant movement for finding a proper place to drop the carried pixel. 3. limiting maximum number of steps the ant can pass to drop a pixel.

Since all solutions based on behavior of ants are inherently parallel, more research can be done on parallelizing the proposed methods. Also as it has been shown in data clustering, for example [6] and [7], combining ant-based clustering with other methods like fuzzy methods or C-means improves the execution time.

References

1. Dorigo M., Maniezzo V., Colorni A.: Ant System: Optimization by a Colony of Cooperating Agents, IEEE Trans. on Systems, Man, and Cybernetics, 26(1) (1996) 29–41.
2. Deneubourg J.L., Goss S., Franks N., Sendova-Franks A., Detrain C., Chretien L., The Dynamics Of Collective Sorting Robot-like Ants and Ant-like Robots, In Proc. First Conf. on Simulation of Adaptive Behavior (1990) 356–363.
3. Lumer E.D. , Faieta B.: Diversity and Adaptation in Populations of Clustering Ants, Proc. of the Third Int. Conf. on Simulation of Adaptive Behavior: From Animals to Animats 3 (1994) 501–508
4. Kuntz P., Snyers D.: New Results On An Ant-Based Heuristic for Highlighting The Organization Of Large Graphs , In Proc. of the 1999 Congress on Evolutionary Computation (1999) 1451–1458
5. Handl J., Meyer B.: Improved ant-based clustering and sorting in a document retrieval interface , In Proc. of the Seventh Int.Conf. on Parallel Problem Solving from Nature (PPSN VII), LNCS 2439 (2002) 913–923
6. Monmarché N.: On Data Clustering With Artificial Ants (AAAI-99 & GECCO-99). Workshop on Data Mining with Evolutionary Algorithms (1999) 23–26
7. Kanade P. M., Hall L. O.: Fuzzy Ants As A Clustering Concept , 22nd Int. Conf. of the North American Fuzzy Information Processing Society NAFIPS (2003) 227–232
8. Labroche N., Monmarché N., Venturini G.: Visual Clustering With Artificial Ants Colonies, Seventh Int. Conf. on Knowledge-Based Intelligent Information & Engineering Systems (2003) 332–338
9. Scheunders, P.: A comparison of clustering algorithms for color image quantization, Pattern Recognition Letters (1997) 1379–1384
10. Bing, Z., Junyi, S., Qinke, P.: An Adjustable Algorithm For Color Quantization, Pattern Recognition Letters, 25 (2004) 1787–1797

An Orthogonal Search Embedded Ant Colony Optimization Approach to Continuous Function Optimization

Jun Zhang*, Wei-neng Chen, and Xuan Tan

Department of Computer Science, Sun Yat-Sen University, Guangzhou, P.R. China
junzhang@ieee.org

Abstract. Ant colony optimization has been one of the most promising meta-heuristics since its appearance in early 1990s but it is specialized in discrete space optimization problems. To explore the utility of ACO in the filed of continuous problems, this paper proposes an orthogonal search embedded ACO (OSEACO) algorithm. By generating some grids in the search space and embedding an orthogonal search scheme into ACO, the search space is learned much more comprehensively with only few computation efforts consumed. Hence, solutions are obtained in higher precision. Some adaptive strategies are also developed to prevent the algorithm from trapping in local optima as well as to improve its performance. Moreover, the effectiveness of this algorithm is demonstrated by experimental results on 9 diverse test functions for it is able to obtain near-optimal solutions in all cases.

1 Introduction

Ant Colony Optimization (ACO) was first proposed by Marco Dorigo in the early 1990s in the light of how ants manage to establish the shortest path from their nest to food sources [1]. By now, the idea of ACO has been used in a large number of intractable combinatorial problems and become one of the best approaches to traveling salesman problem [2], quadratic assignment problem [3], data mining [4], and network routing [5].

In spite of its great success in the field of discrete problems, the uses of ACO in continuous problems are not significant. G. Bilchev and I. C. Parmee [6] first introduced an ACO metaphor for continuous problems in 1995. Later, several ant algorithms were proposed. E.g., Mathur *et al.* [7] improved Bilchev and Parmee's idea and proposed a bi-level search scheme using global ants and local ants. In this algorithm, crossover and mutation operators derived from GA were also adopted for global search. Based on some other behaviors of ants, two algorithms called API [8] and CIAC [9] were proposed, but they did not follow the framework of ACO strictly. In another approach described in [10], the probability density

* Corresponding author. This work was supported in part by NSF of China Project No.60573066 and NSF of Guangdong Project No. 5003346 and Guangdong Key Lab of Information Security.

M. Dorigo et al. (Eds.): ANTS 2006, LNCS 4150, pp. 372–379, 2006.

functions are applied for pheromone maintenance. Overall, though some good performances have been achieved, there is still a long way before truly success in exploiting the utility of ACO in continuous space optimization problems.

This paper aims at proposing an orthogonal search embedded ACO (OSEACO) algorithm for continuous function optimization problems. Different from the above algorithms, a number of stochastic and dynamic grids are generated in the search space. The main idea of ACO is used to handle the global search procedure of the algorithm. By embedding the orthogonal design technique, the grids could be learned more comprehensively by the ants. Also, an elitist scheme and an adaptive radius scheme are designed to avoid stagnation situations as well as to improve performance. The experimental results from 9 test functions demonstrate the effectiveness of the algorithm as near-optimal solutions could be obtained in all test cases.

2 Background

2.1 Ant Colony Optimization

The idea underlying ACO algorithms is to simulate the autocatalytic and positive feedback process of the forging behavior of real ants. Once an ant finds a path, pheromone is deposited to the path. By sensing the pheromone ants can follow the path discovered by other ants. This collective pheromone-laying and pheromone-following behavior of ants has become the inspiring source of ACO.

Informally, the mechanism underlying ACO can be considered as the interplay of the following three procedures [11]: 1) Solution Construction, in which solutions are built by ants using pheromone (and heuristic information). 2) Pheromone Management, in which pheromone is deposited or evaporated. 3) Daemon Actions, in which centralized controls are executed, e.g., local search procedure or the procedure that finds the best-so-far ants. Daemon actions are optional.

2.2 Orthogonal Experimental Design

Experiments are carried out to study the effects of different factors. However, it takes lots of time and resources to accomplish a complete factorial experiment which makes measurements at each of all possible combinations. The goal of the orthogonal design is to perform a minimum number of tests but acquire the most valuable information of the considered problem [12][13]. It performs by judiciously selecting a subset of level combinations using a particular type of array called the orthogonal array (OA). As a result, well-balanced subsets of level combinations will be chosen as representative members. Each column in the OA represents a specific factor (variable) in the considered problem and each row corresponds to a single test. The OA is notated as $L_n(S^m)$, where n is the number of rows and also the number of tests, m is the number of columns, S represents the number of levels in each factor. More details about the OA can be found in [12][13]. Fig.1 gives two examples of orthogonal arrays.

$$L_4(2^3)=\begin{matrix}1&1&1\\1&2&2\\2&1&2\\2&2&1\end{matrix}\qquad L_9(3^4)=\begin{matrix}1&1&1&1\\1&2&2&2\\1&3&3&3\\2&1&2&3\\2&2&3&1\\2&3&1&2\\3&1&3&2\\3&2&1&3\\3&3&2&1\end{matrix}$$

Fig. 1. Orthogonal arrays $L_4(2^3)$ and $L_9(3^4)$

Based on the OA, orthogonal design can be implemented as follows. For example, we consider a function optimization problem:

$$\max f(x_1, x_2, x_3) = x_1 + 10x_2 - 100x_3,\ x_1 \in \{1,2\}, x_2 \in \{3,4\}, x_3 \in \{5,6\} \quad (1)$$

The factors of the problem are x_1, x_2, x_3 which are denoted as A, B, C in Table 1. Each factor has two levels, e.g. x_1 has levels $\{1,2\}$. The OA $L_4(2^3)$ is used, so the corresponding values to the array can be shown in Table 1. Fig. 2 shows the placement of the selected four combinations of factors. The complete arrangement of the three factors are shown as $A_iB_jC_k$ $(i = 1,2; j = 1,2; k = 1,2)$ in the cubic. By using the orthogonal design technique, only four vertexes $A_1B_1C_1$, $A_1B_2C_2$, $A_2B_1C_2$ and $A_2B_2C_1$ are chosen, which are marked by solid notes in Fig. 2. Nevertheless, these represented combinations of factor levels are separated evenly so that they can help us to study the effectiveness of each factor and to reduce the cost of the experiment.

Table 1. An example of orthogonal experimental design using $L_4(2^3)$

test	factors			values		
number	A	B	C	x_1	x_2	x_3
1	1	1	1	1	3	5
2	1	2	2	1	4	6
3	2	1	2	2	3	6
4	2	2	1	2	4	5

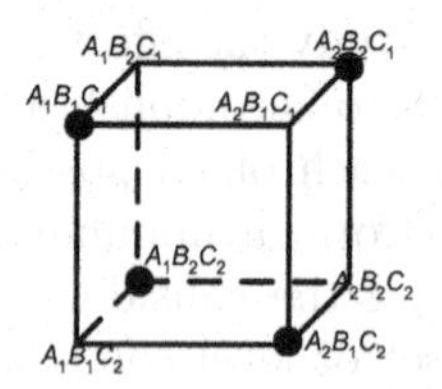

Fig. 2. An example of orthogonal experimental design

3 The Orthogonal Search Embedded ACO Algorithm

The characteristics of OSEACO are mainly in the following four aspects: 1) applying ACO to the global search; 2) incorporating an orthogonal search scheme;

3) generating stochastic and dynamic grids; and 4) employing an elitist strategy and an adaptive radius scheme.

Informally, its procedural steps are summarized as follows.

Step 1) *Initialization*: Grids are created and pheromone values are initialized.

Step 2) *Grid Selection*: Each ant chooses a grid to build its solution according to a grid selection rule in which pheromone values in each grid are made use of.

Step 3) *Orthogonal Search*: Once an ant chooses a grid to explore, this grid will be improved by an orthogonal search strategy.

Step 4) *Elitist Set Construction*: After all ants have built their solutions, the best grids are protected in an elitist set and the other grids will be regenerated.

Step 5) *Pheromone Updating*: Pheromone values are updated on elitist grids.

Step 6) *Termination Test*: If the test is passed, stop; otherwise go to step 2).

Below, the algorithm will be described in detail. To facilitate understanding and explanation of the proposed algorithm, we take the optimization work as minimizing an d-dimension object function $f(X)$, $X = (x_1, x_2, \ldots, x_d)$. The lower and upper bounds of variable x_i are low_i and up_i. Nevertheless, without loss of generality, this scheme can also be applied to other continuous space optimization problems.

3.1 Initialization

At first, N grids $(G_1, G_2, \ldots, G_N)$ are created in the search space randomly. Here, a grid $G_i (1 \le i \le N)$ is defined as a hyper-geometric solid with several attributes, including its center point $X_i = (x_{i1}, x_{i2}, \ldots, x_{id})$, search radius of each factor $R_i = (r_{i1}, r_{i2}, \ldots, r_{id})$, and its fitness value $v_i = f(X_i)$. There are three main characteristics of these grids. First, they are stochastic, for they are initialized randomly. Second, they are adaptive, for the radiuses of their variables can be changed in terms of need. Third, they are movable, for they can move to areas with higher potential. In other words, the center point of G_i can vary from X_i to X_i' if $f(X_i') \le f(X_i)$. Grids are also the carriers of pheromone. The pheromone values on grid i is notated as $\tau_i (1 \le i \le N)$. Therefore, we should set $\tau_1 = \tau_2 = \quad = \tau_N = \tau_{initial}$ in this step where $\tau_{initial}$ is a constant value we set previously as the initial pheromone values on all grids.

3.2 Grid Selection

Let it be M ants in the colony. Each ant builds a solution to the considered problem by applying the grid selection rule. The rule is given by (2), which is the stochastic greedy selection scheme. A random number $q \in [0, 1]$ is generated and is compared to a parameter $q_0 (0 \le q_0 \le 1)$. If q is not larger than q_0, the grid with the largest pheromone value will be selected. Otherwise, a stochastic proportion rule S according to the probability distribution given by (3) is performed. This rule is always referred to as the roulette wheel selection rule, in which the probability of selecting a grid is in direct proportion to the value of the grid's pheromone trails.

$$s = \begin{cases} \arg\max\{\tau_i\}, 1 \le i \le N, \text{if } q \le q_0 \\ S, \text{otherwise} \end{cases} \tag{2}$$

$$p_s = \tau_s / \sum_{i=1}^{N} \tau_i \tag{3}$$

3.3 Orthogonal Search

Once a grid is selected, orthogonal search is applied to improve this grid. In order to introduce the orthogonal design technique to this case, we first consider that each variable of the object function corresponds to a single factor of the experiment. In other words, d factors can be acquired from $(x_1, x_2, \ldots, x_d)$. Then, l levels are obtained in each factor. Considering the jth $(1 \le j \le d)$ vector of grid G_i, its search range is $[x_{ij} - ran \cdot r_{ij}, x_{ij} + ran \cdot r_{ij}]$, where ran is a random number distributed in $[0, 1]$. We divide this range into $l-1$ segments and acquire l dividing values as l levels. The value of the kth $(0 \le k \le l-1)$ level is $x_{ij} - r_{ij} + 2kr_{ij}/(l-1)$. By doing this, the optimizing task is transformed into a d-factor experiment with l levels in each factor. Hence, the idea of orthogonal design can be used. After choosing an OA in terms of d and l, a small fraction of combinations of factor levels are selected. We call these combinations as the orthogonal points. A simple example of orthogonal search has been shown previously in Fig. 1. In a 3-dimension grid, we obtain 2 levels from each vector simply by using two ends of each edge. Then OA $L_4(2^3)$ is applied and 4 orthogonal points are selected.

Soon after all orthogonal points have been selected, they are evaluated in the object function f. Assume that X_i' is the orthogonal point that has the best fitness value. If $f(X_i') < f(X_i)$, X_i' will be the new center point of the grid. That is, the grid moves from X_i to X_i' if better points are found during the orthogonal search procedure.

Another characteristic of a grid is adaptive, that is, the radiuses of all variables $(r_{i1}, r_{i2}, \ldots, r_{id})$ in grid G_i would adjust themselves during the algorithm by applying (4), where $\delta (0 \le \delta \le 1)$ is a parameter.

$$\text{for all } j (1 \le j \le d) \; r_{ij} = \begin{cases} r_{ij}/\delta, \text{if the center point of } G_i \text{ is replaced} \\ r_{ij} \cdot \delta, \text{otherwise} \end{cases} \tag{4}$$

This can effectively help us to improve the grid by deciding whether to move it faster or to let it shrink. If the center point of the grid is not substituted, it is probably that the best region of the grid is around its center so we decrease its radius to obtain a more precise solution. Otherwise, the grid may be out of the trough of a function. In this case, we enlarge the grid to make it move faster to a better area.

3.4 Elitist Set Construction

After all ants have built their solutions, an elitist strategy is employed to reserve the most valuable grids. N is the number of grids. We only reserve μN grids for further exploitation in an elitist set, where $\mu (0 \le \mu \le 1)$ is a parameter.

All grids are ranked according to their fitness values and only the best N grids are preserved. All other grids would be deserted and regenerated randomly. All attributes of a deserted grid should be reinitialized, including its center point, radiuses, and pheromone values. Additionally, redundant identical members in the elitist set should be deleted. That is, if two or more grids in the elitist set appear to be located in the same area of the search space (the Euclid distances between the center points of two grids become smaller than a small enough value), we preserve one of them and desert the others.

3.5 Pheromone Updating

As the pheromone trails on non-elitist grids have been reinitialized, we only apply this step to the grids in the elitist set. A fraction of pheromone would evaporate and the ants which visit the elitist grids are allow depositing pheromone on the grids using a rank-based scheme. Grids have been ranked at the elitist set construction step.

Pheromone values of elitist grids are updated by applying rule (5) where τ_i is the pheromone values on grid i, $\tau_{initial}$ is the initial pheromone values to all grids, $\rho(0 \leq \rho \leq 1)$ is a parameter, μN is the size of the elitist set, $rank_i$ is the rank of grid i, and $count_i$ is the number of ants that choose grid i in that iteration.

$$\tau_i \leftarrow (1-\rho)\tau_i + \rho\tau_{initial}(\mu N - rank_i + count_i), \text{if grid}_i \in \text{Elitist Set} \quad (5)$$

In terms of this rule, a better grid that visited by more ants will receive larger amount of pheromone.

4 Computational Results and Discussing

To demonstrate the effectiveness of the proposed algorithm, 9 test functions in table 2 are selected from [14] where we can obtain more information about these functions (e.g., search domains and minimum function values). $f_1 - f_4$ are unimodel functions and $f_5 - f_9$ are multimodel functions. The performances of OSEACO are compared with two other heuristic approaches, i.e., PSO [15] and API. Parameters of OSEACO are set as: $M = 100, N = 30$(in unimodel cases) or 200(in multimodel cases), $\tau_{initial} = 0.001, \delta = 0.9, \mu = 0.1, q_0 = 0.3, \rho = 0.2$ and the initial radius of the jth vector of grid G_i is set to $r_{ij} = (up_j - low_j)/10$. These configurations are based on a series of experiments. The parameter V_{max} in PSO is set to different values in different functions and the one that has the best results are used in the comparison. Parameters of API are set as: $n = 20, T = 50$, and $P_{local(\alpha_i)} = 50$, $(i = 1, 2, \ldots, n)$. These configurations are based on [8]. The maximum number of function evaluations is set to 100000 in all algorithms. The solutions averaged over 50 independent runs and the best solutions among all 50 runs are shown in Table 2. Obviously, the performances of OSEACO are much better than the performances of PSO and API in both unimodel and multimodel functions. The effectiveness of OSEACO can also be seen in Fig. 3. OSEACO manages to obtain the precision of 0.3 on f_6 in less than 2000 evaluations and obtain the precision of 10^{-10} on f_1 in less than 17500 evaluations. These reveal that solutions with higher-precision can be obtained by OSEACO with fewer evaluations compared with PSO and API.

Table 2. The comparison between PSO, API, and OSEACO on 9 Test Functions

test functions	d	PSO	API	OSEACO
$f_1(x)=\sum_{i=1}^d x_i^2$	4	$2.2e^{-5}$ ($2.8e^{-6}$)	0.0326 (0.072)	$\mathbf{1.8e^{-88}}$ ($\mathbf{9.0e^{-95}}$)
$f_2(x)=\sum_{i=1}^d \lvert x_i\rvert+\prod_{i=1}^d \lvert x_i\rvert$	4	$7.62e^{-4}$ ($2.59e^{-4}$)	0.0939 (0.0525)	$\mathbf{1.1e^{-61}}$ ($\mathbf{2.6e^{-69}}$)
$f_3(x)=\sum_{i=1}^d(\sum_{j=1}^i x_j)^2$	4	$2.72e^{-5}$ ($2.87e^{-6}$)	1.944 (0.153)	$\mathbf{4.9e^{-99}}$ ($\mathbf{1.7e^{-122}}$)
$f_4(x)=\max_i(\lvert x_i\rvert, 1\le i\le d)$	4	$5.12e^{-4}$ ($1.58e^{-4}$)	0.449 (0.211)	$\mathbf{2.6e^{-69}}$ ($\mathbf{2.7e^{-76}}$)
$f_5(x)=\sum_{i=1}^d -x_i\sin(\sqrt{x_i})$	4	−1393.52 (−1671.55)	−1324.22 (−1674.88)	**−1621.87** (**−1675.93**)
$f_6(x)=\sum_{i=1}^d[x_i^2-10\cos(2\pi x_i)+10]$	4	0.570 (0.022)	5.062 (1.231)	**0.1052** (**0**)
$f_7(x)$ Ackley's function	4	0.506 (0.0219)	0.868 (0.397)	**0.0338** ($\mathbf{5.9e^{-16}}$)
$f_8(x)$ Generalized Penalized function	4	$1.84e^{-4}$ ($3.22e^{-5}$)	$3.61e^{-3}$ ($2.68e^{-4}$)	$\mathbf{8.45e^{-7}}$ ($\mathbf{1.6e^{-32}}$)
$f_9(x)=-\sum_{i=1}^5[(x-a_i)(x-a_i)^T+c_i]^{-1}$	4	−7.414 (−10.037)	−6.1048 (−10.1365)	**−10.1518** (**−10.1532**)

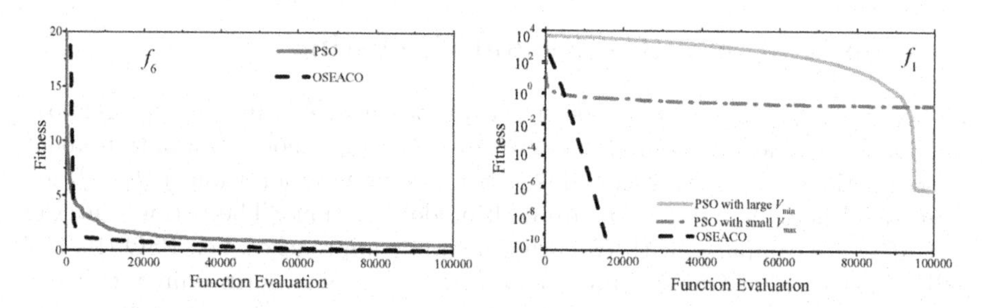

Fig. 3. The best fitness found by OSEACO and PSO in function of evaluation times

5 Conclusion

The OSEACO algorithm for continuous space function optimization problems has been proposed. The general idea underlying this algorithm is to use the orthogonal design scheme to improve the performance of ACO in the filed of continuous optimization problems. An Elitist strategy and an adaptive radius scheme are also proposed. Experimental results on 9 diverse functions demonstrate the effectiveness of the algorithm compared with two other heuristic approaches.

References

1. Dorigo, M., Maniezzo, V., Colorni, A.: Ant system: Optimization by a colony of cooperating agents. IEEE Transactions on Systems, Man, and Cybernetics - part B: Cybernetics **26** (1996) 29–41
2. Dorigo, M., Gambardella, L.M.: Ant colony system: A cooperative learning approach to tsp. IEEE Transactions on Evolutionary Computation **1** (1997) 53–66
3. Dorigo, M., Di Caro, G., Gambardella, L.M.: Ant algorithms for discrete optimization. Artificial Life **5**(2) (1999) 137–172
4. Parpinelli, R.S., Lopes, H.S., Freitas, A.A.: Data mining with an ant colony optimization algorithm. IEEE Transactions on Evolutionary Computation **4** (2002) 321–332
5. Sim, K.M., Sun, W.H.: Ant colony optimization for routing and load-balancing: Survey and new directions. IEEE Transactions on Systems, Man, and Cybernetics - part A: System and Humans **33** (2003) 560–572
6. Bilchev, G., Parmee, I.C.: The ant colony metaphor for searching continuous design spaces. In: Proceedings of the AISB Workshop on Evolutionary Computation. Volume 993 of LNCS., Springer-Verlag (1995) 25–39
7. Mathur, M., Karale, S.B., Priye, S., Jayaraman, V.K., Kulkarni, B.D.: Ant colony approach to continuous function optimization. Ind. Eng. Chem. Res. **39** (2000) 3814–3822
8. Monmarché, N., Venturini, G., Slimane, M.: On how pachycondyla apicalis ants suggest a new search algorithm. Future Generation Computer Systems **16** (2000) 937–946
9. Dréo, J., Siarry, P.: Continuous interacting ant colony algorithm based on dense heterarchy. Future Generation Computer Systems **20**(5) (2004) 841–856
10. Socha, K.: Aco for continuous and mixed-variable optimization. In: Proceedings of the 4th International Workshop on Ant Colony Optimization and Swarm Intelligence (ANTS'2004). Volume 3172 of LNCS., Springer-Verlag (2004) 25–36
11. Dorigo, M., Stützle, T.: Ant Colony Optimization. MIT Press (2004)
12. Fang, K.T., Wang, Y.: Number-Theoretic Methods in Statistics. New York: Chapman & Hall (1994)
13. Hedayat, A.S., Solane, N.J.A., Stufken, J.: Orthogonal Arrays: Theory and Applications. New York: Springer-Verlag (1999)
14. Yao, X., Liu, Y., Lin, G.: Evolutionary programming made faster. IEEE Transactions on Evolutionary Computation **8** (2004) 456–470
15. Elberhart, R.C., Kennedy, J.: A new optimizer using particle swarm theory. In: Proc. 6th Int. Symp. Micromachine Human Sci., Nagoya, Japan (1995) 39–43

Ant Based Mechanism for Crisis Response Coordination

Bogdan Tatomir[1,2] and Leon Rothkrantz[1,2]

[1] Delft University of Technology, Delft, The Netherlands
[2] DECIS Lab, Delft, The Netherlands
{B.Tatomir, L.J.M.Rothkrantz}@ewi.tudelft.nl

Abstract. Many emergency situations involve injured people who need medical help and evacuation to a safe area. Usually there is not enough time to provide medical help to all the victims. So medical doctors have to make choices to help as much victims as possible, taking care of the distribution of victims in the crisis area and the priority of the victims related to the severity of the injuries. This paper describes an ant colony optimization algorithm to route medical doctors along the victims in a crisis area. We tested the algorithm in a simulated crisis environment. Two different routing strategies were implemented and compared with our algorithm.

1 Introduction

Recent terrorist attacks and natural disasters have forced humanity to respond to crisis situations in the most effective ways [1]. In these situations usually many victims are distributed over a large area and doctors entering the field are faced with the problem to provide medical help in an optimal way. They don't want to waste time in finding the victims and spending time to less injured victims while other victims are dying because of deprivation of medical help. Because of the heavy time constraints they are able to handle only a limited number of victims. So they have to make terrible choices.

Immediately after the onset of a disaster there are calls for medical help. A common procedure is that emergency persons enter the field to localize the victims. They use a simple method, called triage, of quickly identifying victims who have immediately life-threatening injuries and who have the best chance of surviving. There are several triage algorithms applied in different parts of the world. A comparison between such algorithms can be found in [2].

In this paper we base our assumptions on the START system [3]. The triage portion of START, relies on making a rapid assessment (taking less than a minute) of every patient, determining which of four categories patients should be in:

- Green: Minor delayed care / can delay up to three hours
- Yellow: Delayed urgent care / can delay up to one hour
- Red: immediate care / life-threatening
- Black: victim is dead / no care required

M. Dorigo et al. (Eds.): ANTS 2006, LNCS 4150, pp. 380–387, 2006.

After the victims are classified in one of the four categories, this information is send to a central point. Based on this information a multiple route is designed for the medical doctors entering the field to provide medical help. The goal is of course to help as much victims as possible, taking care of the priority of the casualties. The problem we have to solve is the following:

- how to route the doctors along the victims,
- how to take care of the priority of the victims,
- how to handle the dynamic aspects (new victims are localized in the course of the time; different medical doctors are entering or leaving the area). So the routes have to be adapted continuously,
- how to handle the time limitations (some victims need help in the shortest time).

1.1 Related Work

Our proposed solution is the Ant Based Control algorithm, which uses a stigmergy-based coordination. This strategy is inspired by the coordination mechanisms found in the ants communities.

The First Aid Case described in this paper can be perceived as a special class of the vehicle routing problem, the Dynamic Vehicle Routing Problem with Time Windows (DVRPTW) [4]. Each medical doctor acts as a vehicle, trying to serve as many customers (casualties) as possible. Mobile agents imitate the trail-laying behavior of ants and try to find the longest cycle which contains each medic and the maximum number of casualties. Each agent constructs a partial solution where each casualty is visited at most once. The ant-based coordination strategy to this problem is very similar to the ant-based routing approach described in [5], [6] and [7] where the static VRPTW is discussed. Still our problem differs from the ones mentioned before. The number of casualties (customers) is not fixed and is changing continuously. The number of doctors (vehicles) is a constraint that also changes in time. In [8] and [9] we found the closest problem to our case. Here an ant heuristic algorithm was applied to a Dynamic Traveling Salesman Problem (DTSP) with insertion and deletion of the cities.

2 The First Aid Problem

In our Ant Based Control approach model to the First Aid problem we use the following concepts:

- Casualties $C = \{1, 2, ..., n\}$. Every casualty has a fixed location, a triage level and, depending on this level, a precise time of death Td_i.
- Triage level: is a representation of the diagnosis of a casualty, i.e. red, yellow, green. Every level corresponds to a treatment time Tt_i (e.g. red takes 30 minutes). At regular times the triage goes to the next level (e.g. if yellow untreated for 40 minutes - switch to red) and as a consequence the time necessary for the treatment of the patient also grows.

- Medics $M = \{1, 2, ..., m\}$: they need to give the first aid to the casualties. Their main objective is to minimize the number of deaths. A second objective is to reduce the total effort time of the medics: the time travelled between patients plus the time spent with their treatment. This is because they have a limited effort capacity E.
- Ants: ants are generated by the medics and travel between the nodes in the graph of casualties. They try to optimize the path followed by the doctors to help the casualties. As a quality of their generated solution, they leave pheromone trails in the form of routing tables.

As already mentioned, we represent the environment as a graph G=(V,A), where V denotes the set of all vertices in the graph; it consists of the subsets C and M (the initial position of the medics) (Figure 1). A travel time T_{ij} is associated with each arc in A. T_{ij} is the time necessary for a medic to move from the casualty i to the casualty j.

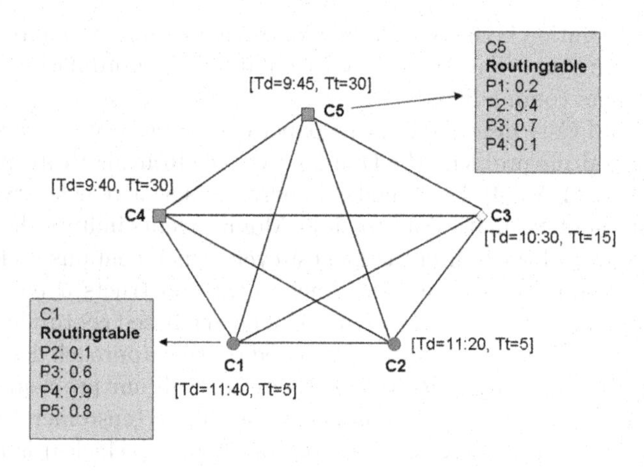

Fig. 1. Graph representation of the problem

G is a fully connected graph. At each casualty node a routing table is maintained. These routing tables are updated by the ants on the basis of pheromone trails. The higher the level of pheromone the more attractive, for example see C1 and C5.

3 Ant Colony System Model

Ants are generated periodically at each node where a medic is. Each ant tries to find the longest cycle which contains each medic and the maximum number of casualties. Each ant constructs a partial solution where each casualty is visited at most once.

Travelling between nodes in G the ant selects the next node to go using the following formula:

$$j = \begin{cases} \max \tau_{iu}\eta_{iu}, q \leq q_0 \\ \\ J, q \geq q_0 \end{cases} \tag{1}$$

$$P_{ij} = \frac{\tau_{iJ}\eta_{iJ}}{\sum_{l \in J} \tau_{il}\eta_{il}} \tag{2}$$

where J represents the set of unvisited nodes and $q_0 = 0.5$.

η is the 'urgency' function. Tt_j is the treatment necessary for the casualty j at the time when the medic is expected to arrive according to this partial solution. In the formula below we tried not to make use of the time Td when the patient will die. This is because although the triage is known, the exact time the patient will die can't be known in advance.

$$\eta_{ij} = \frac{Tt_j}{Tt_j + T_{ij}} \tag{3}$$

Each time an ant makes a step, it removes some quantity of pheromone lay down along the link:

$$\tau_{ij} = (1 - \rho)\tau_{ij} \tag{4}$$

In this way the selected path becomes less attractive for the other ants helping the exploration search for new and better solutions.

After no more casualties can be inserted in the partial solution, this is compared with the best solution found so far. We tried to maximiz the number of total rescued victims. This is the sum of the already rescued victims and the number of victims expected to be rescued (L the length of the cycle). In case the partial solution is better it replaces the global solution. A better solution is considered also if the number of the rescued victims is constant but they have a lower triage level score.

In case we found a new or same global solution an update is done for all the links on the paths:

$$\tau_{ij} = \tau_{ij} + \frac{\rho|L|}{|C_a|} \tag{5}$$

C_a represents the total number of alive casualties which haven't received the first aid yet. L is the length of the partial solution and represents the estimate number of victims that can still be rescued. We choose $\rho = 0.1$.

When a medic has to move to another patient, the selection is made following the path stored in the global solution.

3.1 Reacting to a Change

An important issue in our algorithm is how to deal with the insertion of a new doctor or casualty. When a new node in the graph is introduced the pheromone table in this node is initialized with values proportional with the distance to the other nodes. A new element corresponding to the new node is introduced in the pheromone table of each of the other nodes already existent in the graph G. The pheromone value set on this field is equal with the maximum value found in the routing table. This is because we want that new solutions containing the new element to be fast generated. When casualty dies or its treatment is completed, the links corresponding to this node are removed from the graph.

4 Experiment Environment

Consider a disaster area of 1 km x 1 km (Figure 2). The simulation starts at T=0 and will stop at T=360 minutes (6 hours). In this area casualties will appear at random locations, at random intervals in a period of three hours. After three hours, there are no new casualties.

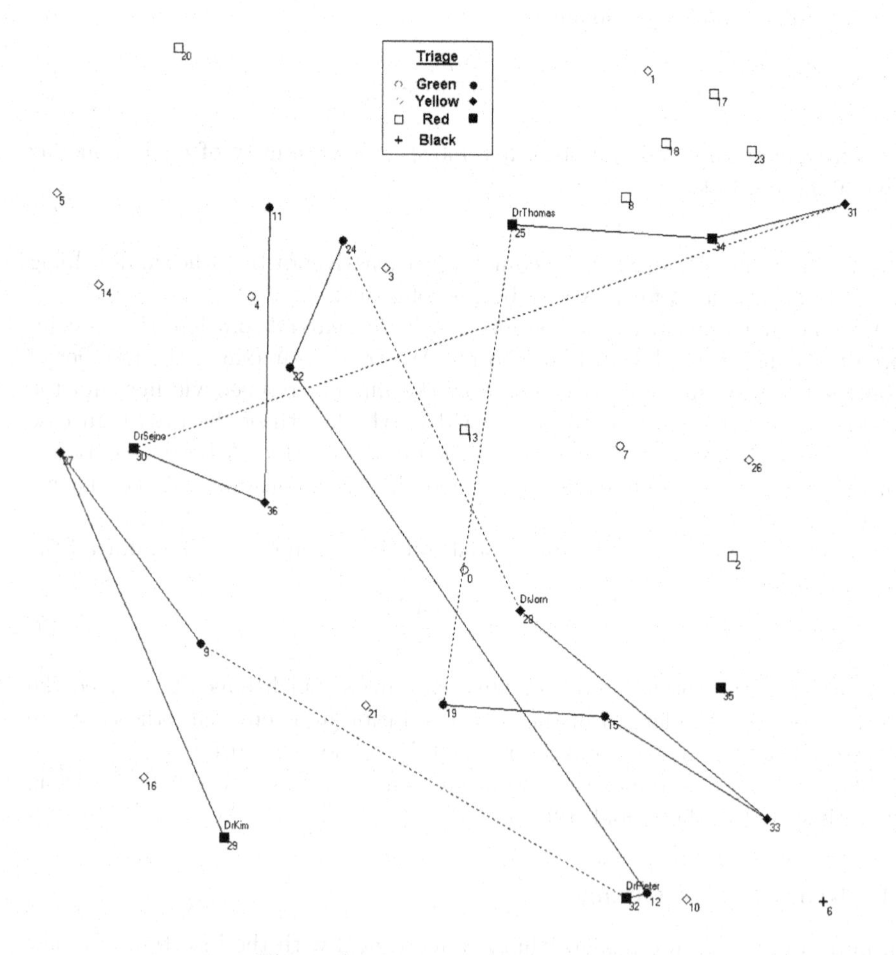

Fig. 2. Area with casualties and doctors

At the moment a new casualty appears, the triage level is known and will either be green (minor), yellow (delayed) or red (immediate). New casualties do not have triage level black (morgue). Casualties will stay in one place. Medics can enter the area from the up or the left border and will move around in the disaster area by foot. It will take 15 minutes a medic to walk 1 kilometer (i.e. 4 km/h). All the medics have a fixed effort time of 3 hours. After this time passes they have to leave the area.

In Figure 2 the dark signs represent victims which are waiting for help or are under the treatment. The empty figures represent patients that already have been treated. The death of patient 6 can be noticed by the cross sign "+". We used circle for green, diamond for yellow and square for red. The best available schedule is displayed. The solid lines draw the paths the medics will follow. The dotted lines are connecting the scheduling paths of different medics. The patient 35 is not part of the solution and eventually will be let to die. In our case the schedule plan is:

- for Dr Kim: 29, 27, 9.
- for Dr Pieter: 32, 12, 22, 24.
- for Dr Jorn: 28, 33, 15, 19.
- for Dr Thomas: 25, 34, 31.
- for Dr Seine: 30, 36, 11.

The condition of casualties will get worse in time if they are not treated. Degeneration intervals are as follows:

- Triage level green to yellow after 120 minutes.
- Triage level yellow to red after 40 minutes.
- Triage level red to black after 20 minutes.

As soon as a medic starts paying attention to a casualty, the condition of that casualty will remain stable (i.e. the triage level will remain the same over time). Each medic will spend a certain amount of time to each casualty, depending on the triage level:

- Triage level green: treatment time = 5 minutes
- Triage level yellow: treatment time = 15 minutes
- Triage level red: treatment time = 30 minutes

In order to compare our Ant Based Control algorithm (ABC) we implemented two greedy approaches. The first selects the next victim in the nearest neighbourhood (nearest neighbour NN). The second approach treats the victims based on the triage level (reeds first RF).

5 Tests and Results

First we run a "middle of the road" scenario. An area with 25 casualties was simulated: 5 green, 10 yellow and 10 red. Two medics were sent into the field. The results in Figure 3 show that using the ABC algorithm 2 more patients can be saved.

For the second "escalation scenario" we raised the number of casualties to 50: 10 green, 20 yellow and 20 red. We tested now a case with 5 medics in the

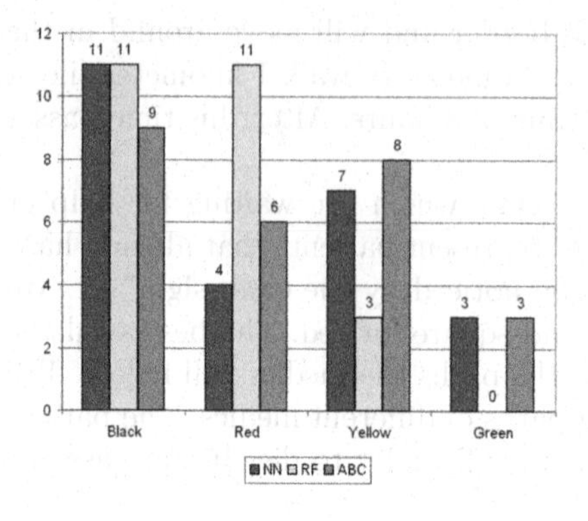

Fig. 3. Middle of the road scenario: 25 casualties and 2 doctors

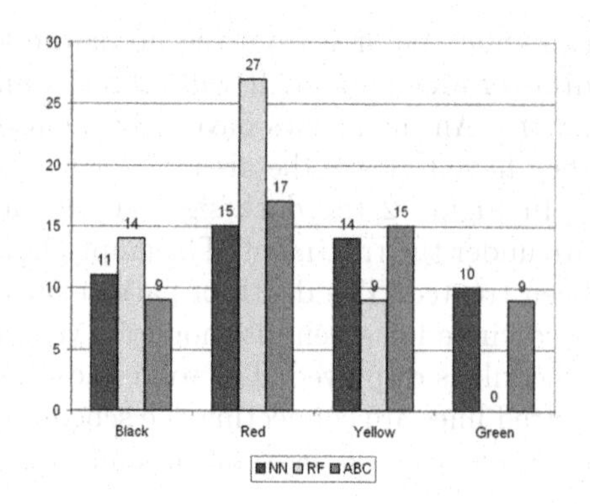

Fig. 4. Escalation scenario: 50 casualties and 5 doctors

field. The results are shown in Figure 4. Again the ABC algorithm saves 2 more victims than NN and 5 more than the RF strategy.

The third scenario we used was a "saturation scenario". 100 casualties are discovered during the first three hours after the disaster. 20 have the green triage level, 40 yellow and 40 red. This time we tested 2 situations, one with 10 medics available and one with 20 medics available. The results are displayed in Figure 5 and Figure 6.

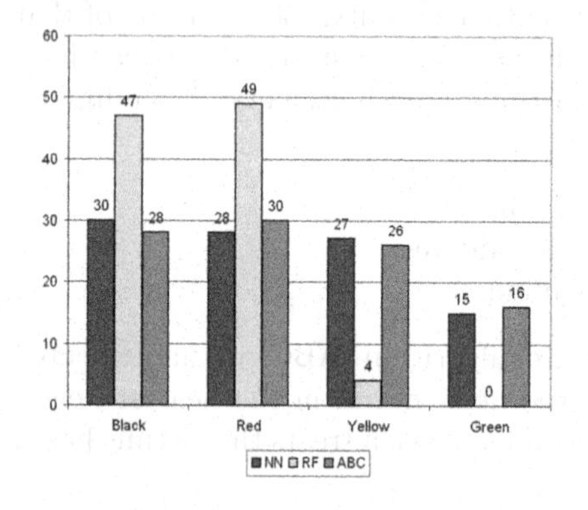

Fig. 5. Saturation scenario: 100 casualties and 10 doctors

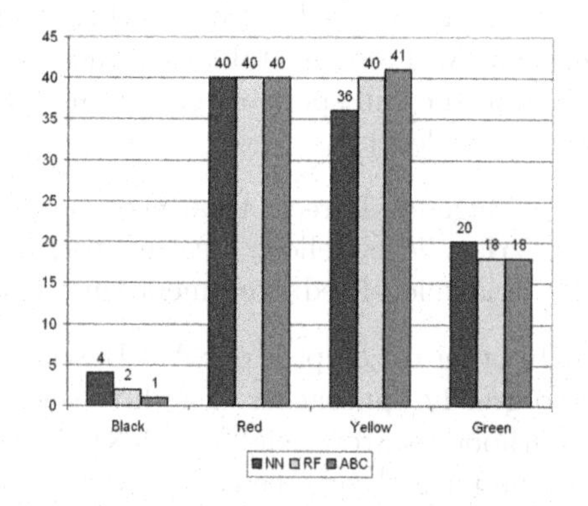

Fig. 6. Saturation scenario: 100 casualties and 20 doctors

The results of the Ant Based Control algorithm are better then the other two in all the cases. Choosing always the nearest neighbour proved to be more efficient than making the choice based on the triage level especially when there is a lack of rescuers. In the current version of our algorithm, once a medic is assigned to a patient, it stays there until the complete treatment is done. Allowing the medics to leave the patients, without finishing their treatment, and to switch to a new discovered patient might improve the results.

6 Conclusion

In this paper we applied an ant based algorithm to solve the problem of scheduling doctors to casualties in a crisis area. The approach proved to be suitable to solve such a dynamic optimization problem. The First Aid Case enables us to conduct comparative experiments, but also allows us to illustrate how coordination strategies can help crisis response organizations to deal with complex, dynamic crisis situations. In the near future we are planning to compare our algorithm with the Self Managing Distributed System (SMDS) and with a negotiation-based approach described in [1].

References

1. van Veelen, J., Storms, P., van Aart, C.: Effective and Efficient Coordination Strategies for Agile Crisis Response Organizations. In: International Workshop on Information Systems for Crisis Response and Management ISCRAM 2006. (2006)
2. Garner, A., Lee, A., Harrison, K., Schultz, C.H.: Comparative analysis of multiple-casualty incident triage algorithms. Annals of Emergency Medicine **38**(5) (2001) 541–548
3. Benson, D., Keonig, K., Schultz, C.: Disaster triage: START then SAVE - a new method of dynamic triage for victims of a catastrophic earthquake. Prehosp Disaster Med **11** (1996) 117–124
4. Montemanni, R., Gambardella, L., Rizzoli, A., Donati, A.: A new algorithm for a Dynamic Vehicle Routing Problem based on Ant Colony System. In: Proceedings of Third International Workshop ANTS 2002. (2002) 111–122
5. Dorigo, M., Maniezzo, V., Colorni, A.: Ant System: Optimization by a colony of cooperating agents. IEEE Transactions on Systems, Man, and Cybernetics Part B: Cybernetics **26**(1) (1996) 29–41
6. Dorigo, M., Gambardella, L.: Ant Colony System: A Cooperative Learning Approach to the Traveling Salesman Problem. IEEE Transactions on Evolutionary Computation **1**(1) (1997) 53–66
7. Ellabib, I., Basir, O.A., Calamai, P.: An Experimental Study of a Simple Ant Colony System for the Vehicle Routing Problem with Time Windows. In: Proceedings of Third International Workshop ANTS 2002. (2002) 53–64
8. Guntsch, M., Middendorf, M., Schmeck, H.: An Ant Colony Optimization Approach to Dynamic TSP. In: Proceedings of the Genetic and Evolutionary Computation Conference (GECCO-2001). (2001) 860–867
9. Guntsch, M., Middendorf, M.: Applying Population Based ACO to Dynamic Optimization Problems. In: Proceedings of Third International Workshop ANTS 2002. (2002) 111–122

Autonomous Gossiping of Information in a P2P Network with Artificial Ants

Christophe Guéret, Nicolas Monmarché, and Mohamed Slimane

Laboratoire d'Informatique, Université François Rabelais, Tours, France
{gueret, monmarche, slimane}@univ-tours.fr

Abstract. They appeared in our life some years ago with the awakening of the PC and now the are everywhere : computers have become ubiquitous and, almost, irreplaceable. Classical ways of creating, managing and exchanging information have been progressively replaced by electronic means. In spite of this plebiscite, computer supported collaborative work (CSCW) softwares can be blamed for requiring the user to do an effort to use them. This paper describes an artificial ant based autonomous information dissemination algorithm. It constitutes the communication layer of our framework PIAF ("Personal Intelligent Agent Framework") intended to help users transparently sharing information. The algorithm uses message gossiping strategy to transfer information items between users.

1 Introduction

Computer based technology occupy an important place in our daily life and now are considered to be ubiquitous [1]. During the last decades, using computers modified users habits. Electronic documents have changed the way to write, archive and diffuse content while Internet has changed the way we collaborate. Now, it's possible to work on a same project, exchanging documents or chatting regardless the physicals positions of the co-workers. In this collaboration context, information gathering may consist in "ask a program" or "ask a person" [2]. Programs are able to deliver indexed content (data files) while persons may give some recommendations (knowledge). Supposing it is possible to represent knowledge as a data file to share, from now on, we will use the generic term of "resource" to design both shared data and knowledge.

Resource sharing can be implicit or explicit. Sending an email is an explicit act while using a software to share idle CPU time is implicit for the user. Explicit sharing is the most challenging task for the user. Let us suppose a user finds an interesting website and wants to have all other, potentially interested, peers know about it. The strategy may be either 1) within the set of known peers, inform the subset of peers more likely to be interested by the website 2) inform all peers and let them decide if they are interested or not. In the first case, the risk for the sender is to omit some interested peers while in the second, the risk is to Spam (*ie*: sending unwanted messages to some peers). Such a simple task belongs to Computer Supported Collaborative Work (CSCW) and

M. Dorigo et al. (Eds.): ANTS 2006, LNCS 4150, pp. 388–395, 2006.

highlight the 3 main problems related to their use [3]. First is the need for a mutual awareness so every peer is aware of other's interests. Depending on the size of the collaboration team, this can be difficult to maintain. Secondly, users must be motivated enough in using a software helping them sharing resources. The user has to make an effort to send emails or learn how to use a new software dedicated to CSCW. This may weaken their motivation and dissuade them from diffusing resources they have. The third and last problem is the difficulty for users to define precisely what they are interested in. If we take the example of web browsing, users are most likely to jump from page to page looking for interesting links rather than follow a precise and predetermined path.

We believe a resource sharing system based on implicit sharing could cope with those problems. The PIAF software we design is based on this idea. PIAF stands for Personal Intelligent Agent Framework, this framework is divided in two main layers: communication and dialog. The communication layer takes in charge the information flows within the network. The dialog layer is the interface between the user and the network. Personal agents are settled in this layer, along with sensors softwares to create new information items from users' activities and help them use the network. An information is a data structure used to inform peers of an available resource. This shared resource may be located on the source of the information or elsewhere, the network itself is only seen as a space where information items are exchanged. Resources exchanges are performed in the dialog layer. In this paper, we focus on the communication layer.

The reminder of this paper is organized as follows. In section 2, we discuss some existing solutions for communications in P2P networks. The following section 3 contains a description of the different algorithms we developed. Finally, we present experimental results in sections 4 and 5 before concluding on the perspectives for this work in section 7.

2 Communications in a P2P Network

Information flows are generated by exchanges between a server having the information and a client asking for it. Considering a group of collaborators, every one may have resources he wants to share or fetch. Hence, the tricky task for a client is mainly to find a relevant server to contact. Supposing the client asks for something, several solutions have already been proposed to perform queries routing in structured [4] and unstructured P2P networks [5]. To inform a group of peer, publish/subscribe mechanisms like social web portals [6] or event based systems [7] can be considered. This paper deals with the problem of automatic dissemination. Hence we consider no queries are specified by user nor they get subscribed on publication service. Therefore, we choose a gossip-based scheme for exchanging information items [8].

2.1 Gossip-Based Diffusion

Gossips protocols consists in sending a message to a subset of the connected peers, according to a probability p. Flooding is in fact a particular type of Gossip with $p = 1$. Gossiping is particularly suited for diffusing information within a defined group of peers. Unlike solutions aimed at globally replicate a data structure [9], that is, ensure all peers recover all existing information items, our algorithm does directed and focused diffusion. A given information is gossiped to peers more likely to be interested of it.

To estimate neighboors' interest in an information item, the *de facto* strategy consists in using similarity between profiles. a data structure summarizing peer's interests. In order for each peer to keep up-to-date peer's profiles, two strategies can be used: 1) Whenever a peer changes its profile, it pushes this update to other peers [10]. 2) Peers periodically browse the network to fetch profiles [11]. Both solutions have pros and cons but share two main drawbacks. First, since they are supposed to summarize what a peer is interested in, profiles are not easy to define. Although we can have them automatically constructed [12] problems can appear if a user has changing or spurious points of interest. Secondly, a profile is related to the unique peer it describes whereas, in real life communications, a given person can act has a referral and suggest an other peer to contact. Hence, one peer profile also compromise profiles of other peers he knows.

We propose another new solution to profiles managing inspired by the ideas of overhearing [13] and use of information trails [14] in a network. We consider that whenever a peer sends a message over the network, he gives an hint about what he is interested in. Hence, instead of inquiring about the profiles of one's peers, we guess them from traffic they generate over the network. The use of estimated profiles differentiate our work from other solutions proposed to perform autonomous gossiping of information [15].

2.2 Topology Management

It has been observed that a network of collaborators exhibits small world properties: the network is made of many dense groups loosely connected to each other. Those groups appear when individuals congregate as they found themselves having shared interests. Considering a peer x, a clustering coefficient γ_x quantifies how dense its neighborhood is. If $\gamma_x \simeq 1$, x is part of a dense group. Dynamically adjusting the topology of the P2P network in order to make it similar to the underlying small world can improve sharing efficiency. Using estimated profiles, we propose an algorithm aimed at identifying such group and dynamically adjust connections. A criteria is used to decide if two peers have similar interests or not. Depending on this criteria, a given connection may be dropped or kept.

3 Artificial Ant for Information Dissemination

One can view the P2P network as a directed graph $G(V, E)$. Each node of V is associated to a peer. An edge $e_{xy} \in E$, with $E \subseteq V \times V$, represents a connection from a peer x to a peer y. We define the neighborhood $\Gamma(x)$ of a peer x as the subset of peers x is connected to: $\Gamma(x) = \{y \in V \backslash \{x\} \mid e_{xy} \in E\}$. Although $e_{xy} \neq e_{yx}$ we will consider $\forall e_{xy} \in E \Rightarrow e_{yx} \in E$. This model defines a social network, that is a network which edges define relations between nodes. The relation here is some shared interests concerning some information items.

Artificial ants are used to move information items. We also use artificial pheromones defined on a vector space $\mathbb{R}^n$. The dimension n depends on the semantic of pheromones. For instance, if pheromones results from a TFIDF encoding, n would be the dictionnary size. In this paper, no assumption is made concerning this semantic. Each information X has an associated pheromone vector τ_X. An other vector τ_{xy} is associated to a connection e_{xy}. We suppose the existence of a similarity s defined on this vector space $s : \mathbb{R}^n \times \mathbb{R}^n \mapsto [0, 1]$. Thus, it is possible to evaluate the similarity between two connections, as well as the similarity of a connection and an information. For the simulations, the similarity used is the standard cosine.

Pheromones associated to connections are used as a global network memory and store traces about information exchanged. Peers stores pheromones related to incoming connections so they "remembers" what other peer sends to them over the time. This is why an ant going from a peer x to a peer y, pheromones on the link from y to x are updated.

3.1 Ant's Gossiping Activity

Ants work as follows when disseminating an information X. Every T_g unit of time, the ant will push this information from its nest (a peer x) to another nest (peer y) chosen randomly in the neighborhood of x. A stochastic algorithm is used to select y among $\Gamma(x)$. According to a similarity threshold λ, neighbors are first sorted in two groups : interested peers V_I and non interested peers $V_{\overline{I}}$. In this step, already visited peers ψ are ignored. Two counters $I(y)$ and $\overline{I}(y)$, respectively used to store good and bad evaluations for a connection, are also adjusted. If y goes into V_I, $I(y)$ is incremented.

$$V_I(x, X) = \{y \in \Gamma(x) \setminus \psi \mid s(\tau_{xy}, \tau_X) \geq \lambda\} \qquad V_{\overline{I}}(x, X) = \Gamma(x) \setminus V_I(x, X) \quad (1)$$

For an interested peer, its chances to being chosen as a destination are proportional to its relative similarity. Meanwhile, non interested peers may be equiproportionally chosen. A strategy aimed at having an optimal information flow would lead to always sending an information only to the most interested peer. On the other hand, in order to find new peers to connect to, network exploration involves trying to send information items to some other neighbors even if they does not seem to be interested. A trade-off hence must be found. We grant ants with a notion of freewill : when it is about to send an information, an ant toss a coin to choose whether it will contact a peer interested or not (η being the probability of choosing a peer within $V_I(x, X)$).

$$\forall y \in \Gamma(x),\ p_{x \to y}(X) = \begin{cases} \eta \dfrac{s(\tau_{xy}, \tau_X)}{\sum_{z \in V_I(x,X)} s(\tau_{xz}, \tau_X)} & \text{if} \quad y \in V_I(x, X), \\ (1 - \eta) \dfrac{1}{|V_{\overline{I}}(x, X)|} & \text{if} \quad y \in V_{\overline{I}}(x, X) \\ 0 & \text{if} \quad y \in \psi. \end{cases} \quad (2)$$

Whenever an information is carried from a peer x to a peer y, pheromones vector of e_{yx} is updated. The amount of pheromones laid depends on the activity on the link and information's origin. The more information items are transfered through a connection, the more pheromones deposit will be important. Pheromones deposit should also decrease as the information is farther away from its source. We have chosen to model this using two Gaussian (see equation 3). As seen in equation 4, ρ is used to both evaporation and deposit of pheromones. ρ_{max}, α and σ are regulation factors. $r(X)$ is the number of peers X has crossed by from its origin up to y. $t^{'}$ designs present time and t last time an information was transfered through this connection.

$$\rho = \rho_{max} \cdot \exp^{-\alpha r(X)} \cdot \exp^{-(\frac{t'-t}{\sigma})^2} = \rho_{max} \exp^{-(\frac{t'-t}{\sigma})^2 - \alpha r(X)} \tag{3}$$

$$\tau_{yx}^{(t')} = (1-\rho)\tau_{yx}^{(t)} + \rho \tau_X \tag{4}$$

3.2 Topology Management

Moving the nest consists in modifying is neighborhood by adding or removing some links. To establish a new connection a peer x grab an other peer y from his address book $D(x)$. Only not already connected nor in standby peers may be picked up from this directory.

A peer may have a maximum of k_{max} opened connections. Hence if $|\Gamma(x)| = k_{max}$ a connection must be dropped before contacting y. For each connection, a quality score $f(y)$ is defined as the ratio between the number of time the neighbor was estimated to be interested $I(y)$ and the total number of estimations performed by ants, $I(y) + \overline{I}(y)$. If $f(y)$ falls under a given threshold β, the connection is not efficient enough and may be dropped with a probability $p_d(y)$ (see equation 5). The higher the difference between β and $f(y)$ is, the higher $p_d(y)$ is.

$$\forall y \in \Gamma^{'}(x)\ ,\ p_d(y) = \frac{\beta - f(z)}{\sum_{z \in \Gamma'(x)} \beta - f(z)} \tag{5}$$

$\Gamma^{'}(x) = \{y \in \Gamma(x) \mid f(y) < \beta\}$ is the subset of peers not efficient enough.

If x was not able to find a peer in is directory, it asks one of its neighbors to send him a suggestion. x sends to y a message with a copy of is own directory. y browses its own directory and answers sending back the address of the most similar peer w that x does not already know. To choose which y of $\Gamma(x)$ will be interrogated, a rank-based selection is applied on the scores $f(y)$. Using a rank based selection provides us with a lower risk of always sending suggestions requests to the same peer.

4 Simulation Environment and Performance Criteria

A discrete event simulator was used in order to implement and test our information dissemination algorithm. We have chosen to use the OmnetPP simulator [1].

[1] OmnetPP discrete event simulator, `www.omnetpp.org`

The tested network is made of 20 peers allowed to maintain a maximum of $k = 4$ connections. Initially, no connection is established and in their address book peers have the address of a unique randomly chosen peer. To simulate the presence of a user, each peer periodically (every 100 unit of time) publish an information item grabbed from a global dataset made of 400 documents distributed in 4 topic. In this dataset, average intraclass and interclass distances are respectively of 0.75 and in $[0.08, 0.25]$. Ant's λ parameter is set to 0.7 in order to have high probability to correctly recognize elements of a same class. Each peer is supposed to be interested in only one topic. Topics are equally distributed in order to form 4 groups of peers having same interests. It has been proved in [16], that in this case the most clustered graph is a "caveman graph" with a clustering value of 0.6 bounded by an amount of $O(1/k^3)$.

The performance of the diffusion algorithm is evaluated through 3 estimators: the clustering coefficient γ, completeness and efficiency. Theses values are computed for each peer and then averaged over all network members. The clustering coefficient is defined as the number of connections present within a peer's neighboorhood divided by the number of possible connections. Completeness is defined as the number of received information items related to his interest divided by the total number of information items related to his topic present in the network. Precision quantify the Spam ratio and is measured as the number of interesting information items received divided by the total amount of information items received.

5 Results

Several tests have been performed to estimate the impact of each parameter on the overall system but, because of space constraints, only tests for the minimum of quality for a connection are presented here. The test goal is to find when it is better considering a peer not to be efficient: is it when $\overline{I} > I$? when the value for I is twice ones of $\overline{I}$? The answer is statued from simulation results averaged over 50 executions of the algorithm, using different values of β. Five condition are tested : $I < 4\overline{I}$, $I < 2\overline{I}$, $I < \overline{I}$, $2I < \overline{I}$ and $4I < \overline{I}$. They respectively corresponds to $\beta = 0.8$, $\beta = 0.67$, $\beta = 0.5$, $\beta = 0.33$ and $\beta = 0.2$.

Considering the figure 1, the best solution is allowing to drop a connection when $\overline{I} \geq I$ since, for $\beta = 0.5$, $\gamma \simeq 0.6$. Higher values of β leads to drop many connections, therefore almost no cluster can be formed and $\gamma \simeq 0.15$. On the other hand, if β is set to lower values, only very bad connections are dropped. Nevertheless, the clustering value is higher and the resulting topology is a network of isolated caves. Since β is low, fewer connections may satisfy $f(y) < \beta$. Hence a peer candidate for disconnection has more chances to be actually disconnected and less time to exchange information items.

The figure 2 shows how information dissemination evolutes during the simulation. On the figure 2a it can be noted that peers tends to gather more information they are interested in when $\beta = 0.2$. This was predictable since, as seen in figure 1, when $\beta = 0.2$ peers are connected almost only to peers sharing similar

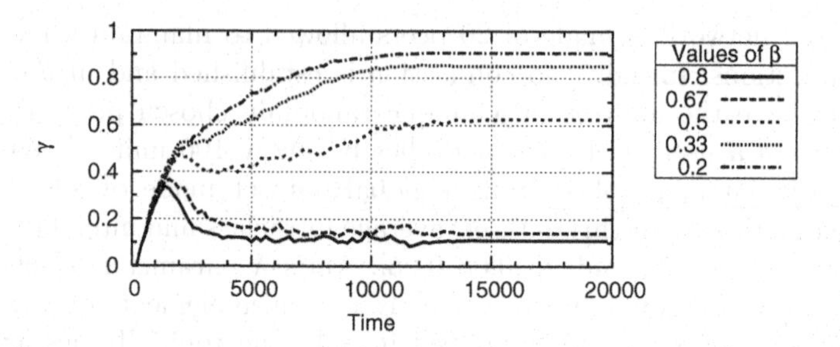

Fig. 1. Evolution of clustering coefficient

interests. Setting $\beta = 0.5$ leads to middle-range performance with a precision of $\simeq 0.64$. On the figure 2b, we can remark that for a $\beta \leq 0.5$, peers' information storages contains about 85% of the total number of information they are interested in present in the network. This convergence of the results for $\beta = 0.5$ and $\beta = 0.2$ proves that the diffusion algorithm is as efficient with connected caves as with isolated caves.

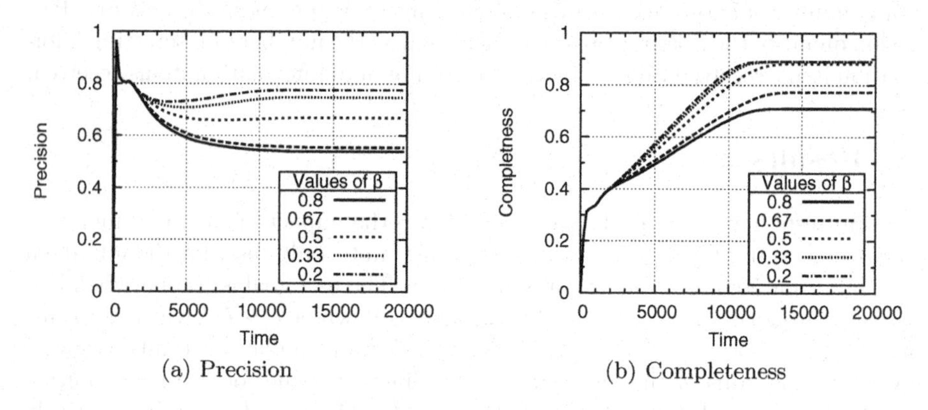

(a) Precision

(b) Completeness

Fig. 2. Evolution of completeness and precision

6 Conclusion and Future Directions

We presented in this paper an algorithm using artificial ants to diffuse information in a P2P network. Using a collective memory based on pheromones and a gossip based diffusion strategy, this algorithm allows to share information without having to define a profile for users nor sending request over the network. The information diffusion is proactive and transparent for the user.

Without being really changed, the algorithm presented in this paper has been progressively improved since it's earlier versions [17]. Future developments will include trying to improve it further. Particularly, we are looking forward using Bloom Filter to summarize storages and directories contents.

References

1. Weiser, M., Gold, R., Brown, J.S.: The origins of ubiquitous computing research at parc since the late 1980s. IBM Systems Journal **38**(4) (1999) 693–696
2. Kautz, H., Selman, B., Milewski, A.: Agent amplified communication. In: Proceedings of the Thirteenth National on Artificial Intelligence (AAAI-96), Portland, OR (1996)
3. Grudin, J.: Why cscw applications fail: problems in the design and evaluation of organization of organizational interfaces. In: Proceedings of the 1988 ACM conference on Computer-supported cooperative work, Portland, Oregon, United States (1988) 85–93
4. Stoica, I., Morris, R., Karger, D., Kaashoe, M.F., Balakrishnan, H.: Chord: A scalable peer-to-peer lookup service for internet applications. In: SIGCOMM'01. (2001)
5. Di Caro, G., Dorigo, M.: AntNet: Distributed stigmergetic control for communications networks. Journal of Artificial Intelligence Research **9** (1998) 317–365
6. Google: Orkut online community. `www.orkut.com` (2003)
7. Leeb, M., Sua, S.Y., Lama, H.: Event and rule services for achieving a web-based knowledge network. Knowledge-Based Systems **17** (2004) 179–188
8. Burmester, M., Le, T.V., Yasinsac, A.: Adaptive gossip protocols: Managing security and redundancy in dense ad hoc networks. Ad Hoc Networks (2006)
9. Demers, A.J., Greene, D.H., Hauser, C., Irish, W., Larson, J., Shenker, S., Sturgis, H.E., Swinehart, D.C., Terry, D.B.: Epidemic algorithms for replicated database maintenance. Operating Systems Review **22**(1) (1988) 8–32
10. Haase, P., Siebes, R.: Peer selection in peer-to-peer networks with semantic topologies. In: Proceedings of the 13th International World Wide Web Conference, New York City, NY, USA (2004)
11. Schmitz, C.: Self-organization of a small world by topic. In: First International Workshop on Peer-to-Peer Knowledge Management (P2PKM). (2004)
12. Kautz, H., Selman, B., Shah, M.: Referral web: combining social networks and collaborative filtering. Communications of the ACM **40**(3) (1997) 63–65
13. Busetta, P., Serafini, L., Singh, D., Zini, F.: Extending multi-agent cooperation by overhearing. Technical Report 0101-01, Istituo Trentino di Cultura (2001)
14. Payton, D.W.: Discovering collaborators by analyzing trails through an information space. In: AAAI Fall Symposium on Artificial Intelligence and Link Analysis. (1998)
15. Datta, A., Quarteroni, S., Aberer, K.: Autonomous gossiping: A self-organizing epidemic algorithm for selective information dissemination in wireless mobile adhoc networks. LNCS **3226** (2004) 126–143
16. Watts, D.J.: Small Worlds - The Dynamics of Networks between Order and Randomness. Princeton University Press (2003)
17. Guéret, C., Monmarché, N., Slimane, M.: Sharing resources with artificial ants. In: Proceedings of the 9th International Workshop on Nature Inspired Distributed Computing (NIDISC'06), Rhodes Island, Greece (2006)

Cooperative VLSI Tiled Architectures: Stigmergy in a Swarm Coprocessor

Gianmarco Angius, Cristian Manca, Danilo Pani, and Luigi Raffo

DIEE - Department of Electrical and Electronic Engineering
University of Cagliari, Cagliari, Italy
{g.angius, pani, luigi}@diee.unica.it

Abstract. Stigmergy is a form of indirect interaction for coordination and communication purposes that can be found in many swarm systems. In this paper we present a tiled coprocessor for computation-intensive applications that explicitly exploits stigmergy to achieve adaptability avoiding the usual time-consuming handshakes involved in direct interactions. This adaptability, without any centralized control, directly implies architectural scalability at design time, flexibility in multitasking environment, adaptive load balancing and fault-tolerance at run-time. A CMOS 0.13μm implementation of such architecture for simple array processing operations is presented and evaluated. Obtained results show the potentiality of the proposed approach.

1 Introduction

Many studies have been carried out to improve the performances of uniprocessor systems for computation-intensive applications. CMOS technology scaling allows higher operating frequencies and higher integration density. However, novel approaches are required to face the new challenges of digital design, e.g. wire delay and fault tolerance. With the upcoming of Moore's Law, scientists are trying to figure out what kind of the so called "emerging technologies" could provide spatial scalability with good performances [1]. Emerging technologies are characterized by undependable manufacturing processes leading to many structural defects. Today's design styles are not suited for this kind of implementation neither in terms of architectural structure nor in terms of fault tolerance. Tiled architectures (architectures composed of identical computational tiles arranged in a regular grid) have always been considered a good solution to address the issues of fault tolerance, multitasking and wire delay.

The most fine-grained tiled architectures are Cellular Automata (CA), where every tile in the system is a simple automaton. Different studies in the field, aimed to exploit some of the properties of CA on silicon [2] or on future emerging technologies [1], were presented. A bio-inspired fine-grained fault tolerant platform is Embryonics [3], where every element is not an automaton but instead a multiplexer-based cell. The self-repair mechanism, implemented at two levels, involves a significant cost in terms of wasted hardware resources, and the number of recoverable faults is limited.

M. Dorigo et al. (Eds.): ANTS 2006, LNCS 4150, pp. 396–403, 2006.

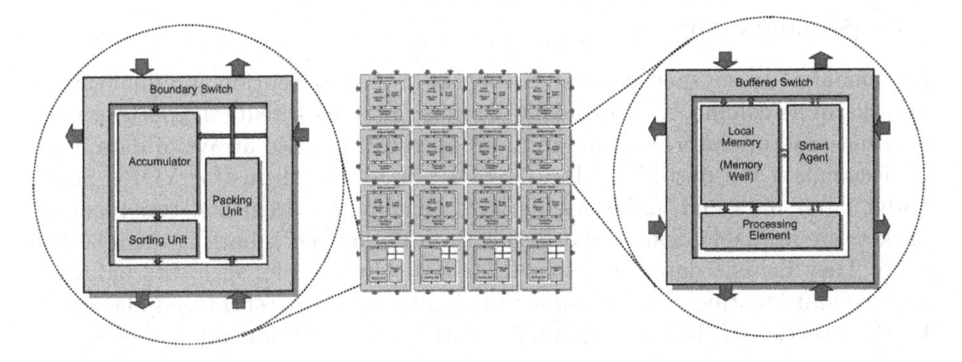

Fig. 1. The fabric structure of identical tiles with the boundary tiles

In this paper, following the original studies about swarm architectures presented in ANTS2004 [4], we applied Swarm Intelligence to design a cooperative tiled architecture with inherent properties of scalability, flexibility and fault tolerance in multitasking environments. The system is composed of *hardware agents* with limited capabilities placed in an artificial and mutable computational environment. To avoid direct communications between adjacent agents, which require handshake protocols and synchronization, we exploited *stigmergy*. This powerful mechanism provides for workload diffusion from overloaded zones towards unloaded ones to achieve improvements in computational performances and fault-tolerance. An absolutely scalable system was conceived, and a 4×8 fabric composed of 32 hardware agents was implemented in a standard cell CMOS 0.13μm technology and tested on simple array processing algorithms.

The remainder of this paper is organized as follows. Section 2 deals with the main characteristics of the proposed architecture, whereas Section 3 explains how stigmergy was used in such architecture. Some simulation results are presented in Section 4. Section 5 presents the conclusions.

2 The Proposed Tiled Architecture

The proposed architecture consists of a fabric of identical elementary tiles locally interconnected in a two-dimensional grid. The interface between the fabric and any external entity is managed by an additional row of boundary tiles (Fig. 1). The granularity of a tile is coarser than those of CA. A smart hardware agent placed in every cell is responsible for stigmergic behaviors. The adoption of a simple packet switching network allows the realization of a loosely structured collection of tiles even within a regular mesh, compared with a configurable network. The routing is performed exploiting only relative displacements to allow scalability and fault tolerance. The architecture is able to support the array element-by-element operations of ADD, SUB, MUL, shift (SHL, SHR) and compare (CMP), producing a vectorial result (VOP), and multiply-accumulate (MAC) and accumulation (ACC) operations producing a scalar result (SOP).

2.1 Boundary Tile

A boundary tile is a special tile able to perform the distribution and finalization of tasks, in its column. The boundary tile must be able to create the packets with operands and operator, starting from an operator and two arrays of data, and to accumulate the partial results (for SOP) or to sort them (for VOP). Since 16-bit input data and 32-bit results have different requirements, two types of packets are available: Source and Result Data Packet (DP), both of them 50-bit wide. They include data and other fields basically used for elements indexing and dynamic routing. The boundary tile implements a decentralized algorithm for the selection of the best column in multitasking environments.

2.2 Computational Tile

The computational tile is basically composed of three elements, a *processing element* (PE), a *local memory* (MW), and *a buffered switch* (BS). To these elements, able to ensure the correct functionality of the architecture in a way similar to other tiled architectures, we added a *smart agent*, which implements the stigmergic behaviors, as depicted in Fig.1.

The *Processing Element* is an arithmetic unit that processes the packets fetched from the MW. It is equipped with two self-test structures for fault detection. The latency of the different operations, i.e. the number of clock cycles required to carry out them, varies with the operations and with the complexity of data (only for multiplications, as in [4]). ADD, SUB and CMP require 1 cycle, shifts require 2 cycles, ACC requires 3 cycles and MUL and MAC require 3-8 cycles.

The *Local Memory* (also called *Memory Well*) is a little storage area composed of dual-ported FIFOs, one for the input data (WI) and other two (WO) for the output ones. A memory manager is responsible for the correct packets management, even in case of faults, and for the workload monitoring. Workload information is continuously evaluated during the activity of the tile, and consists in the number of packets stored in WI.

The *Buffered Switch* is the module which performs data movements into the fabric. Every input port of the switch is buffered by means of small FIFOs to break long connections and to ensure a buffering in case of high network traffic. The routing rules are very simple, since the switch doesn't require any absolute address. During task loading, this module performs data movement filling its MW with incoming SourceDPs until it is full, and then passing the new incoming SourceDPs to the neighboring tile on the north.

The *Smart Agent* is a module which transforms a computational tile in an interactive hardware agent able to carry out the computations cooperatively and with fault tolerance support in multitasking environments. It consists of some comparators and a subtractor, and some finite state machines. The smart agent is responsible for the coordinated activity within the small colony of tiles in the architecture, exploiting the stigmergy. It should be noted that every tile is able to work even without this module, but in this case no cooperation can take place.

3 Stigmergy on VLSI Platforms

The massively parallel structure of swarms allows huge parallelization exploiting the available individuals at every instant of time. The individuals are able to coordinate without a coordinator complex activities communicating by means of the environment. This is an example of *stigmergy*: one individual modifies the environment and another one responds to the new environment at a later time [5]. An example of stigmergy is the clustering behavior exhibited by some species of ants. Indirect interactions do not require neither synchronization on the same communication channel nor handshakes, since environment and agents are different entities. Since agents communication overhead does not increase with the size of the group, stigmergy implicitly allows scalability. It should be noted that stigmergy doesn't explain how the indirect communication takes place: it provides a general mechanism that relates individual and colony-level behaviors.

3.1 Agent Intelligence and Colony Behavior

Like in every swarm system, the colony behavior descends from the interactions among the individuals of a colony. Our tiled architecture can be conceived as an environment composed of a set of fillable wells locally connected by dedicated pipes. The wells present some leakages so that normally they will be progressively emptied. The wells are our local MWs, which can be filled with SourceDPs; the leakage corresponds to the activity of the PE, and the pipes among these wells are the BSs and their channels. Without any cooperative behavior, MWs are filled from the southern border of the fabric (Fig.1) and are emptied only due to the single PE processing (horizontal links are not useful). The agent intelligence is related to the behavior of the agent with respect to the environment. The agent is able to sense the environment locally (the workloads of the 4 neighboring wells and the one of its tile). It takes decisions about the amount of SourceDPs to transfer towards adjacent wells, if the workload difference is over a minimal threshold of convenience, and the best direction. It can move data only *from* its well; this indirect interaction is a form of stigmergy used to coordinate the operations into the array. More precisely we have an *active* stigmergy, which is related to explicit data movements (performed by the agents of a tile with the mechanism described above) and to the normal activity of the PEs (which consume the input data stored in the attached MWs). The workload is spread across the fabric so that the largest number of available tiles can be involved in a task, thus exploiting at the most the computational platform every time, reducing the computational latency even in presence of faulty tiles.

A comparative analysis of the behavior of the system with or without stigmergy in the case of only two tasks (130-element MACs) is presented in Fig. 2. Different snapshots are taken at the same instants of time in the two cases. It should be noted, on the time axis, that with stigmergy the latency is considerably reduced (47% of the case without stigmergy). It should be noted that, in large architectures, it would be impossible to efficiently perform such a run-time analysis and to take the relative decisions in a centralized way.

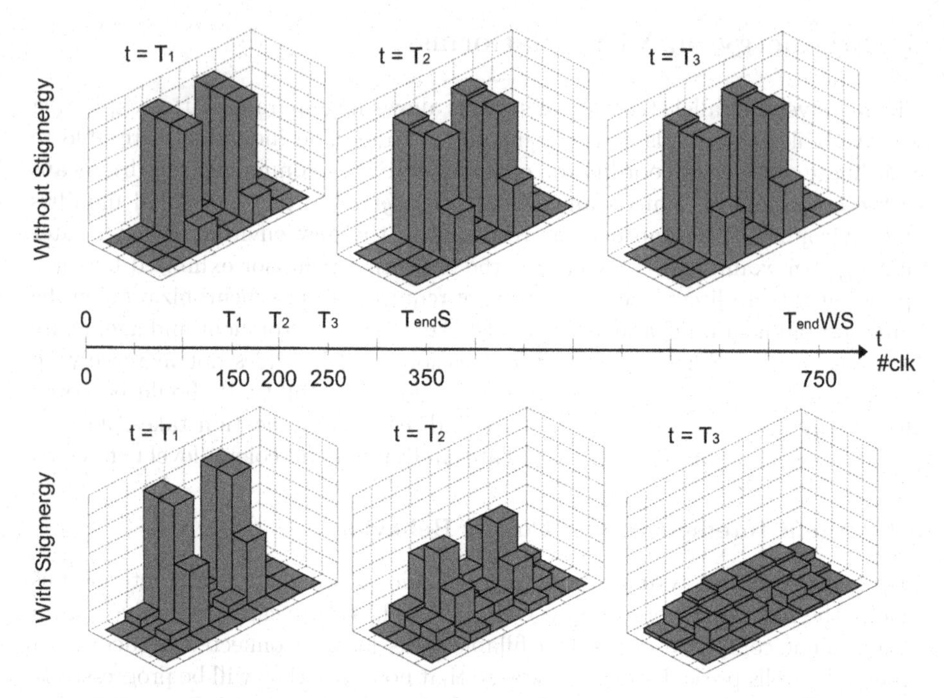

Fig. 2. Workload distribution evolution in different instants of time in a 4 × 8 fabric with 2 tasks loaded with or without stigmergy

3.2 Cell Health and Fault Tolerance

The swarm approach allows the introduction of fault tolerance. The chosen scheme was the so called *cell exclusion*. At this time, we implemented two simple mechanisms applied to the PE to detect faults, by means of fixed-operands operations with known results during inactivity periods, and of a smart watchdog timer to verify the correctness of the latency of the current operation during normal activity. If a fault is detected, the smart agent resets the PE and reloads the last operation. If the fault comes again, the smart agent deactivates the PE and plugs the MW, so that no other packets can be loaded in that tile.

Two mechanism were implemented for cell exclusion. A first mechanism is the bypass: the faulty tile becomes transparent and the packets from other tiles can traverse it but cannot be loaded into its MW. The faulty tile exports to its neighbors the workload information of the opposite tile, to allow the correct workload spreading as if the tile was not present. During bypass, the BS must be perfectly functional. If this would not be true, a second mechanism is activated: the faulty tile is completely isolated, with the I/O ports blocked. In this case the smart agent directly activates a block signal which is propagated to the 4 neighboring BSs, and from them to the respective smart agents to prevent requests of data movement towards that tile. In these cases, the stigmergic behavior is very useful to exploit all the reachable healthy tiles in the architecture, overcoming the difficulties of the reduced number of available tiles in a column.

3.3 Keeping It Simple to Improve Performances: The Bubble Effect

In principle, the workload was conceived as the sum of the latencies of the operations in the MW (WI) of a tile. This way, the workload difference between two tiles is the difference in terms of processing time based on the *number* and *weight* of the SourceDPs. Such a solution doesn't take into account the fixed size of the MWs which is not related to the weight of the operations. We called the strange effect descending from this drawback "bubble effect". This effect was only observable when tasks with different latencies were contemporarily loaded into the fabric (e.g. ACC and MAC), and when the sizes of such tasks were large enough to saturate the MWs of the pertinent column. The stigmergic behavior tries to spread the workload into the fabric, sending packets to the tiles with less workload regardless the space available. The "heavy" packets saturates some FIFOs of the BSs so that the distribution of the "light" task is delayed after the execution of the "heavy" ones. This way the "light" operations seems to emerge like bubbles at the end of the computation since the "heavy" tasks terminate before. To avoid this situation, we move back to the natural swarm abstraction: the weight of a packet must influence its execution speed, not the decision about how many packets should be moved. This means that the workload must reflect only the number of SourceDPs in the MWs and not the type of them. This way, stigmergy doesn't produce stalls, the lighter tasks being carried out faster than the heavier ones (as natural). Furthermore, the workload monitor circuitry is dramatically simplified compared to the previous solution.

4 Synthesis and Experimental Results

The architecture presented in this paper was implemented in standard cell CMOS 0.13μm technology. The operating frequency is 900MHz, with the only exception of the PE that runs at 450MHz. Thanks to the swarm approach, the architecture is fully scalable, i.e. the area grows linearly with the number of tiles, and the frequency is not affected by it. The area of a tile is about 42.8K equivalent gates, with 80% of such area occupied by the memories.

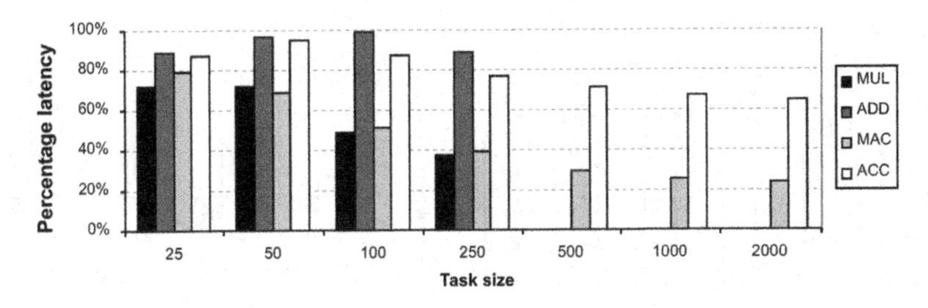

Fig. 3. Percentage of latency for a single task of different size loaded in one of the two central columns on a 4×8 fabric in case of enabled stigmergy with respect to the case of disabled stigmergy

4.1 Exploring System Potentialities: Single Task Simulations

In this case, we obtain the highest performances since the number of the tiles involved in the task execution is the highest possible. It should be reminded that a task is allocated to a column, and all the stigmergic interactions take place at run-time without any predefined scheme. In Fig. 3 the percentages of latency in case of stigmergy, with respect to the case without stigmergy, for 4 significative tasks are presented. As can be seen, for long-latency operations (MULs and MACs) the effect of cooperation is well visible with significant performance improvements. For low-latency operations (ADDs and ACCs, in this case) the latency reduction is less sharp since the latency of data transfers due to stigmergic interactions is comparable to those of the single operations.

4.2 Performances in Case of Faults

The system was tested with a fixed set of 40 tasks whose size and operands were randomly generated. The operators for the task were chosen randomly within a set of 4 operators (MUL, MAC, SUB, SHL) to equilibrate latencies. The starting times were chosen randomly in a timing interval such that the overall simulation was able to engage the fabric for about 50% of its resources on average. For every number k of faults the possible faults locations are $\binom{32}{k}$. Starting with the same set of tasks with the same starting times, we performed 5 simulations for every number of faults (randomly placed), calculating for each one the percentage of latency in case of enabled stigmergy with respect to the case of disabled stigmergy. The histogram in Fig. 4(a) highlights the average of such values for every fixed number of faults (bypassed tiles). As can be seen, increasing the number of faulty elements, the stigmergic approach leads to better performances compared to the non-cooperative case. The cooperative approach, allowing the involvement of the largest number of computational tiles in every computation, shows an outstanding adaptability that cannot be found on traditional architectures with statically scheduled operations. Compared to related works, no rerouting

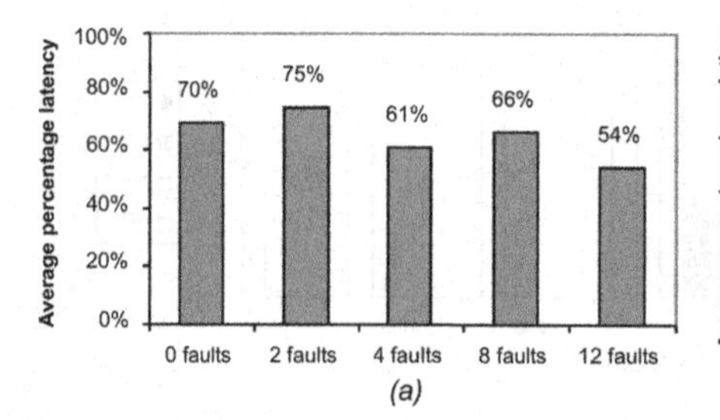

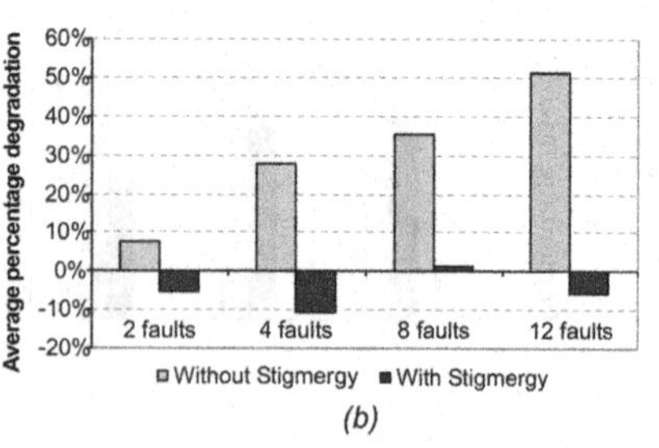

Fig. 4. Percentage of the average task latency in case of enabled stigmergy with respect to the case of disabled stigmergy for 40-task simulations with variable number of faults *(a)*. Average percentage degradation with or without stigmergy when increasing the number of faults *(b)*.

and no spare resources are needed to "repair" a fault. From Fig. 4(b) it is possible to see the performance degradation for the same set of simulations, varying the number of faults but always using stigmergy or not. As can be seen, whilst without stigmergy the performances degradation is always increasing, the adaptability exhibited by the stigmergic approach prevents such degradation showing also some performance improvements. This is mainly due to a more intensive exploitation of the upper part of the array (Fig.1), which is usually less used, due to the presence of bypassed cells in the middle.

5 Conclusions

In this paper an innovative cooperative tiled architecture was presented and evaluated in a simulated multitasking environment. The choice to adopt stigmergy to coordinate the interactions between the smart agents embedded in the tiles leads to a flexible computational platform able to involve in the computation the largest number of available tiles. Such an adaptation to environmental changes in run-time without any centralized control is an unique feature that was made possible by the adoption of a decentralized cooperative approach inspired by the Swarm Intelligence paradigm. Beyond stigmergy, some other interesting aspects typical of swarm system can be found in the proposed architecture, namely an absolute scalability, flexibility in tasks execution, adaptability, parallelism, simultaneous multitasking and fault tolerance. From the simulations performed it is possible to see how the adoption of cooperative behaviors based on stigmergic interactions enables the achievement of better performances in normal processing and significantly masks the effects of the presence of faulty tiles, compared to a non-cooperative approach.

References

1. Abelson, H., Allen, D., Coore, D., Hanson, C., Rauch, E., Sussman, G.J., Weiss, R.: Amorphous computing. Communications of the ACM **43**(5) (2000) 74–82
2. Gruau, F., Lhuillier, Y., Reitz, P., Temam, O.: Blob computing. In: Computing Frontiers 2004 ACM SIGMicro. (2004)
3. Mange, D., Sipper, M., Stauffer, A., Tempesti, G.: Toward robust integrated circuits: the embryonics approach. In: Proc. of the IEEE. Volume 88. (2000) 516–541
4. Pani, D., Raffo, L.: A VLSI multiplication-and-add scheme based on swarm intelligence approaches. In: Proc. of the 4th International Workshop on Ant Colony Optimization and Swarm Intelligence - ANTS2004, Brussels, Belgium (2004) 13–24
5. Bonabeau, E., Dorigo, M., Theraulaz, G.: Swarm Intelligence: From Natural to Artificial Systems. Oxford University Press (1999)

Distributed Shortest-Path Finding by a Micro-robot Swarm

Marc Szymanski, Tobias Breitling, Jörg Seyfried, and Heinz Wörn

Institute for Process Control and Robotics (IPR)
Universität Karlsruhe, Karlsruhe, Germany
{szymanski, seyfried, woern}@ira.uka.de

Abstract. This paper describes a distributed algorithm for solving the shortest path problem with a swarm of JASMINE micro-robots. Each robot is only connected via infra-red communication with its neighbours. Based on local information exchange and some simple rules the swarm manages to find the shortest path (shortest path in the number of robots on the path) in a labyrinth with dead-ends and cycles. The full algorithm and simulation results are presented in this paper.

1 Introduction

In swarm robotics an often needed behaviour is to search for interesting spots within the workspace and to form a communication/transportation line between the found spot(s) and another area of interest or another object.

Several researchers proposed algorithms based on signalling wavefronts to solve shortest path problems in sensor and communication networks or the multi robot domain. E.F. Moore described in [1] four wavefront algorithms to find the shortest path in a maze. And also the Bellman-Ford algorithm computes the smallest spanning tree in a maze. O'Hara and Balch described in [2] an algorithm that guides robots with the help of fixed communication nodes exploiting Payton's pheromone algorithm. In Payton's algorithm described in [3] a robot close to the source will broadcast a hop count pheromone message through the swarm. If a robot close to the target gets this message, it will send a second hop count pheromone message in the opposite direction. All robots know the direction vector to the target and the source now. Adding those direction vectors leads to the shortest path. Two problems could occur with this algorithm. Firstly the robots do not know if they are on the shortest path or not. If the robots follow the gradient, they will be guided to the source or the target, but they do not keep up a path between the source and the target. And secondly if two robots in the swarm become source robots at the same time, the algorithm will be confused by different directions. Inspired by those algorithms we tried to overcome this problem. The pheromone used in our algorithm counts the hops up and down. This enables the robots to know wether they are on the shortest path or not despite of cycles in the maze.

We implemented a distributed algorithm, that finds the shortest path between two initiating source robots within a labyrinth and afterwards gathers the other

M. Dorigo et al. (Eds.): ANTS 2006, LNCS 4150, pp. 404–411, 2006.

robots around those initial robots while keeping a communication path between the two sources. The shortest path is found by an unidirectional negotiation algorithm, that sends information waves through the robot swarm.

The algorithm was evaluated in the Breve simulation environment [4] based on a model of the real swarm robot JASMINE[1], see Fig. 1.

2 The JASMINE Swarm Micro-robot

The underlying swarm micro-robot JASMINE was developed especially for swarm robot research and swarm robot games. Despite its small size of about $27 \times 27 \times 35$ mm^3, it has excellent local communication abilities and a far distance scanning and distance measuring sensor. The excellent communication abilities result from six infra-red sensors and emitters arranged around the robot with a displacement of 60°. Those sensors could also be used for short distance measurements. The far distance measuring sensor is hooked to the front of the robot. Two differentially driven wheels give this micro-robot a high manoeuvrability at a high speed. Optical encoders allow odometric measurements in the mm-range. Different from many other swarm robots JASMINE supports only local communication. Long distance communication via radio frequency is not implemented and does not correspond with the views of the construction team about swarm robot capabilities. Figure 1 shows the JASMINE swarm micro-robot and the sensor placement.

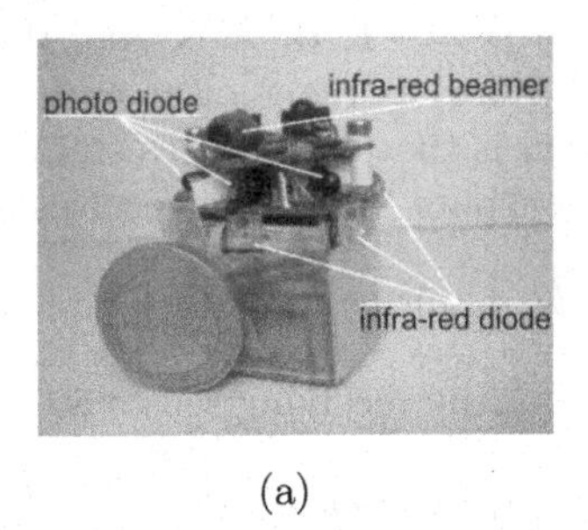

(a)

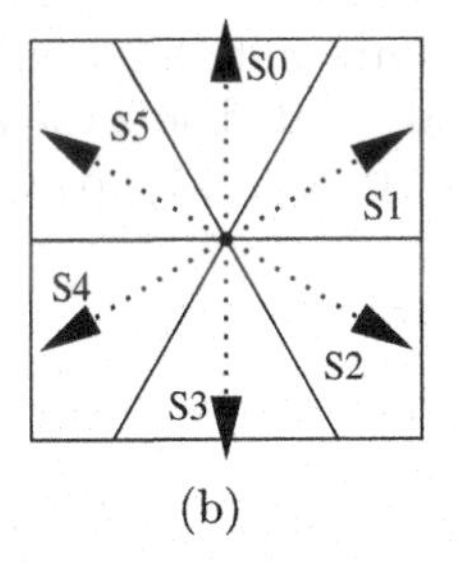

(b)

Fig. 1. (a) The micro-robot JASMINE; (b) Numbering and directions of the sensors. Starting with sensor S0 at the robot's front.

3 Distributed Shortest-Path Algorithm

The whole process of searching the shortest path could be separated into three phases:

[1] JASMINE is designed by the *Micromechatronics and Microrobotics Group at the Institute for Process Control and Robotics at the Universität Karlsruhe, Germany,* and the *Collective Micro-Robotics Team at the Institute of Parallel and Distributed Systems at the University of Stuttgart, Germany,* for the I-SWARM project. For more details see *http://www.swarmrobot.org.*

1. uniform distribution phase and search phase,
2. shortest path negotiation phase,
3. and aggregation phase.

Whereas the *uniform distribution phase* ensures that all robots are uniformly distributed through the whole maze and the *aggregation phase* leads to a contraction of the robots at the nearest sources. However, the most interesting phase is the *shortest path negotiation phase*, where the shortest path finding takes place. This paper will focus on the second phase. How a swarm could be dispersed can be found in [5].

4 Shortest Path Negotiation Phase

During the *shortest path negotiation phase* each sensor has a memory for the expected distance *from* and *to* the source $\mathbf{d}_f$ and $\mathbf{d}_t$. It also has to remember from which source it received the pheromone message. This is saved in the *pheromone* vector $\mathbf{p}$.

$$\mathbf{d}_f = (d_f^0, d_f^1, \ldots, d_f^5) \in \mathbb{N}^6 \quad (1)$$

$$\mathbf{d}_t = (d_t^0, d_t^1, \ldots, d_t^5) \in \mathbb{N}^6 \quad (2)$$

$$\mathbf{p} = (p^0, p^1, \ldots, p^5) \in \{0, 1\}^6 \quad (3)$$

The upper index always denotes the sensor in respect to Fig. 1(b).

The robots' values are initially set to $\mathbf{p} = -\mathbf{1}$, $\mathbf{d}_t = \infty$ and $\mathbf{d}_f = \infty$. The source robots are initialised with $\mathbf{p} = \mathbf{0}$ for source 0 or $\mathbf{p} = \mathbf{1}$ for source 1, $\mathbf{d}_t = \infty$ and $\mathbf{d}_f = \mathbf{0}$. The basic algorithm is that the sources starts sending the message $\mathbf{m} = (0, \infty, \{0, 1\})$ over all six outputs. If a robot receives a message

$$\mathbf{m} = (\bar{d}_f, \bar{d}_t, \bar{p}) \quad (4)$$

it will store those values in the memory of the receiving sensor s

$$d_t^s = \bar{d}_t,\ d_f^s = \bar{d}_f,\ p^s = \bar{p}, \quad (5)$$

and afterwards sends the message

$$\mathbf{m} = (\bar{d}_t - 1,\ \bar{d}_f + 1,\ \bar{p}), \quad (6)$$

over sensors $((s+2) \bmod 6)$, $((s+3) \bmod 6)$ and $((s+4) \bmod 6)$, on the opposite side of the receiving sensor s. This ensures the wave like dispersion of the pheromones. If the source gets a message from the other source it sets $\mathbf{d}_t = \bar{d}_f$ if $(\bar{d}_f < d_t)$ and continues to send the new message. If it gets a message with $\bar{d}_t = 0$ and $\bar{d}_f \cdot \mathbf{1} = \mathbf{d}_t$ the source knows that it got a message over the shortest communication path. It will wait for an arbitrary number of such messages before it sends the shortest path found signal (found_signal) which will start the aggregation phase. This delay ensures that the message was really send via the shortest path and not coincidentally via a second path.

Table 1. Shortest path algorithm for a source robot

```
initialise:
d_f := 0; d_t := ∞; p := {0,1};
receive_count := 0;

begin:
while (receive_count < receive_threshold) {
   send( −1, d_f + 1, d_t − 1, p);
   if (received message m := (d̄_f, d̄_t, p̄) on sensor  s) {
      if (p̄ ≠ p) {
         if (d̄_f < d_t)
            d_t := d̄_f;
         if (d̄_t = 0 ∧ d_t = d̄_f)
            receive_count := receive_count + 1;
      }
   }
}

while (¬stop_condition)
   send( found_signal, 0);
```

```
send(s, d̄_f, d̄_t, p̄) {
   if (s ≠ −1)
      send the message m := (d̄_f, d̄_t, p̄) over sensors
      (s + 2) mod 6, (s + 3) mod 6 and (s + 4) mod 6.
   else
      send the message m := (d̄_f, d̄_t, p̄) over all sensors
}
```

However, this basic algorithm is too simple and would lead to problems with cycles in the connection graph. Table 2 shows an improved algorithm for a robot, that is not a source. As long as it does not get a signal from the sources, that the shortest path was found (found_signal) each robot does the following: After receiving a message on sensor s the robot will test if $\bar{d}_t = 0$ which implies the robot is not on the shortest path and will not commit this message any further, because only a source robot could receive a $\bar{d}_t = 0$ as the distance to itself. If $\bar{d}_t > 0$ the robot will update the memory on the receiving sensor if it has not been updated before or if the received distance

$$\bar{d}_f \leq \min(d_f^i \mid p^i = \bar{p};\ i \in \{0, \ldots, 5\}) \tag{7}$$

is smaller as or equal to all other distances to the goal received from the sending source. If the memory was updated the robot will send the message updated by (6) over the sensors on the opposite of the receiving sensor s otherwise the

Table 2. Shortest path algorithm for a normal (non-source) robot

```
initialise:
d_f := ∞; d_t := ∞; p := −1;
on_shortest_path := false;

begin:
while(received message m ≠ found_signal) {
   if (received message m := (d̄_f, d̄_t, p̄) on sensor s) {
      if (d̄_t ≠ 0) {
         if ( p^s = −1
            ∨ d̄_f ≤ min(d^i_f | p^i = p̄, i ∈ {0,...,5})
            ∨ d^s_f ≥ min(d^i_f | p^i = p^s, i ∈ {0,...,5}\s) ) {
            d^s_t := d̄_t; d^s_f := d̄_f; p^s := p̄;
            send(s, d̄_f + 1, d̄_t − 1, p̄);
         }
      }

      if (∃ i, j ∈ {0,...,5} : d^i_t = d^j_f ∧ d^j_t = d^i_f ∧ i ≠ j)
         on_shortest_path := true;
   }
   wait_for_next_message();
}

goto_source();
```

message will be blocked. A problem that could occur due to communication problems is that a message from one source could be propagated around a loop in the labyrinth and a robot has values from the same source on opposite sensors. If we just use the obvious condition in (7) the robot would not accept any messages from the other source anymore. This would lead to a live-lock and the shortest path would never be found. To solve this, another constraint has to be introduced:

$$d_f^s \geq \min(d_f^i \mid p^i = p^s;\ i \in \{0, \ldots, 5\} \backslash s) \tag{8}$$

Equation (8) allows to overwrite a value from one source by the other one in case, that there is a better d_f^i value from the overwritten source on another sensor i.

This behaviour will iteratively lead to the case, that each robot on the shortest path will have at least two sensors $i,\ j,\ i \neq j$, with data from different sources that have cross-over the same values $d_t^i = d_f^j$ and $d_t^j = d_f^i$. The shortest path condition is:

$$\exists\, i, j \in \{0, \ldots, 5\} : d_t^i = d_f^j \wedge d_t^j = d_f^i \wedge i \neq j. \tag{9}$$

The length of the shortest communication path can be computed as $d_{SP} = d_t^i + d_f^j$.

It is important for the robots to know if they are on the shortest path or not. Because the robots on the shortest path will behave as beacons for the other robots that gather near the closest source or transport objects along this path similar to [6]. The knowledge being or not being on the shortest path triggers different behaviours during the aggregation phase.

5 Experiments

The simulation for evaluating the proposed algorithm is based on a model of JASMINE. The simulation environment Breve described in [4] has been used for the simulation of the JASMINE robots. The distributed shortest-path algorithm was implemented in Steve[2] and MDL2ϵ which is our further development from MDLe by Manikonda et al., see [7].

Many simulation experiments have been performed with several source positions. Figure 2 shows an experiment[3]. Hundred robots have been equally distributed in Fig. 2(a) with a communication range of 20 cm. Figure 2(b) shows the robots in the aggregation phase the white robots stand on the shortest path, the sources are green and the black robots are moving towards the nearest source.

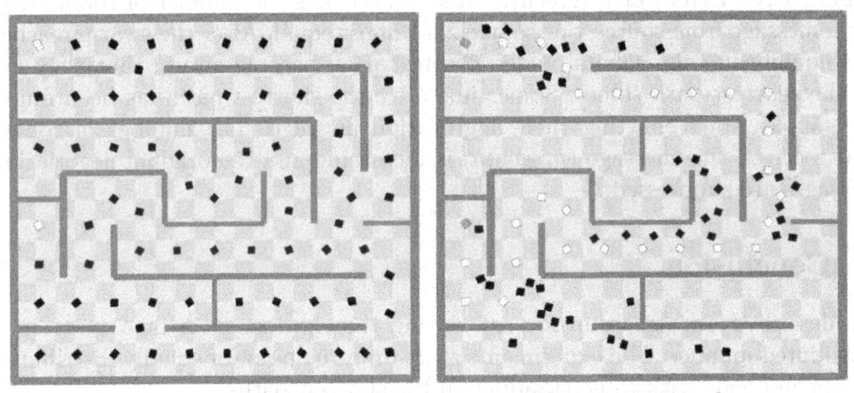

(a) Beginning of the negotiation phase. White robots are the sources.

(b) Aggregation phase.

Fig. 2. Experiments with the simulation environment Breve. Black robots are normal robots, white robots are on the shortest path.

The experiments showed that the algorithm is stable with respect to communication problems that occur randomly distributed over the robots. If a robot cannot transmit or receive any messages it will be detected as an obstacle and the algorithm searches another path around this robot if possible.

[2] Programming language for Breve.

[3] A video showing the whole experiment can be found at `http://wwwipr.ira.uka.de/~szymansk/video/SlimeMould.avi`.

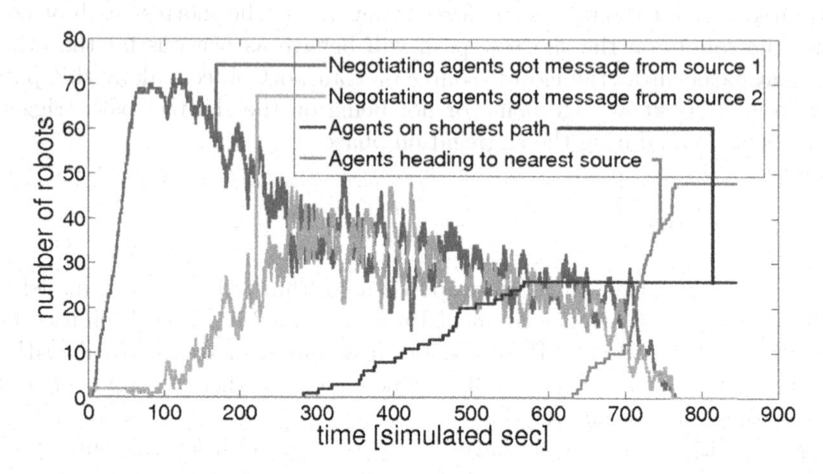

Fig. 3. State distribution during an experiment

Figure 3 shows the state distribution during an experiment. One can see that in the beginning all robots are in the negotiation state and get a message from either source 1 (red) or 2 (green). After 300 seconds the number of robots that got a message from source 1 or source 2 is almost equal. After the swarm reaches this equilibrium the robots on the shortest path become aware, that they are on the shortest path (blue). Some time after the shortest-path has been found the sources start emitting their aggregation message. The number of robots that received this message (magenta) is increasing and the original negotiation messages are suppressed.

6 Conclusion

We described an algorithm for calculating the shortest path in a maze in a distributed manner. The experiments start from the point of equally distributed robots and ends with robts standing on the shortest path.

We showed in several experiments that the algorithm is stable regarding communication errors. This algorithm is currently not scalable in the sense that we add more sources that should be all connected. Experiments showed if a source with the same source identification is added the shortest path between two different sources will also be found. One problem for this algorithm is if an agent is "dead" and does not respond to any input this agent will be treated as a wall. This problem could also be seen as an advantage, because a path around this "dead" robot will be found. This "dead" robot is nothing more than an obstacle.

7 Future Work

In the future we are going to implement the algorithm including the dispersion part on JASMINE robots to evaluate the performance on real robots. It could also be compared with other algorithms that find the shortest path between two sources in a workspace and build a communication path. And a theoretical proof of the stability of this algorithm could be derived.

References

1. Moore, E.F.: The shortest path through a maze. In: Proc. of the International Symposium on the Theory of Switching, Harvard University Press (1959) 285–292
2. O'Hara, K.J., Balch, T.R.: Distributed path planning for robots in dynamic environments using a pervasive embedded network. In: AAMAS '04: Proceedings of the Third International Joint Conference on Autonomous Agents and Multiagent Systems, Washington, DC, USA, IEEE Computer Society (2004) 1538–1539
3. Payton, D., Daily, M., Estowski, R., Howard1, M., Lee, C.: Pheromone robotics. In: Special Issue on Biomorphic Robotics. Volume 11 of Autonomous Robots., Springer Netherlands (2001) 319 - 324
4. Klein, J.: breve: a 3D Environment for the Simulation of Decentralized Systems and Artificial Life. In: ICAL 2003: Proceedings of the eighth international conference on Artificial life, Cambridge, MA, USA, MIT Press (2003) 329–334
5. Hsiang, T.R., Arkin, E.M., Bender, M.A., Fekete, S., Mitchell, J.S.B.: Online dispersion algorithms for swarms of robots. In: SCG '03: Proceedings of the nineteenth annual symposium on Computational geometry, New York, NY, USA, ACM Press (2003) 382–383
6. Li, Q., Rosa, M.D., Rus, D.: Distributed algorithms for guiding navigation across a sensor network. In: MobiCom '03: Proceedings of the 9th annual international conference on Mobile computing and networking, New York, NY, USA, ACM Press (2003) 313–325
7. Manikonda, Krishnaprasad, Hendler: Languages, Behaviors, Hybrid Architectures, and Motion Control. Mathematical Control Theory (1998)

Fleet Maintenance Scheduling with an Ant Colony System Approach

Fernando Teixeira Mendes Abrahão[1] and Nicolau Dionísio Fares Gualda[2]

[1] Instituto de Logística da Aeronáutica - ILA, Divisão de Pesquisa, Guarulhos, Brasil
ferabr@ig.com.br
[2] Escola Politécnica da Universidade de São Paulo, São Paulo, Brasil
ngualda@usp.br

Abstract. The paper presents highlights of a doctoral research addressed to model and solve the problem of preventive maintenance scheduling of fleets of vehicles. The problem is formulated and several Ant Colony based approaches were proposed and tested considering different instances of the maintenance scheduling problem, including applications related to the preventive maintenance of an aircraft fleet belonging to the Brazilian Air Force. The most successful approach has shown to be the one based on ACS with a specific local search procedure and inspired on the application of ACO to the timetabling problem. Details on the problem formulation, on the proposed approaches, and on the performed tests and applications are presented.

1 The Fleet Preventive Maintenance Scheduling Problem - FPMSP

The objective of the FPMSP is to find a feasible chronological sequence for vehicle preventive maintenances. The solution involves the maximization of the available life cycle of the vehicles (c_{ij}) searching for a feasible schedule that minimizes the waste of available hours between preventive maintenance intervals. Sarac, Batta and Rump (2004) explain the measures used by air companies related to the maximization of available hours for fleets of aircrafts [1]. Another measure to be maximized is the availability of vehicles for the operator (b_i). While a vehicle is under preventive maintenance, it is not available for the operator. In such a case, a tolerance exists until a minimum number of available vehicles is reached (EAv). In the same vain, workshops capacities act as restrictions to the problem (a_i). Simultaneous preventive maintenances could be prohibitive due to workshop capacity constraints. The FPMSP presents dynamic characteristics and whenever a vehicle j goes to preventive maintenance at period i, all of its remaining life cycle is altered to adjust to the new set of available hours m given by the maintenance action. The remaining of the maintenance system also changes as maintenance resources get busy in those specific periods, which affects other periods as well as the values of the fleet's availability. The FPMSP presents similarities with assignment and subset problems. The Aircraft Maintenance Routing Problem - AMRP presented by Sarac; Batta and

M. Dorigo et al. (Eds.): ANTS 2006, LNCS 4150, pp. 412–419, 2006.

Rump (2004) presents similarities with the FPMSP and shows that the problem is *NP-hard* since it contains a *NP-complete* covering problem as a subproblem. All similar problems studied present exponential complexity and *NP - Hardeness* with authors suggesting the use of meta-heuristics to solve practical cases [2,3,4,5]. For a simple case, with just one vehicle type and only one workshop, the solution is to find:

$$x_{ij} = \begin{cases} 1 & \text{if vehicle } j \text{ is assigned to maintenance at period } i \\ 0 & \text{otherwise.} \end{cases} \quad (1)$$

Objective:

$$Min \sum_{i=1}^{I} \sum_{j=1}^{J} (a_i + b_i + c_{ij)} \quad (2)$$

Where:

$$a_i = \begin{cases} [(\sum_{1}^{J} x_{ij}) - Cap_i] * wa & \text{if } \sum_{1}^{J} x_{ij} > Cap_i \\ [Cap_i - (\sum_{1}^{J} x_{ij})] * wa' & \text{if } \sum_{1}^{J} x_{ij} < Cap_i \end{cases} \quad (3)$$

$$b_i = \begin{cases} (1 - \frac{VerAv_i}{EAv}) * wb & \text{if } VerAv_i < EAv \\ 0 & \text{otherwise.} \end{cases} \quad (4)$$

$$c_{ij} = \begin{cases} (\frac{(m - mAv_{(i-1)j})}{m}) * wc * x_{ij} & \text{if } mAv_{(i-1)j} > 0 \\ (1 - x_{ij}) * wc' & \text{if } mAv_{ij} = mAv_{(i-1)j} = 0 \\ o & \text{otherwise.} \end{cases} \quad (5)$$

Penalties related to the mismatch of the workshop capacity (a_i) are calculated by multiplying the parameter wa (or wa') by the observed excess (or lack) of capacity for each period. Penalties related to availability (b_i) are calculated when the expected availability (EAv) is not reached for each period by multiplying the parameter wb by the percentage not reached. Variable (mAv_{ij}) indicates the remaining available time for each vehicle at each period. Penalties related to the waste of available hours (c_{ij}) are calculated by multiplying the parameter wc by the percentage of life cycle wasted when a vehicle is assigned to maintenance with available hours still remaining. If a vehicle has no available hours and is not assigned to maintenance at period i, the penalty is equal to wc'. The wa, wa', wb, wc and wc' values correspond to the weight of applied penalties and are adjusted according to each instance nature. The research of Abrahão (2006) presents details of the formulation of the FPMSP including the complete description of the problem constraints [6].

2 Solution Strategies to Solve the FPMSP

All procedures begin building an entirely connected graph formed by all vehicles and periods ($I \times J$). Each node x_{ij} represents possible maintenance task assignments. Penalties are computed for each period and the solution cost is the sum of the costs computed for all periods. While maintenance tasks are being scheduled, penalties are computed for each period (i) as constraints (a_i, b_i and c_{ij}) are affected.

2.1 Constructive Heuristic for the FPMSP - *CH*

In this case, preventive maintenance tasks are scheduled exclusively according to the proximity of the depletion of available hours. The *CH* objectives a *shortest path* of scheduled preventive maintenances based on the above mentioned rule.

2.2 Local Search Heuristic for the FPMSP - *LS*

The *2-opt* [7] method tests possible exchanges of any 2 arcs, rebuilding connections when improvements are found. It finishes when is not possible to accomplish changes that result in improvements. For the FPMSP, the exchanges are performed on maintenance tasks scheduled for each vehicle (j). The procedure advances maintenance tasks by one period and checks if an improvement is verified. Next, the procedure delays maintenance tasks by one period and checks for improvements. Finally, maintenance tasks are inserted or removed to the solution schedule to verify if improvements could be still found. The procedure stops when, for two consecutive iterations, no improvement is observed.

2.3 Ant System for the FPMSP - *ASmnt*

Among studied problems, the University Course Timetabling Problem - UCTP seems to present the most promising configuration to be adapted to the FPMSP. So, the *ASmnt* is somewhat inspired on the implementation of *ACO* procedures to the UCTP. The agents (ants) use the proximity of the total depletion of remaining hours for the assignment of maintenance tasks as heuristic information (η) and β as a parameter. This, combined with the α parameter and the pheromone trails (τ), which are deposited on nodes (x_{ij}) instead of arcs, provides the required data for ants to move in accordance with the *AS random proportional rule*. The amount of available hours mAv_{ij} is used instead of the distance for the heuristic information calculation and the neighborhood of node x_{ij} is limited to avoid consider periods too far from period i. Its reasonable assume that maintenance actions would be taken close to the period where remaining available hours are expected to be depleted. This limit is set according to fleet's characteristics and the maintenance environment found on each instance. Formulation of pheromone trail update and other *AS* parameters are the same for the *ASmnt*.

2.4 Ant Colony System for the FPMSP - *ACSmnt*

The *Ant Colony System - ACS* adaptation to the *ACSmnt* is similar to the *ASmnt*. Even though *ACS* has been the first *ACO* procedure benefitting from local search procedures, no local search is used for this specific case. The pheromone update strategy applies only to nodes of the best solution of each iteration.

2.5 Constructive and Local Search Heuristic for the FPMSP - *CLSH*

The *CLSH* method is the composition of a constructive heuristic (*CH*) with a local search procedure (*LS*). This method doesn't count with so much diversification compared with *ACO* in the search for solutions, but it is significantly faster and simple to implement.

2.6 Ant System with Local Search for the FPMSP - *ASmntLS*

The *ACSmntLS* is inspired on the adaptation presented in Abrahão and Gualda (2004) and incorporates local search heuristics among the *ASmnt* iterations [8]. When applied at each group of θ iterations, local search procedures are successively repeated until no improvement is observed. The procedure continues their iterations with *ASmnt* until the local search procedure is called again. After local search iterations, pheromone updates only happen if improvements are found, otherwise, pheromone trails remain the same as before, avoiding too much intensification. With *ASmntLS*, *ASmnt* acts as a constructive heuristic while *LS* works as a solution improver. After the first iterations, *ASmnt* and *LS* work as improvement mechanisms. However, *ASmnt* diversifies while *LS* intensifies the search for better solutions.

2.7 Ant Colony System with Local Search for the FPMSP - *ACSmntLS*

The *ACSmntLS* is also inspired on the adaptation presented in Abrahão and Gualda (2004) and incorporates *LS* inserted at each group of θ iterations completed by each ant using *ACSmnt*.

3 Tests and Applications

A set of 18 instances and *ACO* parameters (Table 1) were generated for the tests applied to aircraft fleets. Table 2 shows the results for the *A25P25sq3h2*, one of the 18 instances generated that encompasses 25 aircraft, 25 periods of one month, 3 different squadrons and initial remaining hours randomly generated between 0 and 400 hours for each aircraft. Table 3 shows the comparative tests with all instances and methods.

Tests to compare performances were developed using non parametric tests as suggested by Golden and Stewart (1985) [9]. Friedman's tests [10] were used

Table 1. ACO Parameters

	Parameters	*Values*
α	heuristic parameter	1; 1.5; 2
β	pheromone parameter	1; 1.5; 2
ρ	pheromone update	0.1; 0.4; 0.7; 0.9
q0	pseudo-random ACS rule	0.5; 0.7; 0.9
Iter	Nr of iterations	30000
Tries	Nr of runs	5

Table 2. Obtained results for *A25P25sq3h2*

Method	*Penalties*	*Time (sec)*	*% worst*
CH	997	0,04	72%
LS	985	0,08	70%
ASmnt	1352	339,05	133%
ACSmnt	808	159,77	39%
ASmntLS	645	196,85	11%
ACSmntLS	**580**	**149,83**	**0%**
CHLS	647	6,29	1%

Table 3. Comparative results

	Ranked results for all instances																		
Method	1	2	3	4	5	6	7	8	9	10	11	12	13	14	15	16	17	18	*Total*
CH	6	3	6	6	6	6	6	6	6	6	5	6	7	5	5	5	5	5	100
LS	7	4	7	7	7	7	5	5	5	5	6	5	6	7	6	7	6	7	109
ASmnt	5	5	5	5	5	5	7	7	7	7	7	7	5	6	7	6	7	6	109
ACSmnt	3	6	3	4	4	2	4	4	4	4	3	4	4	4	4	4	4	4	69
ASmntLS	2	7	2	3	3	3	3	3	3	3	4	2	3	2	2	1	1	2	49
ACSmntLS	1	1	1	1	1	1	1	1	1	1	2	1	1	1	3	2	3	3	**26**
CHLS	4	2	4	2	2	4	2	2	2	2	1	3	2	3	1	3	2	1	42

to verify if there is statistical significance to affirm that different performances among methods exist. If the null hypothesis is rejected, multiple comparisons are used to determine which procedures present different performances from the others. Table 4 presents the results of the non-parametric tests. The results so far indicate that the *ACSmntLS* is the most promising procedure to solve the FPMSP. However, *ACO* allows adjustments and is necessary to adjust the parameters in order to get the best of each method and results. The next study compares 108 configurations of parameters (Table 1) applied to the 18 instances with the use *ACSmntLS* procedures. The tests were conducted to find the best set of parameters to be used in real instances of the same kind. Results confirm

Table 4. Friedman's Test and Multiple Comparisons

Parameters	*Values*	
T_2	60.27	(*T statistics*)
F (a, K1, K2)	2.98	(*critical value*)
"P-value"	0.01	-
Reject Ho ?	**Yes**	($T_2 > F$) *for Friedman's Test*
a	0.95	-
$t_{1-a/2,k2}$	1.98	-
T_{critic}	12.40	-
Method	*Ranking*	$R_i - R_{(i-1)} > T_{critic}$
ACSmntLS	26	A
CHLS	42	B
ASmntLS	49	B
ACSmnt	69	C
CH	100	D
ASmnt	109	D

that there are differences among configuration of parameters, but they are only significant between the group formed by the first 14 configurations and the rest. Nevertheless, configuration 79 was found to be the best among the tested ones. More details on the parameters settings, comparisons and tests could be found in Abrahão (2006). With all these results, it is possible to apply the *ACSmntLS* to a practical instance of the FPMSP and An application to a real world preventive maintenance problem related to a fleet of 20 aircrafts of 2 squadrons belonging to the Brazilian Air Force (FAB) was conducted. The time horizon was set to 25 periods ($i = 25$) of one month each (two year span) and each aircraft was supposed to fly 50 hours per period (t). *2nd* level maintenance actions and *3rd* level maintenance actions were to be scheduled. *2nd* level maintenance tasks are expected to take 2 periods of time to be accomplished and should occur at each 400 hours interval (m_{2nd} = 400 hr = *2nd* level maintenance interval). *3rd* level maintenance tasks are expected to take 5 periods of time to be accomplished and should occur at each 1200 hours interval (m_{3rd} = 1200 hr = *3rd* level maintenance interval). The first squadron was composed of used aircraft with remaining available hours randomly distributed between 0 and 1200 . The *2nd* squadron was composed by brand new aircrafts. This difference provides a more comprehensive test. Workshops capacities were set to 4 aircraft per period for *3rd* level and 5 for *2nd* level ($Cap_i = 4$ for *3rd* level and $Cap_i = 5$ for *2nd* level maintenance). Penalty weights were set to 20 for the waste of life cycle time (wc) and 10 for idle aircraft not in maintenance (wc'). Penalties for lowering the fleet availability was set to 5 (wb) and capacity penalties were set to 5 (wa) and 1 (wa'). The expected availability of the fleet is 55% (EAv=55%). Table 5 presents a preliminary application of all methods to FAB problem and confirms

Table 5. Global application of the procedures to the practical case

Method	*Penalty*	*Processing Time (sec)*	*Aircraft Availability (hr)*
CH	465.00	2.37	21431
ASmnt	147.56	158.77	20699
ACSmnt	105,11	86.70	20478
ASmntLS	95,69	178.40	20014
CHLS	95,06	21.66	20092
ACSmntLS	92,60	155.27	19845

ACSmntLS as the best performing procedure. The final solution strategy was to apply the *ACSmntLS* to the problem with configuration 79 of *ACO* parameters. 5 tries with 20 ants were performed, given a total of 3000 iterations. Local search procedures were set to run at every 10 iterations until no improvements were found by two consecutive iterations. The method was applied first to solve the *3rd* level problem. The solution found for the *3rd* level maintenance is considered and updates the graph for the application of the method for the *2nd* level maintenance. So, the graph for the *2nd* level maintenance problem considers as tabu moves the moves that involve time slots ($x_{ij} = 1$) occupied by the *3rd* level solution. The same parameters used for the *3rd* level solution were used for the *2nd* level, except workshops capacities and maintenance lead times. Details on the application and on its results are found in Abrahão (2006). The final results of this application provided aircraft availabilities of 67% and 66% for the first squadron during the first and the second year respectively and of 76% and 73% for the second squadron on the same years. The fleet overall availabilities during the first and the second year were of 71% and 69%. These figures are far beyond the minimum practical availability level of 55% setup by the fleet operator.

4 Conclusions and Recommendations

Several *ACO* based approaches have been proposed and tested to solve the FPMSP and the most successful was the *ACSmntLS*, which combines an *ACS* structure with a specific local search heuristic. This *ACSmntLS* was successfully applied to solve a real problem of a Brazilian Air Force (FAB) fleet of aircrafts. Distinctive aspects of the research include the mathematical formulation and implementation of procedures based on *ACO* approaches to the UCTP to solve the FPMSP and non parametric tests to compare the different proposed solution methods and to adjust *ACSmntLS* parameters. Further research should include: the use of *CHLS* as a constructive heuristic for the *ACSmntLS*; studies to systemically adjust the penalty parameters (*wa, wa, wb, wc* and *wc*) to different fleets and environments; consider more than one maintenance shop; and comparisons of the proposed approaches with the *Max-Min AS* and other other meta-heuristics such as *Tabu Search*, *Genetic Algorithm* and *GRASP*.

References

1. Sarac, A., Batta, R., Rump, C.M.: A branch-and-price approach for operational aircraft maintenance routing. European Journal of Operational Research. Available online `http://www.sciencedirect.com/science`
2. Dorigo, M., Stützle, T.: Ant colony optimization. The MIT Press, Cambridge, Massachusetts (2004)
3. Even, S., Itai, A., Shamir, A.: On the complexity of timetabling and multicommodity flow problems. SIAM Journal of Computation **5** (1976) 691–703
4. Rossi-Doria, O., Sampels, M., Birattari, M., Chiarandini, M., Dorigo, M., Gambardella, L.M., Knowles, J., Manfrin, M., Mastrolilli, M., Paechter, B., Paquete, L., Stützle, T.: A comparison of the performance of different metaheuristics on the timetabling problem. In: 4th International Conference on Practice and Theory of Automated Timetabling - PATAT 2002. (2002)
5. Rardin, R.L., Uzsoy, R.: Experimental evaluation of heuristic optimization algorithms: a tutorial. Journal of Heuristics **7**(3) (2001) 261–304
6. Abrahão, F.T.M.: A meta-heurística colônia de formigas para solução do problema de programação de manutenção preventiva de uma frota de veículos com múltiplas restrições: aplicação na Força Aérea Brasileira. PhD thesis, EPUSP, Departamento de Engenharia de Transportes, Sao Paulo (2006)
7. Lin, S., Kernighan, B.W.: An effective heuristic algorithm for the traveling salesman problem. Operations Research **21** (1973) 498–516
8. Abrahão, F.T.M., Gualda, N.D.F.: Aplicação da metaheurística colônias de formigas e das heuríisticas 2-opt e 3-opt na solução de problemas logísticos da força aérea brasileira. In: XXVI SBPO, São João Del Rei, MG. 2004, Rio de Janeiro, SOBRAPO, Anais: Trigésimo sexto congresso da sociedade brasileira de pesquisa operacional (2004) 856–866
9. Golden, B.L., Stewart, W.R.: Empirical analysis of heuristics. In: LAWLER, E. L. et al. The traveling salesman problem: a guided tour of combinatorial optimization. Wiley Series in Discrete Mathematics & Optimization, New York, John Wiley & Sons (1985) 207–250
10. Conover, W.J.: Practical Nonparametric Statistics. John Wiley & Sons, Inc., New York - NY (1980)

Geoacoustic Inversion and Uncertainty Analysis with $\mathcal{MAX} - \mathcal{MIN}$ Ant System

Vincent van Leijen[1] and Jean-Pierre Hermand[1,2]

[1] Combat Systems Department, Royal Netherlands Naval College
Den Helder, The Netherlands
av.vanLeijen@kim.nl
[2] Environmental hydroacoustics lab, Université libre de Bruxelles, Brussels, Belgium
jhermand@ulb.ac.be

Abstract. Inverse problems in ocean acoustics are generally solved by means of matched field processing in combination with metaheuristic global search algorithms. Solutions that describe acoustical properties of the bed and subbottom in a shallow water environment are typically approximations that require uncertainty analysis. This work compares Ant Colony Optimization with other metaheuristics for geoacoustic inversion, particularly Genetic Algorithms. It is demonstrated that a $\mathcal{MAX}$-$\mathcal{MIN}$ Ant System can find good estimates and provide uncertainty analysis. In addition, the algorithm can easily be tuned, but proper tuning does not guarantee that every run will converge given a limited processing time. Another concern is that a single optimization run may find a solution while there is no clear indication on the accuracy. Both issues can be solved when probability distributions are based on parallel $\mathcal{MAX} - \mathcal{MIN}$ Ant System runs.

1 Introduction

Coastal waters allow for a high concentration of human activities and an expanding exploration of the underwater environment takes place mostly by acoustic techniques. The propagation of sound through the ocean medium is well understood and since no other energy propagates as efficiently, sonar has been the most effective sensor for many years. In a shallow water environment [1] sound interacts with the bottom in complex ways and propagation modeling requires a detailed knowledge of the bottom acoustic properties in addition to the sound speed structure in the water column and the sea surface scattering conditions.

Geoacoustic inversion is a technique that aims to describe a shallow water environment by acoustical parameters which in turn are related to other geophysical parameters. An introduction to geoacoustic inversion is given in Section 2. The concept of matched field processing is explained together with the optimization problem that is part of an inversion process. Section 3 describes a benchmark of geoacoustic inversion based on the Yellow Shark experiments in Mediterranean shallow waters. The section also provides a brief introduction to metaheuristics that, according to literature, have been applied to inversion before. Ant Colony Optimization (ACO) [2] is demonstrated to be feasible as well,

M. Dorigo et al. (Eds.): ANTS 2006, LNCS 4150, pp. 420–427, 2006.

when the world of the ants is regarded as an analogy for the geoacoustic environment. A $\mathcal{MAX}$-$\mathcal{MIN}$ Ant System implementation is tuned for the benchmark and it is shown how accurate results are found. Tuning is important but does not guarantee that every inversion will converge to the same solution. In addition there is no clear indication on the accuracy of solutions. Section 4 adresses these issues with uncertainy analysis.

2 Introduction to Geoacoustic Inversion

Geoacoustic inversion is principally based on matched field processing (MFP) described in this section. The focus is on a representative benchmark of geoacoustic inversion, based on real data from the 1994 Yellow Shark experiments.

2.1 Inversion Based on Matched Field Processing

Inversion is a technique that derives a physical model from a measurable quantity like a sound pressure field. Sound can be measured for a particular frequency and at a certain distance from the sound source. A physical model of a medium can be used to make predictions of pressure fields at a number of positions and frequencies. These predictions are called forward calculations. When the predictions match the measured fields it is fair to say that the model is *acoustic equivalent* with the real world. In other words: when sound propagates through a medium, the sound pressure field upon reception will be the same as it would have been when the sound had propagated through the physical model. The whole process of matching reception with prediction, in order to obtain a physical model, is called *matched field inversion*. An objective function defines mismatch and needs to be minimized to find the best acoustic equivalent model.

2.2 Inversion for Bottom Geoacoustic Parameters

In a typical shallow water environment[1], the medium consists of several layers. The top layer is the water column where most acoustic quantities can be measured or otherwise accurately be predicted. But high-frequency echo sounders used for bathymetric survey do not deeply penetrate into the sedimentary layers and do not reach the underlying hard rock basement. Low-frequency sound however does penetrate these layers and in many cases inversion helps to find acoustic characteristics like sound speed and absorption. A profound understanding of the medium enhances acoustic sensing capabilities and allows for accurate predictions of sonar detection ranges. A fast inversion scheme that obtains reliable geoacoustic parameters is a prerequisite for Rapid Environmental Assessment (REA) for sonar.

[1] *Shallow* waters are usually found at the continental shelf and bound by a depth-contour line of 200 m. In these waters, sound typically propagates with multiple interaction with the sea bottom and surface.

For an arbitrary environment, geoacoustic inversion counts four steps [3]: discretization of the environment, efficient and accurate forward modeling, efficient optimization procedures and finally uncertainty analysis. The first step is commented in the next paragraph for the Yellow Shark experiments. The forward modeling of sound propagation falls beyond the scope of this article. Steps 3 and 4 are addressed with ACO in sections that follow.

2.3 The Yellow Shark Experiments

One of the interesting features of the Yellow Shark experiments is that cores of the bottom material have been taken and the analysis of these samples provide ground truthing for the solutions of geoacoustic inversion. Details on the experiments can be found in Hermand and Gerstoft [4]. For our preliminary study we will use a benchmark with geometric and geoacoustic parameters and an objective function based on phase-coherent processing of pressure time series (wave-forms) received on a vertical array using a model-based matched filter (MBMF), details can be found in [5]. The benchmark is in the form of a pre-computed objective function [6] based on measured sound speed profiles in the water column, averaged over range, and a geoacoustic model of eight parameters that describe a single sediment layer over a half-space sub-bottom.

The objective function $f : S \to \mathbb{R}^+$ needs to be minimized. In YS94, S is an 8-dimensional search space, where every real parameter x_i is subject to $a_i \leq x_i \leq b_i$, with constants a_i and b_i based on general *a priori* information. If each parameter is sampled by just 10 samples, there are already 10^8 possible combinations. Considering that each forward call to the objective function depends on a non-linear propagation model that is computationally demanding, geoacoustic inversion clearly benefits from a metaheuristic approach.

3 Ant Colony Optimization for Inversion

Various metaheuristics have been applied to the optimization part of inversion. Ant Colony Optimization is unique in being a population based method with a short term memory. A $\mathcal{MAX}-\mathcal{MIN}$ Ant System has been applied on the Yellow Shark benchmark and a rule of thumb for the performance parameters has been derived. Results of a tuned run are shown to match with reference solutions for the benchmark.

3.1 ACO and Other Metaheuristics for Inversion

Early environmental inverse problems were solved with an exhaustive search on a limited search space [7]. The introduction of Simulated Annealing and the Genetic Algorithm (GA) [3], [8] made it possible to invert more parameters on a wider range of samples. Only recently, other methods as Tabu Search and Differential Evolution entered the field.

Ant Colony Optimization has most in common with Genetic Algorithms. Both types are population based algorithms that search a discrete search space and that are capable of providing uncertainty analysis. The main difference between the methods are the mechanisms that handle and recombine components of better candidate solutions (pheromones trails versus genetic operators). ACO is further different in having a form of memory (the pheromone trails), while GA's are without memory. When pheromones evaporate, identifiers of paths with above average quality are fading out. High rates of evaporation make that only recent information can be retrieved, as is typical for a short term memory. Low evaporation rates allow recollection of much older information and correspond to a long term memory.

3.2 Application of $\mathcal{MAX}-\mathcal{MIN}$ Ant System

When ACO is applied to inversion, the world of the ants acts as an analogy for the geoacoustic environment. Acoustic parameters that describe a sea bottom or water column are objects between the nest and a food source. Paths that bridge such objects are the sampled values each parameter may take. Ants communicate by depositing pheromones, a measure for the mismatch between predicted and actually measured sound pressure fields.

In previous work [9] we have argued that ACO is a feasible optimizer for the geoacoustic inversion problem and this was demonstrated with a $\mathcal{MAX}-\mathcal{MIN}$ Ant System ($\mathcal{MM}$AS) as introduced by Stützle and Hoos [10]. $\mathcal{MM}$AS has four characteristics [2]. First, only the best-so-far or iteration-best ant is allowed to deposit pheromones. Secondly, pheromone trail values are restricted to the $[\tau_{min}, \tau_{max}]$ interval. All paths are initialized with τ_{max}. And finally, in case of stagnation the system is reinitiated.

The $\mathcal{MM}$AS implementation focussed on here is called *Geoacoustic Inversion with ANTs* (GIANT). Some technical issues:

- Initially, pheromones are equally distributed. Each sample has pheromone value $\tau_0 = 1/f_{min}$, with f_{min} the minimal mismatch from the first iteration.
- Only the best-so-far ant x^* is allowed to deposit pheromones, or (optional) the iteration-best ant.
- Pheromone update $\Delta\tau = \frac{1}{f(x^*)}$ is deposited on each of the n samples of x^*.
- Division by zero does not occur for $\Delta\tau$ as full convergence ($f = 0$) is a stopping criterion. Some small $\epsilon \geq 0$ is defined as a threshold for convergence.
- At each iteration the pheromone trails are scaled to fit an upper bound of $\tau_{max} = n/f(x_{iteration-best})$ and a lower bound of $\tau_{min} = c/\phi n^2$, with $c > 0$ a constant and ϕ the average f of the iteration [10]. With $c = 1$ it follows that $\tau_{min} > 0$ and therefore ants have access to the complete search space at any iteration.

3.3 Tuning of $\mathcal{MAX}-\mathcal{MIN}$ Ant System

The efficiency of an average optimization run strongly depends on some algorithm-specific parameters, unique for a particular problem. A metaheuristic runs on a pair (S, f), generally called an *instance* [11]. The solution space S is determined by problem specific parameters x_i with upper and lower boundaries a_i and b_i that are based on general *a priori* knowledge. For combinatorial search methods S is usually discretized by sampling the (a_i, b_i) intervals. The sampled geoacoustic parameters strongly influence the search space and the instance. Therefore, sampling has not been regarded as part of the tuning.

For geoacoustic inversion problems, the objective function f is usually computationally demanding as for each forward call a mathematical propagation model needs to be calculated. The ideal solution with $f = 0$ is not always contained in S due to simplifications in the models, the presence of ocean noise or limitations in the sampling of the search space. Parameters that have been tuned for GIANT are colony size N and pheromone evaporation factor ρ.

In a typical application of geoacoustic inversion, many signals are transmitted and many instances are to be considered. As the objective functions are computationally time consuming, *rapid* assessments do not permit extensive tuning for separate instances. Tuning results and their sensitivity in general can not directly be transferred to other instances.

3.4 Results for Yellow Shark

Good procedures exist for tuning metaheuristics [12], [11]. For the benchmark however, an exhaustive search was carried out to find a simple rule of thumb for optimal values for N and ρ. This approach is time consuming and not recommended for practical use, but does provide a good understanding of the interaction between performance parameters and the speed of convergence. The method further indicates how sensitive tuning results are and just how bad non-optimal settings work out. Results are shown in Fig. 1 for a maximum number of 10^5 forward calls. Averages are taken over 30 runs for various combinations of N and ρ. A benchmark specific convergence threshold of $\epsilon = 1.743$ has been adapted.

The left diagram shows the average final mismatch. In most runs GIANT did not converge to ϵ, plotted in black. The shades of grey point out that N and ρ should carefully be chosen to find the best solution with the available forward calls. For 10^5 available calls, there exists a range of good settings, briefly characterized as N large and $\rho < 0.5$.

The diagram on the right shows how many forward calls were needed to get the minimal mismatch in the left diagram. The number of calls corresponds with computing time for the best-so-far solution. For a minimal number of forward calls, N must be as small as the minimal mismatch allows for, ρ is less bounded.

Combined, the two plots reveal a small area of interest: $\rho < 0.4$ and $N \approx 250$. The low evaporation rates make $\mathcal{MAX} - \mathcal{MIN}$ Ant System act like it has a long term memory. With small ρ, ants start to choose the same solution components and are reducing the intensity of exploration. For the benchmark a rule of thumb for colony size is $N \approx \sqrt{k}$, with k the available number of forward calls [9].

Table 1 lists solutions for a run with $\rho = 0.1$ and $N = 250$. The results are almost identical to the reference solutions [4]. Since the search space was discretized in 50 samples, S does not contain all the values of the reference

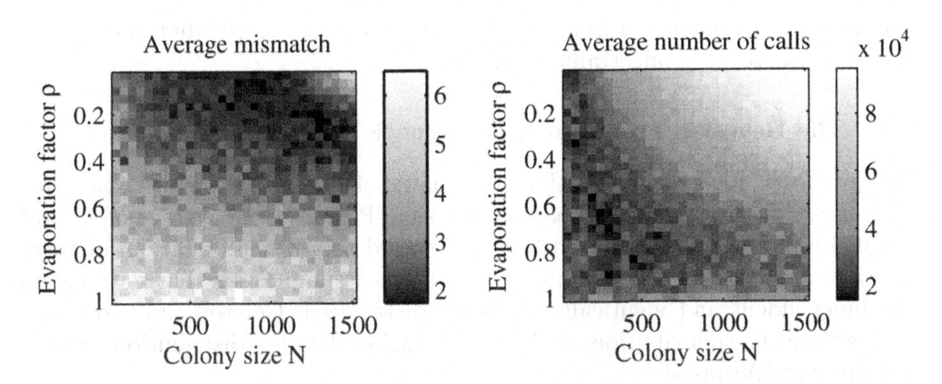

Fig. 1. Tuning results for GIANT on the YS94 benchmark. Left: final mismatch over 30 runs. Right: average number of forward calls that were needed to find these minimal mismatches. In both plots the best settings are found in the darkest areas.

Table 1. Geoacoustic parameters and results for tuned GIANT on YS94 ($\rho = 0.1$ and $N = 250$). Listed are parameters x_i with their physical meaning, a_i and b_i are lower and upper bounds, s^* is the reference solution and s the solution found by GIANT.

x_i Physical meaning	a_i	b_i	s^*	s
ρ_2 sediment density in g/cm^3	1	2	1.5	1.5
α_2 sediment absorption in dB/λ	0	0.175	0.03	0.0315
c_2 sediment sound speed in m/s	1440	1540	1470	1470
g_2 sediment sound speed gradient in s^{-1}	1	9	2	1.96
d_2 thickness of sediment layer in m	0	11	7.5	7.48
ρ_3 bottom density in g/cm^3	1	2.5	1.8	1.81
α_3 bottom absorption in dB/λ	0.05	0.375	0.15	0.18
c_3 bottom sound speed in m/s	1450	1600	1530	1531

solution. This can be noticed for g_2, where neighboring samples are 1.96 and 2.12: the exact solution $g_2 = 2$ s^{-1} is out of reach.

Good settings have been found after tuning, Fig. 1 reveals that on average, still some 40.000 forward calls are needed to get the desired solution. Another issue is that not every run of a metaheuristic is guaranteed to converge to the same solution. An alternative approach is to have less forward calls for each run, stops before final convergence and then let uncertainty analysis point out what the intermediate results show.

4 Uncertainty Analysis

Metaheuristics find solutions of above-average quality and tuning will speed up the average inversion process. Still, there is no guarantee that for a single run the convergences are fast enough and the final solution lacks a clear indication on the degree of confidence in the results. Both issues are solved when probability distributions provide uncertainty analysis.

4.1 The Bayesian Framework for Genetic Algorithms

In geoacoustic inversion codes that use genetic algorithms, it has become common use to base *posterior* probability density (PPD) on the average of several parallel inversions [3]. In the absence of detailed *a priori* information, uniform *a priori* distributions are assumed [8]. Another assumption is that data errors are independent and identically Gaussian distributed. Gerstoft did show how the average gene distributions and the marginal probability distribution became similar over 20 runs [3].

4.2 Uncertainty Analysis with $\mathcal{MAX} - \mathcal{MIN}$ Ant System

Without a Bayesian argumentation but in line with the genetic approach, the probability distributions in Fig. 2 are based on the averages over 10 GIANT runs. For this multi-start procedure equal settings for performance were used: $N = 50$, $\rho = 0.05$ and a maximum of 200 iterations. In total, the cost of acquiring the distributions did not exceed 10^5 forward calls. Arrows mark the reference solutions and are found at central positions within the distributions. Figure 2 not only gives correct central values, it adds an indication on the uncertainty. Intervals of likely values narrow down when more processing is allowed.

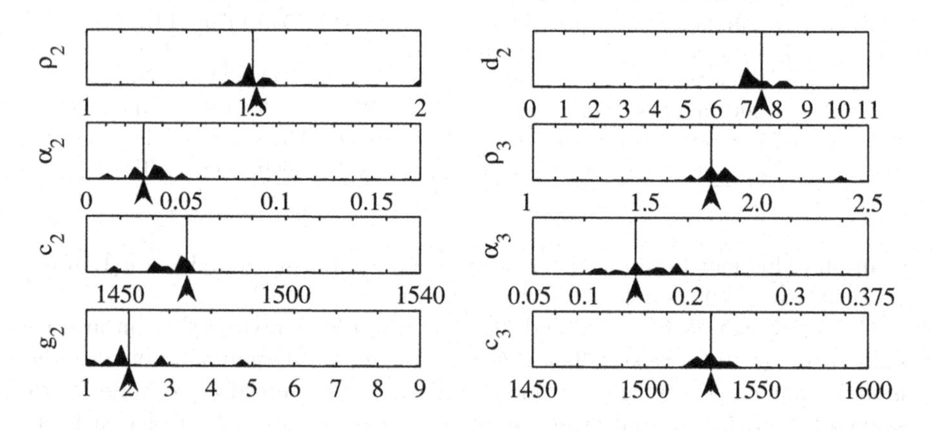

Fig. 2. GIANT probability distributions on the YS94 benchmark. Plotted are averages over 10 parallel runs with a total of 10^5 forward calls. Arrows mark reference solutions.

5 Conclusion

With geoacoustic inversion, a marine application has been presented for Ant Colony Optimization. A basic $\mathcal{MAX} - \mathcal{MIN}$ Ant System was implemented to search for geoacoustic properties of a shallow water environment. The algorithm has been tuned on a benchmark based on real world data from the 1994 Yellow Shark experiments. Accurate solutions were found within the given processing time. Runs that do not reach full convergence are shown to be useful in a multi-start approach, when $\mathcal{MM}$AS provides uncertainty analysis by combining final results of parallel runs into probability distributions. The next step is to improve convergence for geoacoustic inversion by customizing $\mathcal{MM}$AS with *a priori* knowledge on the search space.

References

1. Caiti, A., Chapman, N., Hermand, J.P., Jesus, S., eds.: Acoustic Sensing Techniques for the Shallow Water Environment Inversion Methods and Experiments. Springer (2006)
2. Dorigo, M., Stützle, T.: Ant Colony Optimization. MIT press (2004)
3. Gerstoft, P.: Inversion of seismoacoustic data using genetic algorithms and a posteriori probability distributions. J. Acoust. Soc. Am. **95**(2) (1994) 770–782
4. Hermand, J.P., Gerstoft, P.: Inversion of broad-band multitone acoustic data from the yellow shark summer experiments. IEEE Journal of Oceanographic Engineering **21**(4) (1996) 324–346
5. Hermand, J.P.: Broad-band geoacoustic inversion in shallow water from waveguide impulse response measurements on a single hydrophone: Theory and experimental results. IEEE Journal of Oceanographic Engineering **24**(1) (1999) 41–66
6. Hermand, J.P., Holland, C.: Geoacoustic characterisation of fine-grained sediments using single and multiple bottom reflection data. In: Shallow Water Environments. Volume Marine Geophysical Researches, Special Volume on Subsurface Imaging and Sediment Characterization of 26. (2005) 267–274
7. Tolstoy, A.: Matched Field Processing for Underwater Acoustics. World Scientific (1993)
8. Gerstoft, P., Mecklenbräuker, C.F.: Ocean acoustic inversion with estimation of *a posteriori* probability distributions. J. Acoust. Soc. Am. **104**(2) (1998) 808–819
9. van Leijen, A., Hermand, J.P.: Geoacoustic inversion with ant colony optimisation. In Jesus, S., Ródriguez, O., eds.: Proceedings of the Eighth European Conference on Underwater Acoustics, Carvoeiro, Portugal, 8th ECUA (2006)
10. Stützle, T., Hoos, H.: The $\mathcal{MAX} - \mathcal{MIN}$ ant system and local search for the traveling salesman problem. In Bäck, T., Michalewicz, Z., Yao, X., eds.: Proceedings of the 1997 IEEE International Conference on Evolutionary Computation (ICEC'97), Piscataway, NJ, IEEE Press (1997) 309–314
11. Birattari, M.: The problem of tuning metaheuristics as seen from a machine learning perspective. PhD thesis, Université Libre de Bruxelles (2004)
12. Birattari, M., Stützle, T., Paquete, L., Varrentrapp, K.: A racing algorithm for configuring metaheuristics. In et al, W.L., ed.: Proceedings of the Genetic and Evolutionary Computation Conference, San Fransisco, CA, USA, Morgan Kaufmann (2002) 11–18

Higher Order Pheromone Models in Ant Colony Optimisation

James Montgomery

Faculty of Information & Communication Technologies
Swinburne University of Technology, Melbourne, Australia
jmontgomery@ict.swin.edu.au

Abstract. Ant colony optimisation is a constructive metaheuristic that successively builds solutions from problem-specific components. A parameterised model known as pheromone—an analogue of the trail pheromones used by real ants—is used to learn which components should be combined to produce good solutions. In the majority of the algorithm's applications a single parameter from the model is used to influence the selection of a single component to add to a solution. Such a model can be described as first order. Higher order models describe relationships between several components in a solution, and may arise either by contriving a model that describes subsets of components from a first order model or because the characteristics of solutions modelled naturally relate subsets of components. This paper introduces a simple framework to describe the application of higher order models as a tool to understanding common features of existing applications. The framework also serves as an introduction to those new to the use of such models. The utility of higher order models is discussed with reference to empirical results in the literature.

1 Introduction

Ant colony optimisation (ACO) is a constructive metaheuristic that belongs to the model-based search (MBS) class of optimisation algorithms [1]. In an MBS algorithm, successive solutions are built using a parameterised probabilistic model, the parameters of which are revised over time using the solutions produced by the algorithm in order to direct its search towards promising areas of the solution space. The model used in ACO is referred to as a *pheromone model* in reference to the trail pheromones laid down by real ants to mark paths from their nest to a food source. In essence, the model describes relationships between components in the solution, such as one component succeeding another or whether a component is in a solution or not. The problem being solved thus partially dictates what can be modelled. Although the initial application of ACO to the well-known travelling salesman problem (TSP) used a model that very closely resembles the environment in which real ants move—there is a clear similarity between a Hamiltonian cycle in an edge-weighted graph and alternative routes between nest and food—as the range of problems to which it is applied has grown so too has the range of models [2].

M. Dorigo et al. (Eds.): ANTS 2006, LNCS 4150, pp. 428–435, 2006.

During a single iteration of a typical ACO algorithm, each artificial ant constructs a solution by successively adding problem-specific solution components. The relative utility of alternative components is given by the parameters of the pheromone model. When the utility of adding a candidate component to a partial solution is described by a single parameter, the model is said to be first order [3,4,2]. A higher order model is one in which the utility of adding a candidate component to a partial solution is described by several parameters, requiring that the information be aggregated before judging that component's utility.

This paper introduces a simple framework for higher order pheromone models that serves both as a tool to understand existing applications of such models and as a guide for their future application. The utility of such models is also discussed with reference to their empirical performance. These topics are organised as follows. Section 2 formalises the discussion of pheromone models, introducing some necessary notation, before Section 3 describes the framework. The empirical performance of higher order models is compared to that of first order alternatives in Section 4. Section 5 provides some concluding remarks.

2 Pheromone Models

An ACO algorithm consists of a number of iterations of solution construction, within which each (artificial) ant builds its solution by successively selecting a solution component to add to its sequence. Solution components are typically selected probabilistically, biased by the parameters of the pheromone model, which provide an estimate of the utility of adding a solution component to an ant's partial solution. It should be noted that the term *solution component* is somewhat overloaded in the ACO literature, at times being used to refer to components of the model rather than the components from which solutions are built (sometimes referred to as *natural solution components* [3]). As both kinds of "component" are discussed in this paper, the term *solution characteristic* [2] is introduced to describe components of the model. A pheromone model thus consists of a set of solution characteristics, and is denoted by C. To each solution there corresponds a set of solution characteristics that is a proper subset of that defined by the model.

In essence, the model describes relationships between components or components and the solution. For instance, the model used with the TSP describes a relationship between exactly two components, the candidate being considered and the previous component added. The model commonly used with knapsack problems represents a relationship between a candidate and the solution as a whole, indicating the utility of including a component at all. The relationships described by a pheromone model may also make reference to aspects of solutions other than the components from which they are built. For instance, some models used in the literature represent the absolute position of components in a solution, while others represent the assignment of items from one set of entities to another (with the solution components being drawn from only one of the two sets). When a model represents a relationship between a candidate solution component and at most one other component in the partial solution, as in these examples, that model is said to be first order. In other words, the utility of adding a candidate component is described by a single pheromone value (a non-zero, real-valued number denoted by τ). When the relationships modelled relate to multiple solution components, the pheromone model is higher order.

Higher order pheromone models implicitly define two sets of solution characteristics: one set relates to decisions to include individual solution components based on the current state of a partial solution, while the other set relates to higher order decisions, i.e., whether a single solution should exhibit two or more particular solution characteristics at the same time. Given a first order model C, an nth-order model may be a set of n-tuples of solution characteristics, denoted by C^n. Higher order models may be contrived by modelling combinations of n solution characteristics from some first order model or, when the solution characteristics modelled relate many parts of the solution to each other, will form naturally as a consequence of having to combine information from each relationship. In the latter case there is typically no related first order model.

3 Using Higher Order Models

The two main issues that arise when using a higher order model are how the higher order information is used, and the trade-off between the computational overhead associated with the larger pheromone model versus the benefits of using the extra information it provides. The former is discussed in this section, while the latter is discussed in Section 4 below.

Although higher order pheromone models have been used in a number of ACO algorithms, there is no single approach to their use. Nevertheless, there are common features of each of the approaches currently described in the literature that allow a general framework to be proposed.

When using a first order pheromone model, each constructive step is a competition between individual solution characteristics (and hence between the solution components they implicitly represent). When using a higher order model, the pheromone associated with adding a particular solution component is an aggregate of a number of pheromone values. Given a first order solution characteristic $c \in C$, denote the set of all other single solution components or characteristics to which c is related, and which consequently should be used to inform the decision to include c in the current partial solution, by C_c. In general, for an nth order pheromone model it is important to know to which tuples of $(n-1)$ solution components or characteristics a first order characteristic c is related, denoted C_c^{n-1}. Given an appropriate definition for C_c^{n-1}, a suitable aggregation function must also be defined, as well as an alternative when $C_c^{n-1} = \varnothing$.

Assuming that a solution characteristic c corresponds to a single constructive step (e.g., the addition of a single solution component or a single assignment), and denoting the pheromone associated with adding c to the partial solution s^p using an nth order pheromone model by $\tau(s^p, c, n)$, a generic function for

$\tau(s^p, c, n)$ where $n \geq 2$ is given by

$$\tau(s^p, c, n) = \begin{cases} f(s^p, c, \tau_n) & \text{if } \exists\, \tau_n \text{ and } |C_c^{n-1}| > 0 \\ \tau(s^p, c, n-1) & \text{otherwise} \end{cases} \qquad (1)$$

where $\tau_n : C \times C^{n-1} \rightarrow \mathbb{R}^+$ is a function from collections of n solution characteristics to pheromone values, and $f(s, c, \tau_n)$ is an aggregation function over the pheromone values associated between c and the elements of C_c^{n-1}, which is discussed in more detail below. Note that the equation is recursive; if C_c^{n-1} is empty or τ_n does not exist then a lower order pheromone model is sought. To ensure that the recursion defined by Equation 1 is well-founded, τ_1 must be defined, either to be a constant value or a separate first order pheromone model. In pheromone models where the elements of C_c^{n-1} are taken from s^p, early in solution construction s^p contains few solution characteristics and it is likely that $C_c^{n-1} = \varnothing$, and hence a lower order pheromone model such as τ_1 must be used until the nth order model τ_n can be used. Conceivably, for $n > 1$, if an nth order model is used, $n - 1$ other pheromone models may also be employed to deal with the first $n - 1$ steps of solution construction. In practice, most higher order pheromone models are only second order, so at most two pheromone models may be required.

Instances of higher order models are specified by providing definitions for the three components of this general framework, C_c, f and τ_1. The definition of C_c is highly problem specific and closely tied to the way solutions are constructed.

A number of options are available for the aggregation function f, four of which are to take the minimum, maximum, mean or sum of the different pheromone values involved. These four alternative definitions of f are denoted by $min(\tau_n)$, $max(\tau_n)$, $mean(\tau_n)$ and $sum(\tau_n)$ respectively.

The definition of C_c^{n-1} also determines which pheromone values from τ_n are updated by a solution s. For instance, using a second order pheromone model that represents the learned utility of having pairs of solution characteristics $(c_i, c_j) \in C^2$ copresent in a solution, pheromone is updated for all pairs (c_i, c_j) such that $c_i, c_j \in s, c_i \neq c_j$. Alternatively, when using a second order model that represents the utility of placing a solution component before certain other solution components, the value of C_c^{n-1} *when the solution was constructed* must be used to identify which pheromone values to update.

3.1 Defining C_c, f and τ_1 for a Problem

The definition of C_c is problem specific and typically apparent from the higher order solution characteristics being modelled. For instance, if a second order model is used to learn whether pairs of components should be part of the same solution, then intuitively C_c should contain those components already in the partial solution. Alternatively, given a different second order pheromone model that models pairs of components that should not be part of the same solution (or where there is a relationship based on the relative order of the pairs of components as in [4]), and faced with a first order decision about whether to include

Table 1. Sample of customisations of Equation 1. τ_{max} is an upper bound on pheromone values imposed in the $\mathcal{MAX}-\mathcal{MIN}$ Ant System algorithm [6].

C_c	f	τ_1	Example(s)
$\in s^p$	*sum*	unknown	[7,8]
		equivalent to 1	[9]
		1st order model	[10]
	mean	1	[11,12]
		1	[13]
		1	[5]
		1st order model	[5]
$\notin s^p$	*min*	τ_{max}	[14]
		∞	[4]

a candidate characteristic, intuitively C_c should contain only those components that have not yet been added to the partial solution.

The definitions of f and τ_1 can be somewhat separate from the solution characteristics modelled and so may appear to be arbitrary choices. Nevertheless, a number of observations may be made concerning existing applications of higher order models. Table 1 categorises the usage of higher order pheromones found in the literature, based on whether the elements of C_c come from the partial solution or its complement, the aggregation function used and definition of τ_1. Full details of the retrospective application of the framework to the works cited are given in [5].

With regards to the aggregation function f, all the examples in Table 1 use *min*, *mean* or *sum*, while none uses *max*. The use of the *min* function can be characterised as a cautious approach—any single low pheromone value can in effect veto the first order decision being considered. Conversely, *max* allows any single high pheromone value to make the decision more likely. The functions *sum* and *mean* allow each higher order solution characteristic's pheromone value to influence the first order decision, with the choice of whether to use *sum* or *mean* dependent on the number of higher order solution characteristics available for each candidate first order characteristic (or component). In the examples cited, *sum* is used in all cases where $|C_c| = |C_{c'}| \; \forall \; c \neq c'$ for a fixed partial solution size, while *mean* is used in those cases where this is not the case (or where the magnitude of observed pheromone values must be kept constant).

Notably, *min* is used only in those cases where the elements of C_c are not present in the partial solution. Blum and Sampels [4] describe an ACO algorithm for shop scheduling problems in which each solution characteristic indicates the relative order of operations that must be processed on the same machine. In this application, the rationale for using *min* is that if any pheromone value is low then there must exist at least one related operation that should be scheduled before the one being considered. The *min* function is also used in an ACO algorithm for a university timetabling problem developed by Socha, Knowles and Sampels [14], where higher order pheromone values are used to learn which events should *not* be placed in the same timeslot. Conceivably, taking the minimum value between the current event and those already assigned a timeslot might produce similar results to considering unscheduled events. However, this approach may allow an event to be placed in a timeslot that suits another unscheduled event better and which may increase solution cost if that other event were later placed in the same timeslot. Consequently, taking the minimum value may avert such undesirable actions. Thus, in both examples, using *min* in relation to those solution components or characteristics that have yet to be added to a partial solution appears to avoid making decisions that may force the algorithm to make an inferior decision later in solution construction. In contrast, the use of *sum* and *mean* with pheromone values associated with solution components or characteristics already in a partial solution appears to promote the selection of a solution component that is well suited to the existing partial solution.

In those examples where τ_1 is clearly defined and the elements of C_c are taken from the partial solution, the first solution characteristic is chosen either randomly or using a first order pheromone model when used in conjunction with *sum*, while it is assigned a constant value when used with *mean*. Where the elements C_c do not come from the partial solution, τ_1 is set to either a high value (τ_{max}) or a candidate component is chosen as if it had a high value (such as ∞).

4 Utility of Higher Order Pheromones

When implemented, higher order pheromone models require greater computational resources than their first order counterparts. While storage overhead is typically not problematic—most higher order models are second order, representing a squaring in size—higher order models necessarily take longer to process as multiple pheromone values must be considered for each solution characteristic. This increased computational overhead must be weighed against any potential improvements to the quality of solutions produced by the algorithm, as the following examples show.

A comparative study of first and second order pheromone models for the k-cardinality tree problem found that, given the same amount of execution time, the latter produces fewer solutions and thus the algorithm makes less progress towards good solutions [10]. It was concluded that the first order model is consequently a better choice for this problem.

Roli, Blum and Dorigo [7] compared the performance of an ACO algorithm for constraint satisfaction using three alternative pheromone models, including a first order pheromone model that represents which assignments should be made and a second order pheromone modelling which pairs of assignments should be made. Both models performed similarly well. However, again due to the increased computational overhead for the second order model, the first order model is promoted as the best-suited to that problem.

Montgomery [5] compared first and second order pheromone models for the knapsack problem, also finding that the two gave equivalent performance in terms of solution quality, with the second order model increasing the required computation time for an equivalent number of solutions produced.

Montgomery [5] also compared a first and two second order pheromone models for a car sequencing problem in which different car models must be assigned positions in a production sequence such that the separation penalty between cars of the same model is minimised (i.e., it is desirable keep cars of the same model apart). The penalty varies between models. The first order pheromone model represents the assignment of a car model to a sequence position. One of the higher order models represents pairs of sequence positions assigned the same car model, similar to the model used by Costa and Hertz [11] for the graph colouring problem in which nodes in a partially connected graph are partitioned into colour groups. The other higher order model represents pairs of sequence positions assigned the same car model plus which model is assigned. The study found that the first order model and the second order model that includes the actual car model assigned both outperformed the model inspired by that used by Costa and Hertz. This finding is commensurate with suggestions by Montgomery, Randall and Hendtlass [2] regarding appropriate pheromone models. However, the first order model performed best overall.

These four studies would appear to suggest that higher order models have little or no utility. However, there are some combinatorial problems, such as the maximum-clique [9] and graph colouring [11] problems, for which first order models may not be appropriate—both of these problems involve the assignment of components to groups. Indeed, potential problems have been identified with first order models for the graph colouring problem [2]. Furthermore, the best performing model for shop scheduling problems is the second order model developed by Blum and Sampels [4,15], a key feature of which is that it is not derived from a first order model. Taken together, these results suggest that if a simpler pheromone model is "appropriate" for a given problem, it is unnecessary to use a higher order (typically, a second order) model. Montgomery, Randall and Hendtlass [2] put forward a number of qualities of a model that make it "appropriate" for a particular problem, the chief one being that it represents characteristics of solutions that directly impact on solution cost. Using this guiding criterion, the use of higher order pheromone models is implicated in a number of problem domains.

5 Conclusions

The majority of pheromone models used in ACO algorithms are first order, with a single parameter of the model representing the learned utility of adding a single solution component to a partial solution. A number of higher order models have also been developed which give more detailed information about the utility of adding a single component. This paper has introduced a simple framework for describing higher order pheromone models, which serves as a tool to understand common features of existing applications and may assist in the future development of new higher order models for other problems.

A review of studies that compare first and second order models suggests that higher order models will often give equivalent performance to first order counterparts, but at the expense of greater computation times. However, there are some problems where the use of higher order models appears necessary. Therefore, if a problem appears to require the use of a higher order model then a first order model should also be developed and its performance examined.

References

1. Zlochin, M., Dorigo, M.: Model-based search for combinatorial optimization: A comparative study. In: 7th International Conference on Parallel Problem Solving from Nature. (2002) 651–662
2. Montgomery, J., Randall, M., Hendtlass, T.: Automated selection of appropriate pheromone representations in ant colony optimisation. Artificial Life **11** (2005) 269–291
3. Blum, C.: Theoretical and practical aspects of ant colony optimization. Volume 282 of Dissertations in Artificial Intelligence. IOS Press, Nieuwe Hemweg, The Netherlands (2004)
4. Blum, C., Sampels, M.: Ant colony optimization for FOP shop scheduling: A case study on different pheromone representations. In: 2002 Congress on Evolutionary Computation. (2002) 1558–1563
5. Montgomery, E.J.: Solution Biases and Pheromone Representation Selection in Ant Colony Optimisation. PhD thesis, Bond University (2005)
6. Stützle, T., Hoos, H.: $\mathcal{MAX} - \mathcal{MIN}$ ant system. Future Gen. Comp. Sys. **16** (2000) 889–914
7. Roli, A., Blum, C., Dorigo, M.: ACO for maximal constraint satisfaction problems. In: 4th Metaheuristics International Conference, Porto, Portugal (2001) 187–192
8. Solnon, C.: Ants can solve constraint satisfaction problems. IEEE Trans. Evol. Comput. **6** (2002)
9. Fenet, S., Solnon, C.: Searching for maximum cliques with ant colony optimization. In Raidl, G., et al., eds.: EvoWorkshops 2003. Volume 2611 of LNCS. Springer-Verlag (2003) 236–245
10. Blum, C.: Ant colony optimization for the edge-weighted k-cardinality tree problem. In Langdon, W., ed.: Genetic and Evolutionary Computation Conference, Morgan Kaufmann Publishers (2002) 27–34
11. Costa, D., Hertz, A.: Ants can colour graphs. J. Oper. Res. Soc. **48** (1997) 295–305
12. Schoofs, L., Naudts, B.: Solving CSPs with ant colonies. In: Abstract Proceedings of ANTS2000, Brussels, Belgium (2000)
13. Ducatelle, F., Levine, J.: Ant colony optimisation and local search for bin packing and cutting stock problems. J. Oper. Res. Soc. **55** (2004) 705–716
14. Socha, K., Knowles, J., Sampels, M.: A $\mathcal{MAX} - \mathcal{MIN}$ ant system for the university course timetabling problem. In Dorigo, M., Di Caro, G., Sampels, M., eds.: Ant Algorithms. Volume 2463 of LNCS, Springer-Verlag (2002) 1–13
15. Blum, C., Sampels, M.: An ant colony optimization algorithm for shop scheduling problems. J. Math. Model. Algorithms **3** (2004) 285–308

Hybrid Particle Swarm Optimization: An Examination of the Influence of Iterative Improvement Algorithms on Performance

Jens Gimmler[1], Thomas Stützle[2], and Thomas E. Exner[1]

[1] Theoretische Chemische Dynamik, Universität Konstanz, Konstanz, Germany
{Jens.Gimmler, Thomas.Exner}@uni-konstanz.de
[2] IRIDIA, CoDE, Université Libre de Bruxelles, Brussels, Belgium
stuetzle@ulb.ac.be

Abstract. In this article, we study hybrid Particle Swarm Optimization (PSO) algorithms for continuous optimization. The algorithms combine a PSO algorithm with either the Nelder-Mead-Simplex or Powell's Direction-Set local search methods. Local search is applied each time the PSO part meets some convergence criterion. Our experimental results for test functions with up to 100 dimensions indicate that the usage of the iterative improvement algorithms can strongly improve PSO performance but also that the preferable choice of which local search algorithm to apply depends on the test function. The results also suggest that another main contribution of the local search is to make PSO algorithms more robust with respect to their parameter settings.

1 Introduction

PSO is a population-based optimization technique introduced by Kennedy and Eberhart in 1995 [1]. Today there exist many variants of PSO like PSO with inertia weights [2, 3], a constricted version of the PSO [4], the usage of various topologies for the particles neighborhood [5], etc. These variants try to introduce variations on how to balance diversification and intensification of the search to improve over the performance of the basic PSO algorithm.

Another option, which is very commonly used in hybrid stochastic local search (SLS) algorithms [6], has less frequently been exploited in PSO research: the combination of a global exploration mechanism with a local search algorithm. (Examples for combinations of PSO with local search for continuous function optimization can be found in [7, 8].) The main focus of this article is an experimental study of a hybrid Particle Swarm Optimization (HPSO) algorithm that combines PSO with local search for continuous optimization problems of the form $f : D \to \mathbb{R}$ where $D \subseteq \mathbb{R}^n$, $n \in \mathbb{N}^+$. In this article, we examine the extension of a standard PSO algorithm with either the Nelder–Mead Simplex (NMS) or Powell's Direction-Set (PDS) local search methods [9] as well as with a reinitialization process and we study the performance of the resulting algorithms.

M. Dorigo et al. (Eds.): ANTS 2006, LNCS 4150, pp. 436–443, 2006.

2 Basic Algorithms

Particle Swarm Optimization. PSO algorithms maintain a population of particles, each particle i having associated a current position $\vec{x}_i \in D$, a velocity vector $\vec{v}_i \in \mathbb{R}^n$ and the position $\vec{p}_i \in D$ of the best objective function value it found so far. Starting from some initial population, PSO cycles through the following steps. First, all particles update their velocity and next each particle i updates its position $\vec{x}_i$ using its new velocity vector and computes the function value $f(\vec{x}_i)$. PSO repeats these steps until some termination criterion is met.

The basic PSO extended with a parameter called *inertia weight* [2] performs the velocity and position update for a particle i according to

$$\vec{v}_i(t+1) := w(t)\vec{v}_i(t) + \phi_1 \vec{r}_{1i}(\vec{p}_i(t) - \vec{x}_i(t)) + \phi_2 \vec{r}_{2i}(\vec{p}_g(t) - \vec{x}_i(t)) \qquad (1)$$

$$\vec{x}_i(t+1) := \vec{x}_i(t) + \vec{v}_i(t+1), \qquad (2)$$

where $w(t)$ is the (time varying) inertia weight, $\phi_1 \in \mathbb{R}^+$ and $\phi_2 \in \mathbb{R}^+$ are positive constants, $\vec{r}_{i1}$ and $\vec{r}_{i2}$ are random numbers with each dimension being uniformly distributed in $[0,1]$ and g is the index of the particle with the best previous position in the neighborhood of particle i at iteration t. Usually, each component j of the velocity vector $\vec{v}_i$ is bounded to $-v_{max} \leq v_{ij} \leq v_{max}$; $v_{max} \in \mathbb{R}^+$ is a positive constant. Another important variant of PSO uses the *constriction coefficient* χ [4] and the velocity update becomes

$$\vec{v}_i(t+1) := \chi(\vec{v}_i(t) + \phi_1 \vec{r}_{1i}(\vec{p}_i(t) - \vec{x}_i(t)) + \phi_2 \vec{r}_{2i}(\vec{p}_g(t) - \vec{x}_i(t))). \qquad (3)$$

The constriction coefficient is often used as a function of ϕ_1 and ϕ_2 and thus could be considered as a special case of the variant with inertia weights [10].

Apart from the above mentioned parameters, the neighborhood topology can have a significant influence on PSO performance [11]. Usually, the velocity update equation of every particle i only depends on $\vec{p}_i$ and $\vec{p}_g$. The basic version of PSO uses the *gbest* neighborhood topology, i.e., all particles are neighbored to each other and, hence, $\vec{p}_g$ is the best position found so far by the algorithm.

Iterative improvement algorithms. The NMS and the PDS method are deterministic iterative-improvement algorithms for multidimensional continuous optimization. Both methods use only function evaluations for the optimization process and are easy to implement. Starting from an initial solution $\vec{x}_{start}$, they iteratively improve upon their current solution until their termination criterion is met. We use the implementation of the NMS and PDS methods proposed in [9].

3 Hybrid Particle Swarm Optimization

An advantage of iterative improvement methods is their good exploitation quality; however, they have no mechanism for a robust exploration of the search space. In contrast, PSO can offer such exploration capabilities. Hence, it may be a good idea to combine the two techniques in one algorithm.

In this article, we examine such a combination, in which we apply iterative improvement algorithms only occasionally. In particular, we introduce a convergence criterion to the PSO algorithm that is based on the distance between the current positions of the particles. Once this convergence criterion is met, we apply an iterative improvement algorithm to the best solution seen so far. Upon termination of the local search, the particles are reinitialized to randomly chosen positions and the PSO part is restarted. This cycle of repeated applications of a PSO algorithm until convergence and subsequent local search is applied until the termination criterion of our HPSO algorithm, typically a maximum number of function evaluations or computation time, is met and the algorithm then returns the best solution found.

The main difference to a standard PSO loop is the usage of the convergence criterion and the usage of the iterative improvement algorithm. To enforce a fast convergence of the PSO algorithm and to, hence, apply the local search a significant number of times, we use as the underlying PSO algorithm the *gbest* PSO with Equation 1 for the velocity update. For $w(t)$ we use the nonlinear function

$$w(t) := (1 - g(t))^{w_e} \cdot (w_{initial} - w_{final}) + w_{final}, \tag{4}$$

which was proposed in [3] to calculate the inertia weight at iteration t, where $g(t)$ returns a number in the range $[0, 1]$ and $w_e, w_{initial}, w_{final} \in \mathbb{R}$ are constants. The function $g(t)$ returns 0 at the beginning of a search process and increases every iteration until it reaches 1 at the end of the search. The constant $w_{initial}$ is the initial value of the inertia weight and w_{final} is the desired value of the inertia weight at the end of the search process. A constant inertia weight is obtained if $w_{initial}$ and w_{final} are equal. As the convergence criterion in our algorithm we use the distance of the particles to the best solution found so far $\vec{x}^*$. In particular, the convergence criterion is met, if $d_{max}/d_{domain} < 0.001$, where $d_{max} \in \mathbb{R}^+$ is the maximum Euclidean distance to $\vec{x}^*$ over all current positions of the particles and $d_{domain} = \sqrt{n(b_u - b_l)^2}$ is the diameter of the search space (b_u and b_l are the upper and lower bound of one dimension of the search space). As iterative improvement methods we explore the usage of either NMS or PDS. For the reinitialization of the population P, each particle in P is moved to a random position $\vec{x}$ in the search space and $\vec{p}$ is set to $\vec{x}$. The components of the velocity vectors receive new, uniformly distributed values in the range $[-v_{max}, v_{max}]$.

4 Experiments on PSO Variants

Experimental setup. As our test-bed we use five test functions taken from the literature [4, 12, 13]; the functions are given in Table 1 together with their particular search domain and their globally optimal function value. All functions were tackled using 30 and 100 dimensions.

The swarm size was set to 30 particles. All algorithms use $v_{max} = b_u - b_l$. Standard parameters for ϕ_1 and ϕ_2 and the constant inertia weight were taken

Table 1. The functions of our test-bed. See the text for details.

Name	Domain $[b_l, b_u]^n$	f_{opt}	Formula
Ackley	$[-32, 32]^n$	0	$f_1(x) = -20\exp\left(-0.2\sqrt{n^{-1}\sum_{i=1}^n x_i^2}\right) - \exp\left(n^{-1}\sum_{i=1}^n \cos(2\pi x_i)\right) + 20 + e$
Griewank	$[-600, 600]^n$	0	$f_2(x) = 1 + \frac{1}{4000}\sum_{i=1}^n x_i^2 - \prod_{i=1}^n \cos\left(\frac{x_i}{\sqrt{i}}\right)$
Rastrigin	$[-5.12, 5.12]^n$	0	$f_3(x) = 10n + \sum_{i=1}^n [x_i^2 - 10\cos(2\pi x_i)]$
Rosenbrock	$[-30, 30]^n$	0	$f_4(x) = \sum_{i=1}^{n-1} [100(x_{i+1} - x_i^2)^2 + (x_i - 1)^2]$
Sphere	$[-100, 100]^n$	0	$f_5(x) = \sum_{i=1}^n x_i^2$

Table 2. Parameter sets for the standard algorithms; see the text for more details. The "C" and "D" at the end of algorithm identifiers refer to the usage of either constant or decreasing inertia weights.

Type of inertia weight	Configuration	ϕ_1	ϕ_2	Inertia weight
constant	PSO1, RPSO1C, HPSO1C	1.7	1.7	0.6
	PSO2, RPSO2C, HPSO2C	1.494	1.494	0.729
decreasing	RPSO1D, HPSO1D	1.7	1.7	$(0.8, 0.2, 2)$
	RPSO2D, HPSO2D	1.494	1.494	$(0.8, 0.2, 2)$

from the literature [12]. An experiment with an n-dimensional function was terminated after $1000n$ function evaluations and each experiment was repeated 100 independent times. The initial positions of the swarm are distributed uniformly at random in the search space.

The algorithmic variants include the basic *gbest* PSOs, where the velocity update is calculated according to Equation 1. We also study an RPSO algorithm that is the same as our HPSO, except that it does not use any iterative improvement method. Hence, RPSO only differs in the usage of the occasional restarts from the basic PSO. By including it into the analysis, we get a better impression of the impact the local search and the occasional restarts have on the overall performance. RPSO and HPSO were tested with constant and decreasing inertia weights. All algorithmic variants and the individual parameters of each algorithm variant in a first batch of experiments are shown in Table 2, which is split in dependence of the usage of constant or decreasing inertia weights. The decreasing inertia weight parameters are given as a triple $(w_{initial}, w_{final}, w^e)$.

Experimental comparison. The results for our HPSO are given for either the usage of NMS or of PDS. In Table 3, we give the average function values obtained; all entries smaller than 10^{-10} were rounded to zero.

Our experiments show that RPSO reaches at least comparable but often much better solution quality than basic PSO on the multimodal functions (Ackley, Griewank, Rastrigin), while the reinitialization process leads to no improvement for the unimodal functions (Rosenbrock, Sphere). Second, almost all results of the HPSO configurations are better than the results of PSO1 and PSO2. Again, the 30-dimensional Sphere function is an exception where HPSO with NMS is

Table 3. Average function values for PSO, RPSO and HPSO variants on the test functions after $1000n$ function evaluations. All configurations of HPSO that are in bold-face reach lower averages than the best among all PSO and RPSO variants. The best configuration for each function is marked with an asterisk. See Table 2 for the meaning of the algorithm identifiers.

Configuration	Ackley	Griewank	Rastrigin	Rosenbrock	Sphere
		$n = 100$, 100000 function evaluations			
PSO1	5.8161	0.2590	267.465	231.779	0.0003
PSO2	5.9400	0.3734	276.049	279.856	1.0557
RPSO1C	0.3064	0.0129	114.134	295.935	0.0055
RPSO2C	1.2262	0.0153	132.128	319.547	0.0062
RPSO1D	1.7313	0.0057	167.863	263.382	0.0023
RPSO2D	2.7645	0.0354	234.573	367.831	0.0933
HPSO1C PDS	***0**	0.1435	***86.290**	**150.011**	***0**
HPSO2C PDS	***0**	0.1351	**87.131**	**168.364**	***0**
HPSO1D PDS	**0.2510**	0.1437	143.900	**134.895**	***0**
HPSO2D PDS	0.6713	0.2148	159.969	**130.187**	***0**
HPSO1C NMS	2.1163	***0.0040**	**113.872**	**153.167**	0.0003
HPSO2C NMS	2.2404	0.0075	142.284	**176.449**	**0.0002**
HPSO1D NMS	3.0200	0.0068	150.047	**130.198**	0.0003
HPSO2D NMS	4.3143	0.0193	171.966	***82.794**	0.0003
		$n = 30$, 30000 function evaluations			
PSO1	2.1911	0.0326	56.0857	42.1194	*0
PSO2	1.9050	0.0314	57.2996	48.9864	*0
RPSO1C	0.0081	0.0213	30.8356	56.9872	0.0003
RPSO2C	0.0230	0.0244	35.8105	60.0397	0.0002
RPSO1D	0.0090	0.0248	39.6500	54.7927	0.0004
RPSO2D	0.0886	0.0342	47.3336	69.9956	0.0040
HPSO1C PDS	***0**	**0.0196**	***17.4212**	**15.7134**	*0
HPSO2C PDS	0.0134	***0.0193**	**19.7989**	**23.1755**	*0
HPSO1D PDS	0.0116	0.0282	**25.1322**	**19.8705**	*0
HPSO2D PDS	0.2095	0.0453	**30.1778**	***14.2611**	*0
HPSO1C NMS	0.0948	0.0248	**29.3964**	**34.5819**	6.614e-10
HPSO2C NMS	0.1384	0.0241	33.5891	44.5354	3.780e-10
HPSO1D NMS	0.3952	0.0215	31.5297	**28.9776**	1.161e-09
HPSO2D NMS	1.1578	0.0231	34.4734	**22.5492**	1.254e-09

outperformed by PSO although especially the Sphere function should be easily solved by an iterative improvement method. Here, apparently NMS may have problems with convergence or the termination criterion used in [9]. Third, the best choice of the iterative improvement method depends somewhat on the test function. For the Ackley function, HPSO with PDS performs clearly better than HPSO with NMS, while on the 100-dimensional Griewank function it is dominated by HPSO with NMS. On the Rastrigin and Rosenbrock functions, HPSO

Table 4. Average function values of HPSO on n dimensions. All configurations that performed better than the best corresponding (HPSOC: HPSO1C, HPSO2C; HPSOD: HPSO1D, HPSO2D) configuration in Table 3 are in bold-face. Configurations that are equal to the best corresponding configuration of HPSO in Table 3 are in brackets.

Configuration	Ackley	Griewank	Rastrigin	Rosenbrock	Sphere
n = 100, 100000 function evaluations					
HPSOC	(0)	0.1303	**79.3003**	**126.509**	(0)
HPSOD	**0**	0.1357	**87.5583**	115.592	(0)
n = 30, 30000 function evaluations					
HPSOC	(0)	0.0267	**17.0231**	**10.3569**	(0)
HPSOD	**0**	0.0267	**18.8141**	14.6747	(0)

Table 5. The entries show the average function values of the ith best parameter setting for HPSO

Inertia weight	i	Ackley	Griewank	Rastrigin	Rosenbrock	Sphere
n = 100, 100000 function evaluations						
	20	0.0188	0.0096	101.994	19.9772	0
constant	100	2.9717	0.2052	158.834	96.7507	0.0003
	200	20.9588	2456.14	1642.66	1.18e+09	272874
	20	0.1892	0.0814	111.136	16.9327	0
decreasing	100	0.9343	0.2192	153.055	58.3007	0
	200	7.7326	0.3141	201.682	173.815	0.0003
n = 30, 30000 function evaluations						
	20	0.0031	0.0065	22.8257	3.9928	0
constant	100	1.0219	0.0344	38.5897	23.7757	5.13e-10
	200	20.3653	509.89	423.832	2.07e+08	54336.2
	20	0.0135	0.0069	24.0287	8.2877	0
decreasing	100	0.5883	0.0182	34.0737	21.1996	0
	200	2.5140	0.0838	45.8425	36.363	1.29e-09

with PDS performs somewhat better than with NMS if we compare the variants using a same parameter setting for the PSO related parameters; on the Sphere function HPSO with PDS outperforms clearly HPSO with NMS. Across all the test functions, the configuration HPSO1C PDS seems to be rather robust. It achieved the best results of all configurations seven times and in the other cases it is typically among the best variants.

Parameter settings. Besides the standard parameters (see Table 2), we examined the influence of the parameter settings on our HPSO. The possible values for ϕ_1 and ϕ_2 that we used were $\{0.75, 1.125, 1.5, 1.875, 2.25\}$, the values for the constant inertia weight w were $\{0.2, 0.4, 0.6, 0.8, 1.0\}$. In the case of the decreasing inertia weight $w(t)$ only $w_{initial}$ varies. Possible values we considered were $\{0.4, 0.6, 0.8, 1.0, 1.2\}$; the parameters $w_{final} = 0.2$ and $w_e = 2$ were fixed. To rank the different parameter settings we used a quality criterion that is calculated as follows: First determine the mean worst and the mean best found function value f_{max} and f_{min} of function f. For the jth parameter setting we compute a quality $q_{f_j} = (f_j - f_{min})/(f_{max} - f_{min})$ where f_j is the mean function value obtained by parameter setting j. The final quality criterion measured is $q_j = |F|^{-1} \sum_{f \in F} q_{f_j}$, where F is the set of test functions.

The highest ranked HPSO configuration according to this quality criterion for the constant inertia weight as well as for the decreasing inertia weight was $\phi_1 = 1.5$, $\phi_2 = 2.25$ with PDS. The best constant inertia weight was $w = 0.2$ and the best initial inertia weight for the decreasing approach was $w(t) = 0.4$. The values reached by these two parameter settings are shown in Table 4, where HPSOC uses the constant inertia weight and HPSOD the decreasing inertia weight. Compared to the standard parameters from Table 3, we can see that there is room for further improvement with parameter tuning, but that a standard parameterization (Table 2) performs reasonably good.

An interesting observation is that the best results for the configurations with decreasing inertia weights are not better than the best results for the constant inertia weight, but that many more parameter settings with decreasing inertia weight find reasonably good quality solutions than when using a constant inertia weight. This is shown in Table 5, where we give the average function values of the 20th, 100th and 200th best parameter setting per test function for constant and decreasing inertia weights. A closer inspection of the algorithm traces suggested an explanation for the more robust behavior: By decreasing inertia weights the algorithm more rapidly and more easily meets our convergence criterion and, hence, also uses more often restarts and the local search algorithms.

5 Conclusions

In this article, we have introduced a new hybrid PSO algorithm that uses occasional restarts and combines PSO with iterative improvement methods. Our computational results clearly show that the inclusion of iterative improvement methods into PSO is highly desirable concerning solution quality reached but especially also concerning the robustness of our overall HPSO algorithm with respect to the parameter settings of the PSO part. For this latter aspect, particularly the variants with decreasing inertia weight appear to be best suited. Future work will comprise a more careful comparison of the hybrid algorithms proposed here to other, recent PSO variants and also to other techniques for tackling continuous optimization problems. Another line of research will study other schemes for implementing hybrid PSO algorithms. In fact, we are convinced that our initial results also indicate that the study of hybrids between PSO and local optimization techniques deserves more attention in future research on PSO.

Acknowledgments. Jens Gimmler thanks Oliver Korb, Patrick Duchstein and Christine Bauer for the support writing this article. Thomas Stützle acknowledges support of the Belgian FNRS, of which he is a research associate.

References

1. Kennedy, J., Eberhart, R.C.: Particle swarm optimization. In: Proceedings of IEEE International Conference on Neural Networks. (1995) 1942–1948
2. Shi, Y., Eberhart, R.: A modified particle swarm optimizer. In: Proceedings of the IEEE Congress on Evolutionary Computation (CEC 1998). (1998) 69–73
3. Chatterjee, A., Siarry, P.: Nonlinear inertia weight variation for dynamic adaptation in particle swarm optimization. Computers and Operations Research **33**(3) (2006) 859–871
4. Clerc, M., Kennedy, J.: The particle swarm - explosion, stability, and convergence in a multidimensional complex space. IEEE Transactions on Evolutionary Computation **6**(1) (2002) 58–73
5. Mendes, R., Kennedy, J., Neves, J.: The fully informed particle swarm: Simpler, maybe better. IEEE Transactions on Evolutionary Computation **8**(3) (2004) 204–210
6. Hoos, H.H., Stützle, T.: Stochastic Local Search-Foundations and Applications. Morgan Kaufmann Publishers, San Francisco, CA, USA (2004)
7. Fan, S., Liang, Y., Zahara, E.: Hybrid simplex search and particle swarm optimization for the global optimization of multimodal functions. Engineering Optimization **36**(4) (2004) 401–418
8. Wang, F., Qiu, Y., Bai, Y.: A new hybrid NM method and particle swarm algorithm for multimodal function optimization. In: IDA. Volume 3646. (2005) 497–508
9. Press, W.H., Flannery, B.P., Teukolsky, S.A., Vetterling, W.T.: Numerical Recipes in C : The Art of Scientific Computing. second edn. (1992)
10. Eberhart, R.C., Shi, Y.: Comparing inertia weights and constriction factors in particle swarm optimization. In: Proceedings of the 2000 Congress on Evolutionary Computation, 2000. Volume 1. (2000) 84–88
11. Kennedy, J.: Small worlds and mega-minds: effects of neighborhood topology on particle swarm performance. In: Proceedings of the 1999 Congress on Evolutionary Computation, 1999. CEC 99. Volume 3. (1999) 1931–1938
12. Trelea, I.C.: The particle swarm optimization algorithm: convergence analysis and parameter selection. Information Processing Letters **85**(6) (2003) 317–325
13. Ali, M.M., Khompatraporn, C., Zabinsky, Z.B.: A numerical evaluation of several stochastic algorithms on selected continuous global optimization test problems. Journal of Global Optimization **31**(4) (2005) 635–672

Introducing a Binary Ant Colony Optimization

Min Kong and Peng Tian

Shanghai Jiaotong University, Shanghai, China
kongmin@sjtu.edu.cn, ptian@sjtu.edu.cn

Abstract. This paper proposes a Binary Ant Colony Optimization applied to constrained optimization problems with binary solution structure. Due to its simple structure, the convergence status of the proposed algorithm can be monitored through the distribution of pheromone in the solution space, and the probability of solution improvement can be in some way controlled by the maintenance of pheromone. The successful implementations to the binary function optimization problem and the multidimensional knapsack problem indicate the robustness and practicability of the proposed algorithm.

1 Introduction

Ant Colony Optimization (ACO) is a stochastic meta-heuristic for solutions to combinatorial optimization problems. Since its first introduction by M. Dorigo and his colleagues [1] in 1991, ACO has been successfully applied to a wide set of different hard combinatorial optimization problems, such as traveling salesman problem[2], quadratic assignment problem[3], and vehicle routing problem[4]. The main idea of ACO is the cooperation of a number of artificial ants via pheromone laid on the path. Each ant contributes a little effort to the solution, while the final result is an emergence of the ants' interactions.

Besides its success in practical applications, there have also been some studies on the theory of ACO[5,6,7], which are mainly focused on the convergence proof of ACO. But the cybernetics of ACO is still unclear, specially on the speed of convergence. This paper tries to reveal some properties of the convergence speed of ACO by a simple ant system, BAS, which works in binary space and whose performance is verified through the binary function optimization problem and the multidimensional knapsack problem.

2 The Binary Ant System

2.1 Solution Construction

In BAS, artificial ants construct solutions by walking on the mapping graph as described in Fig. 1. At every iteration, a number of n_a ants cooperate together to search in the binary solution domain, each ant constructs its solution by walking sequentially from node 1 to node $n+1$ on the routing graph. At each node i, ant either selects the upper path $i0$ or selects the lower path $i1$ to walk to the

M. Dorigo et al. (Eds.): ANTS 2006, LNCS 4150, pp. 444–451, 2006.

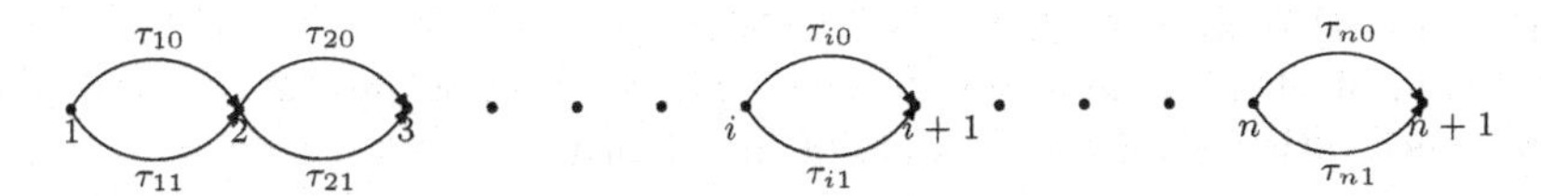

Fig. 1. Routing Diagram for Ants in BAS

next node $i+1$. Selecting $i0$ means $x_i = 0$; and selecting $i1$ means $x_i = 1$. The selecting probability is dependent on the pheromone distributed on the paths:

$$p_{is}(t) = \tau_{is}(t), i = 1, \cdots, n, s \in \{0, 1\} \tag{1}$$

where t is the number of iteration.

The solutions constructed by the ants may not be feasible when tackling the constrained binary optimization problems. A solution repair operator is incorporated to transfer the infeasible solutions to the feasible domain.

2.2 Pheromone Update

Initially, BAS set all the pheromone values as $\tau_{is}(0) = 0.5$, which is the same as that of HCF[8], but uses a simplified pheromone update rule:

$$\tau_{is}(t+1) \leftarrow (1-\rho)\tau_{is}(t) + \rho \sum_{x \in S_{upd} | is \in x} w_x \tag{2}$$

where S_{upd} is the set of solutions to be intensified; w_x are explicit weights for each solution $x \in S_{upd}$, which satisfying $0 \le w_x \le 1$ and $\sum_{x \in S_{upd}} w_x = 1$; ρ is the evaporation parameter, which is set initially as $\rho = \rho_0$, but decreases as $\rho \leftarrow 0.9\rho$ every time the pheromone re-initialization is performed.

S_{upd} consists of three components, they are: the global best solution S^{gb}, the iteration best solution S^{ib}, and the restart best solution S^{rb}.

Different combinations of w_x are implemented according to the convergence status of the algorithm. The convergence status is monitored by a convergence factor cf, which is defined as:

$$cf = \frac{\sum_{i=1}^{n} |\tau_{i0} - \tau_{i1}|}{n} \tag{3}$$

Under this definition, when the algorithm is initialized with $\tau_{is}(0) = 0.5$ for all the paths, $cf = 0$, while when the algorithm gets into convergence or premature, $|\tau_{i0} - \tau_{i1}| \to 1$, thus that $cf \to 1$.

Table 1 describes the pheromone update strategy in different value of cf, where w_{ib}, w_{rb} and w_{gb} are the weight parameters for S^{ib}, S^{rb} and S^{gb} respectively, $cf_i, i = 1, \cdots, 5$ are threshold parameters within the range of [0,1].

In BAS, once $cf > cf_5$, the pheromone re-initialization procedure is performed according to S^{gb}:

$$\begin{cases} \tau_{is} = \tau_H & \texttt{if } is \in S^{gb} \\ \tau_{is} = \tau_L & \texttt{otherwise} \end{cases} \tag{4}$$

where τ_H and τ_L are two parameters satisfying $0 < \tau_L < \tau_H < 1$ and $\tau_L + \tau_H = 1$. This kind of pheromone re-initialization will focus more on the previous search experience rather than a total redo of the algorithm.

Table 1. Pheromone Update Strategy for BAS

	$cf < cf_1$	$cf \in [cf_1, cf_2)$	$cf \in [cf_2, cf_3)$	$cf \in [cf_3, cf_4)$	$cf \in [cf_4, cf_5)$
w_{ib}	1	2/3	1/3	0	0
w_{rb}	0	1/3	2/3	1	0
w_{gb}	0	0	0	0	1

3 Theoretical Analysis

3.1 Pheromone as Probability

Lemma 1. *For any pheromone value τ_{is} at any iteration step t in BAS, the following holds:*

$$0 < \tau_{is}(t) < 1 \tag{5}$$

Proof. From the pheromone update procedure described in the previous section, obviously we have $\tau_{is}(t) > 0$. And because $\sum_{x \in Supd} w_x = 1$, we can calculate the upper limit for any particular path according to equation (2):

$$\begin{aligned} \tau_{is}^{max}(t) &\leq (1-\rho)\tau_{is}^{max}(t-1) + \rho \\ &\leq (1-\rho)^t \tau_{is}(0) + \textstyle\sum_{i=1}^{t} (1-\rho)^{i-1} \rho \\ &= (1-\rho)^t \tau_{is}(0) + 1 - (1-\rho)^t \end{aligned} \tag{6}$$

Since BAS set the initial pheromone value as $0 < \tau_{is}(0) < 1$, we have $\tau_{is}^{max}(t) < 1$ according to the final sum of equation (6).

Theorem 1. *The pheromone values in BAS can be regarded as selecting probabilities throughout the iterations.*

Proof. Initially, since all the pheromone are set to 0.5, it obviously satisfies the statement of the theorem. For the following iterations, what we need to do is to prove $\tau_{i0}(t) + \tau_{i1}(t) = 1$, $0 < \tau_{is}(t) < 1$ holds for every variable x_i under the condition of $\tau_{i0}(t-1) + \tau_{i1}(t-1) = 1$.

From Lemma 1, we can see that $0 < \tau_{is}(t) < 1$ holds for any pheromone. After the pheromone update procedure, all the pheromone values are evaporated, and there must be one and only one of τ_{i0} and τ_{i1} associated with any $x \in S_{upd}$ that receives pheromone intensification, therefore, for any pheromone pair τ_{i0} and τ_{i1}, we have:

$$\begin{aligned} \tau_{i0}(t) + \tau_{i1}(t) &= (1-\rho)\tau_{i0}(t-1) + \rho \sum_{x \in S_{upd} | i0 \in x} w_x \\ &\quad + (1-\rho)\tau_{i1}(t-1) + \rho \sum_{x \in S_{upd} | i1 \in x} w_x \\ &= (1-\rho)(\tau_{i0}(t-1) + \tau_{i1}(t-1)) + \rho \sum_{x \in S_{upd}} w_x \\ &= 1 - \rho + \rho = 1 \end{aligned}$$

3.2 Relation with PBIL

It is interesting that BAS, which is developed from ACO, is much similar to PBIL[9], which is developed from another successful meta-heuristic, the Genetic Algorithm. The main reason lies probably in the fact that both BAS and PBIL incorporate the same reinforcement learning rule represented as equation (2).

The pheromone as selecting probabilities in BAS seems identical to the probability vector in PBIL, but BAS focuses more on the pheromone monitor and is controlled by additional complicated pheromone maintenance methods to guide the search to the direction for a quick convergence. While PBIL only deals with binary function optimization problem, BAS also applies to constrained combinatorial optimization problems.

4 Experimental Results

4.1 Function Optimization Problem

Normally, the function optimization problem can be described as:

$$\begin{array}{ll} \min & f(y), \quad y = [y_1, \cdots, y_v] \\ \text{s.t.} & a_j \leq y_j \leq b_j, \quad j = 1, \cdots, v \end{array} \tag{7}$$

where f is the object function and v is the number of variable.

In BAS, each variable y_j is coded into a binary string $[x_{j1}, \cdots, x_{jd}]$ in BCD code, where d is the coding dimension for every variable. The final solution representation $x = [x_{11}, \cdots, x_{1d}, x_{21}, \cdots, x_{2d}, \cdots, x_{v1}, \cdots, x_{vd}]$ is the combination of all the variables y_j in series, so we have $n = vd$ as the total dimension of the binary representation to the function optimization problem.

Repair Operator. The purpose of the repair operator for the function optimization problem is to decode the binary solution string $x = [x_1, \cdots, x_n]$ into the real variables and make sure that each variable y_j falls into the constrained region $[a_j, b_j]$. The process can be described as:

$$\begin{array}{l} y'_j = \sum_{i=1+d(j-1)}^{dj} x_i 2^{i-1}, \quad j = 1, \cdots, v \\ y_j = \frac{y'_j}{2^d}(b_j - a_j) + a_j \end{array} \tag{8}$$

Local Search. A one-flip local search is applied to S^{ib} and S^{gb}, it checks every bit by flipping the value from 0 to 1 or from 1 to 0, to see whether the resulting solution is better than the original one. If it is improved, the solution is updated, otherwise, the original solution is kept for further flips.

Comparison with Other Ant Systems. For all the tests, we use general parameter settings as: $d = 12$, $n_a = 5$, $\tau_0 = 0.5$, $\tau_H = 0.65$, $\rho_0 = 0.3$, $cf_i = [0.2, 0.3, 0.4, 0.5, 0.9]$, and the algorithm stops until the total function evaluation number exceed 10000 or the search is considered success by satisfying the following condition:

$$|f^* - f_{known\_best}| < \epsilon_1 \cdot |f_{known\_best}| + \epsilon_2 \tag{9}$$

where f^* is the optimum found by BAS, f_{known_best} is the known global optimum. ϵ_1 and ϵ_2 are accuracy parameters, which is set to be: $\epsilon_1 = \epsilon_2 = 10^{-4}$.

Table 2 reports the success rate and average number of function evaluations on different benchmark problems over 100 runs. It is clear that BAS finds the best_known solutions every time for all the benchmarks. Meanwhile, considering the average number of function evaluation, BAS is also very competitive.

Table 2. Comparison of BAS with CACO[10], API[11], CIAC [12], and ACO[13]

f	CACO		API		CIAC		ACO		BAS	
	% ok	evals	% ok	evals	% ok	evals	% ok	evals	% ok	evals
$R2$	100	6842		[10000]	100	11797		2905	100	5505.4
SM	100	22050		[10000]	100	50000		695	100	74.37
GP	100	5330			56	23391		364	100	1255.59
MG	100	1688			20	11751			100	2723.36
St		[6000]			94	28201			100	1044.66
Gr_5				[10000]	63	48402			100	1623
Gr_{10}	100	50000			52	50121			100	1718.43

4.2 Multidimensional Knapsack Problem

The multidimensional knapsack problem (MKP) is a well-known NP-hard combinatorial optimization problem, which can be formulated as:

$$\text{maximize} \sum_{j=1}^{n} p_j x_j \tag{10}$$

$$\text{subject to} \sum_{j=1}^{n} r_{ij} x_j \leq b_i, \quad i = 1, ..., m, \tag{11}$$

$$x_j \in \{0, 1\}, \quad j = 1, ..., n. \tag{12}$$

Repair Operator. In BAS, a repair operator is incorporated to guarantee feasible solutions. The idea comes from Chu and Beasley [14], which is based on the pseudo-utility ratio calculated accodring to the surrogate rate. The general idea of this approach is described very briefly as follows.

The surrogate relaxation problem of the MKP can be defined as:

$$\text{maximize} \sum_{j=1}^{n} p_j x_j \tag{13}$$

$$\text{subject to} \sum_{j=1}^{n} \left(\sum_{i=1}^{m} \omega_i r_{ij}\right) x_j \leq \sum_{i=1}^{m} \omega_i b_i \tag{14}$$

$$x_j \in \{0, 1\}, \quad j = 1, 2, ..., n \tag{15}$$

where $\omega = \{\omega_1, ..., \omega_m\}$ is a set of surrogate multipliers (or weights) of some positive real numbers. We obtain these weights by a simple method suggested by Chu and Beasley [14], in which we solve the LP relaxation of the original MKP and use the values of the dual variables as the weights. The weight ω_i can be seen as the shadow price of the ith constraint in the LP relaxation of the MKP.

Table 3. The results of BAS_MKP on 5.100 instances. For each instance, the table reports the best known solutions from OR-library, the best and average solutions found by Leguizamon and Michalewicz[15], the best solution found by Fidanova [16], the best and average solutions found by Alaya et.al. [17], and the results from BAS_MKP, including the best and average solutions over 30 runs for each instance.

$N°$	Best Known	L.&M.		Fidanova	Alaya et.al.		BAS_MKP	
		Best	Avg.	Best	Best	Avg.	Best	Avg.
00	24381	24381	24331	23984	24381	24342	24381	24380.7
01	24274	24274	24245	24145	24274	24247	24274	24270.7
02	23551	23551	23527	23523	23551	23529	23551	23539.7
03	23534	23527	23463	22874	23534	23462	23534	23524.1
04	23991	23991	23949	23751	23991	23946	23991	23978.5
05	24613	24613	24563	24601	24613	24587	24613	24613
06	25591	25591	25504	25293	25591	25512	25591	25591
07	23410	23410	23361	23204	23410	23371	23410	23410
08	24216	24204	24173	23762	24216	24172	24216	24205.4
09	24411	24411	24326	24255	24411	24356	24411	24405.5
10	42757			42705	42757	42704	42757	42736.2
11	42545			42445	42510	42456	42545	42498.9
12	41968			41581	41967	41934	41968	41966.5
13	45090			44911	45071	45056	45090	42074.8
14	42218			42025	42218	42194	42198	42198
15	42927			42671	42927	42911	42927	42927
16	42009			41776	42009	41977	42009	42009
17	45020			44671	45010	44971	45020	45016.5
18	43441			43122	43441	43356	43441	43408.8
19	44554			44471	44554	44506	44554	44554
20	59822			59798	59822	59821	59822	59822
21	62081			61821	62081	62010	62081	62010.4
22	59802			59694	59802	59759	59802	59772.7
23	60479			60479	60479	60428	60479	60471.8
24	61091			60954	61091	61072	61091	61074.2
25	58959			58695	58959	58945	58959	58959
26	61538			61406	61538	61514	61538	61522.5
27	61520			61520	61520	61492	61520	61505.2
28	59453			59121	59453	59436	59453	59453
29	59965			59864	59965	59958	59965	59961.7

The pseudo-utility ratio for each variable, based on the surrogate constraint coefficient, is defined as:

$$u_j = \frac{p_j}{\sum_{i=1}^{m} \omega_i r_{ij}} \quad (16)$$

The repair operator consists of two phases. The first phase examines each bit of the solution string in increasing order of u_j and changes the bit from one to zero if feasibility is violated. The second phase reverses the process by examining each bit in decreasing order of u_j and changes the bit from zero to one as long as feasibility is not violated.

Local Search. A random-4-flip method is designed as the local search for S^{ib} and S^{gb}, it randomly selects 4 bits to flip, then repairs the solution if necessary, and checks the result. If the resulting solution is better, then update the solution, otherwise, keeps the original solution. This kind of flips are performed for a certain number of $10n$ to S^{ib} and S^{gb} at each iteration, where n is the problem dimension.

Comparison with Other Ant System. Table 3 displays the comparison results of 5.100 from OR library. The parameters for all the tests are: $n_a = 20$, $\tau_0 = 0.5$, $\tau_H = 0.55$, $\rho_0 = 0.3$, $cf_i = 0.3, 0.5, 0.7, 0.9, 0.95$, and the algorithm stops when 2000 iterations are performed. On these instances, BAS_MKP outperforms all the other three algorithms in the results of the average solution found. Actually, BAS_MKP finds 29 best solutions out of the 30 instances tested.

5 Conclusions

This paper presented BAS, a binary version of hyper-cube frame of ACO to handle constrained binary optimization problems. In the proposed version of the system, pheromone trails are put on the selections of 0 and 1 for each bit of the solution string, and they directly represent the probability of selection. Experimental results show that BAS works well on binary function optimization problem and performs excellently in multidimensional knapsack problem. The results reported in the previous experimental sections demonstrate that BAS is capable of solving these various problems very rapidly and effectively.

References

1. Dorigo, M., Maniezzo, V., Colorni, A.: Positive feedback as a search strategy. Technical report, Dipartimento di Elettronica e Informatica, Politecnico di Milano, IT (1991)
2. Dorigo, M., Maniezzo, V., Colorni, A.: Ant System: Optimization by a colony of cooperating agents. IEEE Transactions on Systems, Man and Cybernetics, Part B **26**(1) (1996) 29–41
3. Gambardella, L., Taillard, E., Agazzi, G.: Macs-vrptw: a multiple ant colony system for vehicle routing problems with time windows. In Corne, D., Dorgo, M., Glover, F., eds.: New Ideas in Optimization. McGraw-Hill Ltd. (1999) 63–76

4. Gambardella, L., Taillard, E., Dorigo, M.: Ant colonies for the quadratic assignment problem. Journal of the Operational Research Society **50**(2) (1999) 167–176
5. Gutjahr, W.: A graph-based ant system and its convergence. Future Generation Computer Systems **16**(8) (2000) 873–888
6. Stützle, T., Dorigo, M.: A short convergence proof for a class of ant colony optimization algorithms. IEEE Transactions on Evolutionary Computation **6**(4) (2002) 358–365
7. Kong, M., Tian, P.: A convergence proof for the ant colony optimization algorithms. In: 2005 International Conference on Artificial Intelligence, ICAI2005. (2005) 27–30
8. Blum, C., Dorigo, M.: The hyper-cube framework for ant colony optimization. IEEE Transactions on Systems, Man and Cybernetics, Part B **34**(2) (2004) 1161–1172
9. Baluja, S., Caruana, R.: Removing the genetics from the standard genetic algorithm. In Prieditis, A., Russel, S., eds.: The International Conference on Machine Learning 1995, Morgan-Kaufmann Publishers (1995) 38–46
10. Bilchev, G., Parmee, I.: Constrained optimization with an ant colony search model. In: 2nd International Conference on Adaptive Computing in Engineering Design and Control. (1996) 26–28
11. Monmarche, N., Venturini, G., Slimane, M.: On how pachycondyla apicalis ants suggest a new search algorithm. Future Generation Computer Systems **16**(8) (2000) 937–946
12. Dreo, J., Siarry, P.: Continuous interacting ant colony algorithm based on dense heterarchy. Future Generation Computer Systems **20**(5) (2004) 841–856
13. Socha, K.: Aco for continuous and mixed variable optimization. In Dorigo, M., Birattari, M., Blum, C., eds.: ANTS 2004. LNCS 3172, Springer (2004) 25–36
14. Chu, P., Beasley, J.: A genetic algorithm for the multidimentional knapsack problem. Journal of Heuristics **4**(1) (1998) 63–86
15. Leguizamon, G., Michalewicz, Z.: A new version of ant system for subset problems. In: Congress on Evolutionary Computation. (1999) 1459–1464
16. Fidanova, S.: Evolutionary algorithm for multidimensional knapsack problem. In: the Seventh International Conference on Parallel Problem Solving from Nature (PPSNVII) Workshop. (2002)
17. Alaya, I., Solnon, C., Ghedira, K.: Ant algorithm for the multi-dimensional knapsack problem. In: International Conference on Bioinspired Optimization Methods and their Applications (BIOMA 2004). (2004) 63–72

Kernelization as Heuristic Structure for the Vertex Cover Problem

Stephen Gilmour and Mark Dras

Department of Computing, Macquarie University, Sydney, Australia
{gilmour, madras}@ics.mq.edu.au

Abstract. For solving combinatorial optimisation problems, exact methods accurately exploit the structure of the problem but are tractable only up to a certain size; approximation or heuristic methods are tractable for very large problems but may possibly be led into a bad solution. A question that arises is, From where can we obtain knowledge of the problem structure via exact methods that can be exploited on large-scale problems by heuristic methods? We present a framework that allows the exploitation of existing techniques and resources to integrate such structural knowledge into the Ant Colony System metaheuristic, where the structure is determined through the notion of kernelization from the field of parameterized complexity. We give experimental results using vertex cover as the problem instance, and show that knowledge of this type of structure improves performance beyond previously defined ACS algorithms.

1 Introduction

For solving combinatorial optimisation problems, exact methods accurately exploit the structure of the problem but are tractable only up to a certain size; approximation or heuristic methods are tractable for very large problems but may possibly be led into a bad solution. A third approach could be to combine heuristics and exact methods, which would hopefully still run quickly but the quality of solution would be improved over just regular heuristics. Some examples of combining heuristics with exact methods are discussed in [1]. In the work discussed in this paper, we investigate the use of an already well established body of techniques from the field of parameterized complexity [2,3] for identifying problem structure as part of an exact solution, the extent to which these techniques can be integrated into heuristics, and what advantage this gives over a standard heuristic approach.

In section 2 of this paper, we discuss ant colony system (ACS) for the vertex cover problem. In section 3 we discuss parameterized complexity and kernelization. In section 4 we present our framework for integrating ACS with kernelization. In section 5 we give experimental results for our new algorithms and in section 6 we conclude.

2 ACS for the Minimum Vertex Cover Problem

In this section we will present an algorithm by Gilmour and Dras [4] for ant colony system on the vertex cover problem (VCP). Given the graph $G = (V, E)$,

M. Dorigo et al. (Eds.): ANTS 2006, LNCS 4150, pp. 452–459, 2006.

the minimum VCP is the problem of finding a set of nodes $V' \in V$ such that every edge E is covered by a node from V' and such that the number of nodes in V' is minimised. Our ant colony system algorithm is based upon the ACS algorithm for the TSP [5] and the algorithm by Shyu, Yin, & Lin [6] for the weighted vertex cover problem. In section 2.1 we discuss the problem representation as defined by Shyu *et al.* [6] for ACS on the vertex cover problem. In sections 2.2 and 2.3 we present our adapted random proportional transition rule and pheromone update rules.

2.1 Problem Representation

Shyu *et al.* note that, unlike the Traveling Salesman Problem (TSP) for which the first ACS algorithm was designed, a VCP solution does not constitute a path in the graph. Thus in order to allow our algorithm to find unordered subsets of nodes, we construct a complete graph $G_c = (V, E_c)$. This will "guarantee that there always exists a path in G, a sequence of unrepeated adjacent vertices, which covers exactly and only the vertices in V'" [6]. But we want to solve the minimum vertex cover problem for G and not G_c and therefore we need to preserve the details of the original graph within this new representation. Therefore we define for each ant k a binary connectivity function $\psi_k : E_c \rightarrow \{0, 1\}$ as

$$\psi_k(i,j) = \begin{cases} 1 \text{ if edge } (i,j) \in E; \\ 0 \text{ if edge } (i,j) \in E_c - E, \end{cases} \tag{1}$$

2.2 Random Proportional Transition Rule

Our rule for deciding which node j ant k should place in its vertex cover construction next is:

$$j = \begin{cases} \arg\max_{u \in J^k} \{[\tau_u(t)] \cdot [\eta_u]^\beta\} & \text{if } q \leq q_0; \\ J & \text{if } q > q_0, \end{cases} \tag{2}$$

where q is randomly selected from the distribution $[0, 1]$; q_0 is a tunable parameter such that $0 \leq q_0 \leq 1$; $\tau_u(t)$ is the amount of pheromone on node u; $\eta_u = \sum_{z \in N(u)} \psi_k(u, z)$ is the heuristic goodness of node u; J^k is the set of nodes that ant k may still visit; and $J \in J^k$ is a node that is randomly selected according to the probability:

$$p_{J^k}(t) = \frac{[\tau_J(t)] \cdot [\eta_J]^\beta}{\sum_{l \in J^k} [\tau_l(t)] \cdot [\eta_l]^\beta} \tag{3}$$

After an ant visits a node which it has just placed in its candidate solution, it sets $\psi_k(i, j) = 0$ for every edge (i, j) connected to the node j that it has just visited. This allows the ant to keep track of which edges have been covered. A solution is constructed when all edges are covered. But, before the next cycle can continue, the connectivity values need to be reset according to equation (1).

2.3 Pheromone System

For the vertex cover problem, pheromone is placed on nodes since we are interested in constructing an unordered subset of all the nodes within the graph. Therefore, the global and local pheromone update rules need to be updated to work with nodes. Our global pheromone update rule, which is executed by one ant at the end of each cycle, is defined as:

$$\tau_i(t) \leftarrow (1-\rho) \cdot \tau_i(t) + \rho \cdot \Delta\tau_i(t) \quad (4)$$

where i are the nodes belonging to the current best solution T^+; $\Delta\tau_i(t) = 1/L^+$ such that L^+ is the size of the solution T^+; and ρ is a parameter governing pheromone decay such that $0 < \rho < 1$.

Similarly, our local pheromone update rule, which is executed by every ant on each node as it is placed in the candidate solution, is defined as:

$$\tau_i = (1-\varphi)\tau_i + \varphi\tau_0 \quad (5)$$

where $\varphi \in (0,1)$ is a parameter which simulates the evaporation rate of pheromone; and τ_0 is the amount of pheromone every edge is initially set to before this algorithm starts.

Similar to the TSP, our formulation for τ_0 is $\tau_0 = \frac{1}{(n \cdot L_{nn})}$ where L_{nn} is the size of a solution produced by a simple greedy heuristic.

3 Parameterized Complexity

An overview of parameterized complexity is available in [2,3]; we give a brief outline here. In section 2 we defined what we will now call the general minimum vertex cover problem. A related problem is the k-vertex cover problem: given a graph $G = (V, E)$ and a parameter k, the k-vertex cover problem is the problem of finding a vertex cover of size less than or equal to k. This problem is still NP-complete. A key idea of parameterized complexity is that if some value of a problem is known to be bounded in a certain context, useful algorithms with good complexity properties can be developed. For the k-vertex cover problem, k might be bounded by the maximum size of the vertex cover that we are trying to find; in such a case, there is a best-case algorithm with complexity $O(kn + 1.2852^k)$ [7] that is tractable for k up to 400. Parameterized complexity also allows a more fine-grained analysis of problems classified as NP-complete: some are amenable to this treatment, while others have only brute-force solutions of complexity $O(n^{k+1})$. Those that are amenable to this approach are called fixed-parameter tractable (FPT).

Parameterized complexity contains both a framework of complexity analysis and a corresponding toolkit of algorithm design. One such tool for algorithm design is kernelization. The idea behind kernelization is reducing a problem in polynomial time to its problem kernel such that ideally, even a brute-force attack is an option; however, usually an approach such as bounded search trees is necessary for difficult problems. Kernelization is the core idea behind the successful algorithms for the vertex cover problem.

Many different optimization problems [3] have been analyzed using the notions of parameterized complexity, for example the dominating set problem, traveling salesman problem, and the 3-CNF satisfiability problem, often with several proposed algorithms for each problem. There is thus a wide range of tools available to be used. The aim of this paper is to see whether, and in what way, these can be combined with ACO as a kind of template in the context of the vertex cover problem.

4 Ant Colony System with Structure

The key idea in this paper is that kernelization can be used to give ACS information about the problem. We will present six variant algorithms for combining these. Within the algorithms that we propose, we will be utilizing just one kernelization rule. The rule we have chosen to use is: "If G has adjacent vertices u and v such that $N(v) \subseteq N[u]$, then replace (G, k) with $(G - u, k - 1)$ and place u in the vertex cover" [2]; here $N(v)$ denotes the set of vertices that form the neighbourhood of v and $N[v]$ denotes $N(v) \cup \{v\}$. See [8] for a more detailed discussion on how we came to use this rule.

Kernelized Ant Colony System. This algorithm performs kernelization within the initialisation stage of the algorithm and then runs regular ACS on the resulting kernel graph. We define a set χ that contains all the nodes that are identified by kernelization as belonging to an exact solution. Within the kernelization phase we remove from the graph all the nodes that belong to χ and all edges connected to nodes in χ.

PreKernelized Ant Colony System. This algorithm works similarly to Kernelized ACS except rather than removing nodes from the graph, it sets the pheromone on the selected nodes to be τ_{kern}. This will initially make the nodes in the kernelization set more attractive to ants than any other nodes in the problem.

CycleKernelized Ant Colony System. This algorithm is similar to PreKernelized ACS except that the kernelization information is continually reinforced in pheromone. Therefore we have moved the kernelization component out of the initialisation phase and into the global pheromone update rule. As well as placing pheromone on the current best solution, the ant selected to perform the global pheromone update rule also reinforces the pheromone on the nodes in χ.

KernelAnts Ant Colony System. ACS uses m ants to generate solutions to a problem. However, KernelAnts ACS uses k additional ants to kernelize the graph and place pheromone on the nodes identified through kernelization whilst the original m ants continue to perform regular ant colony system. Each turn, these kernelants set the pheromone on one node each to τ_{kern}; this node is selected by choosing the node in χ with the smallest pheromone value. This occurs in parallel with the regular ants continuing to construct solutions to the problem influenced by the pheromone on the graph.

TransKernelized Ant Colony System. Within this algorithm, we have incorporated the kernelization into the ants' random proportional transition rule. When an ant draws a random number between zero and one that is less than the threshold q_0, the ant will first look in the kernelization set χ to see if there is a node to visit before picking the best of all possible options as with regular ACS. Should the random number be above the threshold, the ant assigns to each potential node a probability and decides where to go next probabilistically, as with regular ACS.

Neighbourhood TransKernelized Ant Colony System. One problem with TransKernelized ACS is that it can involve a lot of kernelization on the fly. Neighbourhood TransKernelized ACS is an alternative algorithm that picks a node using the regular ACS random proportional transition rule (see equation (2)). However, if $q \leq q_0$, this algorithm then tests all the neighbours of node j to ensure that none of them belong to the kernelization set χ and therefore make a better choice. Since either j is in the vertex cover or all of its neighbours are, it is safe to include j into our vertex cover should none of its neighbours be in the kernelization set.

5 Evaluation

5.1 Parameter Investigation

We generated a set of 160 graphs with number of nodes ranging from 100 to 800 and number of edges ranging from 150 to 4000 and performed parameter analysis on ant colony system and our six new algorithms for the vertex cover problem. We timed how long ant colony system took to complete 200[1] iterations on each graph and that was the amount of time given to each algorithm during parameter analysis. We then explored one parameter at a time; table 1 contains the parameters found to be good.

5.2 Challenging Benchmarks

Benchmarks with Hidden Optimum Solutions for Graph Problems[2] is a website with a collection of challenging instances of graph problems constructed by hiding optimum solutions for a specific problem in hard graphs [9]. This website contains forty instances for the VCP of between 450 and 1534 nodes.

We initially timed how long it took ant colony system to run on each graph for 200 iterations. We then set each algorithm to run on each graph for that

[1] We chose 200 iterations because the Mann-Whitney statistical test has shown statistical improvement between 50, 100, 150, and 200 iterations but no improvement between 200 and 250 iterations.

[2] http://www.nlsde.buaa.edu.cn/~kexu/benchmarks/graph-benchmarks.htm

Table 1. Good parameters for ant colony system and our six new algorithms. These parameters are: number of ants m; number of kernelants k; influence of heuristic information β; pheromone trail evaporation ρ; probability of including the best choice in tour construction q_0; and quantity of pheromone to drop on kernelized nodes τ_{kern}.

	ACS	KACS	PKACS	CKACS	KAACS	TKACS	NTKACS
m	50	40	40	40	40	10	40
k	-	-	-	-	10	-	-
β	4	5	5	4	4	5	5
ρ	0.1	0.1	0.1	0.1	0.1	0.2	0.1
q_0	0.9	0.9	0.9	0.9	0.9	0.999	0.9
τ_{kern}	-	-	0.5	0.5	0.95	-	-

Table 2. The sum and average of all results for Shyu *et al.*'s algorithm, ACS, the optimal solution, and our six new algorithms, on benchmark instances

	SYL	ACS	KACS	PKACS	CKACS	KAACS	TKACS	NTKACS	Opt
sum	39219	39177	39110	39116	39035	39037	39077	39092	38690
average	980.475	979.425	977.75	977.9	975.875	975.925	976.925	977.3	967.25

period of time. Table 2 contains the sum and average of the results for each algorithm. We have adopted the approach of Birattari [10], of using a maximum number of instances possible with just one run per instance for all experimentation. See [8] for a more detailed discussion of all experiments and results.

We applied the Mann-Whitney U-test—recommended for use in metaheuristic analysis [11]—to these results and made the following conclusions. Firstly, all algorithms including our ACS algorithm were statistically significant improvements over the algorithm by Shyu *et al.*. Secondly, all kernelization algorithms were a statistically significant improvement over regular ACS. Thirdly, there was no statistical difference between CycleKernelized ACS and KernelAnts ACS but they were statistically better than all other algorithms. Lastly, Kernelized ACS, PreKernelized ACS, TransKernelized ACS, and Neighbourhood TransKernelized ACS all performed roughly the same; there is only a small statistical preference for TransKernelized ACS. The primary conclusion from these results is that the kernelized algorithms do outperform regular ACS algorithms.

5.3 Random Graphs

We constructed two groups of graphs of 500 nodes each. The first group of graphs contains graphs with 100 to 500 nodes, the second 600 to 1000 nodes. Each algorithm ran on each graph for the quantity of time required for ACS to perform 200 iterations on that graph. All random graphs were generated using the algorithm proposed by Skiena [12] and selected to contain a variety of parameters for number of nodes, number of edges, and kernel sizes. Again only one run per graph was performed.

Table 3. The sum and average of all results for Shyu *et al.*'s algorithm, ACS, and our six new algorithms, on random graphs with 100 to 500 nodes

	SYL	ACS	KACS	PKACS	CKACS	KAACS	TKACS	NTKACS
sum	99229	95486	95286	95366	95354	95273	95010	94967
average	198.458	190.972	190.572	190.732	190.708	190.546	190.02	189.934

Table 4. The sum and average of all results for Shyu *et al.*'s algorithm, ACS, and our six new algorithms, on random graphs of 600 to 1000 nodes

	SYL	ACS	KACS	PKACS	CKACS	KAACS	TKACS	NTKACS
sum	264700	255329	254407	254703	254785	254517	253671	253272
average	529.4	510.658	508.814	509.406	509.57	509.034	507.342	506.544

Table 3 contains the sum and average of the results for each algorithm for the first group of graphs. We applied the Mann-Whitney U-test to these results and made the following conclusions. Firstly, all algorithms are a statistical improvement over the algorithm by Shyu *et al.*. Secondly, all kernelization algorithms except PreKernelized ACS are a statistical improvement over regular ACS. Thirdly, TransKernelized ACS and Neighbourhood TransKernelized ACS perform statistically speaking roughly the same, and these two algorithms are a statistical improvement over all other algorithms.

Table 4 contains the sum and average of the results for each algorithm for the second set of graphs. We applied the Mann-Whitney U-test to these results and made the following conclusions. Firstly, all algorithms were a statistical improvement over the algorithm by Shyu *et al.*. Similarly, all kernelization algorithms were a statistical improvement over regular ACS. Secondly, Neighbourhood TransKernelized ACS was an improvement over all other algorithms and TransKernelized ACS a clear second.

6 Conclusion

Our overall conclusion is that kernelization rules from the field of parameterized complexity are a useful and extensive resource for combination with ACO. Specifically, we have found that our six kernelization algorithms are useful for getting better results for both our benchmark problems and random graphs. In the larger, harder benchmark problems, it was found that pheromone based kernelization algorithms performed the best. This is probably because pheromone based algorithms consume less CPU time and so more iterations of ACS can be performed which is benefical for these hard problems. However, the algorithms with kernelization integrated into the random proportional transition rule work better on the random graphs; probably because the random graphs are not quite as hard and so more time can be used performing kernelization. We have further identified a structure through this work that is common enough in both our benchmark problems and our random graphs to significantly affect quality of solutions, and that ant colony system is poor at solving.

There are two broad avenues for future work. Firstly, further experimentation of this kind on the vertex cover problem using different kernelization rules would be useful for getting greater insight into what structures ant colony system is weak at solving. Investigation into why this is the case could also prove fruitful. Secondly, we plan to extend the approach to other optimization problems.

References

1. Blum, C.: Ant colony optimization: Introduction and recent trends. Physics of Life Reviews **2** (2005) 353–373
2. Downey, R., Fellows, M., Stege, U.: Parameterized complexity: A framework for systematically confronting computational intractability. DIMACS Series in Discrete Mathematics and Theoretical Computer Science (1997)
3. Downey, R., Fellows, M.: Parameterized Complexity. Monographs in Computer Science. Springer-Verlag (1998)
4. Gilmour, S., Dras, M.: Understanding the Pheromone System within Ant Colony Optimization. Australian Conference on Artificial Intelligence (2005) 786–789
5. Dorigo, M., Stützle, T.: Ant Colony Optimization. A Bradford Book. MIT Press (2004)
6. Shyu, S.J., Yin, P.Y., Lin, B.M.T.: An Ant Colony Optimization Algorithm for the Minimum Weight Vertex Cover Problem. Annals of Operational Research **131** (2004) 283–304
7. Chen, J., Kanj, I., Jia, W.: Vertex cover: Further observations and further improvements. Journal of Algorithms **41** (2001) 280–301
8. Gilmour, S., Dras, M.: Exactness as Heuristic Structure for Guiding Ant Colony System. Technical report, Department of Computing, Macquarie University, Australia (2006)
9. Xu, K., Boussemart, F., Hemery, F., Lecoutre, C.: A simple model to generate hard satisfiable instances. Proceedings of 19th International Joint Conference on Artificial Intelligence (IJCAI) (2005) 337–342
10. Birattari, M.: On the Estimation of the Expected Performance of a Metaheuristic on a Class of Instances: How many instances, how many runs? IRIDIA Technical Report No. TR/IRIDIA/2004-001 (2005)
11. Taillard, E.D.: Comparison of non-deterministic iterative methods. MIC'2001 - 4th Metaheuristic International Conference (2001) 272–276
12. Skiena, S.S.: The algorithm design manual. Springer-Verlag New York, Inc., New York, NY, USA (1998)

Minimizing Total Earliness and Tardiness Penalties with a Common Due Date on a Single-Machine Using a Discrete Particle Swarm Optimization Algorithm

Quan-Ke Pan[1], M. Fatih Tasgetiren[2], and Yun-Chia Liang[3]

[1] College of Computer Science, Liaocheng University, Liaocheng, China
qkpan@lctu.edu.cn
[2] Department of Management, Fatih University, Istanbul, Turkey
ftasgetiren@fatih.edu.tr
[3] Department of Industrial Engineering and Management
Yuan Ze University, Taiwan
ycliang@saturn.yzu.edu.tw

Abstract. In this paper, a discrete particle swarm optimization (DPSO) algorithm is presented to solve the single machine total earliness and tardiness penalties with a common due date. A modified version of HRM heuristic presented by Hino et al. in [1], here we call it MHRM, is also presented to solve the problem. In addition, the DPSO algorithm is hybridized with the iterated local search (ILS) algorithm to further improve the solution quality. The performance of the proposed DPSO algorithm is tested on 280 benchmark instances ranging from 10 to 1000 jobs from the OR Library. The computational experiments showed that the proposed DPSO algorithm has generated better results, in terms of both percentage relative deviations from the upper bounds in Biskup and Feldmann [2] and computational time, than Hino et al. [1].

1 Introduction

In a single machine scheduling problem with a common due date, n jobs are available to be processed at time zero. Each job has a processing time and a common due date. Preemption is not allowed and the objective is to sequence jobs such that the sum of weighted earliness and tardiness penalties is minimized. That is,

$$f(S) = \sum_{j=1}^{n} (\alpha_j E_j + \beta_j T_j) \ . \tag{1}$$

When the job j completes its operation before its due date, its earliness is given by $E_j = max(0, d - C_j)$, where C_j is the completion time of the job j. On the other hand, if the job finishes its operation after its due date, its tardiness is given by $T_j = max(0, C_j - d)$. Earliness and tardiness penalties of job j are given by parameters α_j and β_j, respectively. It is well-known that for the case of

M. Dorigo et al. (Eds.): ANTS 2006, LNCS 4150, pp. 460–467, 2006.

restrictive common due date with general penalties, there is an optimal schedule given the following properties:

1. No idle times are inserted between consecutive jobs [3].
2. The schedule is V-Shaped. In other words, jobs that are completed at or before the due date are sequenced in non-increasing order of the ratio p_j/α_j. On the other hand, jobs whose processing starts at or after the due date are sequenced in non-decreasing order of the ratio p_j/β_j. Note that there might be a straddling job, that is, a job that its processing is started before its due date and completed after its due date [2].
3. There is an optimal schedule in which either the processing of the first job starts at time zero or one job is completed at the due date [2].

The complexity of the restrictive common due date problem is proved to be NP-complete in the ordinary sense [4]. Feldmann and Biskup [5] applied different meta-heuristics such as evolutionary search (ES), simulated annealing (SA) and threshold accepting (TA). In addition, Hino et al. [1] most recently compared the performance of TS, GA, and their hybridization. PSO was first introduced to optimize continuous nonlinear functions by Eberhart and Kennedy [6]. Authors have successfully proposed a DPSO algorithm to solve the no-wait flowshop scheduling in [7]. Based on the experience above, this study aims at solving the single-machine total earliness and tardiness penalties with a common due date problem by the DPSO algorithm.

Section 2 introduces the modified MHRM heuristic. Section 3 provides the details of the proposed DPSO algorithm. The computational results on benchmark instances are discussed in Section 4. Finally, Section 5 summarizes the concluding remarks.

2 Modified MHRM Heuristic

Consistent with the HRM heuristic in [1], the MHRM heuristic consists of: (i) determining the early and tardy job sets, (ii) constructing a sequence for each set, and (iii) setting the final schedule S as the concatenation of both sequences. In order to ensure that S will satisfy properties (1) and (2), there will be no idle time between consecutive jobs, and the sequences of S^E and S^T will be V-shaped. The following notation consistent with Hino et al. [1] is used:

P : set of jobs to be allocated
g : idle time inserted at the beginning of the schedule
S^E : set of jobs completed on the due date or earlier
S^T : set of jobs completed after the due date
S : schedule representation $S = (g, S^E, S^T)$
e : candidate job for S^E
t : candidate job for S^T
E^e : distance between the possible completion time of job e and the due date

T^t : distance between the possible completion time of job t and the due date
d^T : time window available for inserting a job in set S^T
d^E : time window available for inserting a job in set S^E
p_j : the processing time of job j
H : total processing time, $H = \sum_{j=1}^{n} p_j$

The procedure of the modified $MHRM$ heuristic is summarized as follows:

Step 1: Let $P = 1, 2, , n; S^E = S^T = \phi$,$g = max\{0, d - H \times \sum_{j=1}^{n} \frac{\beta_j}{\alpha_j + \beta_j}\}$; $d^E = d - g$ and $d^T = g + H - d$.
Step 2: Set $e = arg\ max_{j \epsilon p}\{p_j/\alpha_j\}$and $t = arg\ max_{j \epsilon p}\{p_j/\beta_j\}$ (in case of a tie, select the job with the longest p_j).
Step3: Set $E^e = d^E - p_e$ and $T^t = d^T$.
If $E^e \leq 0$, then go to step 5. If $T^t - p_t \leq 0$, then go to step 6.
Step 4: Choose the job to be inserted:
If $E^e > T^t$, then $S^E = S^E + \{e\}$,$d^E = d^E - p_e$ and $P = P - \{e\}$.
If $E^e < T^t$, then $S^T = S^E T + \{t\}$,$d^T = d^T - p_t$ and $P = P - \{t\}$.
If $E^e = T^t$, then if $\alpha_e > \beta_t$ then $S^T = S^T + \{t\}$,$d^T = d^T - p_t$ and $P = P - \{t\}$; else $S^E = S^E + \{e\}$, $d^E = d^E - p_e$ and $P = P - \{e\}$. Go to step 7.
Step 5: Adjustment of the idle time (end of the space before the due date):
If $g + E_e < 0$, then $S^T = S^T + \{t\}$, $d^T = d^T - p_t$ and $P = P - \{t\}$,
else $S^{E'} = S^E$,$S^{T'} = S^T \cup P$,$g' = d - \sum_{j \epsilon S^{E'}} p_j$, $S' = (g', S^{E'}, S^{T'})$;
$S^{E''} = S^E + \{e\}$, $S^{T''} = S^T \cup P - \{e\}$, $g'' = d - \sum_{j \epsilon S^{E''}} p_j$,
$S'' = (g'', S^{E''}, S^{T''})$.
If $f(S^{E'}) \leq f(S^{E''})$, then $S^T = S^T + \{t\}$,$d^E = 0$, $d^T = d^T - p_t + g' - g$, $g = g'$ and $P = P - \{t\}$.
Else $S^E = S^E + \{e\}$,$d^E = 0$, $d^T = d^T + g'' - g$, $g = g''$ and $P = P - \{e\}$.
Go to step 7.
Step 6: Adjustment of the idle time (end of the space after the due date):
If $g < T^t$, then $S^E = S^E + \{e\}$, $d^E = d^E - p_e$ and $P = P - \{e\}$,
else $S^{T'} = S^T$,$S^{E'} = S^E \cup P$,$g' = d - \sum_{j \epsilon S^{E'}} p_j$, $S' = (g', S^{E'}, S^{T'})$;
$S^{T''} = S^T + \{t\}$, $S^{E''} = S^E \cup P - \{t\}$, $g'' = d - \sum_{j \epsilon S^{E''}} p_j$,
$S'' = (g'', S^{E''}, S^{T''})$.
If $f(S^{E'}) \leq f(S^{E''})$, then $S^E = S^E + \{e\}$,$d^T = 0$, $d^E = d^E - p_e + g' - g$, $g = g'$ and $P = P - \{e\}$;
else $S^T = S^T + \{t\}$, $d^T = 0$, $d^E = d^E + g - g''$, $g = g''$ and $P = P - \{t\}$.
Step 7: If $P \neq \phi$ then go to step 2.
Step 8: If there is a straddling job (it must be the last job in), then $S^{E'} = S^E$, $S^{T'} = S^T$, $g' = d - \sum_{j \epsilon S^{E'}} p_j$, $S' = (g', S^{E'}, S^{T'})$. If $f(S') \leq f(S)$ then $g = g'$.

The main difference between HRM and MHRM heuristics is due to the fact that the inserted idle time g is calculated in Step 1 such that :

$$g = max\{0, d - H \times \sum_{j=1}^{n} \frac{\beta_j}{\alpha_j + \beta_j}\} . \tag{2}$$

By doing so, the inserted idle time completely depends on the particular instance. It implies that if the total tardiness penalty of a particular instance is greater than the total earliness penalty of that instance, that is, $\sum \beta_j > \sum \alpha_j$, the inserted idle time would be larger for that particular instance. Hence more jobs would be completed before the due date. In other words, more jobs would be early. Since $\sum \beta_j > \sum \alpha_j$, the total penalty imposed on the fitness function would be less than the one used in the HRM heuristic. In addition, the following modification is made in Step 3. If the distance between the possible completion time of candidate job t and the due date is smaller or equal to zero, both the start time and the completion time of the job t will be before or at the due date, i.e., the job t is not a straddling job. In our algorithm, $T^t - p_t \leq 0$ is employed instead of $T^t \leq 0$ because $T^t - p_t \leq 0$ implies that the job t is a straddling job. In this case, the adjustment of the idle time for the end of the space after the due date through Step 6 should be made. Accordingly, necessary modifications are also made in Step 5, 6, and 8.

3 Discrete Particle Swarm Optimization Algorithm

It is obvious that standard PSO equations cannot be used to generate a discrete job permutation since position and velocity of particles are real-valued. Instead, Pan et al. [7] proposed a new method to update the position of particles as follows:

$$X_i^t = c_2 \oplus F_3(c_1 \oplus F_2(w \oplus F_1(X_i^{t-1}), P_i^{t-1}), G^{t-1}). \tag{3}$$

Given that λ_i^t and δ_i^t are temporary individuals, the update equation consists of three components: The first component is $\lambda_i^t = w \oplus F_1(X_i^{t-1})$, which represents the velocity of the particle. F_1 indicates the binary swap operator with the probability of w. In other words, a uniform random number r is generated between 0 and 1. If is less than w, then the swap operator is applied to generate a perturbed permutation of the particle by $\lambda_i^t = F_1(X_i^{t-1})$, otherwise current permutation is kept as $\lambda_i^t = X_i^{t-1}$. The second component is $\delta_i^t = c_1 \oplus F_2(\lambda_i^t, P_i^{t-1})$ where F_2 indicates the one-cut crossover operator with the probability of c_1. Note that λ_i^t and P_i^{t-1} will be the first and second parents for the crossover operator, respectively. It is resulted either in $\delta_i^t = F_2(\lambda_i^t, P_i^{t-1})$ or in $\delta_i^t = \lambda_i^t$ depending on the choice of a uniform random number. The third component is $X_i^t = c_2 \oplus F_3(\delta_i^t, G^{t-1})$ where F_3 indicates the two-cut crossover operator with the probability of c_2. It is resulted either in $X_i^t = F_3(\delta_i^t, G^{t-1})$ or in $X_i^t = \delta_i^t$ depending on the choice of a uniform random number. The pseudo code of the DPSO algorithm is given in Fig.1.

A binary solution representation is employed for the problem. The x_{ij}^t, the jth dimension of the particle X_i^t, denotes a job; if $x_{ij}^t = 0$, the job j is completed before or at the due date, which belongs to the set S^E; if $x_{ij}^t = 1$, the job j is finished after the due date, which belongs to the set S^T.

After applying the DPSO operators, the sets S^E and S^T are determined from the binary representation. Then every fitness calculation follows property (2).

Note that the set S^T might contain a straddling job. If there is a straddling job, the first job in the early job set is started in time zero. After completing the last job of the early job set, the straddling job and the jobs in the tardy job set are sequenced. On the other hand, if there is no straddling job, the completion time of the last job in the early job set is matched with the due date and the processing in the tardy job set is followed immediately.

```
Initialize parameters
Initialize population
Evaluate
Do{
  Find the personal best
  Find the global best
  Update position
  Evaluate
  Apply local search(optional)
}While (Not Termination)
```

Fig. 1. DPSO algorithm with a local search

```
s0 = G^t
s=LocalSearch(s0)
Do{
  s1=perturbation(s)
  s2=LocalSearch(s1)
  s=AcceptanceCriterion(s, s2)
}While (Not Termination)
 if f(s) < f(G^t) then G^t = s)
```

Fig. 2. Iterated local search algorithm

At the end of each iteration, the ILS algorithm is applied to the global best solution G^t to further improve the solution quality. The ILS algorithm in Fig.2 was based on the simple binary swap neighborhood. The perturbation strength was 5 binary swaps to avoid getting trapped at the local minima. In the *LocalSearch* procedure, the binary swap operator was used with the size of $min(6n, 600)$ and the size of the $do - while$ loop was 10. The binary swap operator consists of two steps: (i) generate two random integers, a and b, in the range $[1, n]$; (ii) if $x^t_{ia} = x^t_{ib}$, then $x^t_{ia} = (x^t_{ia} + 1) mod2$; else $x^t_{ia} = (x^t_{ia} + 1) mod2$ and $x^t_{ib} = (x^t_{ib} + 1) mod2$.

4 Computational Results

The DPSO algorithm is coded in Visual C++ and run on an Intel P IV 2.4 GHz PC with 256MB memory. Regarding the parameters of the DPSO algorithm, the crossover probabilities were taken as $c_1 = c_2 = 0.8$, respectively. The swap probability was set to $w = 0.95$. One of the solutions in the population is constructed with the MHRM heuristic, the rest is constructed randomly. The proposed DPSO algorithm was applied to the benchmark problems developed in Biskup and Feldmann [2]. 10 runs were carried out for each problem instance and the average percentage relative deviation was computed as follows:

$$\Delta_{avg} = \sum_{i=1}^{R} \left(\frac{(F_i - F_{ref}) \times 100}{F_{ref}} \right) / R \tag{4}$$

where F_i,F_{ref}, and R were the fitness function value generated by the DPSO algorithm in each run, the reference upper bounds generated by Biskup and Feldmann [2], and the total number of runs, respectively. The maximum number of iterations was fixed to 50 and the algorithm was terminated when the global best solution was not improved in 10 consecutive iterations. The computational results of the MHRM heuristic are given in Table 1 where the MHRM heuristic is superior to its counterpart HRM heuristic in terms of relative percent improvement.

Most recently, Hino et. al. [1] developed a TS, GA and hybridization of both of them denoted as HTG and HGT. They employed the same benchmark suite of Biskup and Feldmann [2]. Table 2 summarizes the computational results of the DPSO and those in Hino et al. [1]. As seen in Table 2, the DPSO algorithm outperforms all the metaheuristics of Hino et al. [1] in terms of the minimum percentage relative devia-tion since the largest improvement of -2.15 on overall mean is achieved. Besides the average performance of the DPSO algorithm, it is also interesting to note that even the worst performance of the DPSO algorithm, i.e., the maximum percentage relative deviation, was better than TS, GA, HGT and HTG algorithms of Hino et al. [1]. Regarding the CPU time requirement of the DPSO algorithm, the maximum CPU time until termination was not more than 1.33 seconds on overall average whereas Hino et al. [1] reported that their average CPU time requirement was 21.5 and 7.8 seconds for TS and hybrid

Table 1. Statistics for the MHRM Heuristic

	h	10	20	50	100	200	500	1000	**Mean**
	0.2	1.53	-3.97	-5.33	-6.02	-5.63	-6.32	-6.68	-4.50
HRM	0.4	8.68	0.46	-3.87	-4.42	-3.51	-3.46	-4.26	-1.48
	0.6	19.27	9.78	7.59	4.69	3.71	2.53	3.23	7.26
	0.8	22.97	13.52	8.10	4.70	3.71	2.53	3.23	8.39
	Mean	13.11	5.17	1.62	-0.26	-0.43	-1.18	-1.12	2.42
	h	10	20	50	100	200	500	1000	**Mean**
	0.2	1.00	-3.57	-5.45	-6.02	-5.62	-6.32	-6.69	**-4.67**
MHRM	0.4	5.91	-0.49	-4.03	-4.27	-3.52	-3.45	-4.27	**-2.02**
	0.6	2.77	2.02	1.51	1.50	1.71	1.41	1.55	**1.78**
	0.8	3.95	4.07	2.13	1.43	1.71	1.41	1.55	**2.32**
	Mean	**3.41**	**0.51**	**-1.46**	**-1.84**	**-1.43**	**-1.74**	**-1.97**	**-0.65**

strategies, respectively. To sum up, all the statistics show and prove that the DPSO algorithm was superior to all the metaheuristics presented in Hino et al. [1]. Note that the best results so far in the literature are reported in bold in Table 2.

Table 2. Statistics for the DPSO Algorithm

		DPSO				*TS*	*GA*	*HTG*	*HGT*
	h	Δ_{min}	Δ_{max}	Δ_{avg}	Δ_{std}	Δ_{min}	Δ_{min}	Δ_{min}	Δ_{min}
10	0.2	**0.00**	0.00	0.00	0.00	0.25	0.12	0.12	0.12
	0.4	**0.00**	0.00	0.00	0.00	0.24	0.19	0.19	0.19
	0.6	**0.00**	0.00	0.00	0.00	0.10	0.03	0.03	0.01
	0.8	**0.00**	0.00	0.00	0.00	**0.00**	**0.00**	**0.00**	**0.00**
20	0.2	**-3.84**	**-3.84**	**-3.84**	0.00	**-3.84**	**-3.84**	**-3.84**	**-3.84**
	0.4	**-1.63**	**-1.63**	**-1.63**	0.00	-1.62	-1.62	-1.62	-1.62
	0.6	**-0.72**	**-0.72**	**-0.72**	0.00	-0.71	-0.68	-0.71	-0.71
	0.8	**-0.41**	**-0.41**	**-0.41**	0.00	**-0.41**	-0.28	**-0.41**	**-0.41**
50	0.2	-5.68	-5.67	-5.68	0.01	**-5.70**	-5.68	**-5.70**	**-5.70**
	0.4	**-4.66**	-4.58	-4.64	0.03	**-4.66**	**-4.60**	**-4.66**	**-4.66**
	0.6	**-0.34**	**-0.34**	**-0.34**	0.00	-0.32	-0.31	-0.27	-0.31
	0.8	**-0.24**	**-0.24**	**-0.24**	0.00	**-0.24**	-0.19	-0.23	-0.23
100	0.2	**-6.19**	-6.16	-6.18	0.01	**-6.19**	-6.17	**-6.19**	**-6.19**
	0.4	**-4.94**	-4.88	-4.92	0.02	-4.93	-4.91	-4.93	-4.93
	0.6	**-0.15**	**-0.15**	**-0.15**	0.00	-0.01	-0.12	0.08	0.04
	0.8	-0.18	**-0.18**	**-0.18**	0.00	-0.15	-0.12	-0.08	-0.11
200	0.2	**-5.78**	-5.73	-5.76	0.02	-5.76	-5.74	-5.76	-5.76
	0.4	-3.74	-3.67	-3.72	0.03	-3.74	**-3.75**	**-3.75**	**-3.75**
	0.6	**-0.15**	**-0.15**	**-0.15**	0.00	-0.01	-0.13	0.37	0.07
	0.8	**-0.15**	**-0.15**	**-0.15**	0.00	-0.04	-0.14	0.26	0.07
500	0.2	**-6.42**	-6.39	-6.41	0.01	-6.41	-6.41	-6.41	-6.41
	0.4	-3.56	-3.49	-3.53	0.02	-3.57	**-3.58**	**-3.58**	**-3.58**
	0.6	**-0.11**	**-0.11**	**-0.11**	0.00	0.25	**-0.11**	0.73	0.15
	0.8	**-0.11**	**-0.11**	**-0.11**	0.00	0.21	**-0.11**	0.73	0.13
1000	0.2	**-6.76**	-6.73	-6.75	0.01	-6.73	-6.75	-6.74	-6.74
	0.4	-4.38	-4.32	-4.36	0.02	-4.39	**-4.40**	-4.39	-4.39
	0.6	**-0.06**	**-0.06**	**-0.06**	0.00	1.01	-0.05	1.28	0.42
	0.8	**-0.06**	**-0.06**	**-0.06**	0.00	1.13	-0.05	1.28	0.40
	Mean	**-2.15**	**-2.13**	**-2.15**	0.01	-2.01	-2.12	-1.94	-2.06

5 Conclusions

A modified version of the HRM heuristic with much better results is developed along with the discrete version of the PSO algorithm. The DPSO algorithm is hybridized with the ILS algorithm to further improve the computational results. The proposed DPSO algorithm was applied to 280 benchmark instances of Biskup and Feldmann [2]. The solution quality was evaluated according to the reference upper bounds generated by Biskup and Feldmann [2]. The computational results show that the proposed DPSO algorithm generated better results than those in Hino et. al. [1].

References

1. Hino C.M., Ronconi D.P., Mendes A.B.: Minimizing earliness and tardiness penalties in a single-machine problem with a common due date. European Journal of Operational Research, **160** (2005) 190–201
2. Biskup D., Feldmann M.: Benchmarks for scheduling on a single machine against restrictive and unrestrictive common due dates. Computers & Operations Research, **28** (2001) 787–801
3. Cheng T.C.E., Kahlbacher H.G.: A proof for the longest-job-first policy in one-machine scheduling. Naval Research Logistics, **38** (1990) 715–720.
4. Hall N.G., Kubiak W., Sethi S.P.: Earliness-tardiness scheduling problems II: weighted- deviation of completion times about a restrictive common due date. Operations Research, **39**(5) (1991) 847–856
5. Feldmann M., Biskup D.: Single-machine scheduling for minimizing earliness and tardiness penalties by meta-heuristic approaches. Computers & Industrial Engineering, **44** (2003) 307–323
6. Eberhart R.C., Kennedy J.: A new optimizer using particle swarm theory. In: Proceedings of the Sixth International Symposium on Micro Machine and Human Science, Nagoya, Japan (1995) 39–43
7. Pan Q.K., Tasgetiren M.F., Liang Y.C.: A discrete particle swarm optimization algorithm for the no-wait flowshop scheduling problem with makespan criterion. In: Proceedings of the International Workshop on UK Planning and Scheduling Special Interest Group (PLANSIG2005), City University, London, UK (2005) 34–43

Model Selection for Support Vector Machines Using Ant Colony Optimization in an Electronic Nose Application

Javier Acevedo, Saturnino Maldonado, Sergio Lafuente, Hilario Gomez, and Pedro Gil

University of Alcala, Teoría de la señal, Alcala de Henares, Spain
javier.acevedo@uah.es

Abstract. Support vector machines, especially when using radial basis kernels, have given good results in the classification of different volatile compounds. We can achieve a feature extraction method adjusting the parameters of a modified radial basis kernel, giving more importance to those features that are important for classification proposes. However, the function that has to be minimized to find the best scaling factors is not derivable and has multiple local minima. In this work we propose to adapt the ideas of the ant colony optimization method to find an optimal value of the kernel parameters.

1 Introduction

Electronic noses are defined as an array of sensors and a pattern recognition (PARC) system [1]. Over the past years these systems have been applied to many different applications with a considerable success. One of the aspects gaining in importance in the electronic nose field is feature extraction [2]. The importance of this stage for the PARC system lies in the need to enhance those features that have more importance for classification. In fact, this method is quite similar to the way the brain processes the information in the olfactory bulb, giving more importance to those signals coming from the nose that are useful to identify the target odor. However, most of the work developed about feature extraction is done using the filter approach, that is, the data are transformed independently of the learning machine employed. This approach is straightforward but it is not as coherent as the wrapper approach, which makes the data transformation depending on the result of the classification machine.

Support vector machines (SVM) have demonstrated to be a powerful learning method [3] and its use in the electronic nose field is getting more importance [4]. In particular, best results are achieved with radial basis function (RBF) kernels [5]. When using such kernels, some hyperparameters must be tuned, but this process is usually done picking up some values and testing with an external dataset. In [6] the proposal was to use a multi-gamma kernel, where every feature had its own gamma parameter. In this way, the tuning of the hyperparameters can be used as a feature extraction with a wrapper approach. This

M. Dorigo et al. (Eds.): ANTS 2006, LNCS 4150, pp. 468–475, 2006.

tuning is usually done testing several values of the parameters and measuring a classification error, such as cross-validation. However, when the dimension of the problem is not very small, as it is usual in the electronic nose problems, the number of possible solutions makes it impossible to test all the combinations.

It has to be noted that the functions employed to test the classification error, and so to select the scaling factor values, are not derivable. As a matter of fact, these functions have multiple local minima, so the use of optimization methods based on a gradient descent is less than appropriated. Ant Colony Optimization (ACO) [7],[8] is a recent meta-heuristic search method, based in swarm intelligence, that is providing good results to solve hard combinatorial problems. In this work, we have adapted this method to search the scaling factors values that minimize the classification error.

2 SVM Classification Error

Given a problem with a set of training vectors $\boldsymbol{x_i} \in \mathrm{R}^n, i = 1, ..., l$ and a vector of labels $\boldsymbol{y} \in \mathrm{R}^l, y_i \in \{-1, 1\}$ the training of the SVM implies to solve the following optimization problem:

$$\begin{aligned} &\min_{\alpha} \; W(\boldsymbol{\alpha}) = \tfrac{1}{2}\boldsymbol{\alpha}^T \boldsymbol{Q} \boldsymbol{\alpha} - \boldsymbol{e}^T \boldsymbol{\alpha} \\ &\text{subject to } 0 \le \alpha_i \le C, i = 1, ..., l \\ &\qquad\qquad \boldsymbol{y}^T \boldsymbol{\alpha} = 0 \, . \end{aligned} \tag{1}$$

where $\boldsymbol{e}$ is the unity vector, C is a regularization parameter and $Q_{ij} = y_i y_j K(\boldsymbol{x_i}, \boldsymbol{x_j})$ is a symmetric matrix, with $K(\boldsymbol{x_i}, \boldsymbol{x_j})$ being the kernel function. With certain kernels the Q matrix is positive definitive and so, the problem described in (1) has a unique solution of $\boldsymbol{\alpha}$ that can be found quickly using a gradient descent method or some other decomposition method like the one proposed in [9]. Once the solution of $\boldsymbol{\alpha}$ is found, there will be only a number of training patterns $\boldsymbol{x_i}$ with α_i different to zero. These patterns are known as support vectors. Then, for a new incoming pattern $\boldsymbol{x}$ we have a decision function:

$$f(\boldsymbol{x}) = \operatorname{sgn}\left(\sum_i \alpha_i y_i K(\boldsymbol{x}, \boldsymbol{x_i}) + b \right), \; \boldsymbol{x_i} \in \boldsymbol{V} \, . \tag{2}$$

being V the set of support vectors. As it has been mentioned, in the electronic nose best results have been achieved with a RBF kernel. In this work, the proposal is to use the following kernel:

$$K(\boldsymbol{x}, \boldsymbol{y}) = e^{-\sum\limits_{i=1}^{n} \gamma_i (x_i - y_i)^2} \, . \tag{3}$$

Being $\gamma_i \in [0, 1]$ the scaling factors associated to each feature. These scaling factors have to be tuned to improve the performance of the SVM. Now, the question is how the classification error can be measured. Most of the works that use SVM with a simple RBF kernel use an external test set to adjust the values that minimize the classification error. It is especially interesting the measure obtained with the leave-one-out procedure because it gives us an unbiased estimator of the classification error. However, this procedure is very expensive from a computational point of view, since it requires l trainings. It is well known that only support vectors are able to introduce error in the leave-one-out procedure. Moreover, in [10] it was demonstrated that if a support vector fulfils (4), then that support vector does not introduce error in the leave-one-out process:

$$2\alpha_i^0 R^2 - y_i f^0(x_i) < 0 \ . \tag{4}$$

With R being an upper bound of the kernel used and α_i^0, $f^0(x_i)$ are the solution of the optimization problem described in (1) and the decision function respectively. In our case, the kernel described in (3) has an upper bound of 1. So, there is an upper bound of the leave-one-out error that can be calculated as:

$$\widehat{LOO} = \frac{1}{l} \sum_i u\left(2\alpha_i^0 - y_i f^0(x_i)\right) \ i \in V \ . \tag{5}$$

where $u\,(.)$ is the step function.

3 The Proposed ACO Procedure

For a given dataset $\boldsymbol{x_i} \in \mathrm{R}^n$, $i = 1, ..., l$ we have to find the best combination of $\boldsymbol{\gamma} \in \mathrm{R}^n$, $\gamma_i \in [0, 1]$ that minimizes the leave-one-out error. The proposed ACO procedure has the following elements:

- A number of artificial ants. Each ant travels across a path associated to a solution of the γ vector.
- For each feature i, there are m possible states of γ_i. The continuous space [0,1] is divided into m discrete values. S_j^i is the j-th state associated to the feature i. There is an initial state S^0 where all the ants start the travel.
- From one state j to another state z, there is a path that contains the following elements:
  - The propability p_{jz} of that path to be chosen by an ant.
  - A pheromone value τ_{jz} that depends on the amount of ants that have travel across this path.

Every ant begins its travel at the initial state S^0 where it can choose between m possible paths that will arrive to a state S_j^1. With this initial movement the ant has selected the value of γ_1. The choice is done with a probability p_{0j}. Once an ant has reached a state S_j^i it can only move to a state S_z^{i+1} and it will take that path with a probability p_{jz}. When the ant is in a state S^n it has reached the end of the travel and it has a possible solution of the γ vector. At every instant of time t, there are k concurrent ants that are traveling across the states. In Fig.(1) it is shown a problem with 4 features, where the scaling factor values have 5 steps. Then, an ant starting in the initial node can choose in every state between the paths drawn in dashed line, and in this example it has selected the

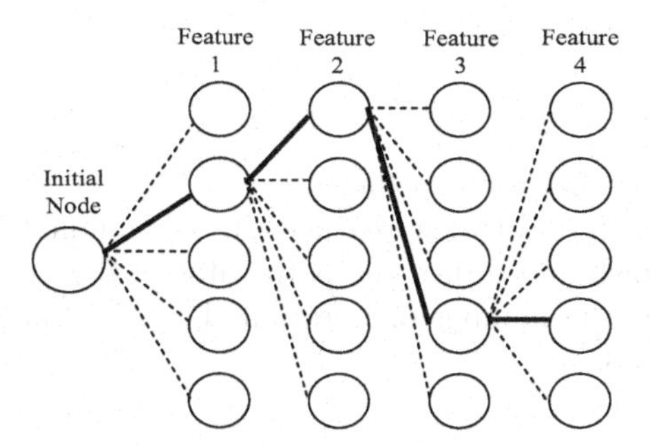

Fig. 1. An example of the possible paths an ant can choose in a 4-feature problem

continuous line. Once all the ants have finished their routes, the estimator of the leave-one-out described in (5) is evaluated for each ant.

It is important to note that we have proposed in this work to take into account only discrete values of the continuous space. The main reason for doing this, is that similar values of $gamma_i$ should give similar results.

The basic algorithm is summarized in the following steps:

1. Create the available states. In the first iteration set a uniform distribution for all probabilities.
2. Set the same amount of pheromone τ_{jz} in all the possible paths.
3. Set k ants at the initial node. Each ant will select a path to the state j with a probability p_{0j}. Once the ant is in the state j the choice of the next path is done depending on the probability of each path.
4. Once the k ants have finished their travel, compute for each of them the estimator of the leave-one-out $\widehat{LOO}_k$, as is defined in (5). Select the B ants that have achieved the lowest values in this iteration. If there are two ants with the same evaluation function value, it is first selected the one with less number of support vectors.
5. Increment the pheromone value of the tail following:

$$\Delta\tau_{jz} = \begin{cases} \frac{Q}{\widehat{LOO}_k} \; if \; k \in B \\ 0 \; if \; k \notin B \end{cases} . \tag{6}$$

   where Q is a constant that has to be adjusted depending on the problem under study.
6. If the winning path until this moment has not be included in the previous point, include it now.
7. Recalculate all the path probabilities:

$$p_{jz} = \frac{\tau_{jz}^{\lambda}}{\sum \tau_j^{\lambda}} . \tag{7}$$

   Where λ is a constant to be adjusted.

8. Evaporate pheromone values:

$$\tau_{jz}^{it+1} = (1 - \rho)\,\tau_{jz}^{it} \; . \tag{8}$$

   Where ρ, is the evaporation coefficient.
9. If the number of iterations has reached the maximum number of iterations allowed, then finish and return the path value of the ant with lowest leave-one-out value. If the number of iterations has not reached the maximum, repeat from step 3.

4 Results and Discussion

4.1 Standard Datasets

Before testing the proposed method on electronic nose datasets, we have tested it on some datasets obtained from the UCI repository database. The main reason for doing this is the high number of features the electronic nose datasets have. In order to have some control on the convergence rate, number of ants per population and some other parameters like the described constants (Q, λ, ρ) it is very useful to test on some datasets with less features in such a way that it is possible to calculate the absolute minima of the problem.

In Fig.(2) it is shown the mean of the standard deviation of the paths selected by the ants against the number of generations.It is clear that after a number of iterations, all the ants in the generation follow the same path. If this number of iterations is very low, there is a high risk of falling into a local minima. However, if the standard deviation of the solution keeps high after a number of generations, it would mean that the algorithm is not searching towards the solution, but it searches in a randomize way. In this figure it is shown different curves for different values of Q and it can be appreciated the importance of choosing an adequate value of Q. To make this figure, we have executed the algorithm one hundred of times per each value of Q, drawing the mean of the standard deviations calculated. Repeating this process for several datasets, we found that a good value of Q can be obtained calculating the mean of the the evaluation function in the first generation and dividing it by two.

One important measure for our experiments is the number of generations needed to reach the absolute minima and how many times is it reached. A measure related to this is the indicator M, that is defined as the number of generations needed to have at least a success in 80% of the tests. This indicator measures how fast the algorithm converges to the optimal solution. In Fig.(3) is represented the percentage of success against the number of generations used for the Indian-Diabetes dataset with 60 ants per generation, in continuous line, and 15 ants per generation in dashed-dot line. For this test the constants have the following values: $\rho = 0.8$, $\lambda = 2$, and $Q = 100$. It can be appreciated that $M = 45$ in the case of 60 ants per generation but in the case of 15 ants per generation this value is not obtained, since many times the algorithm falls into a local minima and the maximum success obtained is a 63%.

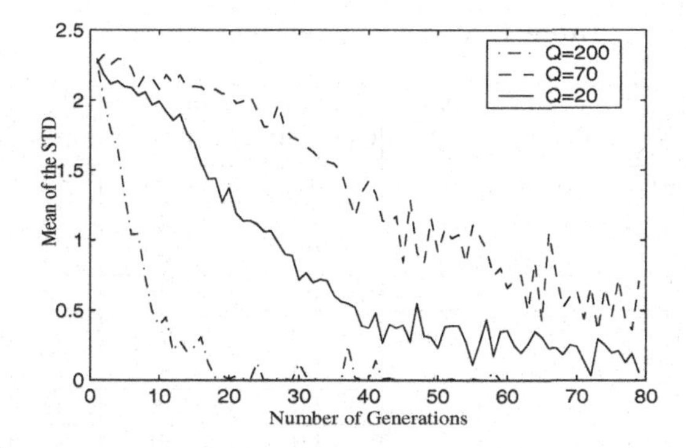

Fig. 2. Standard Deviation against number of generations for different values of Q

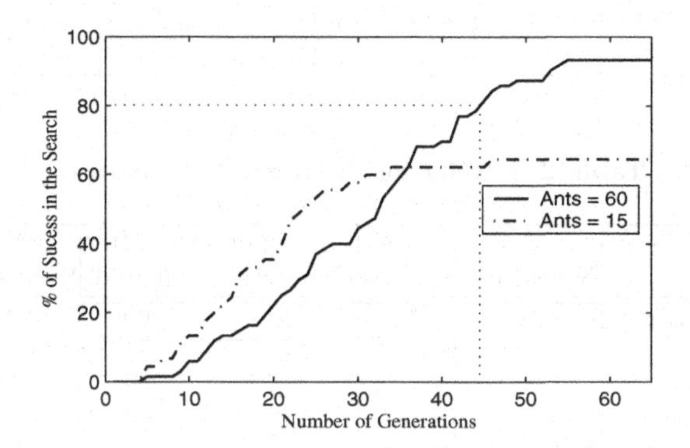

Fig. 3. Indian-Diabetes Dataset. Success for two different size populations.

In Table(1) are shown the values for several standard datasets. The Q parameter was always adjusted as explained, with $\rho = 0.9$ and $\lambda = 2$. In all cases, the exact minima was obtained through exhaustive search since although the number of combinations is big it is possible to obtain it, due to the reduced number of features.

It can be appreciated how the algorithm has a good behavior in the first and the third datasets, whereas in the second case the percentage of success is so low that the M parameter could not be measured although the test was done increasing a lot the number of generations. This can be explained because the leave-one-out estimator gives for this dataset almost plain values and the search is not optimized. These measures are interesting to work with the algorithm in a real system.

4.2 Electronic Nose Datasets

Once we have tested the method with standard datasets we can focus on the electronic nose datasets. In this case we have tested 3 datasets that were obtained

Table 1. M value for different datasets

DATASET	N Features	N of γ steps	Ants per Generation	Max. Iter Allowed	% Sucess	M Parameter
Indian-Diabetes	8	4	30	200	89	60
			30	50	75	NA
			60	200	94	45
			60	50	85	45
Breast Cancer	9	4	30	200	43	NA
			30	50	35	NA
			60	200	55	NA
			60	50	35	NA
Ecoli	5	5	30	200	90	29
			30	50	83	29
			60	200	94	27
			60	50	86	27

Table 2. M value for Electronic Nose datasets

DATASET	N Features	N of γ steps	Ants per Generation	Max. Iter Allowed	% Sucess	M Parameter
RonVsWhisky	88	10	30	200	95	31
			30	50	89	31
			60	200	97	28
			60	50	92	28
EthanolVsAlcohols	88	10	30	200	83	94
			30	50	67	NA
			60	200	85	78
			60	50	71	NA
COVsNO_2	88	10	30	200	78	NA
			30	50	58	NA
			60	200	84	150
			60	50	65	NA

in our laboratory using SnO_2 sensors with thermomodulation [4]. The first one is composed on some samples from a whisky and some others from Ron. The second one, is composed from composed from Ethanol and some other alcohols like methanol and propanol. The third one takes samples from CO and NO_2. Table (2) shows the results obtained for these datasets.

One of the relevant issues from the information obtained with these measures is that the algorithm works better for a population of 60 ants rather than for 30 ants. The main reason is that ants works under a cooperation way and, especially at the beginning there are more solutions explored. However, we can not conclude that the higher the number of ants the faster the convergence is.

5 Conclusion

In this work we have adapted the ACO method to optimize modified RBF kernels. This procedure is extremely important to know what features should be enhanced in electronic nose applications. The proposed method reach to a global minima with a high level of success in most of the datasets tested if the necessary parameters are well adjusted. Future work will explore possible attractiveness functions, new datasets and modifications of the ACO algorithm to work with continuous spaces.

Acknowledgement. This work was supported by Comunidad of Madrid project CAM-UAH 2005/031.

References

1. Hines, E., Llobet, E., Gardner, J.: Electronic noses: a review of signal processing techniques. IEE Proc. Circuits Dev.and Systems **146** (1999) 297–310
2. Distante, C., Leo, M., Siciliano, P., Persaud, K.: On the study of feature extraction methods for an electronic nose. Sensors and Actuators B: Chem. **87** (2002) 274–288
3. Vapnik, N.V.: The Natureof Statistical Learning Theory. Springer-Verlag, New York. (2000) 1ed: 1998.
4. Al-Khalifa, S., Maldonado, S., Gardner, J.: Identification of co and no2 using a thermally resistive microsensor and support vector machine. IEE Proc. Science Meas. and Tech. **150**(6) (2003) 11–14
5. Pardo, M., Sberveglieri, G., Gardini, S., Dalcanale, E.: Classification of electronic nose data with support vector machines. Sensors and Actuators B: Chem. **107** (2005) 730–737
6. Chapelle, O., Vapnik, V., Bousquet, O., Mukherjee, S.: Choosing multiple parameters for support vector machines. Machine Learning **46**(1) (2002) 131–159
7. Dorigo, M., Stützle, T.: Ant Colony Optimization. MIT Press (2004)
8. Dorigo, M., Di Caro, G.: The ant colony optimization meta-heuristic. In Corne, D., Dorigo, M., Glover, F., eds.: New Ideas in Optimization. McGraw-Hill, London (1999) 11–32
9. Platt, J.: Fast training of svms using sequential minimal optimization. In Schölkopf, B., Burges, C., Smola, A., eds.: Advances in Kernel Methods – Support Vector Learning. MIT Press (1998) 185–208
10. Joachims, T.: Estimating the generalization performance of a SVM efficiently. In Langley, P., ed.: Proc. of ICML-00, Morgan Kaufmann , San Francisco, US (2000) 431–438

On the Popularization of Artificial Insects: An Interactive Exhibition for a Wide Audience to Explain and Demonstrate Computer Science and Robotic Problem Solving Taking Inspiration of Insects

Pierre Lebocey[1], Julie Fortune[1], Arnaud Puret[1], Nicolas Monmarché[1], Pierre Gaucher[1], Mohamed Slimane[1], and Didier Lastu[2]

[1] Université François Rabelais, Laboratoire d'Informatique, Tours, France
nicolas.monmarche@univ-tours.fr
[2] Muséum d'Histoire Naturelle de la ville de Tours, Tours, France

Abstract. From October 2005 to August 2006, the museum of natural history of the city of Tours is displaying a temporary exhibition designed by the members of the "handicap and new technologies" research team from the computer science lab of the University of Tours. This paper describes this exhibition, the goals and means that have been involved to popularize recent researches in the field of artificial insects. Different robots and computer simulations have been used to explain how our community builds new paradigms inspired by insect models and swarm intelligence concepts.

1 Introduction

Since the beginning of 90's, the amazing world of insects has inspired computer scientists and robots makers in their research activities [1]. The exhibition called "Artificial Insects" offers a visit inside this strange world populated with virtual ants and autonomous robots whose capacities make it possible to solve problems like robot's navigation, data clustering, optimization... With the help of computer simulations and interactive devices designed by the researchers of the Computer Science Lab in Tours, it is possible to apprehend these new concepts, to handle and see evolving/moving these singular and enthralling creatures.

The project of this exhibition was born from the meeting between the current director of the natural history museum, Didier Lastu, and the members of the research team HaNT (Handicap and New Technologies [1]). At University of Tours, works on artificial insects has been started in 1997, at the starting time of PhD work of Nicolas Monmarché [2]. Since then, artificial ants have been the subject of many publications, studies or projects and their use in various fields of computer science has been constantly developed in the Lab [3,4,5,6,7,8,9, for instance].

[1] www.hant.li.univ-tours.fr

M. Dorigo et al. (Eds.): ANTS 2006, LNCS 4150, pp. 476–483, 2006.

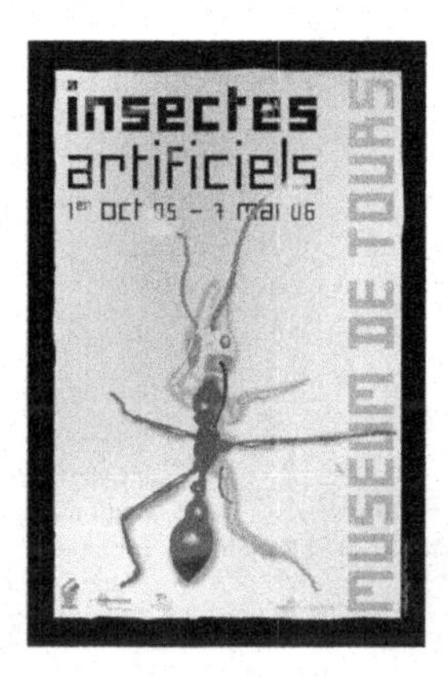

Fig. 1. Poster exhibition (Graphics: Alexandre Saint-Pol)

From the researchers' point of view, the goal of this exhibition is to popularize their work, to communicate the fascination that biological models exert on them for the resolution of difficult problems. For the computer science teachers, it is a question of showing that computer science studies can lead to unexpected considerations and that innovation can take advantage of interdisciplinarity, here with biology, ethology or even sociobiology. It should be underlined that majority of simulations or robots were conceived, developed or programmed in the context of students' projects during their engineering courses. Lastly, for the Natural history museum, it is way of considering a possible future, of showing that natural history can carry out original applications.

In this paper, we present the various modules that have been built for the exhibition. As a web site has been written to help the educational visit preparation, more informations can be extracted from there[2] such as panels (in French) and videos.

2 General Description of the Exhibition

The exhibition is installed in the room of the temporary exhibitions, which is approximately 200 m^2 of surface. The visit is organized in a sequence of modules, with an increasing level of abstraction (see the general map in fig. 2). The first demonstrations are based on the copy of an individual capacity of insects (for example navigation in their environment), whereas aspects related to the collective behavior are introduced later. Each workshop (or modules) focuses on a particular problem or topic and clarifies a concept, most often with two approaches: (1) the first approach focuses on autonomous robotics and presents the animation of one or more robots with possibilities of interactions; (2) the second approach is based on computer simulations and presents by the intermediary of a computer various possibilities of simulation.

[2] `www.hant.li.univ-tours.fr/museum`

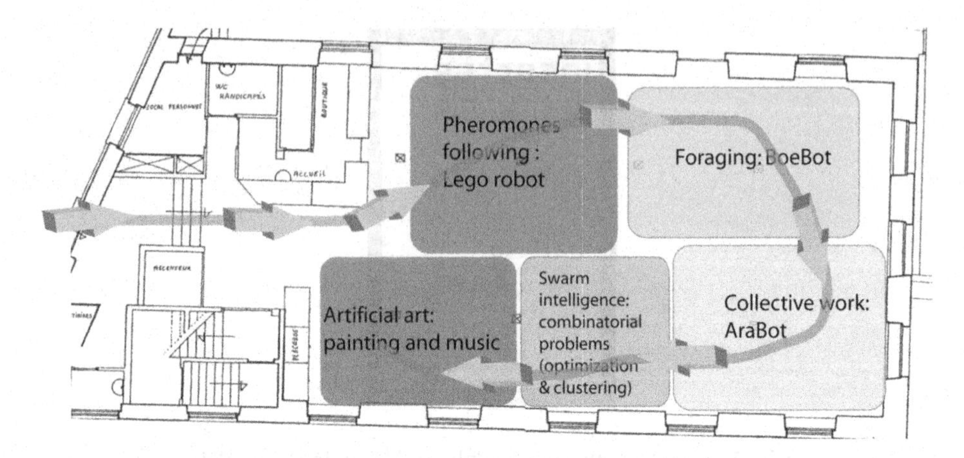

Fig. 2. General map of the exhibition

For each module, we have designed panels that give explanations, other examples taken in the field, and possible applications of the idea presented in the corresponding module.

3 Description of the Modules

3.1 Module 1: Following Pheromone Trails

The ants frequently use pheromones to locate themselves in their environment, for example to find their nest on the way back from a food source. These chemical and volatile substances make it possible to consider minimal strategies of displacement (i.e. consuming only very little energy in communication or calculation).

Follow-Up of Pheromones by a Lego Robot. The goal is to show that with general audience commercial equipment (MindStorm Lego), one can already build a robot that displays a pheromone like following behavior. The robot moves in an autonomous way in a closed area and follows the white lines, which are representing the pheromones trails. The robustness of the strategy of the robot (move straight ahead when the white paint is detected and turn when it detects another color, in our case: green) can be tested by the visitor: the play-ground of the robot can be modified in the same way than the game of the labyrinth where tiles of wood can be inserted laterally to modify a line or a column (see figure 3). This robot gives a simple introduction to robots that can use real odors [10].

Following Digital Pheromones. Even if the Lego robot environment is variable, there is not any pheromone evaporation, as it can be observed in nature. In order to show the impact of this evaporation on the movement of a colony of ants, we have proposed a simulation of ants on a tactile screen. The visitor

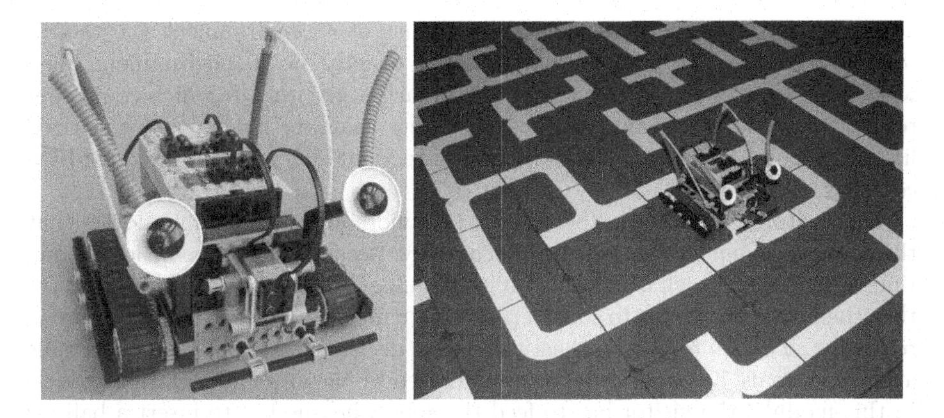

Fig. 3. Lego robot (left) and inside its labyrinth (right)

can play the role of one leading ant by drawing with his/her finger a pheromone trail or let one of the ants deposit one. The main parameters such as the number of ants, and the evaporation speed of pheromones, can be modified during the simulation. This game is very useful for the youngest who can play with these ants and is then one of the favorite display.

3.2 Module 2: The Foraging Boe-Bot

The objective of this module is to show that one can take as a starting point the the navigation capacities of insects to design robots able to move in a variable and dubious environment. We wanted to show that in this case, with a voluntarily simple behavior, a insect-robot is apparently not very effective but finally enough robust to perform the expected task.

To illustrate this idea, we have chosen the daily concern of many insects: finding food and to bringing it back to their nest. The chosen robot for this demonstration, the Boe-Bot one (Parallax) (fig. 4(a)), uses simple behaviors,

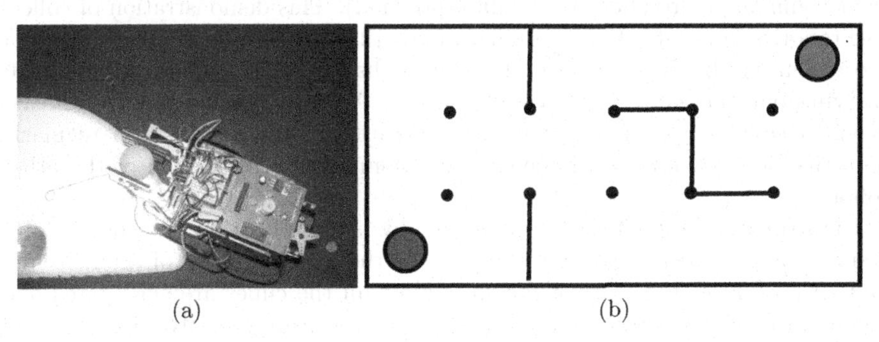

(a) (b)

Fig. 4. Boe-Bot with its grip and compass (a) and its environment (food source in the lower left corner and nest in the upper right corner)

according to the very limited knowledge it has of its environment via its sensors. It is located in a labyrinth, or more exactly in an environment filled with the obstacles (fig. 4(b)) and tries to reach a given area (which could represent the nest for an ant). As many insects that are able to use polarized light, the Boe-bot robot can benefit from directional information provided by a compass.

The visitors can observe the robot's behaviors and for example notice that the robot does not have any learning capacity: it can spend more or less a long time to achieve its task depending on its "chance". The public can thus perceive that there are other ways to solve a problem than trying to reproduce a complex reasoning. Thanks to the stochasticity of its decisions, a robot-insect can thus achieve its goals, its patience being only limited by the autonomy of its batteries. In this module, the visitor has to feed the robot: he/she has to insert a ball for table tennis in the feeding area.

3.3 Module 3: Team Work

The objective of this module is to show the importance of collective behaviors in insect societies for the resolution of robotics problems. We have conceived and realized, a mobile and autonomous robot: ArABot (for Artificial Ant Robot). The constraints that we have decided to verify are the following: the robot must be easily re-programmed, easy to build and at a cost not exceeding 80 EUR. Until march 2006, we have built 12 robots (fig.5(a)).

A Colony of ArABots. A colony of robots is installed at the museum (fig. 5(b)). ArAbots move randomly in their environment in which cubes made of wood are dispersed. As they meet cubes they push them but if the load is too heavy (there are too many cubes or one of the limit of the play ground is reached), the robot changes direction and leaves the cubes in their place. From a global point of view, one can very quickly assist to the agglomeration of the cubes: although the robots do not communicate directly between them, they communicate by the intermediary of the modifications that they cause on their common environment (the cubes' position). This demonstration of collective robotics illustrates the possibility of creating apparently simple robots, not too expensive, but with an overall liability that is better than a single robot, carrying out the same work, could provide. Actually, in the event of breakdown, a collective system will remain operational (the visitors can regularly note that a robot out of service does not compromise the action of the other robots).

Unfortunately, the efficiency of the colony is a problem in the context of a demonstration which must be ran during all the opening hours of the museum: half an hour after the start of robots, almost all the cubes are aggregated and an employee of the museum needs to open the furniture to scatter the pieces of wood. Then we have introduced a special robot which spends all its time turning around the area and driving the cubes toward the center.

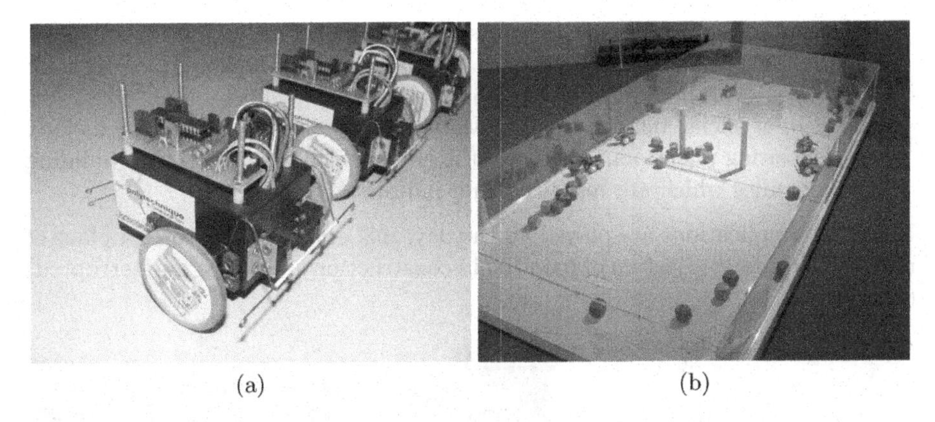

(a) (b)

Fig. 5. AraBots Robots (a) and general vue of the module number 3 with ArAbots inside

Simulations of Collective Behaviors. Two simulations of collective robotics are also presented on computers (inside special furnitures where the screen can only be visible for the visitor):

- the first one is a simplification of ArAbots. We wanted to show here, that robotics also needs a phase of simulation to check the algorithms before establishing them in the robots. The advantages of simulation are visible (speed, reliability, reproducibility) but the limits of the model also (discretization of space...);
- the second simulation models robots able to imitate weaver ants: they gather plates (leaves) to build a larger structure (their nest). They form chains by clinging the ones to the others and try to catch another chain coming from another plate in order to draw the two plates one toward the other.

The objective of these simulations is to show that simultaneous work with several robots can make it possible to reach more efficient results than by using only one more vulnerable and more expensive robot.

3.4 Module 4: Resolution of Combinatorial Problems

This module explains with two panels both combinatorial optimization and data clustering with artificial ants. A simplified example of the TSP is used to explain that ants can lay down artificial pheromones onto the arcs of a graph and by this mechanism they can find good paths.

3.5 Module 5: Artistic Design

The last module of the exhibition proposes an original use of the collective intelligence of ants: the generation of abstract paintings [6] and the generation of music [5]. In both cases, the paradigm of pheromones is used:

- painting of artificial ants: the ants move on a white canvas and deposit colored pheromones. The various characteristics of the ants produce abstract paintings in perpetual evolution (Fig. 6(a));
- music of artificial ants: the ants move between the notes and build a melody by depositing pheromones between the notes which they meet, thus building the melodies which are repetitive and infinite.

These two applications are played all the day, this module constitutes a place of relaxation (video projection of painting in construction and music uninterrupted) (Fig. 6(b)).

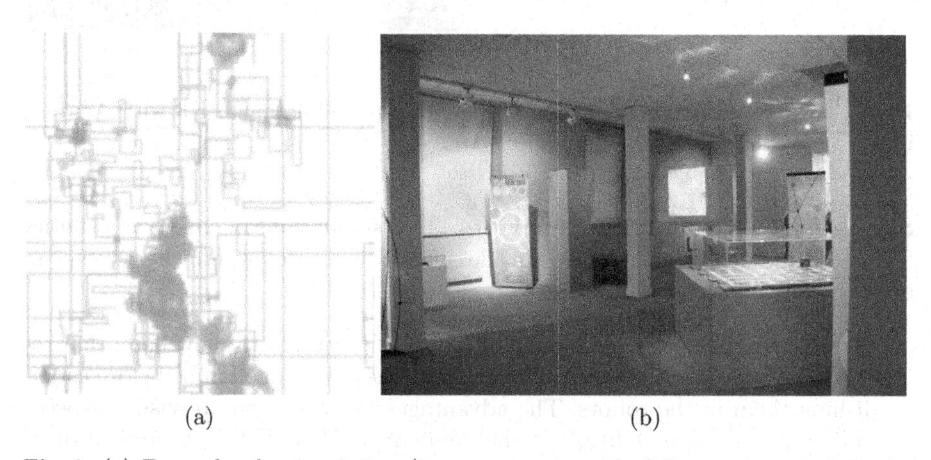

(a) (b)

Fig. 6. (a) Example of ant painting (two green ants with different characteristics in their moves). (b) General vue of the exhibition.

4 Conclusion

This exhibition has necessitated about one year of work (it opened on October 2005, 1st). The main difficulty is to keep the robots in good health, not because of visitors misuses but simply because robots are running 8 hours a day and 6 days per week and it is intensive for small mechanics.

During the opening hours, some of our students had to improve robots (for instance the Lego has been rebuild) and that was one of our goals: having a living exhibition, not only because robots moves alone but also because students, researchers, ... keep on improving the demonstrations.

In our opinion, this kind of exhibition is important to make people aware of the main concepts that underline our researches. Even if robots are more appropriate to catch the attention of visitors, it is necessary to explain what computer science research can imply (even if mathematical aspects have been removed), it is a good way to build vocations within young visitors. As can be observed in other cities, people are found of these experiments and good results in museum-going is our reward: from October to April, about 7,400 visitors have seen the exhibition and 3,000 of them were children accompanied by their professor.

References

1. Bonabeau, E., Dorigo, M., Theraulaz, G.: Swarm Intelligence: From Natural to Artificial Systems. Oxford University Press, New York (1999)
2. Monmarché, N.: Algorithmes de fourmis artificielles : applications à la classification et à l'optimisation. Thèse de doctorat, Laboratoire d'Informatique, Université de Tours (2000)
3. T'Kindt, V., Monmarché, N., Tercinet, F., Laügt, D.: An Ant Colony Optimization algorithm to solve a 2-machine bicriteria flowshop scheduling problem. European Journal of Operational Research **142**(2) (2002) 250–257
4. Monmarché, N., Venturini, G., Slimane, M.: On how *Pachycondyla apicalis* ants suggest a new search algorithm. Future Generation Computer Systems **16**(8) (2000) 937–946
5. Guéret, C., Monmarché, N., Slimane, M.: Ants can play music. In: Fourth International Workshop on Ant Colony Optimization and Swarm Intelligence (ANTS 2004). Volume 3172 of Lecture Notes in Computer Science., Université Libre de Bruxelles, Belgique, Springer-Verlag (2004) 310–317
6. Aupetit, S., Bordeau, V., Monmarché, N., Slimane, M., Venturini, G.: Interactive Evolution of Ant Paintings. In: IEEE Congress on Evolutionary Computation. Volume 2., Canberra, IEEE Press (2003) 1376–1383
7. Labroche, N., Monmarché, N., Venturini, G.: AntClust: Ant Clustering and Web Usage Mining. In Cantu-Paz, E., ed.: Genetic and Evolutionary Computation Conference. Volume 2723 of Lecture Notes in Computer Science., Chicago, Springer-Verlag Telos (2003) 25–36
8. Monmarché, N., Ramat, E., Desbarats, L., Venturini, G.: Probabilistic Search with Genetic Algorithms and Ant Colonies. In Wu, A., ed.: Proceedings of the Optimization by Building and Using Probabilistic Models workshop, Genetic and Evolutionary Computation Conference, Las Vegas, Nevada (2000) 209–211
9. Monmarché, N., Slimane, M., Venturini, G.: On improving clustering in numerical databases with artificial ants. In Floreano, D., Nicoud, J., Mondala, F., eds.: 5th European Conference on Artificial Life (ECAL'99). Volume 1674 of Lecture Notes in Artificial Intelligence., Swiss Federal Institute of Technology, Lausanne, Switzerland, Springer-Verlag (1999) 626–635
10. Russell, A., D., T., Mackay-Sim, A.: Sensing odour trails for mobile robot navigation. In: IEEE International Conference on Robotics and Automation, San Diego, CA (1994) 2672–2677

Solution Representation for Job Shop Scheduling Problems in Ant Colony Optimisation

James Montgomery[1,*], Carole Fayad[2], and Sanja Petrovic[2]

[1] Faculty of Information & Communication Technologies
Swinburne University of Technology, Melbourne, Australia
jmontgomery@ict.swin.edu.au

[2] School of Computer Science & IT, University of Nottingham
Nottingham, United Kingdom
{cxf, sxp}@cs.nott.ac.uk

Abstract. Production scheduling problems such as the job shop consist of a collection of operations (grouped into jobs) that must be scheduled for processing on different machines. Typical ant colony optimisation applications for these problems generate solutions by constructing a permutation of the operations, from which a deterministic algorithm can generate the actual schedule. This paper considers an alternative approach in which each machine is assigned a dispatching rule, which heuristically determines the order of operations on that machine. This representation creates a substantially smaller search space that likely contains good solutions. The performance of both approaches is compared on a real-world job shop scheduling problem in which processing times and job due dates are modelled with fuzzy sets. Results indicate that the new approach produces better solutions more quickly than the traditional approach.

1 Introduction

Ant colony optimisation (ACO) is a constructive metaheuristic in which, during successive iterations of solution construction, a number of artificial ants build solutions by probabilistically selecting from problem-specific solution components, influenced by a parameterised model of solutions (called a pheromone model in reference to ant trail pheromones). The parameters of this model are updated at the end of each iteration using the solutions produced so that, over time, the algorithm learns which solution components should be combined to produce the best solutions. When adapting ACO to suit a problem an algorithm designer must first decide how solutions are to be represented and built (i.e., what base *components* are to be combined to form solutions) and then what characteristics of the chosen representation are to be modelled.

Production scheduling problems consist of a number of jobs, made up of a set of operations, each of which must be scheduled for processing on one of a number of machines. Precedence constraints are imposed on the operations of each

* Corresponding author.

M. Dorigo et al. (Eds.): ANTS 2006, LNCS 4150, pp. 484–491, 2006.

job. The majority of ACO algorithms for these problems represent solutions as permutations of the operations to be scheduled (operations are the base components of solutions), which determines the relative order of operations that require the same machine (see, e.g., [1,2]). A deterministic algorithm can then produce the best possible schedule given the precedence constraints established by the permutation. This approach is more generally referred to as the *list scheduler algorithm* [1]. An alternative approach is to assign different heuristics to each machine which determine the relative processing order of operations, thereby searching the reduced space of schedules that can be produced by different combinations of the heuristics [3]. Building solutions in this manner may offer an advantage by concentrating the search on heuristically good solutions. This paper compares these two solution representations by using a real-world job shop scheduling problem (JSP).

A formal description of the JSP is given in Section 2, including further details of the two solution construction approaches. Section 3 describes the real-world JSP instance to which both approaches are applied, in which processing times and due dates are modelled by fuzzy sets to reflect the uncertain nature of these in industrial settings. Details of the ACO algorithms developed for the fuzzy JSP are given in Section 4, followed by analysis of their empirical performance in Section 5. Section 6 describes the implications of the results for the future application of ACO to such problems. An extended version of this paper, including more extensive empirical analyses, is presented in [4].

2 Job Shop Scheduling and Solution Construction

The JSP examined in this study consists of a set of n jobs $J_1, \ldots, J_n$, with associated release dates $r_1, \ldots, r_n$ and due dates $d_1, \ldots, d_n$. Each job consists of a sequence of operations (determined by the processing requirements of the job) that must each be scheduled for processing on one of m machines $M_1, \ldots, M_m$. Only one operation of a job may be processed at any given time, only one operation may use a machine at any given time and operations may not be preempted. Two criteria have to be minimised simultaneously, the average tardiness of jobs C_{AT} and the number of tardy jobs C_{NT}, calculated as follows:

$$C_{AT} = \frac{1}{n} \sum_{j=1}^{n} T_j \tag{1}$$

where $T_j = \max\{0, C_j - d_j\}$ is the tardiness of of job J_j and C_j is the completion time of job J_j.

$$C_{NT} = \sum_{j=1}^{n} u_j \tag{2}$$

where $u_j = 1$ if $T_j > 0$, 0 otherwise.

It is common in ACO applications for the JSP and related scheduling problems to generate a permutation of the operations, which implicitly determines the relative processing order of operations on each machine. These algorithms are restricted to creating permutations that respect the required processing order of operations within each job, which can consequently be called *feasible permutations.* A deterministic algorithm transforms the relative processing order into an actual schedule.

Different solution construction approaches produce different search spaces. The space of feasible permutations of operations for a JSP is very large (a weak upper bound is $O(k!)$, where k is the number of operations) and is certainly much larger than the space of actual solutions. This space also has a slight bias towards good solutions, which can be exploited by some pheromone models and proves disastrous for others [5]. Another notable feature of this search space is that while all solutions can be reached, solutions (schedules) are represented by differing numbers of permutations.

An alternative approach to building solutions is to assign different *dispatching rules* (i.e., ordering heuristics) to each machine, which subsequently build the actual schedule [3]. The search space then becomes the space of all possible combinations of rules assigned to machines, which is $O(|D|^m)$ where D is the set of rules and m the number of machines. Given a small number of dispatching rules (this study uses four, described in Section 4) it is highly probable that this search space is a subset of the space of all feasible schedules. However, assuming the dispatching rules are individually likely to perform well it is expected that this reduced space largely consists of good quality schedules.

The performance of these two approaches is compared on a real-world JSP instance, described in the next section.

3 A Real-World JSP

The data set used has been provided by a printing company, Sherwood Press, in Nottingham, United Kingdom [6]. There are 18 machines in the shop floor, grouped within seven work centres: printing, cutting, folding, card-inserting, embossing and debossing, gathering, stitching and trimming, and packaging.

Due to both machine and human factors, processing times of jobs are uncertain and due dates are not fixed but promised instead. Therefore, fuzzy sets are used to model these uncertain values. A triangular membership function $\mu_{\tilde{p}_{ij}}(t) = (p_{ij}^1, p_{ij}^2, p_{ij}^3)$ is used to model the fuzzy processing time $\tilde{p}_{ij}$ of job J_j on machine M_i, $i = 1, \ldots, m$, $j = 1, \ldots, n$, where p_{ij}^1 and p_{ij}^3 are lower and upper bounds of the processing time, while p_{ij}^2 is the so-called modal point [7]. An example of fuzzy processing time is shown in Fig. 1(a). A trapezoidal fuzzy set (d_j^1, d_j^2) is used to model the due date $\tilde{d}_j$ of each job, where d_j^1 is the crisp due date and the upper bound d_j^2 of the trapezoid exceeds d_j^1 by 10%, following the policy of the company. An example of a fuzzy due date is given in Fig. 1(b).

The objective function takes into account both the average tardiness of jobs and the number of tardy jobs. As these are measured in different units they are

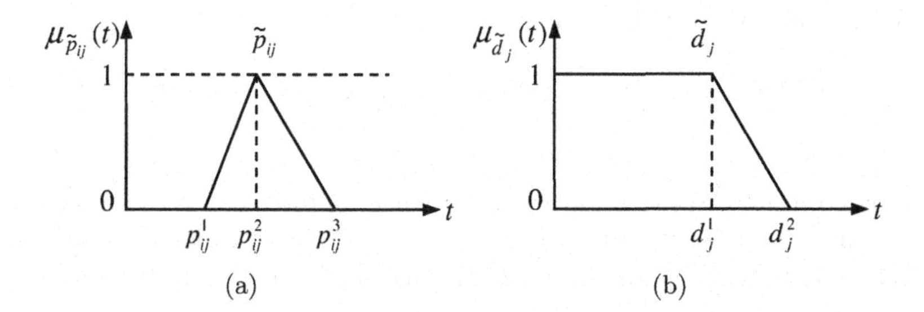

Fig. 1. Fuzzy (a) processing time and (b) due date

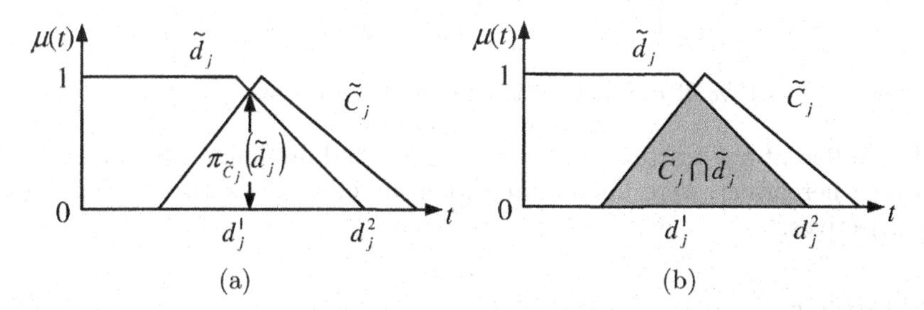

Fig. 2. Satisfaction grade of tardiness using (a) possibility measure and (b) area of intersection

mapped onto satisfaction grades in the range [0, 1], which are then combined in an overall satisfaction grade. Two approaches used to measure tardiness in [6] are investigated:

1. The possibility measure $\pi_{\tilde{C}_j}(\tilde{d}_j)$, used by Itoh and Ishii [8] to handle tardy jobs in a JSP, measures the satisfaction grade of a fuzzy completion time $SG_T(\tilde{C}_j)$ of job J_j by evaluating the possibility of a fuzzy event $\tilde{C}_j$ occurring within the fuzzy set $\tilde{d}_j$ [8] (illustrated in Fig. 2(a)):

$$SG_T(\tilde{C}_j) = \pi_{\tilde{C}_j}(\tilde{d}_j) = \sup \min\{\mu_{\tilde{C}_j}(t),\ \mu_{\tilde{d}_j}(t)\} \qquad j = 1, \ldots, n \tag{3}$$

   where $\mu_{\tilde{C}_j}(t)$ and $\mu_{\tilde{d}_j}(t)$ are the membership functions of fuzzy sets $\tilde{C}_j$ and $\tilde{d}_j$ respectively. This measure is referred to as *poss* hereafter.
2. The area of intersection measure (denoted *area* hereafter), introduced by Sakawa and Kubota [9], measures the proportion of $\tilde{C}_j$ that is completed by the due date $\tilde{d}_j$ (illustrated in Fig. 2(b)):

$$SG_T(\tilde{C}_j) = (area\ \tilde{C}_j \cap \tilde{d}_j)/(area\ \tilde{C}_j) \tag{4}$$

The satisfaction grades of tardiness defined in (3) and (4) are used in two objectives:

1. To maximise the satisfaction grade of *average tardiness* S_{AT}:

$$S_{AT} = \frac{1}{n}\sum_{j=1}^{n} SG_T(\tilde{C}_j) \tag{5}$$

2. To maximise the satisfaction grade of *number of tardy jobs* S_{NT}: A parameter λ is introduced such that a job J_j, $j = 1, \ldots, n$, is considered to be tardy if $SG_T(\tilde{C}_j) \leq \lambda$, $\lambda \in [0, 1]$. After calculating the number of tardy jobs $nTardy$, the satisfaction grade S_{NT} is calculated as:

$$S_{NT} = \begin{cases} 1 & \text{if } nTardy = 0 \\ (n'' - nTardy)/n'' & \text{if } 0 < nTardy < n'' \\ 0 & \text{if } nTardy > n'' \end{cases} \tag{6}$$

where $n'' = 15\%$ of n, where n is the number of jobs.

Two different aggregation operators, average and minimum (denoted *average* and *min* hereafter), were investigated for combining the satisfaction grades of the objectives.

4 ACO for a Fuzzy JSP

Two ACO algorithms were developed based on the $\mathcal{MAX}-\mathcal{MIN}$ Ant System ($\mathcal{MM}$AS), which has been found to work well in practice [10]. The first of these, denoted $\mathcal{MM}\text{AS}_{perm}$, constructs solutions as permutations of the operations, while the second, denoted $\mathcal{MM}\text{AS}_{rules}$, assigns dispatching rules to machines. The set of dispatching rules D consists of the following four rules: Early Due Date First, Shortest Processing Time First, Longest Processing Time First and Longest Remaining Processing Time First.

The two solution representations require different pheromone models. The models chosen have been found to produce the best performance for their respective solution representations [11]. For $\mathcal{MM}\text{AS}_{perm}$, a pheromone value, denoted $\tau(o_i, o_j)$, exists for each directed pair of operations that use the same machine, and represents the learned utility of operation o_i preceding operation o_j [1]. At each step of solution construction, the set of unscheduled operations that require the same machine as a candidate operation o is denoted by O_o^{rel}. Blum and Sampels [1] take the minimum of the relevant pheromone values. Thus, the probability of selecting an available operation o to add to the partial permutation p is given by

$$P(o, p) = \frac{\min_{o_r \in O_o^{rel}} \tau(o, o_r)}{\sum_{o' \notin p} \min_{o_r \in O_{o'}^{rel}} \tau(o', o_r)}. \tag{7}$$

The last operation on each machine is scheduled as soon as it becomes available.

For $\mathcal{MM}\text{AS}_{rules}$, a pheromone value $\tau(M_k, d)$ is associated with each combination of machine and dispatching rule $(M_k, d) \in M \times D$, where M is the set of machines. At each step of solution construction, a machine is assigned a dispatching rule. The probability of assigning a dispatching rule $d \in D$ to machine M_k is given by

$$P(M_k, d) = \frac{\tau(M_k, d)}{\sum_{d' \in D \setminus \{d\}} \tau(M_k, d')}. \tag{8}$$

In both algorithms, after each iteration all pheromone values are reduced in proportion to $(\rho - 1)$ while those for the iteration best solution s^{ib} are increased by $\rho \cdot F(s^{ib})$, where ρ is the pheromone evaporation rate and F is the overall satisfaction grade of s^{ib} (given by either the *average* or *min* aggregation operator).

5 Computational Results

The performance of the algorithms was compared on one month's data collected from Sherwood Press (the March set used by Fayad and Petrovic [6]). The resulting JSP instance consists of 549 operations partitioned into 159 jobs.

The algorithms were implemented in the C language and executed under Linux on a 2.6GHz Pentium 4 with 512Mb of RAM. The $\mathcal{MM}$AS control parameters used were: 10 ants per iteration; 3000 iterations; $\rho = 0.1$; $\tau_{max} = 1$; $\tau_{min} = 1 \times 10^{-3}$ in $\mathcal{MM}\text{AS}_{rules}$ and $\tau_{min} = 1 \times 10^{-4}$ in $\mathcal{MM}\text{AS}_{perm}$. The values of τ_{min} and τ_{max} were chosen to approximate those suggested by Stützle and Hoos [10] based on the size of the solution representation and pheromone update.

Both algorithms were executed with different combinations of parameter values for solution evaluation: *poss* and *area* tardiness measures, and *average* and *min* aggregation operators. The value of λ was fixed at 0.7. Each combination was run with 10 different random seeds.

5.1 Solution Quality

The results revealed that when using the *min* aggregation operator, $\mathcal{MM}\text{AS}_{perm}$ is unable to find a solution with a non-zero objective value. This is because the algorithm, facing a large number of solutions with $S_{NT} = 0$, searches randomly until a subset of pheromone values is updated. Further testing confirmed that a random search of permutations is unlikely to produce solutions with $S_{NT} > 0$. A second version of the algorithm, named $\mathcal{MM}\text{AS}_{perm}^{min}$, was developed in which the pheromone update was modified such that, if all solutions in an iteration have an objective value of zero, the best solution in terms of S_{AT} is used to update pheromone values using the *average* aggregation operator. Such a modification was not necessary for $\mathcal{MM}\text{AS}_{rules}$ as random assignments of dispatching rules to machines typically produced solutions with $S_{NT} > 0$.

Table 1 summarises the satisfaction grades of tardiness measures according to the aggregation operator used for each algorithm. It is evident that $\mathcal{MM}\text{AS}_{perm}^{min}$ is much more successful than its original form when using the *min* aggregation operator. Further investigation revealed that it required the use of the *average* aggregation operator in up to 33% of iterations. Across solution evaluation measures, $\mathcal{MM}\text{AS}_{rules}$ clearly outperforms $\mathcal{MM}\text{AS}_{perm}$.

Table 1. Performance of the algorithms. The best result for each measure is given with the mean value in parentheses. Bold items are best within each solution quality measure.

Algorithm	F	S_{AT}	S_{NT}	C_{NT}
Using *poss* and *average*				
$\mathcal{MM}AS_{perm}$	0.69 (0.62)	0.91 (0.91)	0.46 (0.34)	13 (15.9)
$\mathcal{MM}AS_{rules}$	**0.73 (0.73)**	**0.93 (0.93)**	**0.54 (0.53)**	**11 (11.2)**
Using *poss* and *min*				
$\mathcal{MM}AS_{perm}$	0	0.72 (0.71)	0	48 (51.2)
$\mathcal{MM}AS^{min}_{perm}$	0.42 (0.35)	0.89 (0.88)	0.42 (0.35)	14 (15.5)
$\mathcal{MM}AS_{rules}$	**0.54 (0.53)**	**0.93 (0.93)**	**0.54 (0.53)**	**11 (11.3)**
Using *area* and *average*				
$\mathcal{MM}AS_{perm}$	0.62 (0.59)	0.90 (0.90)	0.33 (0.28)	16 (17.3)
$\mathcal{MM}AS_{rules}$	**0.71 (0.70)**	**0.93 (0.93)**	**0.50 (0.48)**	**12 (12.5)**
Using *area* and *min*				
$\mathcal{MM}AS_{perm}$	0	0.70 (0.69)	0	49 (52.1)
$\mathcal{MM}AS^{min}_{perm}$	0.42 (0.32)	0.88 (0.87)	0.42 (0.32)	14 (16.4)
$\mathcal{MM}AS_{rules}$	**0.50 (0.48)**	**0.93 (0.92)**	**0.50 (0.48)**	**12 (12.5)**

5.2 CPU Time

An order of magnitude difference was observed between the CPU time of the two algorithms, with $\mathcal{MM}AS_{perm}$ taking more than 1400 seconds compared to approximately 100 seconds for $\mathcal{MM}AS_{rules}$. This is to be expected given the respective number of components each must consider at each constructive step; $\mathcal{MM}AS_{perm}$ considers approximately 40 operations on average, while $\mathcal{MM}AS_{rules}$ considers only four. Moreover, $\mathcal{MM}AS_{rules}$ finds its best solutions very early in each run (often within 1 second) while $\mathcal{MM}AS_{perm}$ does not converge until quite late.

6 Conclusions

Typical ACO algorithms for production scheduling problems such as the JSP build solutions as permutations of the operations to be scheduled, from which actual schedules are generated deterministically. An alternative approach when the problem in question has multiple machines and various criteria upon which to judge the urgency of competing operations is to assign different dispatching rules to each machine. The chosen dispatching rules are then responsible for determining the relative processing order of operations on each machine. This paper compared both approaches on a multi-objective real-world JSP, modelled with fuzzy operation processing times and job due dates. The results show that assigning dispatching rules to machines produces higher quality solutions in far less time than building a permutation of the operations. This supports the claim that the assignment of dispatching rules restricts the search space to an area of good quality solutions.

As this study focused on a single, real-world JSP instance (albeit using a variety of solution quality measures) future work is required to determine if these results hold for other production scheduling instances. Additionally, it is now common practice in most ACO algorithms to use a local search procedure to improve the solutions produced, something not done in this study so that differences between the two solution construction approaches could be observed. While the addition of local search to a permutation-based ACO algorithm for these problems may allow it to perform better, it is potentially more useful in the new approach, where it can explore solutions that combinations of dispatching rules would otherwise never produce.

References

1. Blum, C., Sampels, M.: An ant colony optimization algorithm for shop scheduling problems. J. Math. Model. Algorithms **3** (2004) 285–308
2. Colorni, A., Dorigo, M., Maniezzo, V., Trubian, M.: Ant system for job-shop scheduling. JORBEL **34** (1994) 39–53
3. Dorndorf, U., Pesch, E.: Evolution based learning in a job shop scheduling environment. Comput. Oper. Res. **22** (1995) 25–44
4. Montgomery, J., Fayad, C., Petrovic, S.: Solution representation for job shop scheduling problems in ant colony optimisation. Technical Report SUTICT-TR2006.05, Faculty of Information & Communication Technologies, Swinburne University of Technology, Melbourne, Australia (2006)
5. Montgomery, J., Randall, M., Hendtlass, T.: Structural advantages for ant colony optimisation inherent in permutation scheduling problems. In Ali, M., Esposito, F., eds.: Innovations in Applied Artificial Intelligence. Volume 3533 of LNAI. Springer-Verlag (2005) 218–228
6. Fayad, C., Petrovic, S.: A fuzzy genetic algorithm for real-world job shop scheduling. In Ali, M., Esposito, F., eds.: Innovations in Applied Artificial Intelligence. Volume 3533 of LNAI. Springer-Verlag (2005) 524–533
7. Klir, G., Folger, T.: Fuzzy Sets, Uncertainty and Information. Prentice Hall, New Jersey (1988)
8. Itoh, T., Ishii, H.: Fuzzy due-date scheduling problem with fuzzy processing time. Int. Trans. Oper. Res. **6** (1999) 639–647
9. Sakawa, M., Kubota, R.: Fuzzy programming for multiobjective job shop scheduling with fuzzy processing time and fuzzy duedate through genetic algorithms. Eur. J. Oper. Res. **120** (2000) 393–407
10. Stützle, T., Hoos, H.: $\mathcal{MAX}-\mathcal{MIN}$ ant system. Future Gen. Comp. Sys. **16** (2000) 889–914
11. Montgomery, E.J.: Solution Biases and Pheromone Representation Selection in Ant Colony Optimisation. PhD thesis, Bond University (2005)

Some Experiments with Ant Colony Algorithms for the Exam Timetabling Problem

Michael Eley

Logistics Laboratory (LlAb), Faculty of Engineering
Aschaffenburg University of Applied Science, Aschaffenburg, Germany
michael.eley@fh-aschaffenburg.de

Abstract. The exam timetabling problem faces the problem of scheduling exams within a limited number of available periods. The main objective is to balance out student's workload by distributing the exams evenly within the planning horizon. Ant colony approaches have been proven to be a powerful solution approach for various combinatorial optimization problems. In this paper a Max-Min and a ANTCOL approach will be presented. Its performance is compared with other approaches presented in the literature and with modified graph coloring algorithms.

1 Introduction

The exam timetabling problem faces the problem of scheduling exams within a limited number of available periods. The main objective is to balance out student's workload and to distribute the exams evenly within the planning horizon. To evaluate a given schedule Carter et al. [1] proposed a cost function that imposes penalties P_ω whenever one student has to write two exams scheduled within $\omega + 1$ consecutive periods. ω is called the order of the conflict. In particular, conflicts of order 0 should be avoided, i.e. that a student has to write two exams in the same period.

The exam timetabling problem can be formulated as a modification of the well-known graph coloring problem. Each node represents one exam. Undirected arcs connect two nodes if at least one student is enrolled in both corresponding exams. Weights on the arcs represent the number of student enrolled in both exams. The objective is to find a coloring where no adjacent nodes are marked with the same color or to minimize the weighted sum of the arcs that connect two nodes marked with the same color. The exam timetabling problem is a generalization of this problem as in the objective function also conflicts of higher orders are penalized. As the graph coloring problem is already NP-hard [2] several heuristics have recently been developed for solving practical exam timetabling problems, c.f. [3].

Ant colony optimization algorithms represent special solution approaches for combinatorial optimization problems derived from the field of swarm intelligence. They were first introduced by Colorni, Dorigo and Maniezzo in the early nineties [4]. An in depth introduction into ant systems can be found in [5].

The solution approach in ant colony optimization consists of n cycles. In each of these cycles first each of the m ants constructs a feasible solution. If the

M. Dorigo et al. (Eds.): ANTS 2006, LNCS 4150, pp. 492–499, 2006.

optimization problem consists of finding an optimal sequence for some nodes, the probability that an ant ν that has just chosen node i chooses the next node j is determined by the following formula:

$$p_{ij}^{\nu} = \begin{cases} \frac{(\tau_{ij})^{\alpha}(\eta_{ij})^{\beta}}{\sum_{k \in N_i^{\nu}} (\tau_{ik})^{\alpha}(\eta_{ik})^{\beta}} & \text{if } j \in N_i^{\nu} \\ 0 & \text{otherwise} \end{cases} \quad (1)$$

The value η_{ij} is calculated by a constructive heuristic. τ_{ij} is the amount of pheromone trail, that represents the learned desirability of choosing node j when in node i. This information is repeatedly updated by the ants after they have constructed their solutions. α and β are given weighting factors and N_i^{ν} is the set of nodes that have not yet been visited by ant ν currently located in node i. This type of ant colony optimization algorithm is known in the literature as ant systems (AS).

Different variants of ant colony algorithms have been suggested in the literature, like e.g. ant colony systems (ACS) or Max-Min ant systems (MMAS), c.f. [5]. We will compare some of these strategies with respect to their suitability for our problem. In particular, MMAS, which was first proposed by Stützle and Hoos [6], generated significantly better solutions for the travelling salesmen problem. Socha et al. [7] compared the MMAS variant with ACS and found out that MMAS outperformed the ACS approach for the considered timetabling problem.

The main modification of MMAS are related to the way how the matrix τ is initialized and how pheromone values are updated. Additionally, MMAS uses local search to improve the solutions found by the ants. Details will be discussed in the next section.

As far as the author is aware, ant colony algorithms to scheduling problems have only been applied by Colorni et al. [4] and by Socha et al. [7]. The former article focuses on the job shop scheduling problem, the latter one on the timetabling problems for university classes, which are slightly different from the exam timetabling problem considered here. Finally, Costa and Hertz [8] used an ant colony approach to solve assignment type problems, in particular graph coloring problems. Recently, Dowsland and Thomson modified and improved in [9] this graph coloring algorithm with respect to the examination scheduling problem.

2 An Ant Algorithm for the Exam Scheduling Problem

2.1 General Modifications for the Exam Timetabling Problem

The solution approach consists of n cycles. In each of these cycles first each of the m ants constructs a feasible solution using therefore the constructive heuristic and the pheromone trails. These exam schedules are then evaluated according to the given objective function and the experience accumulated during the cycle is used to update the pheromone trails.

Depending on the choice of a constructive heuristic and the way the pheromone values are used, there are different ways how this basic solution approach can be adapted to the exam timetabling problem.

- At each stage of the construction process in the approach of Costa and Hertz [8] called ANTCOL the ant chooses first a node i and then a feasible color according to a probability distribution equivalent to (1). The matrix τ provides information on the objective function value, i.e. the number of colors required to color the graph, which was obtained when nodes i and j are colored with the same color.

  In contrast to elite strategies where only the ant that found the best tour from the beginning of the trial deposits pheromone, all ants deposit pheromone on the paths they have chosen. According to [5] this strategy is called ant cycle strategy.

  Different priority rules were tested as constructive heuristic. Among those chosen in each step, the node with the highest degree of saturation, i.e. the number of different colors already assigned to adjacent nodes, achieved the best results with respect to solution quality and computation times.
- In Socha et al. [7] a pre-ordered list of events is given. Each ant chooses the color for a given node probabilistically similar to the formula (1). The pheromone trail τ_{ij} contains information on how good the solution was, when node i was colored by color t. The constructive heuristic employed in their approach is not described.

For the exam timetabling problem the way the information in matrix τ is used in both approaches is not meaningful. Due to the conflicts of higher orders the quality of a solution does not depend on how a pair of exams is scheduled nor on the specific period an exam is assigned to. For example, assigning two exams i and j with $c_{ij} = 0$ to the same period can either result in a high or in a low objective function value as the quality of the solution strongly depends on when the remaining exams are scheduled. In the following we implemented a two step approach.

Step I: Determine the sequence according to the exams is scheduled. We will assume that an ant located in a node, corresponding to an exam, has to visit all other nodes, i.e. it has to construct a complete tour. The sequence according to this ant constructs the tour corresponds to the sequence in which the exams are scheduled.

Step II: Find the most suitable period for an exam which should be scheduled. Therefore, all admissible periods are evaluated according to the given penalty function.

Following this two step approach probabilities p_{ij}^{ν} for choosing the next node j that has to be scheduled are computed according to (1). Pheromone values τ_{ij} along the ants' paths are updated by the inverse of the objective function value. For the heuristic value η_{ij} the following simple priority rule for graph coloring was implemented. The exam with the smallest number of available periods is selected. A period would not be available for an exam if it caused a conflict of order 0 with another exam that has already been scheduled. This priority rule corresponds to the saturation degree rule (SD) which was tested in [1]. The value η_{ij} is chosen to be the inverse of the saturation degree.

2.2 MMAS Specifications

MMAS approaches mainly differ from AS algorithms in the way they use the existing information (c.f. [6]):

- Pheromone trails are only updated by the ant that generated the best solution in a cycle. The corresponding values τ_{ij} are updated by $\rho\tau_{ij} + 1/f^{best}$ where f^{best} is equal to the best objective function value found so far. For all other arcs (i, j) that are not chosen by the best ant τ_{ij} is updated by $(1-\rho)\tau_{ij}$. $\rho \in [0, 1]$ represents the pheromone evaporation factor, i.e. the percentage of pheromone that decays within a cycle.
- Pheromone trail values are restricted to the interval $[\tau_{min}, \tau_{max}]$, i.e. whenever after a trail update $\tau_{ij} < \tau_{min}$ or $\tau_{ij} > \tau_{max}$ then τ_{ij} is set to τ_{min} or τ_{max}, respectively. The rationale behind this are that if the differences between some pheromone values were too large, all ants would almost always generate the same solutions. Thus, stagnation is avoided.
- Pheromone trails are initialized to their maximum values τ_{max}. This type of pheromone trail initialization increases the exploration of solutions during the first cycle.

The solution quality of ant colony algorithms can be considerably improved when it is combined with additional local search. In hybrid MMAS only the best solution within one cycle is improved by local search. For the exam timetabling problem a hill climber procedure has been implemented. Within an iteration of the hill climber two sub-procedures are carried out in succession. The hill climber is stopped if no improvement can be found within an iteration.

Within the first sub-procedure of the hill climber for all exams the most suitable period is examined. Beginning with the exam that causes the biggest contribution to the objective function value, all feasible periods are checked and the exam is assigned to its best period. The first sub-procedure is stopped if all exams have been checked without finding an improvement. Otherwise the contributions to the objective function value are recalculated and the process is repeated.

The second sub-procedure tries to decrease the objective function value by swapping all exams within two periods, i.e. all exams assigned to period t' are moved to period t'' and the exams of that period are moved to period t'. Therefore all pairs of periods are examined and the first exchange that leads to an improvement is carried out. Again, the process is repeated as long as the objective function value is decreased.

3 Computational Experiments

The proposed Max-Min algorithm was implemented in Borland Delphi 7.0. It will be referred to as M-ET in the sequel. Test runs were carried out on a computer with 3.2 GHz clock under Windows XP.

3.1 Test Cases

To benchmark algorithms test cases of twelve practical examination problems can be found on the site of Carter (c.f. [10]). To make a comparison meaningful all algorithms must use the same objective function. Therefore, Carter proposed weighting conflicts according to the following penalty function: $P_1 = 16, P_2 = 8, P_3 = 4, P_4 = 2, P_5 = 1$, where P_ω is the penalty for the constrain violation of order ω. The cost of each conflict is multiplied by the number of students involved in both exams. The objective function value represents the costs per student. As the proposed M-ET algorithm does not guarantee that no conflicts of order 0 occur, additionally, the penalty P_0 was imposed and set to 10000.

3.2 Adjustment of the Parameters

The required parameters were specified as follows. The number of cycles was set to 50. Within each cycle 50 ants were employed to construct solutions. Several test runs were carried out in order to determine the required parameters appropriately:

- The evaporation rate ρ was set to 0.3. Like in [6] it turned out that this parameter is quite robust, i.e. the parameter ρ does not clearly influence the performance.
- For the restrictions of the pheromone interval values to strategies were tested. Setting $\tau_{max} = 1/\rho$ obtained slightly better results than in the case of variable τ_{max} and τ_{min} as proposed in [6]. Best results were obtained with τ_{min} equal to 0.019.
- Different values for the weighting factors α and β were tested. It turned out that the approach performed best when α was set to one and β was chosen high. Best results were obtained for β equal to 24. But the difference was on the average less than one percent when β was bigger than eight. A high β forces that exams which can be scheduled, due to zero order conflicts, only in a few remaining periods are scheduled first as they are given a much higher probability in (1). Remember that η_{ij} is the inverse of the saturation degree as explained in section 4.1. Thus, a high β value has the same effect like a so called candidate list [5]. Whereas, values for β lower than 5 solutions with zero order conflicts could not always be avoided.
- As the approach is non-deterministic each test case was solved twenty times.

After determining the parameters in such a way, it turned out that less than 2 % of the solutions were generated more than once. Thus, stagnation, that is caused by the fact that many ants generate almost the same solutions, could not be observed.

3.3 Comparison with Other Exam Timetabling Approaches

The proposed M-ET approach was compared with different other approaches. The results of the benchmarks are taken from the literature [11,12]. Table 1 displays the best solution and the average solution achieved when each test case was solved twenty times.

Additionally, the results were compared with a modified version of the ANTCOL graph coloring algorithm of Costa and Hertz [8], called A-ET in the sequel. Within this approach the ANT_DSATUR(1) procedure was used as a constructive method as described in [8]. The objective function was modified in order to consider conflicts of higher order too. In addition the hill climber already incorporated in the M-ET approach was also implemented. The parameter α was set to 1, β to 35. ρ was set equal to 0.3.

Table 1. Best and average solution after twenty test runs for the benchmark test cases from Carter et al.[1,10,12] (Best value and best average value for each instance is written in bold)

test case		[11]	[13]	[14]	[15]	[16]	[17]	[18]	[19]	M-ET	A-ET
car-f-92	best	4.6	5.2	6.0	**4.0**	4.3	-	-	4.4	4.8	4.3
	avg.	4.7	5.6	6.0	**4.1**	4.4	-	-	4.7	4.9	4.4
car-s-91	best	5.7	6.2	6.6	**4.6**	5.1	5.7	-	5.4	5.7	5.2
	avg.	5.8	6.5	6.6	**4.7**	5.2	5.8	-	5.6	5.9	5.2
ear-f-83	best	45.8	45.7	**29.3**	36.1	35.1	39.4	40.5	34.8	36.8	36.8
	avg.	46.4	46.7	**29.3**	37.1	35.4	43.9	45.8	35.0	38.6	36.3
hec-s-92	best	12.9	12.4	**9.2**	11.3	10.6	10.9	10.8	10.8	11.3	11.1
	avg.	13.4	12.6	**9.2**	11.5	10.7	11.4	12.0	10.9	11.5	11.4
kfu-s-93	best	17.1	18.0	13.8	**13.7**	13.5	-	16.5	14.1	15.0	14.5
	avg.	17.8	19.5	13.8	**13.9**	14.0	-	18.3	14.3	15.5	14.9
lse-f-91	best	14.7	15.5	**9.6**	10.6	10.5	12.6	13.2	14.7	12.1	11.3
	avg.	14.8	15.9	**9.6**	10.8	11.0	13.0	15.5	15.0	12.7	11.7
pur-s-93	best	-	-	**3.7**	-	-	-	-	-	5.4	4.6
	avg.	-	-	**3.7**	-	-	-	-	-	5.6	4.6
rye-s-93	best	11.6	-	**6.8**	-	8.4	-	-	-	10.2	9.8
	avg.	11.7	-	**6.8**	-	8.7	-	-	-	10.4	10.0
sta-f-83	best	158.0	161.0	158.2	168.3	157.3	157.4	158.1	**134.9**	157.2	157.3
	avg.	158.0	167.0	158.2	168.7	157.4	157.7	159.3	**135.1**	157.5	157.5
tre-s-92	best	8.9	10.0	9.4	**8.2**	8.4	-	9.3	8.7	8.8	8.6
	avg.	9.2	10.5	9.4	**8.4**	8.6	-	10.2	8.8	9.1	8.7
uta-s-92	best	4.4	4.2	3.5	**3.2**	3.5	4.1	-	-	3.8	3.5
	avg.	4.5	4.5	3.5	**3.2**	3.6	4.3	-	-	3.8	3.5
ute-s-92	best	29.0	29.9	**24.4**	25.5	25.1	-	27.8	25.4	27.7	26.4
	avg.	29.1	31.3	**24.4**	25.8	25.2	-	29.4	25.5	28.6	27.0
yor-f-83	best	42.3	41.0	**36.2**	36.8	37.4	39.7	38.9	37.5	39.6	39.4
	avg.	42.5	42.1	**36.2**	37.3	37.9	40.6	41.7	38.1	40.3	40.4

The results of table 1 can be summarized as follows: Although, the M-ET approach does not generate outstanding results its performance is comparable with other approaches. It finds better solutions than the approaches in [11], [13], [18], [17] and [19] for most test cases. In addition, it is striking that no approach outperforms all other approaches for all test cases. Thus, there are some test cases where M-ET finds better solutions than the approaches [14], [15] and [19], although one must confirm that these three approaches generate better solutions for most of the test cases. For example, M-ET found better solutions than the approach [14] in four out of the 13 test cases.

Surprisingly, the simple AS approach A-ET outperformed the M-ET for almost all test cases. Even without using the hill climber better results were obtained. In particular, this result is contrary to other results presented in the literature where MMAS algorithms obtained better results for various combinatorial optimization problems (c.f. [5,6]).

Computing times for the M-ET approach lay in the range of 10 seconds for the smallest test cases, i.e. hec-s-92, to 2.5 hours for the pur-s-93 problem. Compared to the M-ET approach the computing time of the A-ET combined with the hill climber was on the average 80 % higher. Thus, one can conclude that A-ET takes more time but gets a better solution quality than M-ET. Please note that the same stopping stopping criteria was used for both algorithms, namely, 2500 solutions. Of course one could argue that the time saved by the M-ET approach could be used to generate more solutions. But, increasing the number of ants and the number of cycles to 100 did not result in achieving better solutions.

4 Conclusion

In this paper different strategies for solving exam timetabling problems were tested. Ant colony approaches are capable of solving large real world exam timetabling problems. The implemented algorithms generated comparable results like other high performance algorithms from the literature.

Unlike for other combinatorial optimization problems like the TSP or the QAP for the exam timetabling problem the MMAS approach did not outperform the simpler AS strategy. Of course, it goes without saying but proper adjusting parameters can improve the performance of an algorithm considerably.

A self-evident extension would be to incorporate additional constraints and requirements like e.g. scarce room resources or precedence constraints between exams.

References

1. Carter, M., Laporte, G., Lee, S.: Examination timetabling algorithmic strategies and applications. Journal of the Operational Research Society **47** (1996) 373–383
2. Garey, M., Johnson, D.: Computers and intractability: a guide to the theory of NP-completness. W. H. Freeman and Company, New York (1979)
3. Carter, M., Laporte, G.: Recent developments in practical examination timetabling. Lecture Notes in Computer Science **1153** (1996) 3–21
4. Colorni, A., Dorigo, M., Maniezzo, V.: Distributed optimization by ant colonies. In: Proceedings of the first european conference on artificial life, Amsterdam, Elsevier Science Publishers (1992) 134–142
5. Dorigo, M., Di Caro, G., Gambarella, L.: Ant algorithms for discrete optimization. Artificial Life **5** (1999) 137–172

6. Stützle, T., Hoos, H.: $\mathcal{MAX} - \mathcal{MIN}$ Ant Systems. Future Generation Computer System **16** (2000) 889–914
7. Socha, K., Sampels, M., Manfrin, M.: Ant algorithms for the university course timetabling problem with regard to state-of-the-art. In: Proceedings of 3rd European Workshop on Evolutionary Computation in Combinatorial Optimization (EvoCOP'2003), Essex, UK (2003) 334 – 345
8. Costa, D., Hertz, A.: Ants can color graphs. Journal of the Operational Research Society **48** (1997) 295–305
9. Dowsland, K., Thompson, J.: Ant colony optimisation for the examination scheduling problem. Journal of the Operational Research Society **56** (2005) 426–439
10. ftp://ie.utoronto.ca/pub/carter/testprob
11. White, G., Xie, B., Zonjic, S.: Using tabu search with long-term memeory and relaxation to create examination timetables. European Journal of Operational Research **153** (2004) 80–91
12. http://www.or.ms.unimelb.edu.au/timetabling/ttexp2.html.
13. Di Gaspero, L., Schaerf, A.: Tabu search techniques for examination timetabling Lecture Notes in Computer Science **2079** (2001) 104–117
14. Caramia, M., Dell'Olmo, P., Italiano, G.: New algorithms for examination timetabling. Lecture Notes in Computer Science **982** (2001) 230–241
15. Burke, E., Newall, J.: Enhancing timetable solutions with local search methods. Lecture Notes in Computer Science **2740** (2003) 195–206
16. Merlot, L., Boland, N., Hughes, B., Stuckey, P.: New benchmarks for examination timetabling. (Test problem database, http://www.or.ms.unimelb.edu.au/timetabling.html)
17. Di Gaspero, L.: Recolour, shake and kick: A recipe for the examination timetabling problem. In: Proceedings of the fourth international conference on the practice and theory of automated timetabling, Gent, Belgium (2002) 404–407
18. Paquete, L., Stützle, T.: Empirical analysis of tabu search for the lexicographic optimization of the examination timetabling problem. In: Proceedings of the fourth international conference on the practice and theory of automated timetabling, Gent, Belgium (2002) 413–420
19. Casey, S., Thompson, J.: Grasping the examination scheduling problem. Lecture Notes in Computer Science **2740** (2003) 233–244

A Search Ant and Labor Ant Algorithm for Clustering Data*

Heesang Lee, Gyuseok Shim, Yun Bae Kim, Jinsoo Park, and Jaebum Kim

Sunkyunkwan University, Suwon, Korea
{leehee, gsshim, kimyb}@skku.edu, jsf001@paran.com, kjbnhd@skku.edu

1 Search Ant and Labor Ant Clustering Algorithm (SLAC)

In 1990, Deneubourg et al. [1] developed the first ant clustering algorithm based on mimicking corpse piling process of ants. In his algorithm, an ant picks up and drops the data items based on the similarity of ant's local neighborhoods. Labroche et al. [2] developed a different ant algorithm, AntClust, based on chemical odor and some behavioral rules of ants.

In existing ant-based clustering algorithms [1], [3], [4], ants are usually having long traveling paths and clustering is sometimes taking too long for a large volume of data items. It is also hard to decide when the clustering is completed since the number of clusters can be changed some times when the number of iterations is changed. They also show relatively ambiguous boundaries to distinguish the clusters.

In this paper, we propose a new ant algorithm, search-ant and labor-ant clustering (SLAC) algorithm for clustering of data items utilizing the behavior of ants not in corpse filing but in food searching and storing. We are using two kinds of ants: search ants and labor ants. At an early phase of the algorithm, a fixed number of storage nests which are storage places for data items are formed by search ants. Labor ants are traveling to the storage nests with picked items.

We set a central point of each storage nest on the lattice to form a storage nest around the central point. A search ant drops the data items around the central point of a storage nest. The data items in the same storage nest will be similar to each other, but dissimilar to the data items of other storage nests. The data dropped by search ants is stored in a nest memory. A stored data item in the nest memory will not be moved later on but keep to stay in the storage nest. In SLAC, we use Lumer and Faieta's neighborhood function in [3] for dissimilarity measure and Deneubourg's threshold function in [1] for pick up and drop probabilities of the items.

2 Experimental Result and Concluding Remarks

In our computational experiment, a real data set (Iris data set of the machine learning repository http://www.ics.uci.edu/~mlearn/MLRepository) and

* This paper was supported by Faculty Reaserch Fund, Sungkyunkwan University, 2004.

M. Dorigo et al. (Eds.): ANTS 2006, LNCS 4150, pp. 500–501, 2006.

synthetic data sets generated by two-dimensional Normal distribution N(μ, σ) are used. Iris data set has three cluster and two synthetic data sets are generated to have four clusters. We compare the performance of our proposed algorithm with the ant-based algorithm, ATTA in [4]. ATTA is a much improved algorithm in clustering speed and robustness compared to the existing corpse piling ant algorithms.

The performance of ant-based clustering varies with the magnitude of overlapping among the data items between the clusters. Outputs of ATTA and SLAC for Art1 data set has four clusters clearly since its data are overlapped barely. Both algorithms ATTA and SLAC produce small clustering errors for Art1 data sets in 30 trials. Art2 data set is more overlapped than Art1 data set. ATTA made average of 3.36 clusters even with 1,000,000 iterations, but SLAC made average of 3.87 clusters with only 50,000 iterations in 30 trials for Art2. If we increase the nest memory of SLAC from 20 to 50, the average number of clusters increased from 3.87 to 3.93.

In Iris data set which has 150 data items, ATTA hardly generated correct three clusters in average clustering time of 6.25 seconds but SLAC generated average of 3.0 clusters in average clustering time of 0.15 seconds with 8% mean of clustering errors in 30 trials.

The proposed algorithm SLAC used search ants and labor ants to balance their work loads and reduce the traveling paths of labor ants. SLAC is faster than existing corpse piling ant clustering algorithms and can be applied to various applications. Initial data items assigned by search ants of SLAC contribute a significant influence toward following clustering process. Hence we need to further study optimal setting for the numbers of search ants and the size of nest memory to improve the performance of SLAC.

References

1. Deneubourg, J.-L., Goss, S., Franks, N., Sendova-Franks, A., Detrain, C., Chrétien, L.: The dynamics of collective sorting: Robot-like ants and ant-like robots. In Proceedings of the First International Conference on Simulation of Adaptive Behaviour: From Animals to Animats 1, Cambridge, MA, MIT Press (1991) 356–365
2. Labroche, N., Monmarché, N., Venturini, G.: A new clustering algorithm based on the chemical recognition system of ants. In Proc. ECAI-02, Lyon FRANCE (2002) 345–349
3. Lumer, E., Faieta, B.: Diversity and adaptation in populations of clustering ants. In Proceedings of the Third International Conference on Simulation of Adaptive Behaviour: From Animals to Animats 3. Cambridge, MA, MIT Press (1994) 501–508
4. Handl, J., Knowles, J., Dorigo, M.: Ant-based Clustering and Topographic Mapping. Artificial Life, **12**(1) (2004)

ACO Applied to Switch Engine Scheduling in a Railroad Yard

Jodelson A. Sabino[1,2], Thomas Stützle[2],
Mauro Birattari[2], and José Eugênio Leal[1]

[1] Departamento de Engenharia Industrial
Pontifícia Universidade Católica do Rio de Janeiro, Rio de Janeiro, Brazil
jel@ind.puc-rio.br
[2] IRIDIA, CoDE, Université Libre de Bruxelles, Brussels, Belgium
jsabino@iridia.ulb.ac.be, {stuetzle, mbiro}@ulb.ac.be

This reasearch studies ACO algorithms for the switch engine scheduling in a Railroad Yard. The cars are moved individually or grouped into blocks by a set of locomotives called switch engines which are dedicated to moving around the cars in the yard. The need for moving comes from the fact that the arriving trains are disassembled into blocks of cars, undergo some operations like, loading, unloading and cleaning and finally are assembled into a new train. Each moving request is called a switch order. The decision of which switch engine will execute which switch order and the sequence of that executution is the core of our problem. The optimization of this schedule reduces the overall operational costs of the yard as weel as the time to assemble new trains, thus leading to a more productive railroad system.

The problem can be summarized as follows: Given the information about the railroad yard layout, the switch engines currently located in it and a list containing all pending planned switch orders the goal is to determine an assignment of switch engines to switch orders, and a sequencing of these such that none of the operational constraints are violated and the costs are minimized. The practical goal os the overall project is to develop a switch engine scheduling algorithm for the Tubarao Railroad Yard, located in Brazil, which is the largest railroad yard in Latin America.

The switch engine scheduling problem is strongly connected to the multiple pickup and delivery problem with time windows (m-PDPTW), which is used to model passenger and good transportation. This is a well known NP-hard problem from which heuristic approaches have shown to produce solutions that are easy to implement and have a low sensitivity to changes in the original problem. A first contribution of our work is the adaptation of an ACO algorithm, in this case COMPETants [1] to the problem, introducing a number of railroad yard specific constraints to model the particularities of the railroad yard scenario.

First tests using real-world instances showed that our algorithm might produce huge savings. Hence we decided to continue the research with further analysis and possibly improving the original algorithm. For these tests we have deployed an instance generator to produce additional sets of data based on parameters

M. Dorigo et al. (Eds.): ANTS 2006, LNCS 4150, pp. 502–503, 2006.

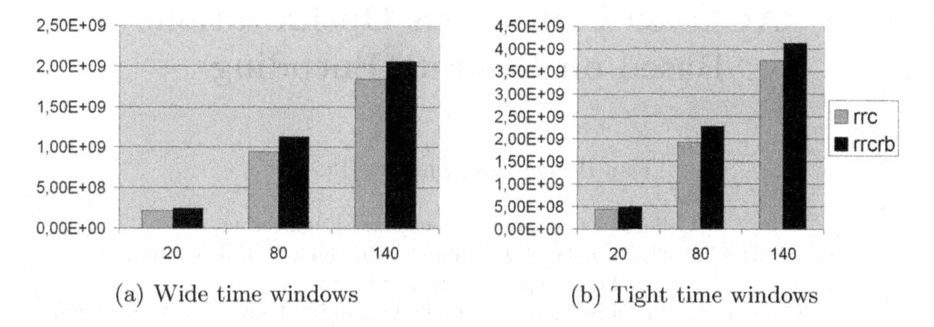

(a) Wide time windows (b) Tight time windows

Fig. 1. RRCRB and RRC solution values comparison for tight and wide time windows, considering plans with 20, 80 and 120 switch orders

that describe the most important operational characteristics of the yard. In our initial tests, we examined the implementations of two different pheromone update rules. The first one, which we called RRC (Railroad COMPETants) was simply the same rule used in the COMPETants algorithm and the second one, called RRCRB (Railroad COMPETants - Rank Based) used the pheromone update rule presented in [2] as the Rank Based Ant System.

An extensive analysis was done comparing the quality of the solutions obtained with both implementations. It showed that the RRC version of the algorithm outperformed the RRCRB version for all the combinations of input parameters and instance type and sizes considered. This can be seen in Figure 1, which illustrates one the results obtained. Paired-t tests showed that both variations deliver results that vary accordingly up and down. We also concluded that further tuning of the algorithm parameters seems not to be very promising to improve the performance of the worse variant.

As the next steps in this research, we plan to try to improve the solution quality obtained by the ACO algorithm through the adoption of more performing ACO techniques or the inclusion of local search procedures. We also plan to extend the scope of this research tackling other problems in the railroad yard operation algorithmically.

References

1. Reimann, M.: Ant Based Optimization in Goods Transportation. PhD thesis, University of Vienna (2002)
2. Dorigo, M., Stützle, T.: Ant Colony Optimization. The MIT Press (2004)

ACO for Continuous Optimization Based on Discrete Encoding

Han Huang[1] and Zhifeng Hao[1,2]

[1] College of Computer Science and Engineering
South China University of Technology, Guangzhou, P.R. China
hanhuang_scut@hotmail.com
[2] StateKey Lab. for Novel Software Technology, Nanjing University, P.R. China

Although ACO has been proved to be an efficient and versatile tool for combinational optimization problems [1,2], it cannot deal with continuous optimization problems directly. Therefore, there are only a few studies on ACO [3] for continuous optimization. This paper presents a novel ACO algorithm (CACO-DE) for continuous optimization based on discrete encoding, which is quite different from other ant methods.

In the CACO-DE, limit-accuracy real number variables are encoded to be strings of integer digits 0-9. For example, given x=397.168, it can be encoded to be a string (0, 3, 9, 7, 1, 6, 8). The first digit in the sequence is 0 for positive numbers and 1 for negative numbers.

The pheromone trail is defined as a 4-dimensional array, in which each element $T[i, j, a, b]$ is defined as follows: i is the parameter index, j indicates the jth digit of the ith parameter, a and b are digits from the set 0 to 9, $T[i, j, a, b]$ reflects the probability of assigning the jth digit of the ith parameter to the digit b conditioned on the $(j-1)$th digit having been assigned to a.

The pheromone trail of CACO-DE is accumulated with an updating rule, in which the increment of pheromone is set according to the quality of the solution obtained by ants. Later, a local strategy is added to optimize the best solution obtained by ants each iteration. The framework of CACO-DE is similar to Ant Colony System (ACS) [2], and the convergence in value [2] can be proved.

The numerical results presented here are based on 100 independent runs of CACO-DE algorithm on each of the test problem. The parameters are set as $\alpha = \beta = 0.1$ and the number of ants is 10 as those in ACS. The accuracy of CACO-DE and ACO [3] is presented in Table 1. Both of the algorithms terminate when the required accuracy is met. And the comparison of average evaluation number between CACO and other methods is presented in Table 2. The comparison presented is not only an indication of the potential of CACO-DE, but also a result showing the advantage of CACO-DE in solution accuracy and evaluation number. Table 1 shows that CACO-DE can find a high-accuracy solution of the test functions and perform better than other ant methods (CACO and ACO) for continuous optimization problem. Table 2 can verify that CACO-DE is the most efficient in all of the test methods.

However, the problems tested here are the ones with a few variables. Moreover, the accuracy of the optimal variables is limited so that the encoding can be

M. Dorigo et al. (Eds.): ANTS 2006, LNCS 4150, pp. 504–505, 2006.

Table 1. Comparison of average accuracy between CACO-DE and other ant methods

No	Test problem	Optimal	Search Domain	Dimension	ACO[3]	CACO-DE
1	De Jong's	3905.93	[-9.999, 9.999]	2	—	0
2	Goldstein, Price	3	[-9.999, 9.999]	2	1.00E-4	0
3	Martin, Gaddy	0	[-9.999, 9.999]	2	—	4.44E-07
4	Rosenbrock	0	[-9.999, 9.999]	2	—	4.44E-07
5	Rosenbrock	0	[-9.999, 9.999]	4	3.00E-3	2.03E-04
6	Sphere model	0	[-9.999, 9.999]	6	1.00E-04	1.00E-06

Table 2. Comparison of average number of function evaluations between CACO-DE and other algorithms in the literature

No	CGA	ECTS	ESA	ACO	CACO	CACO-DE
1	—	—	—	—	6000	1872
2	410	231	783	364	5330	666
3	—	—	—	—	1688	340
4	—	—	—	—	6842	1313
5	960	480	796	2905	8471	624
6	750	338	—	695	22050	270

carried out successfully. Therefore, larger sets of benchmark problems should be tested to indicate the advantage or disadvantage of CACO-DE. Moreover, the performance of CACO-DE should be discussed in detail.

Acknowledgement. This work has been supported by the National Natural Science Foundation of China (60433020, 10471045), Program for New Century Excellent Talents in University (NCET-05-0734), Natural Science Foundation of Guangdong Province (031360, 04020079), Excellent Young Teachers Program of Ministry of Education of China, Fok Ying Tong Education Foundation (91005), Guangdong Key Laboratory of Computer Network Foundation (CN200403), Key Technology Research and Development Program of Guangdong Province (2005B10101010) and State Key Lab. for Novel Software Technology, Nanjing University (200603).

We are grateful to the anonymous reviewers and our colleague Kun Tu.

References

1. Dorigo, M., Di Caro, G., Gambardella, L. M.: Ant algorithms for discrete optimization, Massachusetts Institute of Technology, Artificial Life 5 (1999) 137–172
2. Dorigo, M., Blum, C.: Ant colony optimization theory: A survey, Theoretical Computer Science 344 (2005) 243–278
3. Socha, K.: ACO for continuous and mixed-variable optimization. In Dorigo M. *et al.* (Eds.): ANTS 2004, LNCS 3172 (2004) 25–36

Applying Aspects of Multi-robot Search to Particle Swarm Optimization

Jim Pugh, Loïc Segapelli, and Alcherio Martinoli

Swarm-Intelligent Systems Group
École Polytechnique Fédérale de Lausanne, Lausanne, Switzerland
{jim.pugh, loic.segapelli, alcherio.martinoli}@epfl.ch

1 Introduction

Throughout the history of research, some of the most innovative and useful discoveries have arisen from a fusion of two or more seemingly unrelated fields of study; a characteristic of some method or process is enfused into a completely disjoint technique, and the resulting creation exhibits superior behavior. Some common examples include simulated annealing modeled after the annealing process in physics, Ant Colony Optimization modeled after the behavior of social insects, and the Particle Swarm Optimization algorithm modeled after the patterns of flocking birds.

Particle swarm optimization (PSO) is a promising new optimization technique developed by James Kennedy and Russell Eberhart [1] which models a set of potential problem solutions as a swarm of particles searching in a virtual space for good solutions. The method was inspired by the movement of flocking birds and their interactions with their neighbors in the group. By "flying" the particles through the virtual space, with attraction to positions in the space that yielded the best results, the swarm is able to find optimal solutions.

Within the field of multi-robot systems, one area that has received some attention is collective robotic search, where a group of robots works together to localize one or more targets (e.g., [2,3]). Using a collective robotic approach in search tasks can offer several major benefits over the single robot alternative. Searching can be done massively in parallel, significantly decreasing the time taken to locate the target(s) and improving robustness against failure of single agents. Scalability allows further improvement by only adding more agents, and the extensive amount of sensorial data allows for less error-prone decision-making.

Both PSO and collective robotic search are instances of multi-agent search. For PSO, the search is virtual with no limitations to particle movement, while multi-robot search is situated in the real world with constraints such as inter-robot collisions and limited communication range. However, there may be ideas which can be shared between the two search scenarios to improve one or both; adapting the strategies of PSO particles could yield an effective search technique in multi-robot systems, and the dynamics of the collective robotic search might generate interesting effects in the PSO algorithm.

M. Dorigo et al. (Eds.): ANTS 2006, LNCS 4150, pp. 506–507, 2006.

2 Experiments and Preliminary Results

A first adaptation of the PSO algorithm is to modify particle neighborhoods to be more akin to multi-robot search neighborhoods. In PSO, every particle is a member of some neighborhood of other particles with which it shares information. Typically, these neighborhoods are fixed throughout the algorithm, which may mean that particles in the same neighborhood are very far from each other in the search space. In multi-robot scenarios, communication range is often limited. This restricts information sharing to only between nearby robots. Therefore, to realistically model a multi-robot system, particle neighborhoods should be set in such a way that particles are not required to communicate with other particles outside of some close proximity.

Although proximity-based PSO neighborhoods have been used before in the literature, they have not been thoroughly explored and not with the multi-robot search paradigm in mind. We propose two new models: Model A (PSOA), where the two nearest particles form a particle's neighborhood, and Model B (PSOB), where all particles within some fixed radius r form the neighborhood. These are similar to the neighborhood models used with PSO for multi-robot learning in [4]. We compare these two models to the lbest (PSOL) and gbest (PSOG) techniques for optimizing both low (3) and high (30) dimensional problems.

Table 1. Performance of Algorithms in 3/30 Dimensions

	PSOL	**PSOG**	**PSOA**	**PSOB**
Sphere	0.000/0.297	0.000/0.000	0.000/0.065	0.000/0.000
Rosenbrock	0.023/8.006	0.005/14.31	0.162/147.3	0.005/15.42
Rastrigin	0.042/64.66	0.129/40.66	0.419/76.57	0.070/40.89
Griewank	0.003/0.002	0.007/0.011	0.003/0.012	0.004/0.009

Initial results on standard test functions indicate that Model A does not perform very well in either high or low dimensions. Model B offers some potential benefit in low dimensions, offering a good compromise between PSOL and PSOG, but does not do well in high dimensions.

References

1. Eberhart, R., Kennedy, J.: A new optimizer using particle swarm theory. In: Micro Machine and Human Science 1995. MHS '95., Proceedings of the Sixth International Symposium on, 4-6 (1995) 39-43
2. Gage, D. W.: Randomized search strategies with imperfect sensors. In: Proc. of SPIE Mobile Robots, Boston, **2058** (1993) 270-279.
3. Hayes, A. T., Martinoli, A., Goodman, R.: Distributed Odor Source Localization, Special Issue on Artificial Olfaction, IEEE Sensors, **2**(3) (2002) 260-271
4. Pugh, J., Martinoli, A.: Multi-Robot Learning with Particle Swarm Optimization. In: Proc. of Autonomous Agents and Multi-Agent Systems. Hakodate, Japan (2006) 441-448

Applying Multiple Ant Colony System to Solve Single Source Capacitated Facility Location Problem

Chia-Ho Chen and Ching-Jung Ting

Department of Industrial Engineering and Management
Yuan Ze University, Chung-Li, Taiwan, R.O.C.
s919506@mail.yzu.edu.tw, ietingcj@saturn.yzu.edu.tw

1 Introduction

In the Single Source Capacitated Facility Location Problem (SSCFLP), n customers and m potential facility locations are given. Each customer i has an associated demand d_i that must be supplied only by a single facility j. The cost c_{ij} is the unit shipping cost between customer i and facility j. The capacity of each facility s_j is known, as well as the fixed charge f_j incurred whenever facility j is opened. The objective of the problem is to locate a number of facilities that serve a set of customers at minimum cost.

The Ant Colony Optimization (ACO) has been applied to many combinatorial optimization problems successfully. Descriptions of different ACO algorithms and related literature review can be obtained in [1]. To the best of our knowledge, no applications of ACO for the SSCFLP have been published in the literature. Hence, this research aims to develop a Multiple Ant Colony System (MACS) for the SSCFLP.

2 Multiple Ant Colony System

The MACS coordinates two different solution construction rules: location selection rule and customer assignment rule. One ant colony is used to select the set of facility locations while the other one is applied to allocate customer to each opened location. The heuristic information are the ratio of capacity (s_j) to fixed charge (f_j) for location selection and the distance between customer i and facility j for customer assignment, respectively. The pheromone information are pheromone level of each candidate site j for location selection and the pheromone level of arc (i, j) for customer assignment, respectively. After all ants construct the solutions, two different local search approaches (insertion move and swap exchange) are applied sequentially to improve the best one among all constructed solutions. The framework of the MACS consists in the following steps:

- Step 1: Each ant generates the solution using location selection rule and customer assignment rule. Then update local pheromone information.
- Step 2: Apply the local search to improve the iteration-best solution.
- Step 3: Update global pheromone information according to global-best and iteration-best solutions.

M. Dorigo et al. (Eds.): ANTS 2006, LNCS 4150, pp. 508–509, 2006.

3 Numerical Analysis

In this research, the performance of the MACS is tested by two groups of benchmark instances. The first group of benchmark instances consists of two sets of problems from Holmberg et al. [3]. The number of candidate sites ranges from 10 to 50 while the number of customers ranges from 50 to 150. The computational results show that the MACS can obtain optimal solutions in 37 out of 40 small and median size problems within reasonable run times. Although the MACS cannot find optimal solutions in all medium problems, the average gap from the optimal solutions is about 0.01%.

The second group of benchmark instances, 12 very large size problems, is adopted from the Beasley OR-Library. These problems involve 100 candidate sites and 1,000 customers. We use the best-known solutions of the problems without single source constraint as the lower bounds to measure the gap. Many competitive results of the SSCFLP have been obtained by Lagrangean heuristic (LH). Hence, we compare the performance of the MACS with the LH proposed by Hindi and Pienkosz [2]. The results show that the average gap of the MACS is only 0.27%, which is lower than those obtained by LH (0.77%) in [2].

4 Summary

This research proposes a multiple ant colony system to solve the single source capacitated facility location problem. The computational results show that our approach is competitive with other heuristic algorithms.

Acknowledgment. This research was supported by a grant from the National Science Council of the Republic of China under contract NSC 94-2213-E-155-006.

References

1. Dorigo, M. and Stützle T.: Ant Colony Optimization. Bradford Books. MIT Press (2004)
2. Hindi, K.S., Pienkosz K.: Efficient Solution of Large Scale, Single-Source, Capacitated Plant Location Problems. Journal of the operational Research Society. **50** (1999) 268–274
3. Holmberg, K., Ronnqvist, M., Yuan D.: An Exact Algorithm for the Capacitated Facility Location Problems with Single Sourcing. European Journal of Operational Research. **113** (1999) 544–559

Energy Efficient Sink Node Placement in Sensor Networks Using Particle Swarm Optimization

Kirusnapillai Selvarajah and Visakan Kadirkamanathan

Department of Automatic Control and Systems Engineering
University of Sheffield, United Kingdom
{K.Selvarajah, visakan}@shef.ac.uk

1 Introduction and Problem Formulation

We formulate a non linear optimization problem to find the optimal sink node position for a given wireless sensor networks (WSN) region where sensor nodes generate different amounts of data to send to the sink node [1]. The problem is NP-hard in general and we use the particle swarm optimization technique to solve the optimization problem [2].

For a network with N sensor nodes, where each node i senses and generates data with the rate of g_i, suppose that the initial energy at each node is $E_i (1 \leq i \leq N)$. For the placement problem, we assume that the data rates from node i to node k and to the sink node are f_{ik} and f_{iS}; (x_i, y_i), $1 \leq i \leq N$, are fixed coordinates for the placement of the sensor nodes; (x, y) are the coordinates of the sink node which is to be placed efficiently in the sensor network region $(-L, L) \times (-L, L)$. C_{ik} and C_{iS} are the link cost from node i to node k and to the sink node S, respectively. For each node in the WSN, the following flow balance equation must be met [3]

$$f_{iS} + \sum_{1 \leq k \leq N}^{k \neq i} f_{ik} = \sum_{1 \leq m \leq N}^{m \neq i} f_{mi} + g_i \tag{1}$$

The goal here is to place the sink node in an optimal way to maximize the lifetime of a sensor network consisting of N sensors with the same initial energy deployed in a certain area. According to the problem setup, maximizing the life time is achieved by minimizing total power consumption of N sensor nodes. The power consumption of node i, P_i, can be represented as follows:

$$P_i = \sum_{1 \leq k \leq N}^{k \neq i} c_{ik} f_{ik} + \sum_{1 \leq m \leq N}^{m \neq i} \rho f_{mi} + c_{iS} f_{iS} \tag{2}$$

where P_i is the power dissipation at i, ρ is the power consumption coefficient for receiving data, C_{iS} denotes power consumption cost between nodes i and sink node and f_{iS} denotes the data rate to the sink node from node i. The optimization function that minimizes the total power consumption of the WSN is as follows:

M. Dorigo et al. (Eds.): ANTS 2006, LNCS 4150, pp. 510–511, 2006.

$$\min \sum_{i=1}^{N} \left(\sum_{1 \leq k \leq N}^{k \neq i} C_{ik} f_{ik} + \sum_{1 \leq m \leq N}^{m \neq i} \rho f_{mi} \right) + \sum_{i=1}^{N} C_{iS} f_{iS}) \qquad (3)$$

2 Energy Efficient Sink Node Placement Using Particle Swarm Optimization

To find the optimal location of the sink node in the sensor network region, we have to perform search algorithms such as genetic algorithm or particle swarm optimization as the problem is NP-hard in general. Here we choose particle swarm optimization because its implementation is simple and gives better results in most cases than genetic algorithm [4]. To solve the optimization problem, we need to know the optimal connection pattern for every single search point in the WSN region as the optimization function depends on the multi-hop optimal path connection pattern of every single node to the sink node. If we perform the routing algorithm after every iteration (every possible point for the sink node in the region) online, it can be a computationally expensive process and add more complexity to the optimization algorithm. We consider the WSN networks region as several small clusters to reduce the computational complexity and we perform the optimal multi-hop routing algorithms off-line by assuming the sink node is placed in each cluster. Then we calculate the cost function using the optimal routing connection pattern for each cluster. We assume that if the sink node is placed anywhere in the given cluster it has the same optimal routing path (realistic assumption to calculate near optimal position). After each iteration, our optimization algorithm identifies the appropriate cluster, then selects the cost function for its cluster which depends on the multi-hop optimal routing connection from each sensor node to the sink node.

The simulation results show that the proposed optimal strategy has significant benefit over other existing placement techniques where energy consumption is vital in wireless sensor networks. The future direction of our work is in finding the optimal positions for the multiple sink nodes for large scale sensor networks based on the residual energy of the sensor nodes.

References

1. Akyildiz, I.F., Su, W., Sankarasubramaniam, Y.: A survey on sensor networks. IEEE Communication Mazazine **40**(8) (2002) 102–114
2. Eberhart, R.C., Kennedy, J.: A new optimizer using particle swarm theory. In: Sixth International Symposium on Micromachine and Human Science. Volume 1. (1995) 39–43
3. Ordonez, F., Krishnamachari, B.: Optimal information extraction in energy-limited wireless sensor networks. IEEE Transactions on Selected Areas in Communications **22**(16) (2004) 1121–1129
4. Shi, Y., Eberhart, R.C.: Empirical study of particle swarm optimization. In: Proc. IEEE Congress on Evolutionary Computation. (1999) 1945–1950

Evolution in Swarm Intelligence: An Evolutionary Ant-Based Optimization Algorithm

Christopher Roach and Ronaldo Menezes

Department of Computer Sciences
Florida Institute of Technology, Melbourne, FL, USA
croach@vt.edu, rmenezes@cs.fit.edu

Swarm Intelligent (SI) algorithms draw their inspiration from the interaction of individuals of social organisms. One such algorithm, Ant Colony Optimization (ACO) [1], utilizes the foraging behavior of ants to solve combinatorial optimization problems. Although ACO performs well in a static environment, it has been pointed out that ACO does not perform as well as other heuristics in dynamic situations such as routing. This paper proposes a new algorithm, entitled Evolutionary Ant Colony Optimization (EACO), that combines ACO with elements of Genetic Algorithms (GA). By adding evolution, the EACO algorithm allows the individual ants to develop their own characteristics, thereby removing the homogeneity inherent within ACO. Our results demonstrate the potential of this approach in a dynamic environment.

There have been other attempts in the past at combining SI and Evolution, however, most have used the two as discrete events. White *et al.*[2] came the closest to mimicking nature with their approach by moving two of the global ACO parameters to the individual ants and evolving the ants themselves. However, even in White's experiments, the two algorithms (GA and ACO) are run as discrete steps where the GA runs only after all ants have found a solution.

EACO attempts to approximate nature by directly integrating the bio-operators of GA into ACO and evolving the ants in the system on a continuous basis. In EACO, an initial population of ants is created, and as ants die and are removed from the system, new ants are created in their place. This approach allows the algorithm to run continuously as a single system rather than as two discrete events where entire populations are created and destroyed between each iteration of the GA.

The ants in the ACO algorithm proposed by White *et al.*are characterized by two main characteristics: sensitivity to pheremone and sensitivity to link cost. On the other hand, the genotype in EACO contains five attributes: lifespan, pheremone quality, pheremone sensitivity, speed, and reproduction rate. Our current implementation uses only lifespan and speed.

Two different sets of tests were used to test the efficacy of the EACO algorithm versus traditional ACO. In both tests, a simple map was used to test the ability of each algorithm to find the optimal path between two cities. In the first test, a search for the shortest path was done by both algorithms in a purely

M. Dorigo et al. (Eds.): ANTS 2006, LNCS 4150, pp. 512–513, 2006.

static environment. In the second test, the map was changed (after a period that allowed both algorithms to converge on their initial solutions) to make the current "best path" a local optimum, thus testing how well each algorithm performed in a dynamic environment.

In the first test, as expected, ACO performed better than EACO converging on the most optimal path 98% of the time versus 80% for EACO. However, in the dynamic environment, EACO was able to escape from the local optimum and converge on the new "best path" 62% of the time versus 18% for ACO. Interestingly, in the remaining 38% of the tests, EACO converged on the second most optimal path (i.e., the previous "best path") while ACO proceeded to get worse in the remaining 83% of its tests.

Another interesting fact about the EACO algorithm is that by simply reducing the mutation rate to zero, EACO essentially becomes ACO and, therefore, performs just as well as ACO in a static environment. This planned benefit results from the method in which diversity is introduced into the EACO algorithm. In EACO, all ants in the initial population are introduced into the system with identical genotypes. Then, through mutation and crossover, their genotypes become more diverse. Thus, by simply removing the mutation factor, the genotypes in the EACO algorithm are forced to stay uniform and the algorithm becomes ACO. The end result of this behavior is that EACO acts as a superset of ACO, performing well in both static and dynamic situations depending upon how the system is initialized.

In this paper, we have presented a new approach to solving discrete optimization problems that combines ACO and GA. Our experiments have shown that, in dynamic solution spaces, EACO outperforms the traditional ACO metaheuristic, and can become ACO with a simple change in the system's mutation rate. Future plans for EACO include the inclusion of characteristics other than lifespan and speed. We also plan to incorporate the idea of species into the algorithm thereby allowing a fitness to be associated with a group rather than just an individual ant which should align itself more to the strengths of SI. Finally, we are also looking into adding a mechanism by which EACO can select the appropriate mutation rate according to the dynamism of the system rather than relying on the intervention of a user.

References

1. Dorigo, M., Di Caro, G.: The ant colony optimization meta-heuristic. In Corne, D., Dorigo, M., Glover, F., eds.: New Ideas in Optimization. McGraw-Hill, London (1999) 11–32
2. White, T., Pagurek, B., Oppacher, F.: ASGA: Improving the ant system by integration with genetic algorithms. In Koza, J.R., Banzhaf, W., Chellapilla, K., Deb, K., Dorigo, M., Fogel, D.B., Garzon, M.H., Goldberg, D.E., Iba, H., Riolo, R., eds.: Genetic Programming 1998: Proceedings of the Third Annual Conference, University of Wisconsin, Madison, Wisconsin, USA, Morgan Kaufmann (1998) 610–617

Extending the Particle Swarm Algorithm to Model Animal Foraging Behaviour

Cecilia Di Chio[1], Riccardo Poli[1], and Paolo Di Chio[2]

[1] Department of Computer Science, University of Essex, Colchester, United Kingdom
{cdichi, rpoli}@essex.ac.uk
[2] Dipartimento di Sistemi e Istituzioni per l'Economia
University of L'Aquila, L'Aquila, Italy
pdc@ec.univaq.it

1 Introduction

The particle swarm algorithm [1] contains elements which map fairly strongly to the *group-foraging* problem in behavioural ecology: its continuous equations of motion include concepts of social attraction and communication between individuals, two of the general requirements for grouping behaviour [2]. Despite its socio-biological background, the particle swarm algorithm has rarely been applied to biological problems, largely remaining a technique used in classical optimisation problems. In this paper [3], we show how some simple adaptions to the standard algorithm can make it well suited for the foraging problem.

This work introduces a new way to look at the particle swarm algorithm, i.e. using it as a *simulation* tool in the biological field of behavioural ecology. Our research is part of the *XPS*[1] multidisciplinary project which aims, among other things, to explore extensions of the particle swarm algorithm by including strategies from biology. This work on foraging behaviour represents a first step in simulating more complex group behaviour in animals.

2 Approach

The simulation of animal grouping behaviour is extremely complex. Therefore, in order to make progress, we focus here on simulating an abstraction of the group-foraging problem, where: (a) there are no predators or any other source of risk or danger; (b) animals neither give birth nor die; (c) animals are "blind" and have no sense of smell, but they can communicate with everyone else, regardless of the size of the world; (d) the food sources do not deteriorate unless eaten and once a source is exhausted, it does not regenerate.

We propose two approaches to model foraging behaviour: the first uses a standard particle swarm algorithm, with the particles just slowing down in the proximity of food; the second approach modifies the basic algorithm in order to make the particles actually stop on the food source and remain there to eat.

[1] Details of the project can be found at http://xps-swarm.essex.ac.uk.

M. Dorigo et al. (Eds.): ANTS 2006, LNCS 4150, pp. 514–515, 2006.

The general idea behind our two approaches is the following. The particles in the swarm represent the animals looking for food sources. The sources are distributed over the 2-dimensional world and take the form of *patches* which contain a certain amount of food. To explore different situations that can happen in nature, there are three different configurations of food with respect to the number of patches, their size and the amount of food they contain.

The particles move over the food landscape according to the rules of the particle swarm algorithm. When a particle lands on a patch of food, it starts feeding on it, i.e. the amount of food available on that patch decreases (while the size remains the same). Because of the way the algorithm works, other particles are attracted to this patch, and start feeding as well. Eventually, the food on the patch will be exhausted (i.e. it will reach a minimum threshold), and the particles will start foraging again.

The food eaten by the particle is interpreted as the energy that the animal gains when feeding. The food intake, or equivalently the energy gained, represents the fitness of the particle. In the standard particle swarm algorithm, the particles represent the coordinates of points in the search space and their fitness is simply given by the value of the function we want to optimise. In our simulation, the fitness of the particles is evaluated as the amount of food available for the particle to eat (i.e. the amount of food left on the patch), weighted by the amount of food that a particle can eat (i.e. the particle's own intake factor).

3 Conclusion

We have shown how some simple adaptions to the standard particle swarm algorithm can make it well suited for the foraging problem.

The results (see [3]) show that the changes convert the standard algorithm into one which produces qualitatively realistic behaviour for a simplified model of abstract animals and their foraging environment. We have also highlighted sensible envelopes for parameter values and shown that it is important to keep the particles' trajectories "smooth".

With this simulation, we have begun with a simple abstraction of the group-foraging problem. In the near future, we will extend the analysis of this behaviour further, by introducing more realistic features for the food sources, and by refining the parallelism between particles and animals.

References

1. Kennedy, J., Eberhart, R. C.: *Swarm Intelligence*, Morgan Kaufmann Publishers (2001)
2. Krause, J., Ruxton, G. D.: *Living in Groups*, Oxford University Press (2002)
3. Di Chio, C., Poli, R., Di Chio, P.: *Extending the Particle Swarm Algorithm to Model Animal Foraging Behaviour*, University of Essex (2006)

Particle Swarm Optimization for Facility Layout Problems With/Out Department-Specific Restrictions

Muzaffer Kapanoglu[1] and Fehime Utkan[2]

[1] Osmangazi University, Department of Industrial Engineering, Eskisehir, Turkey
muzaffer@ogu.edu.tr
[2] Anadolu University, Department of Industrial Engineering, Eskisehir, Turkey
futkan@anadolu.edu.tr

The facility layout problems in today's batch-to-mass manufacturing systems have gained a whole new face and popularity due to the requirements of mass customization. Facilities layout problem is still the most popular application for the Quadratic Assignment Problem [1]. The facility layout model addressed in this study is also known as quadratic assignment problem and formulated as integer linear programming formulation [2]. PSO performs a population-based search which emulates the social behavior of bird flocking and fish schooling [3]. In PSO, candidate solutions, called particles, search through the problem space by following the current optimal particles. Locations of particles in the search space are determined by their positions and velocities. Eberhart, Shi and Kennedy [2001] introduce the following equations:

$$\bar{v}_i(t) = \bar{v}_i(t-1) + \varphi_1(\bar{p}_i - \bar{x}_i(t-1)) + \varphi_2(\bar{p}_g - \bar{x}_i(t-1)) \tag{1}$$

where

$$\bar{x}_i(t) = \bar{x}_i(t-1) + \bar{v}_i(t) \tag{2}$$

where $\bar{x}_i$ is the position of a particle i, $\bar{v}_i$ is the velocity for particle i, $\bar{p}_i$ is the individual best-so-far position, $\bar{p}_g$ is the global best-so-far position and φ variables are random numbers defined by an upper limit. The algorithm we used is a typical PSO but it is left due to page restriction. The proposed PSO algorithm for QAP-based FLPs has been implemented in C♯. First set of problems were taken from research papers that dealt with QAP (Table 1). Even though we set the maximum number of iterations to as low as 100, the proposed PSO converged to the known optimal solutions. Aiming a gradual challenge, we considered QAP problems with 30 departments or less, taken from Burkard's QAPLIB as a next stage. This set resulted in 77 problems . For each problem, we propose and run three PSO algorithms: (i) the standard PSO (called as PSO-St), (ii) the standard PSO with velocity restriction (PSO-VR), and (iii) modified PSO that regenerates the population periodically to avoid possible local optima (PSO-M). For each problem, all three PSO's average convergence to the optimal solutions turned out to be around 83%. This performance should be considered with the fact that no problem- specific parameter adjustments or tunings were performed

M. Dorigo et al. (Eds.): ANTS 2006, LNCS 4150, pp. 516–517, 2006.

Table 1. Results of standart PSO for first set of problems

Test problems	PSO Solution	Optimal Solution	Convergence Ratio
Rosenblatt(1)	12822	12822	100
Rosenblatt(2)	14853	14853	100
Rosenblatt(3)	13172	13172	100
Rosenblatt(4)	13032	13032	100
Rosenblatt(5)	12819	12819	100
Zimmermann&Sovereign	389	389	100
Gavett	403	403	100

but simply the generally accepted settings were adhered. Even then in 62 single runs from out of 231 converged to the optimum at a 95% or better percentage. Aiming to reveal the impacts of the parameters on the PSO performance, we designed experiments with different settings over the selected problems with 12, 15, 16, 20, 26, and 30 departments, taken from Burkard's QAPLIB. In some problems like Chr12, fine tuning improved the performance from 57% to 86%. The problems with high convergence rates remained in the same vicinity of performance for different parameter settings. In an attempt to analyze the impact of the department-specific restrictions on the performance of PSO-based algorithm, a subset of the test problems taken from QAPLIB have been modified to represent these restrictions at 10%, 20%, and 30% levels to emulate the undesired department assignment to certain locations. The increased restriction levels naturally resulted in the increase of the objective function values as well. If one increase the number of runs per restriction level, the reliability of the observation should be increased.

Our comprehensive experiments on the test-problems and their promising results enabled us to attack a layout problem of a headline cell with 21 workstations in a truck engine plant. Although no any reconfiguration decision is made yet, the expected improvement in the layout efficiency is approximately 7%. As a conclusion, the computational results indicate that our proposed PSO-based algorithms can be applied in solving QA-based FLPs, and showed that PSO can enable the solution efficiencies to facility layout problems. A future research must be issued to unreveal the impact of a possible relation between the sparsity of the distance and flow matrices and the parameter selection and tuning in the domain of QA-based FLPs.

References

1. Loila, M., Abreu, N., Netto, P., Hahn, P., Querido, T.: A survey for the quadratic assignment problem. European Journal of Operational Research (2005) In press.
2. Golany, B., Rosenblatt, M.: A heuristic algorithm for the quadratic assignment formulation to the plant layout problem. International Journal of Production Research **27** (1989) 293–303
3. Kennedy, J., Eberhart, R., Shi, Y.: Swarm Intelligence. Morgan Kauffman-San Diego (2001)

Self-organized and Social Models of Criminal Activity in Urban Environments

Adriano Melo[1], Ronaldo Menezes[2,*], Vasco Furtado[1], and André L.V. Coelho[1,**]

[1] Graduate Program in Applied Informatics, University of Fortaleza, Fortaleza, Brazil
aanmelo@hotmail.com, vasco@unifor.br, acoelho@unifor.br
[2] Computer Sciences Florida Tech, Melbourne, FL, USA
rmenezes@cs.fit.edu

1 Simulation Police Allocation and Criminal Activity

Multi-Agent Systems (MAS) are extensively used as a tool for simulation of dynamic systems. Geosimulation is an urban phenomena approach that uses the multi-agent methodology to simulate discrete, dynamic, and event-oriented systems. Our focus in this paper is to use self-organization, specially strategies *inspired* by solutions from Swarm Intelligence, as well as the idea of social networks, and demonstrate their effect on learning in geosimulation agents.

The allocation of police officers in urban areas in order to perform *preventive policing* is one of the most important tactical management activities in criminality control. Simulation systems come to be an useful tool for supporting decision support. But effectiveness is directly proportional to the dynamism of the criminal model being utilized. The learning factor tends to be essential in that context because it will help in the identification of the trends.

Our simulator makes a society with three types of agents: crime targets, police, and criminals The simulator is made of a 60×60 grid where 41 crime targets are distributed. The simulation uses 16 criminal agents that form 4 communities (clusters). One hub exists in each community, also playing the role of that community's broker. In general, social network hubs are considered to be the nodes with the highest number of connections (degree of centrality). In this simulator, we define a hub as a criminal agent with the highest success rate (efficiency) among the four in the community – successful criminal agents tend to be known in the community and is likely to have a high degree of centrality. These agents communicate with each other at the end of a day period (48 iterations of our simulation) – the simulation has 90 days.

In our models, each criminal has three actions: commit a crime, not commit a crime, and move to a certain location. In order to reach a decision about the crime they utilize different approaches: LAZY (distance is the sole source of decision); SWARM (experience with target is also used); and SOCIAL-SWARM (decision now considers information from other criminals). All three models above are modeled under one general probability equation common in swarm systems [1].

* Partially funded by the National Science Foundation, USA, grant #INT-0337161.
** Sponsored by CNPq/Funcap, Brazil, under DCR grant #23661-04.

M. Dorigo et al. (Eds.): ANTS 2006, LNCS 4150, pp. 518–519, 2006.

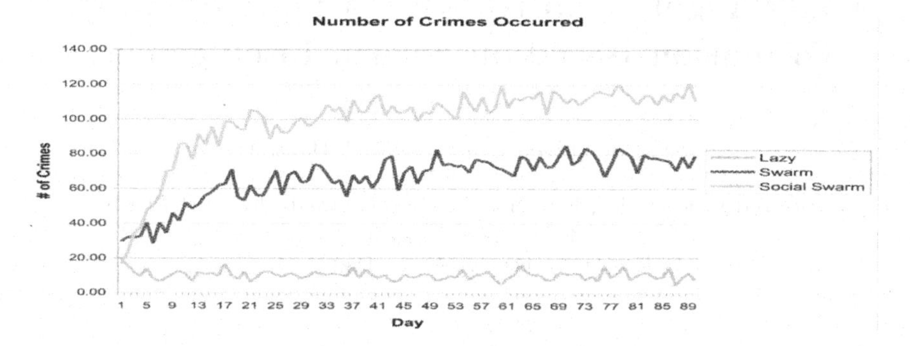

Fig. 1. The evolution of crimes occurred in the Social-Swarm model is the best

2 Evaluation of the Learning Models

First we would like to confirm our previous results showing that criminal agents can make better decisions when they learn from their past criminal activities. In [2] we also show the Swarm model leads to the formation of spatial-temporal structures at the level of criminal agents whose emergent property is their tendency to choose targets that have low or none preventive police patrolling. Second, we would like to investigate the effects of the social connections an agent has on his behavior and learning capability. In all executions we have used 41 crime targets and 38 stationary police patrols in 38 of the targets. This configuration leaves 3 targets without protection. In the experiment in Fig. 1 we demonstrate the learning capability of each model used.

One of the most important results we obtain in this work is the fact that a model to support the study of patrol routes must be consistent with some sociological theories on crime. Our results have demonstrated that learning is essential in the modeling of criminal behavior. We have utilized two models called Swarm and Social-Swarm which indicates the promise of these self-organized approaches in a practical problem. The inclusion of social aspects in the model demonstrates a significant gain over the standard Swarm model when the available points are located next to each other.

The work on modeling the social influence in criminal behavior is in its infancy. Despite our good results, we intend to investigate better the effect that a social network topology has on the social factor considered in this paper.

References

1. Bonabeau, E., Dorigo, M., Theraulaz, G.: Swarm Intelligence: From Natural to Artificial Systems. Oxford University Press (1999)
2. Furtado, V., Melo, A., Menezes, R., Belchior, M.: Using self-organization in an agent framework to model criminal activity in response to police patrol routes. In: Proceedings of the 2006 Florida Artificial Intelligence Research Society Conference, Melbourne, Florida, USA, AAAI (2006)

Traffic Lights Control with Adaptive Group Formation Based on Swarm Intelligence

Denise de Oliveira and Ana L.C. Bazzan*

Instituto de Informática, Univ. Federal do Rio Grande do Sul (UFRGS)
Porto Alegre, Brazil
{edenise, bazzan}@inf.ufrgs.br

1 Introduction

Several traffic control approaches address the problem of reducing traffic jams. A class of them deals with coordination of traffic lights to allow vehicles traveling in a given direction to pass an arterial without stopping. However, in cities where the business centers are no longer located exclusively downtown, this approach is not appropriate: simple offline optimization of the synchronization in *one arterial* alone cannot cope with changing traffic patterns.

This paper is an extension of our previous models: In [1] a decentralized, swarm-based approach was presented, but we have not collected and analyzed information about the group formation. In [2] groups were considered and a technique from distributed constraint optimization was used, namely cooperative mediation. However, this mediation was not decentralized: group mediators communicate their decisions to the mediated agents in their groups and these agents just carry out the tasks. Also, the mediation process may take long in highly constrained scenarios, having a negative impact in the coordination mechanism. Therefore, a decentralized, swarm-based model of task allocation as presented here is necessary. The dynamic group formation without mediation combines the advantages of those two previous works (decentralization via swarm intelligence and dynamic group formation).

2 Approach, Scenario and Results

Our approach seeks to replace the traditional arterial green wave by "shorter green waves" in *segments* of the network. Of course, in some key junctions conflicts may appear because in almost any practical situations, a signal plan does not allow synchronization in more than one traffic direction. However, our approach dynamically deals with the question of which traffic direction shall be synchronized. In this paper, each junction (plus its traffic lights) behaves like a social insect. A signal plan is a unique set of timing parameters comprising basically the cycle length and the split. Due to lack of space, we refer the reader to [1,3] for a detailed explanation about traffic control related concepts. Several plans are normally required for an intersection (or set of intersections in the case

* Authors partially supported by CAPES and CNPq respectively.

M. Dorigo et al. (Eds.): ANTS 2006, LNCS 4150, pp. 520–521, 2006.

of a synchronized system) to deal with changes in traffic flow. As a measure of effectiveness of such systems, one generally seeks to optimize a weighted combination of stops and delays, a measure of the density, or travel time. Here we are focussing on how the coordination is working, so we measure the number of coordinated agents and the number of groups formed.

For the task allocation, we use the mathematical model of division of labor in colonies of social insects [4]. The levels of the stimulus increase if tasks are not performed, or not performed by enough individuals. These concepts are used in our approach in the following way: each agent (traffic light/crossing) has different tendencies to execute one of its signal plans (i.e. an available task), according to the environment stimulus and particular thresholds. The approach also considers that each vehicle leaves a pheromone trace that can be perceived by the agents at the junction. The stimulus s_j of plan j is based on a weighted sum of the accumulated pheromone in each phase of this plan, and on the number of agents in the area of coordination of the signal plan. Every signal plan is associated with a given stimulus according to the direction towards this signal plan is biased.

When there is a change in the traffic flow, there must be an adaptation to the new situation. Traffic lights in the street with intense traffic flow tend to adopt the synchronized plans.

For the experiments, we use a 5x5 Manhattan-like grid, with a traffic light at each junction. We change the insertion rate of vehicles, emulating *unexpected* changes in the scenario. At the beginning of the simulations, all agents have neighbors with different plans, so that no group is formed a priori.

We have performed different experiments. Due to lack of space, these cannot be discussed here. Please go to `http://www.inf.ufrgs.br/~mas/maslab` for the extended description. We measure the number of groups created and the number of agents in the groups. In the experiments, agents were able to create groups of coordination and to coordinate in the direction with the higher traffic flow. In summary, the results show an adaptation to the changes in a fast and independent form, without any hierarchical organization.

References

1. Oliveira, D., Ferreira, P., Bazzan, A.L.C.: Reducing traffic jams with a swarm-based approach for selection of signal plans. In: Proceedings of Fourth International Workshop on Ant Colony Optimization and Swarm Intelligence - ANTS 2004. Volume 3172 of LNCS., Berlin, Germany (2004) 416–417
2. Oliveira, D., Bazzan, A.L.C., Lesser, V.: Using cooperative mediation to coordinate traffic lights: a case study. In: Proceedings of the 4th International Joint Conference on Autonomous Agents and Multi Agent Systems (AAMAS), New York, IEEE Computer Society (2005) 463–470
3. Papageorgiou, M., Diakaki, C., Dinopoulou, V., Kotsialos, A., Wang, Y.: Review of road traffic control strategies. Proceedings of the IEEE **91**(12) (2003) 2043–2067
4. Bonabeau, E., Theraulaz, G., Dorigo, M.: Swarm Intelligence: From Natural to Artificial Systems. Oxford Univ Press (1999)

Using Pheromone Repulsion to Find Disjoint Paths

Peter Vrancx* and Ann Nowé

Computational Modeling Lab, Vrije Universiteit Brussel, Brussels, Belgium
{pvrancx, ann.nowe}@vub.ac.be

We propose an extended ACO algorithm for the Maximum Edge Disjoint Paths (MEDP) problem. In this problem we want to satisfy the largest possible number of request for disjoint paths on a given graph topology. We first proposed our approach in [1]. In that paper a proof of concept was given on a number of quite small graphs. Now we build on that approach and use existing ACO features to develop an algorithm capable of obtaining good results on a set of MEDP benchmark problems.

In our algorithm each ant belongs to a given type. Ants of the same type work together on a common solution, but compete with ants of a different type for the use of resources. In the case of the MEDP each ant type is responsible for a single connection request and the different ant types compete for the use of graph edges. The competition between ant types is modeled by the pheromone system used. Each ant type has its own pheromone. Ants are attracted by pheromone of their own type, but are repulsed by pheromones left by other types. Edges with large amounts of foreign pheromone types will have a lower probability of being used by an ant. To assign a probability to possible next edges e, ants use an adapted version of the Ant System formula:

$$P(e) = \frac{[\tau(e)]^\alpha [\eta(e)]^\beta [1/\phi(e)]^\gamma}{\sum\limits_{e'} [\tau(e')]^\alpha [\eta(e')]^\beta [1/\phi(e')]^\gamma} \quad (1)$$

Every ant now takes into account 2 pheromone values. The value τ_t represent pheromones deposited by its own type t, while ϕ_t is the sum over all pheromone values τ_i associated with the edge, where $i \neq t$. An ant's sensitivity to these foreign pheromones ϕ_t is controlled by the parameter γ. From Formula 1 it is clear that the probability of choosing an edge, decreases as the foreign pheromones increase. This causes ants to explore other edges to find alternatives for edges preferred by other types. Due to a lack of good local heuristics for the MEDP, we do not consider the factor $\eta(e)$ in this algorithm (we set $\beta = 0$). This means that disjoint paths are found based only on the pheromone left by the ants. Initially ants will explore the graph randomly, as each edge has the same probability. As pheromone is deposited on the edges, the different types develop a preference for certain edges, and pheromone repulsion causes types to avoid using the same edges.

* Funded by a Ph.D grant of the Institute for the Promotion of Innovation through Science and Technology in Flanders (IWT Vlaanderen).

M. Dorigo et al. (Eds.): ANTS 2006, LNCS 4150, pp. 522–523, 2006.

We compare our algorithm with the ACO algorithm for the MEDP proposed by Blum and Blesa [2] on a benchmark set described in their paper. In order to make a comparison between the algorithms, both were given the same cpu time. This was done by running the ACO algorithm for a fixed number of iterations (250) on an instance, and giving the MACO algorithm the amount of cpu time used by the ACO algorithm. The average results are summarized in Table 1. Parameter settings for the ACO algorithm were taken from [2].

Table 1. Benchmark results

Instance data		**ACO**		**MACO**		
Topology	#commodities	avg result	σ	avg result	σ	cpu time (s)
graph3	10	9.85	0.35707	9.85	0.35707	4.816
graph3	25	21.3	1.38203	22.1	1.57797	16.0145
graph3	50	33.3	2.14709	33.4	1.772	49.624
graph3	75	38.8	2.56125	38.05	2.83681	89.3765
graph4	10	10	0	9.8	0.50990	9.89
graph4	25	23.9	1.13578	23.25	1.25996	27.4305
graph4	50	39.2	1.93907	39.5	3.13847	91.0965
graph4	75	50.7	3.28786	50.3	3.33317	200.261
bl-wr2-wht2	10	8.75	0.94207	8.4	0.96954	15.0095
bl-wr2-wht2	25	15.8	1.53623	15.7	1.70587	51.0335
bl-wr2-wht2	50	24.75	2.36379	24.9	2.46779	148.036
bl-wr2-wht2	75	29.35	3.16662	29.7	3.13209	316.935

Both algorithms achieve very similar results. The multi-type approach achieves the best result in 5 cases, compared to 6 best results for the ACO algorithm, but differences in solution quality are quite small. The multi-type approach does get the best results on the largest instances (*bl-wr2-wht2*). It should also be noted that the ACO approach requires the calculation of a heuristics matrix. The heuristic stores the shortest distance between node pairs, and is used to guide to ants to short paths for their commodities. This heuristic was calculated once for each topology before the experiment and this calculation time is not included in the above results.

References

1. Nowé, A., Verbeeck, K., Vrancx, P.: Multi-type ant colony: the edge-disjoint paths problem. In: ANTS 2004. Volume 3172 of LNCS., Springer-Verlag (2004) 202–213
2. Blesa, M., Blum, C.: Ant colony optimization for the maximum edge-disjoint paths problem. In: EvoWorkshops 2004. Volume 3005 of LNCS. (2004) 160–169
3. Vrancx, P., Nowé, A.: Using pheromone repulsion to find disjoint paths. Technical report, Computational Modeling Lab, Vrije Universiteit Brussel (2006)

Author Index

Lecture Notes in Computer Science

For information about Vols. 1–4059

please contact your bookseller or Springer

Vol. 4109: D.-Y. Yeung, J.T. Kwok, A. Fred, F. Roli, D. de Ridder (Eds.), Structural, Syntactic, and Statistical Pattern Recognition. XXI, 939 pages. 2006.

Vol. 4108: J.M. Borwein, W.M. Farmer (Eds.), Mathematical Knowledge Management. VIII, 295 pages. 2006. (Sublibrary LNAI).

Vol. 4106: T.R. Roth-Berghofer, M.H. Göker, H. A. Güvenir (Eds.), Advances in Case-Based Reasoning. XIV, 566 pages. 2006. (Sublibrary LNAI).

Vol. 4104: T. Kunz, S.S. Ravi (Eds.), Ad-Hoc, Mobile, and Wireless Networks. XII, 474 pages. 2006.

Vol. 4099: Q. Yang, G. Webb (Eds.), PRICAI 2006: Trends in Artificial Intelligence. XXVIII, 1263 pages. 2006. (Sublibrary LNAI).

Vol. 4098: F. Pfenning (Ed.), Term Rewriting and Applications. XIII, 415 pages. 2006.

Vol. 4097: X. Zhou, O. Sokolsky, L. Yan, E.-S. Jung, Z. Shao, Y. Mu, D.C. Lee, D. Kim, Y.-S. Jeong, C.-Z. Xu (Eds.), Emerging Directions in Embedded and Ubiquitous Computing. XXVII, 1034 pages. 2006.

Vol. 4096: E. Sha, S.-K. Han, C.-Z. Xu, M.H. Kim, L.T. Yang, B. Xiao (Eds.), Embedded and Ubiquitous Computing. XXIV, 1170 pages. 2006.

Vol. 4095: S. Nolfi, G. Baldassare, R. Calabretta, D. Marocco, D. Parisi, J.C. T. Hallam, O. Miglino, J.-A. Meyer (Eds.), From Animals to Animats 9. XV, 869 pages. 2006. (Sublibrary LNAI).

Vol. 4094: O. H. Ibarra, H.-C. Yen (Eds.), Implementation and Application of Automata. XIII, 291 pages. 2006.

Vol. 4093: X. Li, O.R. Zaïane, Z. Li (Eds.), Advanced Data Mining and Applications. XXI, 1110 pages. 2006. (Sublibrary LNAI).

Vol. 4092: J. Lang, F. Lin, J. Wang (Eds.), Knowledge Science, Engineering and Management. XV, 664 pages. 2006. (Sublibrary LNAI).

Vol. 4091: G.-Z. Yang, T. Jiang, D. Shen, L. Gu, J. Yang (Eds.), Medical Imaging and Augmented Reality. XIII, 399 pages. 2006.

Vol. 4090: S. Spaccapietra, K. Aberer, P. Cudré-Mauroux (Eds.), Journal on Data Semantics VI. XI, 211 pages. 2006.

Vol. 4089: W. Löwe, M. Südholt (Eds.), Software Composition. X, 339 pages. 2006.

Vol. 4088: Z.-Z. Shi, R. Sadananda (Eds.), Agent Computing and Multi-Agent Systems. XVII, 827 pages. 2006. (Sublibrary LNAI).

Vol. 4087: F. Schwenker, S. Marinai (Eds.), Artificial Neural Networks in Pattern Recognition. IX, 299 pages. 2006. (Sublibrary LNAI).

Vol. 4085: J. Misra, T. Nipkow, E. Sekerinski (Eds.), FM 2006: Formal Methods. XV, 620 pages. 2006.

Vol. 4084: M.A. Wimmer, H.J. Scholl, Å. Grönlund, K.V. Andersen (Eds.), Electronic Government. XV, 353 pages. 2006.

Vol. 4083: S. Fischer-Hübner, S. Furnell, C. Lambrinoudakis (Eds.), Trust and Privacy in Digital Business. XIII, 243 pages. 2006.

Vol. 4082: K. Bauknecht, B. Pröll, H. Werthner (Eds.), E-Commerce and Web Technologies. XIII, 243 pages. 2006.

Vol. 4081: A. M. Tjoa, J. Trujillo (Eds.), Data Warehousing and Knowledge Discovery. XVII, 578 pages. 2006.

Vol. 4080: S. Bressan, J. Küng, R. Wagner (Eds.), Database and Expert Systems Applications. XXI, 959 pages. 2006.

Vol. 4079: S. Etalle, M. Truszczyński (Eds.), Logic Programming. XIV, 474 pages. 2006.

Vol. 4077: M.-S. Kim, K. Shimada (Eds.), Geometric Modeling and Processing - GMP 2006. XVI, 696 pages. 2006.

Vol. 4076: F. Hess, S. Pauli, M. Pohst (Eds.), Algorithmic Number Theory. X, 599 pages. 2006.

Vol. 4075: U. Leser, F. Naumann, B. Eckman (Eds.), Data Integration in the Life Sciences. XI, 298 pages. 2006. (Sublibrary LNBI).

Vol. 4074: M. Burmester, A. Yasinsac (Eds.), Secure Mobile Ad-hoc Networks and Sensors. X, 193 pages. 2006.

Vol. 4073: A. Butz, B. Fisher, A. Krüger, P. Olivier (Eds.), Smart Graphics. XI, 263 pages. 2006.

Vol. 4072: M. Harders, G. Székely (Eds.), Biomedical Simulation. XI, 216 pages. 2006.

Vol. 4071: H. Sundaram, M. Naphade, J.R. Smith, Y. Rui (Eds.), Image and Video Retrieval. XII, 547 pages. 2006.

Vol. 4070: C. Priami, X. Hu, Y. Pan, T.Y. Lin (Eds.), Transactions on Computational Systems Biology V. IX, 129 pages. 2006. (Sublibrary LNBI).

Vol. 4069: F.J. Perales, R.B. Fisher (Eds.), Articulated Motion and Deformable Objects. XV, 526 pages. 2006.

Vol. 4068: H. Schärfe, P. Hitzler, P. Øhrstrøm (Eds.), Conceptual Structures: Inspiration and Application. XI, 455 pages. 2006. (Sublibrary LNAI).

Vol. 4067: D. Thomas (Ed.), ECOOP 2006 – Object-Oriented Programming. XIV, 527 pages. 2006.

Vol. 4066: A. Rensink, J. Warmer (Eds.), Model Driven Architecture – Foundations and Applications. XII, 392 pages. 2006.

Vol. 4065: P. Perner (Ed.), Advances in Data Mining. XI, 592 pages. 2006. (Sublibrary LNAI).

Vol. 4064: R. Büschkes, P. Laskov (Eds.), Detection of Intrusions and Malware & Vulnerability Assessment. X, 195 pages. 2006.

Vol. 4063: I. Gorton, G.T. Heineman, I. Crnkovic, H.W. Schmidt, J.A. Stafford, C.A. Szyperski, K. Wallnau (Eds.), Component-Based Software Engineering. XI, 394 pages. 2006.

Vol. 4062: G. Wang, J.F. Peters, A. Skowron, Y. Yao (Eds.), Rough Sets and Knowledge Technology. XX, 810 pages. 2006. (Sublibrary LNAI).

Vol. 4061: K. Miesenberger, J. Klaus, W. Zagler, A.I. Karshmer (Eds.), Computers Helping People with Special Needs. XXIX, 1356 pages. 2006.

Vol. 4060: K. Futatsugi, J.-P. Jouannaud, J. Meseguer (Eds.), Algebra, Meaning, and Computation. XXXVIII, 643 pages. 2006.